THE OCEAN BASINS AND MARGINS

Volume 4B
The Western
Mediterranean

THE OCEAN BASINS AND MARGINS

THE OCEAN BASINS AND MARGINS

Edited by

Alan E. M. Nairn and William H. Kanes

Department of Geology
University of South Carolina
Columbia, South Carolina

and

Francis G. Stehli

Department of Geology
Case Western Reserve University
Cleveland, Ohio

Volume 4B
The Western
Mediterranean

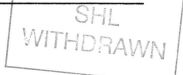
PLENUM PRESS · NEW YORK AND LONDON

Library of Congress Cataloging in Publication Data

Nairn, A   E   M
  The ocean basins and margins.

  Includes bibliographies.
  CONTENTS: v. 1. The South Atlantic.—v. 2. The North Atlantic. v. 3. The Gulf of Mexico and the Caribbean.—v. 4A. The Eastern Mediterranean.—v. 4B. The Western Mediterranean.
  1. Submarine geology. 2. Continental margins. I. Stehli, Francis Greenough, joint author. II. Title. QE39.N27   551.4'608   72-83046
ISBN 0-306-37779-9 (v. 4B)

© 1978 Plenum Press, New York
A Division of Plenum Publishing Corporation
227 West 17th Street, New York, N.Y. 10011

Printed in the United States of America

CONTRIBUTORS TO THIS VOLUME

M. Boccaletti
Istituto di Geologia dell'Università
Florence, Italy
and Centro di Geologia dell'Appennino
Florence, Italy

P. F. Burollet
Compagnie Française des Pétroles
Paris, France

A. Caire
Département de Géologie Structurale
Université de Paris
Paris, France

C. Grandjacquet
Département de Géologie Structurale
Université Pierre et Marie Curie
Paris, France

Michael Robert House
Department of Geology
The University
Hull, U. K.

P. Manetti
Istituto di Mineralogia, Petrografia e
    Geochimica dell'Università
Florence, Italy

G. Mascle
Département de Géologie Structurale
Université Pierre et Marie Curie
Paris, France

Michel Moullade
Centre de Recherche Micropaléontologique
    Jean Cuvillier
Laboratoire de Géologie Structurale
Université de Nice
Nice, France

J. M. Mugniot
Compagnie Française des Pétroles
Paris, France

Hugh Martyn Pedley
Department of Biology and Geology
The Polytechnic of North London
Holloway, London, U. K.

J. M. Ríos
Lagasca 75
Madrid, Spain

J. Salaj
Service Geologique
Tunis, Tunisia
and Geologicky ústav Dionýza Štura
Bratislava, Czechoslovakia

P. Sweeney
Compagnie Française des Pétroles
Paris, France

Brian Waugh
Department of Geology
The University
Hull, U. K.

CONTENTS OF VOLUME 4B

Chapter 3. The Tyrrhenian Sea and Adjoining Regions

M. Boccaletti and P. Manetti

**Chapter 4A.   The Central Mediterranean Mountain Chains in the Alpine
            Orogenic Environment**

A. Caire

Chapter 4B. **The Structure of the Ionian Sea, Sicily, and Calabria–Lucania**

C. Grandjacquet and G. Mascle

PART I: THE STRUCTURE OF THE IONIAN SEA AND SICILY

PART II: THE STRUCTURE OF CALABRIA–LUCANIA

Chapter 5B. **The Geology of the Pelagian Block: The Eastern Tunisian Platform**

J. Salaj

Chapter 5C. **The Geology of the Pelagian Block: The Maltese Islands**

Hugh Martyn Pedley, Michael Robert House, and Brian Waugh

CONTENTS OF VOLUME 4A

Chapter 1

THE MEDITERRANEAN COAST OF SPAIN AND THE ALBORAN SEA

J. M. Ríos

Lagasca 75
Madrid, Spain

I. INTRODUCTION: THE MEDITERRANEAN COAST OF SPAIN

The Mediterranean is a nearly landlocked sea, communicating with the world's oceans only by way of the narrow Straits of Gibraltar (Fig. 1). The western Mediterranean is bounded on the south by the northern coast of Africa. On the north and west it is bounded by the Iberian Peninsula. Consideration of the geology and evolution of the Iberian Peninsula requires first that we take note of its principal features to provide a framework for discussion.

Precambrian rocks, known and inferred, occur in the western and north-western parts of the peninsula, but none have been found along the Mediterranean coast or in the immediate hinterland. They may exist in the Betic Cordillera or in the Pyrenees, but are not known with certainty in either range as yet. During the Paleozoic, the Iberian Peninsula was part of an extensive sedimentary basin, which received Cambrian, Silurian, Devonian, and Carboniferous sediments, predominantly of marine origin. Caledonian orogeny did not strongly affect this region, and is, at most, reflected by a discontinuity in sedimentation in some places. The decisive event in Paleozoic history here was Hercynian orogeny, which resulted in intense deformation and granitic plutonism, and may be said to have consolidated the western half of the pen-

1

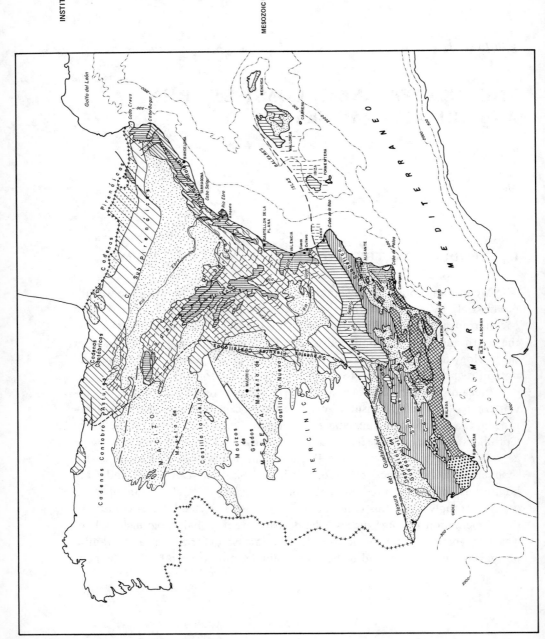

Tectonic Map of Spain
(SLIGHTLY MODIFIED)
MAIN STRUCTURAL UNITS

INSTITUTO GEOLOGICO Y MINERO DE ESPAÑA

—— Madrid 1974 ——

LEGEND

ALPINE CHAINS (S.S.)

PYRENEES (A: Axial zone, mainly Paleozoic)

PREBETIC ZONE

SUBBETIC ZONE

BETIC (A: Nevada–Filabrides complex and equivalents)

Campo de Gibraltar complex (ULTRABETIC?)

BETIC CHAINS

MESOZOIC FRINGE (AND PALEOGENE) OF THE IBERIAN MASSIF

ATLANTIC FRINGE

EASTERN FRINGE (A: deformed originating a chain of intermediate type. B: little or not deformed)

PALEOZOIC MASSIFS OUTCROPPING UNDER THE PLATFORM COVER

Paleozoic of the Catalonian coastal chains

TERTIARY BASINS

Tertiary cover mostly over the Hercynian basement, but also over alpine-deformed areas

TROUGH SYSTEM OF THE EASTERN PART OF THE PENINSULA

Neogene filling of the trough

SCALE 1 : 4,000,000

0   50   100   150   200
Km

Fig. 1. Structural units of the Iberian Peninsula (Instituto Geológico y Minero de España).

insula. Bodies of Hercynian granite are well developed in the central, western and northwestern parts of the peninsula. The eastern region, while intensively deformed at this time, seems to have escaped plutonic intrusion on the scale encountered in the west. The more pronounced Hercynian consolidation of the western regions is reflected in the passive role they have played in the subsequent development of the peninsula. In contrast, the eastern area retained mobility and participated in the subsequent Alpine cycle of sedimentation and tectonism, which is expressed in the Catalonian and Betic Ranges, paralleling the coast, and in the Pyrenean and Iberian Ranges, which are essentially normal to it (Fig. 1).

The sedimentary histories of the Alpine basins, and the tectonic histories of the chains which rose from them, are varied and complex. They have been extensively studied, however, and are the subject of an extensive literature, both Spanish and foreign. Our treatment here must be generalized from this literature, a procedure which of necessity involves some risk of oversimplification. In broad outline, the rocks deformed by Hercynian orogeny are unconformably overlain by continental and paracontinental detrital sediments of Permian and Lower Triassic age. These sediments appear to have resulted from rapid peneplanation of regions uplifted by Hercynian orogeny. The Triassic is generally of "Germanic Facies", but the Permo-Lower Triassic (Bunter) includes extensive andesites in the Pyrenean region, and the Keuper often contains ophites. Between Barcelona and Tarragona, as well as along the coast from Cartagena to Almería and Málaga, ceratite-bearing carbonates intercalated in continental redbeds suggest brief but widespread marine invasions following peneplanation. Regression is suggested by the redbeds and evaporites of the Keuper, which later play a tectonically important role as a zone of décollement and in the roots of diapirs (Ríos, 1968).

Jurassic time began with a general transgression and the deposition of shallow water, mostly dolomitic limestones in the Rhaetic, followed by shales, limestones, and dolomites for the remainder of the period. Within the Alpine basin, the formation of horsts and grabens interrupted the simple sedimentary regime near the end of the Jurassic. The horsts were subjected to erosion and sometimes to the development of a cover of continental sediments, while the grabens continued to receive sediment, partly derived from the adjacent horsts. This tectonic activity destroyed the simple uniformity of the older basin, and there are sharp, local differences in lithology and thickness of the sedimentary rocks. Subsequent to this event parts of the Betic region began to receive deepwater sediments, and its development will be treated separately.

Within the basin, the large and strongly uplifted Ebro Massif and the smaller Ampurdán Massif (Fig. 1) were stripped by erosion of early Jurassic and, in places, Triassic sediments, and remained areas of erosion until submerged by the Cenomanian transgression. Away from the horsts transgression from the north and east is evident as early as Aptian time, though the advancing marine waters do not appear to have joined across the area. A partial regression is indicated by arkosic Albian sedimentation, but relatively uniform marine conditions returned during the Cenomanian and persisted through the remainder of the Cretaceous. Even during the Upper Cretaceous, however, variations in thickness and lithology point to continued crustal instability, and to brief periods of local continental deposition.

Marine conditions persisted into the Cenozoic in some areas, while in others regression resulted in the deposition of continental sediments. Within the basins, these recessions marked the first onset of Alpine tectonic activity. Within the Iberian Ranges, however, Cenozoic deposits are continental, and it therefore appears that this first of the northern Spanish Alpine Chains must have started to rise at the end of the Cretaceous. Folding within the axial zone of the Pyrenees seems to have begun in the Middle Eocene when the range constituted a source of flysch deposits along its southern margin. At this time the Ebro Basin became separated from the Aquitaine Basin by the Pyrenees, and from the Mediterranean by the Catalonian Marginal Chains, while it was limited on the southwest by the Iberian Ranges. These restrictions to inflow of normal marine waters resulted in the deposition of evaporites at the end of the Eocene, locally reaching the level of potash salts, which were covered and preserved by an Oligocene molasse.

Though the Alpine Chains had some topographic relief in Eocene time, a general uplift of the marginal chains did not occur until the Neogene. These younger uplifts resulted in an enlargement northward, eastward, and southeastward of the Spanish Meseta, including the marine platforms which are evidently extensions of the Catalonian Ranges, the Iberian Chains, the Betic Range (and its extension into the Balearic Islands), and to a lesser extent, the eastern end of the Pyrenees.

The post-Hercynian history of the Iberian Peninsula, excluding the Meseta, can thus be seen to be one of sedimentation, until the formation of the principal ranges during the Tertiary. Thereafter, only the eastern side of the peninsula has a Mediterranean history. The Neogene tectonic history of the region appears to be characterized by a conjugate system of tensional faults which drop elongated blocks in the coastal regions to give the coastline its present form. Tectonic uplift persists to the present, and marine and continental Miocene sediments, tens to hundreds of meters above sea level, are known.

The Iberian geology and structures here discussed, of course, continue into or are related to the geology and structures of other Mediterranean countries. These relationships will be reserved for later discussion and will be treated with particular reference to their part in the general theory of plate tectonics and to specific problems posed by the suggested rotation of the Iberian Peninsula.

II. REGIONAL DESCRIPTION OF THE MEDITERRANEAN COAST*

The materials and structures along the Mediterranean coast will be systematically described. For this purpose it is convenient to distinguish five principal zones. These zones are as follows (Fig. 1):

1. The Pyrenean Ranges—Only a short segment of this belt is of interest here, for only the eastern extremity of the Pyrenees intersects the Mediterranean coast between the French border and Cabo Bagur.
2. The Catalonian Coastal Ranges—This zone extends from Cabo Bagur past Barcelona to the Ebro Delta. The orientation of the coast in this section parallels the northeast–southwest trend of the mountains.
3. The Iberian Ranges—These mountains extend west of the Ebro Delta through Castellón and Valencia to Gandía. The coastline in this region is almost perpendicular to the trend of the ranges.
4. The Betic, Pre-Betic and Sub-Betic Ranges—The Pre-Betic and Sub-Betic Ranges extend from Gandía through Alicante to a little north of Cabo de Palos, having a general southwest–northeast trend, and are prolonged by a salient of the continental platform which emerges in the Balearic Islands and then ends abruptly on the east at the steep continental slope. The Betic Ranges proper extend from Cabo de Palos to Cartagena, Almería, and Málaga to the Straits of Gibraltar. The trend of these ranges is roughly parallel to the coast.
5. The Balearic Islands—This zone includes the islands of Menorca, Mallorca, Cabrera, Ibiza, and Formentera, together with the larger submarine plateau from which they rise.

A. The Pyrenean Sector

While structurally complex, the main tectonic trend of the Pyrenees is west to east. At their eastern end the Pyrenees show an Hercynian basement, including rocks of Cambro-Ordovician, Silurian, Devonian, and Carboniferous age, overlain by a thin and incomplete cover of rocks belonging to the Alpine depositional cycle (Ríos, 1961).

The oldest rocks, which are believed to be Cambrian and possibly also Precambrian, occur at the base of a sequence which is 1800–2000 m thick in inland sections, and near the top contains a calcareous horizon which has yielded Caradocian to Ashgillian fossils. The older rocks, once predominantly

---

* The descriptions of the Pyrenean, Catalonian, and Iberian Ranges are based on the texts and maps of the recent series of geological synthesis sheets published by the Instituto Geológico y Mínero de España (see References).

argillaceous, have been metamorphosed to varying degrees and now range from phyllite to gneiss. The degree of metamorphism is lower in the higher beds, so that the sandy shales and occasional conglomerates and the calcareous bands are largely unmetamorphosed. An Ordovician age has been assigned to 700–750 m of beds. There is a transition zone of about 25 m of sandy shale and quartzite between the Ordovician and Silurian. The Lower Silurian has a thickness of about 100 m and contains carbonaceous and amphibolitic shales, which have provided Llandoverian and Wenlockian graptolites. Thin-bedded carbonaceous limestones with a Ludlovian fauna have been found. Rocks assigned to the Devonian have a thickness of 450–600 m and lie discordantly upon the Silurian. The Devonian rocks are predominantly shales, sandy shales, and limestone. Fossils are rare in the Lower and Middle Devonian, but document the presence of rocks of Givetian age. The Upper Devonian, in which limestones are more common, has yielded a richer fauna. The Devonian–Carboniferous boundary is marked by a zone of 8–30 m of black lydites with some slates. The terrigenous detrital component of these rocks increases from east to west and is greater in the upper portion of the section. The Carboniferous age has been established by rare fossils. Discordantly overlying this Culm facies are beds of the higher Carboniferous. They are represented by a detrital–argillaceous sequence of conglomerates, sandstones, and shales, with a few thin coal seams. Only Upper Stephanian beds have been definitely recognized but a wider range of ages may be represented. The sequence of rocks that has just been considered has been regionally affected by the Hercynian orogeny. Folding is widespread and there has been considerable metamorphism. The rocks have been intruded by what appear to be both synorogenic and postorogenic granites and granodiorites.

    The next depositional cycle is represented by Permo-Bunter rocks in red-bed facies, which unconformably overlie the older rocks affected by Hercynian orogeny. The total thickness of the Permo-Lower Triassic (Bunter) is variable, ranging from 100–1000 m. In general, the Permian portion of the section appears to be restricted in extent and is with difficulty, and only rarely, separated from the Lower Triassic. Where a distinction can be made, Permian conglomerates appear to be polygenetic, while the Lower Triassic (Bunter) conglomerates are better sorted and monogenetic. The Triassic appears to be of "Germanic Facies" and the Bunter is succeeded by the yellowish dolomites and gray dolomitic limestones of the Muschelkalk. Above the Muschelkalk come the marls, gypsiferous clays, and the halites of the Keuper. The Keuper is seldom seen in outcrop, but its presence is often marked by saline springs. The presence of evaporites and the mobility which they have given the Keuper make it very difficult to determine original thicknesses, but the section may total a few hundred meters. The Rhaetic, where it is present, is represented by only 15–25 m of platy dolomites.

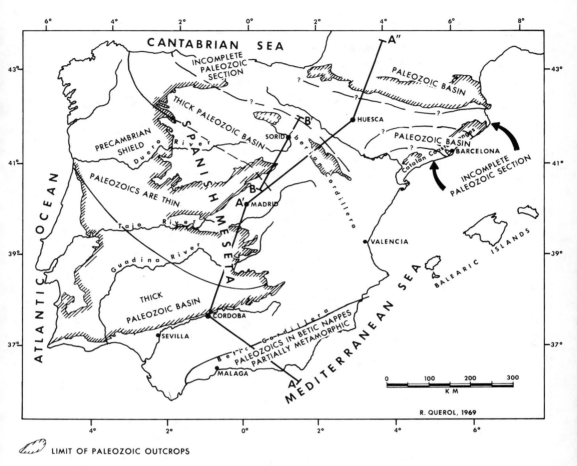

Fig. 2a. Distribution of Paleozoic sediments in Spain (after Querol, 1969).

The Jurassic in this sector is always connected with small overthrusts and slide-nappes. Limestones of Lower and Middle Liassic age are present, but invariably are seen in tectonic, and not in depositional, contact with older beds. The remainder of the Jurassic seems to be absent. The Lower Cretaceous, which is composed of limestones with subordinate calcareous and sandy marls appears, like the Jurassic, to occur only in tectonic contact with the underlying rocks. Aptian, Albian, and Cenomanian rocks have been recognized. The Upper Cretaceous is found either in nappes or lying discordantly on Permo-Triassic or basement rocks. Coniacian and Turonian occur as polygenetic conglomerates with marl and shale intercalations that lie directly on rocks of the Hercynian cycle. The Upper Cretaceous is much thicker than the Lower Cretaceous, and in inland sections reaches 1000 m; however, it thins

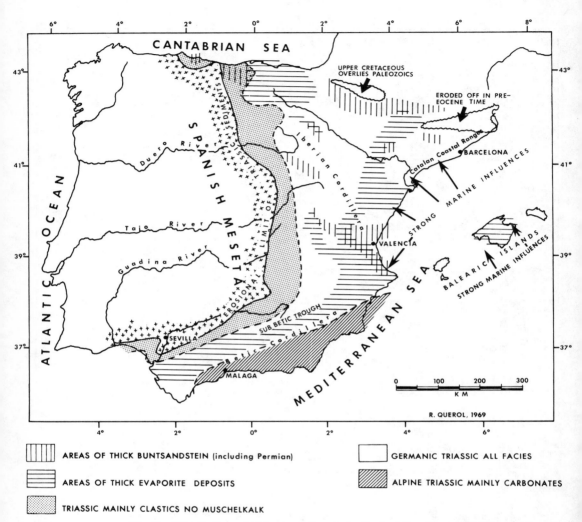

Fig. 2b. Distribution of Triassic sediments in Spain (after Querol, 1969).

in a southeasterly direction and comes finally to be represented by only a few tens of meters of shallow marine or continental deposits.

The Eocene, in the central part of the southern Pyrenean region, is represented by a thick pile of marine beds containing important continental intercalations. Individual formations vary considerably in thickness, and only the Lower Lutetian limestones seem to maintain a roughly constant thickness of 80–100 m. Oligocene and Miocene rocks are poorly represented, but the Pliocene is better developed and begins with marine beds, which change facies upward into a regressive sequence of coarse, detrital, continental sediments.

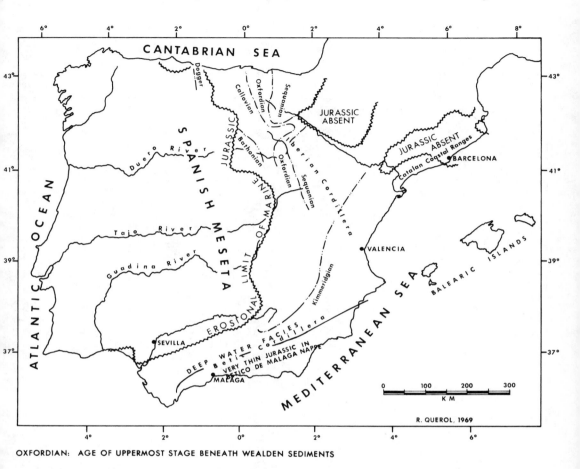

OXFORDIAN: AGE OF UPPERMOST STAGE BENEATH WEALDEN SEDIMENTS

Fig. 2c. Distribution of marine Jurassic sediments in Spain (after Querol, 1969).

Quaternary beds are widely represented. Pliocene to Recent (post-Alpine) volcanism is important west of the coast, where leucite-bearing basaltic and basanitic rocks occur and where Strombolian-type activity reached its greatest intensity in the Middle Quaternary.

The history of the Pyrenean sector can be summarized in terms of two cycles of activity, the Hercynian and the Alpine. The Hercynian cycle is characterized by nearly continuous marine sedimentation of Cambrian to Caradocian age, in which the presence of discordances and of coarse detrital sediments is the only reflection of the more pronounced Caledonian movements of northern Europe. The central Pyrenean sector was a sedimentary trough or furrow in which coarse sediments are absent, but they are found in bordering zones to the north and south. Since the Silurian and Devonian are richer in

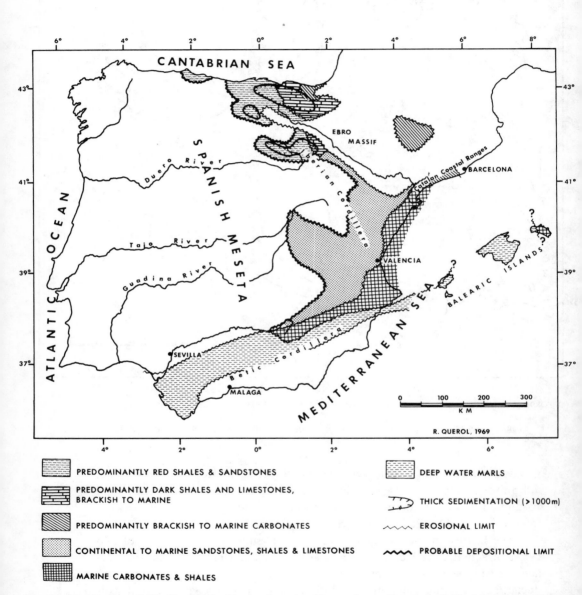

Fig. 2d. Distribution of Wealden–Purbeckian and marine Neocomian sediments in Spain (after Querol, 1969).

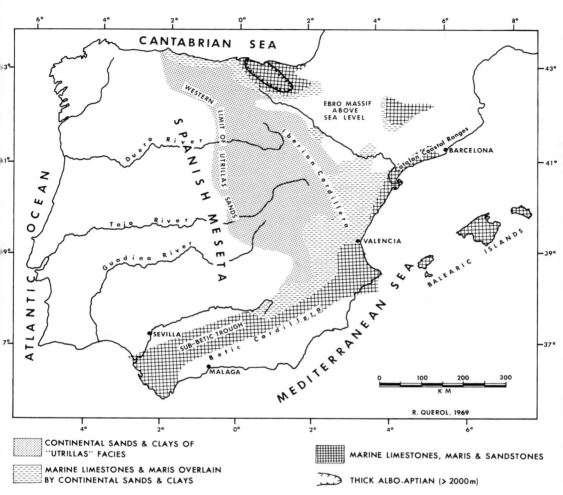

Fig. 2e. Distribution of Aptian–Albian sediments (after Querol, 1969).

limestones, it seems that the supply of terrestrial materials diminished during this time. The Carboniferous was marked by an initial transgression, but coarse, detrital material of the Culm facies was deposited from early in the Tournaisian into the Viséan. There is evidence of weak tectonic movements during the Viséan, but the principal folding began with the Middle Carboniferous (Sudetic phase) and was severe, including isoclinal folding and décollement. Post-tectonic lacustrian sediments of Stephanian age, which include thin coals, were deposited in a number of small, closed, intermontane basins.

Following the close of the Hercynian cycle, active erosion led to pene-planation of the folded region and to the deposition of red, continental Permian

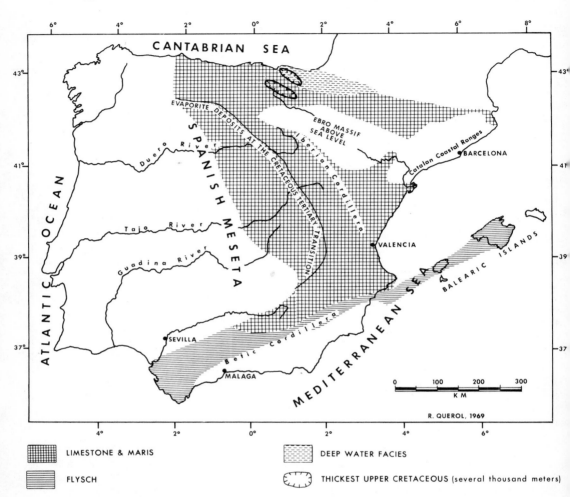

Fig. 2f. Distribution of Upper Cretaceous rocks (after Querol, 1969).

and Triassic sediments, which introduced the second or Alpine cycle characterized by molassic deposits, which in the geographic hinterland display andesitic volcanism. A marine transgression characterized the Middle Triassic,
but was followed by a phase of continental sedimentation and the formation
of evaporites toward the close of Triassic time. Jurassic seas spread over the
peneplaned region, and Jurassic sediments point to a gradual increase in water
depth with the passage of time. Block-faulting at the end of the Jurassic resulted
in the emergence of the Ebro block at the site of the former basin and also of the
Ampurdán block at the eastern end of the Pyrenees, and the northeasternmost
end of the Catalonian Coastal Ranges. This block-faulting episode resulted in

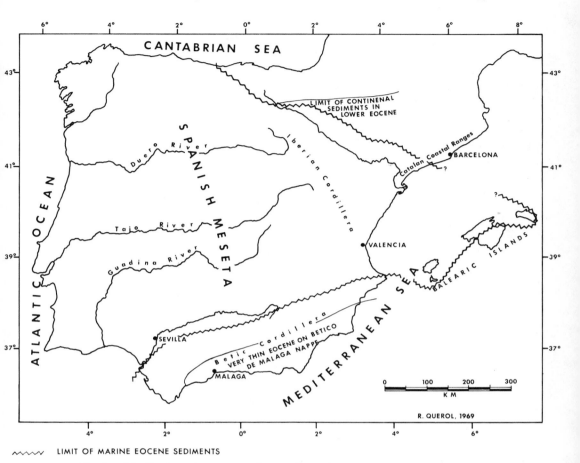

Fig. 2g. Distribution of Eocene sediments in Spain (after Querol, 1969).

a considerable amount of variation in the local patterns of sedimentation. The general picture for the Jurassic and much of the Cretaceous is, however, very difficult to unravel, since the rocks themselves are known only in allochthonous nappes. The early Cenozoic history of the region is a complex one of transgression and regression, reflecting tectonic movements of the Alpine cycle. The folding began in the Eocene and ended in the Oligocene. Oligocene sediments are always continental molasse. The uplift of the Pyrenees was mainly a Miocene event. The Miocene is not significantly developed and was presumably a period of erosion, but the resulting sediments must exist somewhere. The Pliocene, initially marine, ended with a continental phase, indicating further uplift. At the present time the coastal region is rugged, but at Cabo de Rosas an important Quaternary coastal plane occurs through which may be seen inliers of Cretaceous, Eocene, and Pliocene rocks.

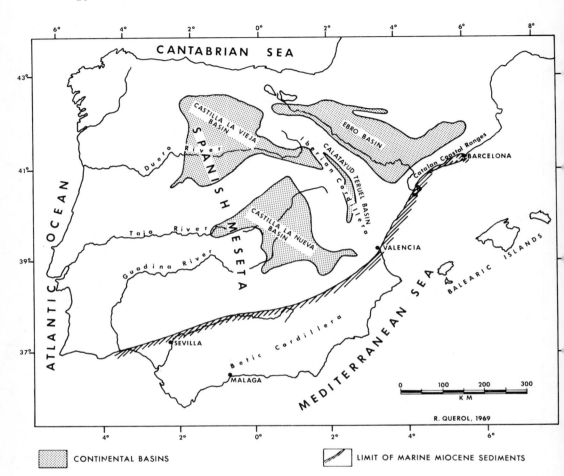

Fig. 2h. Distribution of Oligocene–Miocene sediments in Spain (after Querol, 1969).

B. The Catalonian Coastal Ranges

The Catalonian Coastal Ranges, in which an outer, or littoral, and an inner, or prelittoral, range can be discerned, extend in a northeast–southwest direction from the eastern end of the Pyrenees to the mouth of the Ebro River. The Catalonian Ranges parallel one another and represent a tilted basement block, which has moved in a rather complex manner. Toward the northeast, the granitic terrain of the Hercynian basement is broadly exposed, because the area is almost devoid of Mesozoic cover. The basement surface sinks southwestward, a condition which evidently began in Mesozoic time, because in this direction there is an increasing cover of marine Mesozoic sediments, though

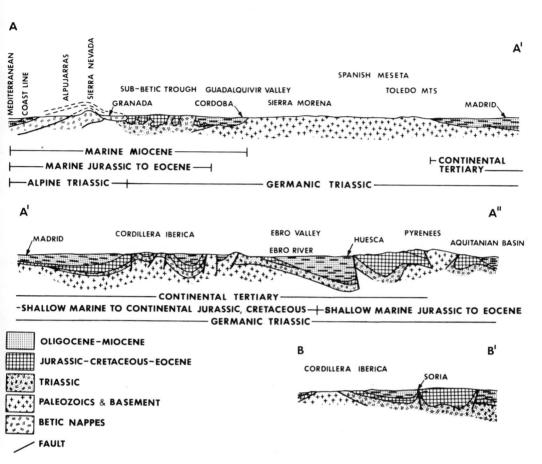

Fig. 2i. Cross sections across Spain A–A′–A″ and across the Pyrenees B–B′; for location see Fig. 2a (after Querol, 1969).

there are also a few continental interludes. In the Cenozoic, this trend was reversed, and there is an increasingly thick and complex series of Cenozoic rocks toward the northeast and the Pyrenean region. The division of the main block by faulting, with resultant formation of the two Catalonian Ranges, was a Miocene event caused by the formation of a median trench. The present coastline along this sector is fault-controlled by modern fractures slightly oblique to the trend of the principal Catalonian Ranges.

The Coastal Ranges are truncated to the north by faults which drop the Hercynian basement, now overlain by a thick Cenozoic cover, to the position it occupies in the Bahía de Rosas. To the southwest, the ranges form the eastern margin of the Ebro Basin, where they disappear beneath the thick and complex

Eocene–Oligocene sedimentary accumulation, which extends from the Pyrenean region to the Iberian Ranges.

To the northeast where the Hercynian basement is well exposed, it resembles that previously described from the Pyrenean Ranges, except that granitic emplacement of batholithic proportions has occurred here. In both areas metamorphism has been most intense in the Lower Paleozoic rocks where gneiss, hornfels, and amphibolites are present, whereas the upper portion of the section (Caradocian and younger) is relatively unaltered. Because of their metamorphism, the older rocks have not been directly dated, and are actually known only to be pre-Caradocian in the Catalonian Ranges. The younger beds, however, are readily identified on the basis of their faunal content. The Silurian, for instance, contains up to 500 m of black, graptolitic shales and a thin terminal limestone. The Devonian is a thin (300 m), dominantly limestone, sequence. The Lower Carboniferous is present in the northern part of the ranges as small, dispersed outcrops in which a Lower Tournaisian, marine paralic sequence with lydites and phosphatic nodules, and an upper, in part Viséan, sequence of continental greywackes, conglomerates, and sandstones can be recognized. The Viséan is also recognized in the southern part of the region, where it likewise is a plant-bearing sequence of sandstones, conglomerates, and sandy shales. The presence of poorly preserved goniatites in the uppermost part of the 600–800 m section suggests that Namurian or even Westphalian beds may be present.

Carboniferous rocks lie discordantly on the Devonian, and represent the last sedimentation of the Hercynian cycle. Permian rocks have not been recognized in this area. The Triassic is represented in the north by a thin, "Germanic Facies" Buntsandstein, followed by the Muschelkalk, but the Keuper is absent, probably as a result of erosion. The Muschelkalk has a middle division of red clays, sandstones, and gypsums, underlain and overlain by calcareous or dolomitic beds. Toward the south, the Triassic is more fully developed, with the Buntsandstein facies being reduced and the Middle and Upper Triassic coming to resemble the Alpine facies. However, the same three divisions of the Muschelkalk recognized in the north are represented in the south.

In the northern part of the Catalonian Chains, both the Jurassic and Cretaceous are absent. Toward the south, both appear and become increasingly thick. The Jurassic first appears, both at the coast and inland, at about the latitude of Tarragona (Cabo Salou), with the occurrence of Rhaetic dolomites and limestones. Farther to the south Liassic, Dogger, and Malm are present. Toarcian and Bathonian have also been found, and possibly Callovian beds are represented. Nevertheless, there are doubts about the continuity of deposition in this area. Still farther south the sequence is thicker and more complete, with both Charmouthian and Aalenian being present. These formations are continuous with those found in the Iberian Chain and progressively thicken in this direction, reaching much greater thicknesses than are observed in the Ebro Region (900 m). The sequence also becomes more complete toward the Iberian Ranges, with the inclusion of members that were missing in the Catalonian Chains.

The earliest Cretaceous deposits, which are marine limestones of Aptian–Albian age, rest upon Triassic and Jurassic in the Prelittoral Chain. In places Cenomanian rocks may also be present. Farther to the south transgression evidently occurred earlier, because both Neocomian and Barremian sediments are encountered. As the Iberian Chains are approached toward the southwest, Upper Cretaceous rocks also appear. In this region the Barremian, Aptian, and Cenomanian all have a transgressive character, while the Neocomian, with its Wealden facies, together with the arkosic Albian and Turonian, represent regression. The continental facies of the Albian, so characteristic of the Iberian Chains, first appears west of the mouth of the Ebro River.

The Eocene, important along the western margin of the Prelittoral Chain, is missing within the Chain itself. The Eocene, along the northeast margin toward the Pyrenees, is a marine sequence of marls and limestones containing occasional clastic beds, some of which are continental. These rocks change facies diachronously toward the southwest into clastics, which extend into the Eocene–Oligocene molassic fill of the Ebro Basin, where the range meets the Iberian Chains. There are evaporite deposits, especially toward the top, where gypsum is developed. Toward the southwest the Eocene and Oligocene are difficult to separate and appear to consist of a single molasse facies, which extends all over the Ebro Basin. The thickness of the marine Eocene sequence decreases from 1750 m in the northeast to less than 100 m north of Tarragona before it disappears completely.

Miocene rocks are found mostly in faulted basins between the two principal Catalonian chains, and in the northeastern sector pass by gradual intercalation from a thin, Oligocene, continental, lacustrine facies (directly overlying granite) into a shallow marine facies. The Pliocene, known only as a continental facies, is restricted to the northeastern part of the area.

Igneous rocks predominate in the northern portion of both Catalonian Chains because of the exposure there of a major batholith. The batholith was emplaced into rocks of the Hercynian cycle, but did not create a broad metamorphic aureole. There is considerable variety in rocks of granitic composition, including aplites, pegmatites, and lamprophyric dikes. Both syntectonic and posttectonic intrusions appear to be present. Because the Hercynian basement of the Catalonian block sinks toward the southwest, granitic outcrops become less frequent in that direction.

Within the Hercynian cycle, folding appears to have begun with the Bretonic phase, but terminated with maximum intensity during the Sudetic phase. There is overthrusting toward the southwest in which the movement occurred along the Silurian graptolitic shales. Evidently uplift terminated in the late Paleozoic and a major erosional cycle began, for Lower Triassic rocks rest upon an already peneplaned mountain system.

The Mesozoic history for the region can be briefly summarized by stating that the northern part of the Catalonian Chain remained above sea level until the beginning of the Paleogene. Toward the south, however, the Hercynian basin was covered by Triassic sediments, and formed an important region of marine deposition following transgression during the Jurassic. Kimmeridgian movements at the end of the Jurassic initiated differential uplift, and several small basins separated by regions of relative uplift can be recognized (Prades, Salou, etc.). The positive region trends northwest–southeast and, in general, the sedimentary sequence is more complete and thicker to the south. The Cretaceous has a history of repeated regression and transgression, ending with a terminal regression in the uppermost Cenomanian. Within this pattern, as in the Jurassic, continental influence is more pronounced to the north while marine influence dominates in the south.

The Catalonian Chains began to rise at the end of the Mesozoic, shedding molasse from their western margin into the Ebro Basin during Eocene and Oligocene time. The Catalonian Massif was tilted so that the northwestern end became submerged and thus received marine sediments. Oligocene sediments are rare within the range. Further tectonic movements resulted in the thrusting of the Mesozoic and Eocene sediments of the western rim of the Prelittoral Chain over the Oligocene molasse of the Ebro Basin. These movements can be dated as having occurred within the Oligocene, for the latest Oligocene sediments of the Ebro Basin cover the thrust structures. The final tectonic phase was tensional and occurred during the Miocene. It resulted in the formation of a parallel-sided trough running northeast–southwest separating the Prelittoral and Littoral Catalonian Chains. The faults that occurred at this time provided channels for the ascent of basic volcanics during the Pliocene and Quaternary. The trough received a fill of Upper Miocene sediments, which follow the usual pattern of passing from dominantly continental at the northeastern end of the Catalonian Chain to dominantly marine at the southwestern end.

C. The Iberian Ranges

As a broad generalization the Iberian Ranges, which reach the coast between the mouth of the Ebro and Valencia, form a coastal link between the Catalonian Littoral Ranges and the Betic Cordillera, and are continuous with both. The Iberian Ranges take the form of a series of subparallel Paleozoic–Mesozoic Sierras, separated by elongate depressions containing a fill of Cenozoic sediments. The grossest simplification is to describe the Iberian Ranges in terms of two parallel structures which, separated by basins filled with Cenozoic sediments, tend to merge to a broad, poorly defined structure toward the Mediterranean.

Paleozoic rocks require no discussion for, with a southeasterly regional dip, outcrops in the Mediterranean sectors are restricted to small local areas. Mesozoic rocks, as previously noted, are continuous with a succession in the southern part of the Catalonian Ranges toward the north, and they are also continuous with the Pre-Betic Ranges toward the south. They are generally rather uniform except for facies changes, especially toward the Mediterranean Sea. Tectonically the northwest–southeast-trending Iberian Ranges are almost perpendicular to the northeast–southwest trend of the Catalonian Chain. The Iberian Ranges reach the coast between Amposta (at the mouth of the Ebro) and Castellón. For convenience the ranges may be discribed in two segments: (1) the coast between Amposta and Castellón, representing a northeastern segment, and (2) the coastal sector between Castellón and Valencia, representing the southwestern sector. Because of their geographical situation a brief discussion of the Columbretes Islands is inserted between considerations of the two principal segments.

1. *The Coast between Amposta and Castellón*

This area of the coast is predominantly composed of Jurassic and Cretaceous rocks. There are only three small outcrops of Hercynian basement which, though usually regarded as exposures of the Devonian, may possibly be of Carboniferous age. These small outliers are overlain by a continental redbed sequence considered to be Permo-Triassic in age. The Triassic itself is of "Germanic Facies," beginning with continental redbeds, exhibiting a middle, Muschelkalk division which displays some redbeds and whose carbonates show some faunal resemblances to the Alpine Triassic, and ending with an upper variegated, evaporitic Keuper sequence. The Keuper provided a décollement surface and its thickness is difficult to determine for this reason and because the immediately overlying beds have been tectonically much disturbed.

Early maps show this area as consisting wholly of Cretaceous, except for sporadic Triassic outcrops. The Jurassic has recently been recognized, however, and depending on the area Pliensbachian to Callovian, Liassic to Oxfordian, Oxfordian to Middle Kimmeridgian, and Upper Kimmeridgian to Portlandian successions have been found. Limestones and dolomites dominate these sequences. An apparent transition from the Jurassic to the Cretaceous is found in a sequence of micrites which extends from the Upper Kimmeridgian to the Berriasian.

At the end of the Jurassic a northwest–southeast-trending swell (the Umbral del Maestrazgo Meridional) was formed. This feature is attributed to the weak Kimmeric phase of orogenic activity. The swell reaches the coast between Oropesa and Benicasim. Over the swell Portlandian rocks are missing, as are those of much of the Upper Cretaceous. In both cases the absence is probably due to erosion. Early Cretaceous deposits appear to be generally regressive, but may be of either Wealdean facies or marine, and they are followed by the marine limestones of a transgressive phase. The transgressive deposits sometimes rest directly on the Triassic, and because they are quite resistant to weathering are responsible for much of the topography in this region. Where it is preserved, the Upper Cretaceous is richly fossiliferous and has provided faunal elements of Turonian and Senonian age. Cenozoic rocks are poorly represented, but continental detrital sequences have been attributed to the Eocene, Oligocene, Miocene, and Pliocene. Quaternary deposits are thick, extensive, and varied. The Ebro Delta, though at present receding at a rate of about 50 m per year, is still an important feature of this area.

The principal sedimentary and tectonic features of the Iberian Ranges may be briefly summarized. The Hercynian basement consists of intensely folded rocks trending northwest–southeast, but cut by conjugate fracture systems, with the same or a northeast–southwest trend, thus defining a number of blocks which are down-faulted toward the Mediterranean coast. Though the faults which define these blocks are seldom exposed, their mobility is reflected throughout the Mesozoic in variations in depositional history. The Hercynian rocks were peneplaned by Permo-Triassic time, and after a marine phase of Middle Triassic age were again transgressed by the seas of the Liassic. The basement blocks were reactivated at the end of the Jurassic, which resulted in the subdivision of a formerly extensive basin. The scarcity of Upper Cretaceous rocks and the continental character of Cenozoic rocks point to an early phase of Alpine activity extending from the end of the Cretaceous into the Oligocene. This activity did not result in acute folding and the area shows little evidence of tectonism, except for Neogene block faulting along the conjugate, and rejuvenated system of basement faults. These faults are still active at the present time.

2. The Columbretes Islands

The Columbretes are a group of small islands, all of volcanic character, lying near the outer edge of the continental platform some 55–65 km east of Castellón. Some of the younger cones still have a clearly recognizable volcanic form. They are composed mostly of pyroclastic material, some of which is found in deposits ascribed to the Würmian. The older volcanic islands are dated as Miocene or Early Pliocene.

3. *The Coast between Castellón and Valencia*

The Iberian Chains, in addition to losing their individuality from north-west to southeast through the disappearance of the intervening Cenozoic basin toward the coast, show notable differences between their southwestern and northeastern flanks near the coast. From Castellón to Valencia, southeast of Teruel, there are extensive outcrops of Triassic and Jurassic rocks in contrast to the situation in the northeastern sector. Common to both regions, however, is the restricted outcrop of Paleozoic rocks. In this southwestern area the only Paleozoic outcrops are quartzites believed to be of Silurian age, and sericitic and muscovitic schists thought to be Devonian.

Along the coast Triassic rocks are well exposed in the facies common to the other areas described, but with the addition of abundant ophitic rocks. The Rhaetic is represented by limestones and dolomites passing upward into a marly, sandy limestone and sandstone sequence. Higher in the section, limestones become more common, first with micrites and bioclastic limestones and then with ferruginous oolites. The end of the Jurassic is marked by regression, and the beds formed at this time show an increased detrital content pointing to a nearby source to the east. A thin, continental, Wealden facies is recognized, so this situation evidently continued into the Cretaceous until terminated by a new transgression and the deposition of marine Aptian limestones. As is the case elsewhere a brief regression marks the Albian prior to the Cenomanian–Turonian(?) transgression.

The region appears to have behaved as a geanticlinal swell which stood above sea level at some stages of the Cretaceous and during the Paleogene. The Neogene is represented by white, continental limestones of Miocene age in the north, passing southward into a detrital molasse facies in the center, and finally in the south into a marine facies followed by sandstones and marls. The Pliocene is poorly represented. The Quaternary deposits form a narrow coastal strip, widening south of Castellón.

In summary, the Hercynian may be ignored since there is so little detail available. The facies of the Triassic differ little from those previously described, except for the greater importance of the evaporites in gravitational and injectional tectonics here. During the Jurassic, rather homogeneous shallow-water carbonates formed. At the end of the period, however, there was a regression which continued into the continental Wealden at the base of the Cretaceous. Transgression began in the Barremian–Aptian, and after a brief regression during the Albian was renewed at the beginning of the Upper Cretaceous. The Upper Cretaceous is poorly represented; nevertheless, it is clear that the end of the period was marked by a regression with the formation of a "Garumnian facies" coinciding with the early phases of Alpine folding. The Eocene and Oligocene are both marked by continental deposition, and marine conditions were not reestablished until the Miocene, and then only in the south.

The whole region was affected by tension and the development of a system of conjugate fractures and associated folding in the late Miocene. The faults have remained active through Recent times and they seem to reflect, though with some distortion due to the soft, sedimentary cover, the fault system of the Hercynian basement.

D. The Betic Ranges

The Betic Ranges occupy a longer stretch of the Mediterranean coast of Spain than any of the other mountain regions described. They reach from south of Valencia to the Straits of Gibraltar. Common subdivision into three elements, the Pre-Betic, Sub-Betic, and Betic Ranges, is partly responsible for the confusion concerning the limits of the Pre- and Sub-Betic divisions, for limits depend upon whether the stratigraphic or tectonic characteristics are being stressed. For the purposes of this discussion, the term Pre-Betic refers to the region from Valencia south to Denia. The term Sub-Betic covers the region from Denia to east of Murcia, and the term Betic refers to the region from Murcia to Estepona. From Estepona to the Straits of Gibraltar there exists a peculiar element, the Cádiz or Campo de Gibraltar Flysch.

The Betic Chains have been formed from rocks deposited in the main sedimentary basin bordering the Meseta, and the formations of the Pre-Betic Chain along the coast are a lateral extension of those seen in the Iberian Chain. The stratigraphic sequences pass from one to the other of the chains without any basic difference.

1. *The Coast between Valencia and Denia: The Pre-Betics*

In this region, the Paleozoic does not outcrop, and the oldest rocks exposed are Triassic of the same general facies as described earlier, though with some variations in detail. The first part of the Pre-Betic Chain between Valencia and Sueca, exhibits a relatively uncomplicated stratigraphic and tectonic pattern and can be considered an extension of the southern portion of the Iberian Chain. The region from Sueca to Denia, although continuous with the previous segment, displays greater complexity due to the imposition of younger Betic-oriented structures upon the earlier Iberian trends. The complexities are enhanced by intensive diapirism, both gravitative and injectional, as well as by consequent sliding on some horizons. Folding between Sueca and Denia was intense, and reconstruction of a sedimentary succession which tends to develop increasing complexity toward the southwest requires considerable care. The

Jurassic is regarded as complete, but here as elsewhere movements between the Jurassic and the Cretaceous have resulted in the presence of an initial, continental, Wealden facies in the Cretaceous. The Aptian is primarily a sequence of marine limestones, but toward the southwest it contains evidence of the presence of a shallow-water detrital facies. The Albian rocks of the Pre-Betic region are generally marine and only in the northwestern part of the area is there evidence of continental deposits (Utrillas) marking an Albian regression. The Upper Cretaceous, in general, is marine and consists of a sequence of limestones, dolomites, and marls. In the Sueca–Cullera area, however, an abundance of detrital material and great thicknesses in the Senonian suggest that perhaps at this time basement was exposed to the east in the area of the present Mediterranean Sea.

The early Cenozoic rocks, Eocene and Oligocene, are marine but the variety of facies points to unstable conditions. In the northeast, now Valencia, the Miocene begins with marine deposits which thicken toward the southwest and end with a phase of post-tectonic continental formations. Microfaunal study shows that the initial marine phase is transgressive and the later Miocene regressive, leading to the terminal Miocene and Pliocene continental deposits mentioned above.

The main decipherable tectonic phases are an older Laramic phase during the Upper Cretaceous, a post-Burdigalian, Styric phase, and a milder terminal, post-Helvetian phase. During the Pliocene, the tectonic history is one of tension with the development of normal faults, which give the coastline its present form (see Figs. 9 and 10). From Valencia to Gandia the structural trends have an Iberian character trending southeast–northwest, which contrasts with the southwest–northeast trend found south of Gandia. The Pre-Betic and Sub-Betic regions are separated by a highly fractured and distorted zone which represents a collision, not a simple passage of one structural trend into another.

2. *The Coast from Murcia to Estepona: The Sub-Betic and Betic (s.s.) Ranges*

Although the Sub-Betic and Betic Ranges arose from sediments deposited in a region adjacent to the Atlantic, their history and materials are related to the old Mediterranean or Tethys. The Betic Ranges (*s.l.*) have been divided into a series of units according to structural and sedimentary history. Only the principal units will be discussed here. These are, according to Fontboté (1957, 1960; Fontboté *et al.*, 1972): *external zones* (Fig. 3): the Pre-Betic, the Sub-Betic, the Guadalquivir Depression; *internal zones:* the Betic (*s.s.*) with the Sierra Nevada as nucleus, and various diverse units like the Alpujár-rides, Maláguides, etc. In addition, there are some singular elements, such as the flysch complex of the Campo de Gibraltar and the post-orogenic volcanism of Cabo de Gata and elsewhere.

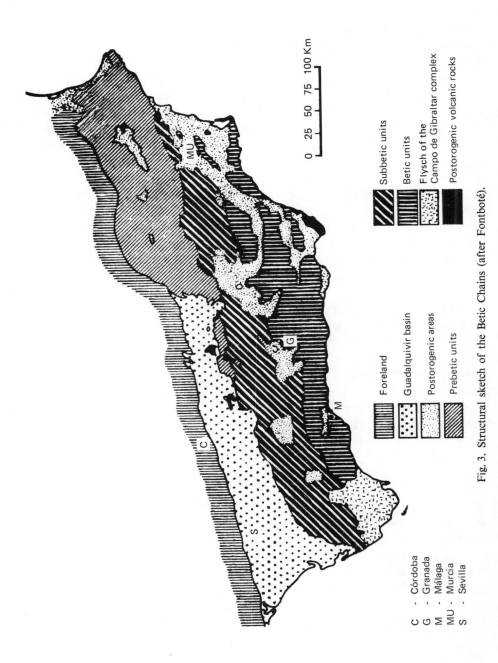

Fig. 3. Structural sketch of the Betic Chains (after Fontboté).

Foreland

Guadalquivir basin

Postorogenic areas

Prebetic units

Subbetic units

Betic units

Flysch of the
Campo de Gibraltar complex

Postorogenic volcanic rocks

0   25   50   75   100 Km

C   -   Córdoba
G   -   Granada
M   -   Málaga
MU -   Murcia
S   -   Sevilla

The rocks forming the Betic may be separated into three groups. The oldest, which have been affected by pre-Alpine orogenies, include possible Precambrian and definite Paleozoic (excluding the uppermost Carboniferous and Permian, which are grouped with the Mesozoic). The second group, composed of Mesozoic and pre-Upper Miocene Cenozoic rocks, has experienced intense deformation during the Alpine orogeny. The last group, which is post-tectogenic, includes the Upper Miocene and younger rocks.

The Structural Elements

1. *The Guadalquivir Depression.* This region is devoid of significant topographic relief and strictly external to the Betic Range. Sandwiched between the Meseta and the Sub-Betic units, the basement of the Guadalquivir Depression tilts toward the southeast. The basin contains a thick succession of young sediments which rest upon Triassic rocks and in which the Miocene alone reaches 1500 m. Along its southern margin, the basin contains olistolites of Mesozoic and Eocene deformed materials for the most part which have slid in from the Sub-Betics and which have been covered by uninterrupted deposition of Miocene sediments.

2. *The Pre-Betic Zone.* The northeastern extremity of this zone has been described already in connection with the coast between Valencia and Gandía. Structurally the zone is relatively simple, although some parts of it display great complexity, e.g., the Sierra de Cazorla. Essentially there is a thin series of platform sediments which in some areas were folded by detachment from, and by gliding over, the basement. The Triassic Keuper series forms the décollement surface. Faulting and disharmonic folding provide local complications. Where Neogene rocks occur, they are thrown into large open folds.

3. *The Sub-Betic Zone.* Folding in this zone is generally more intense than in the Pre-Betic Zone. Disharmonic folding occurs along with the formation of nappes, and from this region olistostromes have slid into the Guadalquivir Depression to the north. The general direction of overthrusting is toward the foreland to the north, and the mechanism appears to have been compression and/or gravitational sliding. The structure observed is the result of several tectogenic phases.

4. *The Betic Zone.* This zone can, to a certain extent, be considered, to show the Metamorphides and Centralides of Kober (1931). The basement appears to have been subjected to three tectonic cycles, pre-Hercynian (?Precambrian), a weak Hercynian phase, and an Alpine phase. The Alpine tectonism is characterized by nappes overthrust toward the north. Regional metamorphism affects all rocks up to the Triassic. The nappes were formed in post-Triassic to pre-Upper Miocene time. Some consider them to have begun in the Jurassic, but a Cretaceous or Eocene age seems more likely. Certainly, they

were formed no later than Lower Oligocene. The development of the Sierra Nevada, the most important orographic feature of the range as a "plis de fond," is a late event. It is a large, simple structure which formed slowly in post-Miocene times, although it may have begun somewhat earlier. Always considered as an autochthon relative to the overlying formations, the Sierra Nevada itself displays, in its core, evidence of old tectonic movements toward the north (F. M. Fontboté, personal communication, 1974). Locally there is young and intense folding associated with the "plis de fond."

The upper nappes (mainly the Alpujarride complex and the Lanjarón unit) have slid perhaps more than 50 km northward beyond and above the Sierra Nevada, with the latter appearing through a tectonic window. The nappes include highly metamorphosed rocks, non-metamorphosed Paleozoic formations, and rocks of Permo-Triassic to Oligocene age. Imbrication and internal detachment are common.

In brief, the Betic Zone is a zone of Alpine nappes, which involve basement rocks. It differs from the external zones in exhibiting strong Alpine metamorphism. Additional allochthonous units occur in the Campo de Gibraltar, where only a relatively thin sedimentary cover ranging from Lower Cretaceous to Lower Miocene in age is involved, thus suggesting that this unit should be separated from the Sub-Betic and Betic units with which it is in contact. These rocks are frequently referred to as the flysch of the Campo de Gibraltar and their emplacement is the result of gravity sliding. The direction of movement is to the northwest, but their original location is still in doubt. One suggestion is that the zone between the external and internal Betics contains the original position of these rocks. They were emplaced at the beginning of the Upper Miocene.

Upper Miocene and younger beds are hardly affected by the Alpine orogeny. The tectonic movement to which they have been subjected is primarily vertical, for instance the opening of the Straits of Gibraltar, due to the subsidence of a segment between the Betic and the Rif Ranges. Such activity continued into the Quaternary. Fontboté (personal communication) stresses that this is an example of a tectonic style superimposed on the Alpine orogenic evolution of the Betics, and is part of a first-order rift pattern, stretching from the lower Rhine to the Alborán Sea and quite independent of the Alpine Chains. The trend of rifting, as well as its extensional nature, suggests a link with the widening of the North Atlantic.

The stratigraphy and structure of the Betic Ranges is already sufficiently well known (Fontboté, 1970) to permit a reconstruction of the preorogenic history of the Betic Basin. This is of particular interest as it concerns the western boundary of the Tethys Sea. In the Pre-Betics a distinction can be made between the neritic facies which is developed toward the border with the Meseta, where it interfingers with continental facies, and the thicker and more complete marine sequence formed in the Pre-Betic–Sub-Betic Subbasin and subjected to notable and prolonged subsidence during the Jurassic and Lower Cretaceous. In the Sub-Betic Zone, the Triassic is represented by the development of a marl–clay–evaporite Andalusian variety of the "Germanic Facies," and is thus not greatly different from that in the Sub-Betic Zone. This similarity continues through the marine facies into the Domerian or Upper Charmouthian. At that time, differential subsidence affects the Sub-Betic Zone, with the formation of radiolarites and turbidites in the more depressed regions. Moreover, during the later Jurassic and early Cretaceous, there was submarine volcanism and there is evidence of synsedimentary breccias and slides. Over most of the region there is a marked pre-Senonian discontinuity. The Senonian is characterized by the development of a uniform, pink marl–limestone unit. Uniformity of facies also characterizes the Paleocene. Differential subsidence was repeated during the Eocene and Oligocene and appears to be related to developing orogenic activity that in some places resulted in the formation of typical flysch facies.

The Aquitanian–Burdigalian, although not well delimited, is the youngest preorogenic sequence in the Pre-Betic Zone. In the Sub-Betic Zone, the Senonian postdates the emplacement of nappes, which may be considered the principal tectonic event without diminishing the importance of the Eocene and Miocene phases of activity.

The sedimentary cover of the internal zones of the Betics is poorly developed, and except around Málaga the Mesozoic is represented only by the Triassic. Some post-Hercynian Paleozoic rocks (Upper Carboniferous or Permian) probably exist but they have not yet been identified definitely. Such rocks as are present appear to represent shallow-water deposition. Around Málaga the section contains Jurassic, Cretaceous, and Paleogene formations, suggesting the existence of conditions similar to those of the Sub-Betic Zone. The main nappes of the Betic Zone appear to have moved in the Cretaceous or Paleogene, although some movements occurred as early as the Jurassic.

Triassic rocks overlap the margin of the Meseta and extend south-southeastward across the Guadalquivir Basin. Thicknesses increase rapidly from the external to the internal Pre-Betic Zones. Farther toward the heart of the Betics thicknesses become variable because of the development of horsts and grabens which affect local sedimentation. The minimum thickness is reached in the Málaga area beyond which the Triassic disappears. The rocks are all of shallow water to very shallow water origin, and include horizons of volcanic and subvolcanic materials. Lower and Middle Liassic rocks are absent on the Meseta and in the Guadalquivir Basin except where they occur as olistolites in the latter. Marine sediments of this age are found in the internal Pre-Betics, where they reach a maximum thickness before becoming thinner and more variable in the external Sub-Betic Zone. However, they thicken once more in the middle Sub-Betic region because of basins formed by fracturing of the basement, and then disappear in the internal Sub-Betic region. The internal zone appears to have remained sediment-free, but beyond it there is developed a facies of shallow water limestone, including some volcanics. The pattern just described continues for the remainder of the Jurassic, but the principal basins continued to subside and became quite deep.

Sedimentation during the Lower Cretaceous expanded to cover a broader area, for a thin sequence covers the Guadalquivir Basin. A thick sequence of sandstones and lutites fills the deep trough of the internal Pre-Betic Zone, but the grabens of the middle Sub-Betic Zone diminish in importance though still receiving pelagic sediments and volcanics. Toward the internal Sub-Betic Zone, as was the case earlier, the sedimentary thickness is reduced and the internal zones remain emergent.

Fontboté (1970a) notes that the history described departs from the typical geosynclinal concept. Instead it appears to suggest a history of extension and differential subsidence from Middle Liassic time onward in the internal zones and volcanic activity associated with fracturing. The internal zones of the Betic (s.s.), which might be expected to be zones of maximum subsidence, are in fact zones of considerable stability at or close to sea level, and only in the Malaguides Zone is there evidence of weak subsidence. These features scarcely seem characteristic of an eugeosyncline. Another unusual feature is that the Paleozoic rocks of the Malaguides Zones were scarcely, if at all, affected by the Hercynian orogeny. Parts of the Pre-Betic internal zones underwent subsidence and can be regarded as mildly geosynclinal during the Jurassic and Lower Cretaceous. The Sub-Betic Zone, because of paleogeographic position, tectonic style, and lack of metamorphism despite some pelagic sedimentation and volcanism, cannot be considered eugeosynclinal. The stability of the internal zones (Betic s.s.) from the Triassic onward suggest that they are best considered as an internal post-orogenic area incorporated later into a Triassic sedimentary basin. Classic geosynclinal development was confined to the external zones and within them can be found features more commonly identified with the internal zones of Alpine chains. Nonetheless, despite the absence of a typical development, the internal zones have been metamorphosed and have generated nappes.

In summary, there was a general thrusting of the Betics over the external zones as postulated by Blumenthal (1930) and Fallot (1948), with subduction apparently occurring along the limit between the external and internal zones and bringing together elements previously far apart. The sediments of the Sub-Betic Zones were deposited some tens of kilometers to the southeast of their present position, but no more exact estimate is possible at this time. From Triassic to Domerian time the Sub-Betic region was a continental platform, subaerial in part during the Triassic, and then marine, which became geosynclinal with many breaks in sedimentation during the Middle Cretaceous.

There is some question whether the deformation of pre-Cenomanian formations was due to normal tectonic processes or was the result of halokinetic movements. At the end of the Cretaceous and particularly in the Paleogene detrital sediments appeared and some areas seem to qualify as flysch. In parts of the Sub-Betic there are important Eocene and Oligocene movements, but these do not appear to affect the region as a whole. The final phase of nappe formation, not always the most important, occurred during the Upper Miocene, imprinting on the Sub-Betic their present character. Subsequent movements were on a small scale.

During the final stages of the Paleozoic, the Betic domain became incorporated into what later became a Triassic sedimentary basin. Thereafter it remained stable at or close to sea level. The hypothesis adopted here is that the materials of the Campo de Gibraltar represent part of the post-Triassic cover lacking in the Betic domain. The original source of the nappes of the Campo de Gibraltar is not known, but it was certainly beyond the Sub-Betic Zone. From the Upper Cretaceous to the Lower Miocene sedimentation was periodically of flysch type. The final stage of nappe emplacement is dated at the beginning of the Upper Miocene.

E. The Balearic Islands

There is still some dissent as to whether Balearic Islands should be regarded as a continuation of the Betic Cordillera. Do they represent a massif separated from the Castilian Massif, or are they structurally related to the Betics? The following survey is based upon the work of Colom and Escandell (1962), both geologists well acquainted with their native islands.

Colom and Escandell have produced a series of paleogeographic maps for the Jurassic, Cretaceous, and some stages of the Cenozoic. These maps indicate the presence, through the Jurassic and Cretaceous, of an Ebro Massif continuous with the Catalonian Massif and lying within the area of the present Mediterranean (Fig. 4a–d). The massif is shown as cutting across the present coast in the vicinity of Barcelona, and extending east of Ibiza, but Mallorca and Menorca lay beyond its eastern margin. The northern margin broadly follows the axis of the Tremp Basin and cuts the coast at the latitude of Palamós. This massif was surrounded by the Mesozoic Tethys. Around the margins of the massif are littoral and neritic deposits which extend to the Iberian Arch in Valderobles, reach the coast at Salou, and extend across Mallorca and Menorca. Subsidence of the massif is indicated by the thick limestones deposited southeast of the Catalonian Massif during the Lower Liassic. These limestones outcrop in the northern sierras of Mallorca and Menorca. There are no indications on either Mallorca or Menorca of sediments of deep-water origin, though they do occur on the island of Cabrera just to the south, and in the Middle

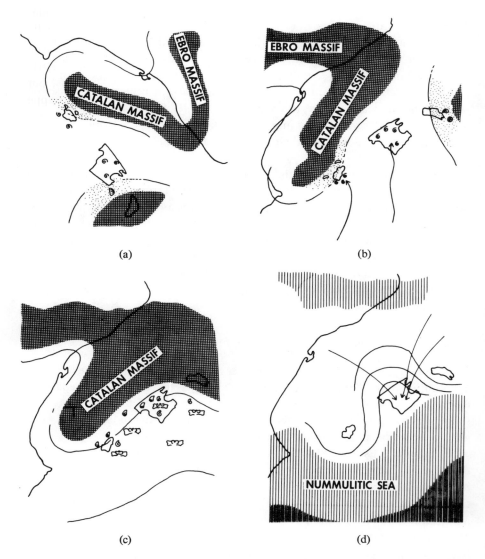

Fig. 4. The Cretaceous and Eocene of the Balearic Islands (after Colom and Escandell, 1962).
(a) Valanginian—Urgonian facies north of Ibiza related to the Catalan Massif and to other
neritic sediments east of Mallorca; bathyal fauna and sediments in Majorca and south of
Ibiza. (b) Hauterivian–Barremian—Maximal development of the geosynclinal trough and
pelagic conditions south of Ibiza and covering Mallorca. Continuing Urgonian facies in
Ibiza and its appearance in Menorca from the Upper Barremian–Aptian. (c) Gault—Emergence of the Catalan Mass, which joins the emergent Menorca Mass, with the southward
migration of the geosyncline. Fine-grained sediments with *Globotruncana* during the Cenomanian–Turonian. End of the geosyncline. (d) Eocene—Total submergence of the Balearic
area with the formation of a large basin over Mallorca. Beginning of limnic associations of
molluscs and mammiferous groups.

Liassic the sediments are of bathyal origin. The Balearic geosyncline appears to have begun to form at about this time, as is suggested by the initiation of deep water deposition and the expansion of the littoral zone at the expense of the Ebro–Catalonian Massif. Concomitant with this development, Mallorca and much of Menorca became incorporated in the bathyal zone with only the northern part of Mallorca remaining in the littoral environment. During the Middle and Upper Jurassic the Catalonian Massif northwest of the Balearic Islands sank, leaving the islands of Ibiza and Mallorca well within the bathyal zone. Menorca, on the other hand, formed a newly emergent massif extending to the east. From Middle Jurassic time onward, a geosynclinal furrow extended in a northeasterly direction towards the south of France. This geosynclinal trough continued through the Sub-Betic region, although true, deep-water conditions and the absence of terrigenous material date from the Lower Cretaceous. The Paleozoic areas of eastern Menorca remained emergent until they were inundated by the sea at the beginning of the Cretaceous.

At the beginning of the Cretaceous the Catalonian Massif expanded somewhat, its margin approaching Ibiza, so that transitional neritic–bathyal facies are found there. Menorca remained bathyal, but it is somewhat affected by the neritic region of the eastern massif containing Menorca. The maximum development of the geosynclinal furrow occurred during Hauterivian and Barremian times, and in Menorca a neritic–littoral facies occurs, suggesting an eastward shift in the margin of that massif. During the Aptian and Albian an important phase of emergence occurred and led to the union of the Ebro–Catalonian and the eastern massifs, with the development upon them of a continental facies. The coast of this enlarged massif apparently lay just north of Ibiza and Menorca, since both islands show neritic–littoral deposits. Menorca itself formed part of the massif at this time. Through the Cretaceous the geosynclinal trough appears to be displaced toward the south, and during the Upper Cretaceous its geosynclinal character comes to an end. Sedimentation may have continued until near the end of the period, but if so, all of the deposits were removed by later erosion. The geosyncline can best be described as an orthogeosyncline of Alpine type, with marine sedimentation occurring principally during the Neocomian. Sedimentary thicknesses within the geosyncline were small (100–200 m) from the Liassic to the Barremian, and of pelagic character during Upper Jurassic to Lower Cretaceous time. The absence of calcareous material between the Tithonian and the Barremian could be as much the result of the depth of the basin (4000 m) as of the lack of planktonic material.

Emergence of the geosyncline was not violent, and the area appears to have become incorporated into the Eocene craton without signs of significant compression. This craton lay between two Nummulitic seas, the Pyrenean to the north, and to the south a sea which lay south of Ibiza but fairly close to the coast of Menorca, and which reaches the coast of present-day Spain at about the latitude of Sueca, extending inland into the Iberian Peninsula. During its emergence the craton itself was subjected to rapid and intense erosion of its relatively soft cover rocks. There were a number of marine transgression phases, however, during the Eocene which particularly affected the region of Menorca.

By Oligocene time, the sea extended farther to the north and the major transgression during the Miocene (Burdigalian) caused the sea to penetrate the center of Menorca and to cover Ibiza. At the same time in Portugal there was a transgression from the Atlantic. The Moroccan–Betic Arch was emergent, but the Sub-Betic and Pre-Betic Zones were submerged. This transgression was short-lived, however, and regression began toward the end of the Burdigalian.

Folding began, mildly at first, in central–southern Menorca in post-Lattorffian but pre-Burdigalian time. A major compressional phase in the central part of Menorca was followed by complete emergence. In both Ibiza and Menorca the final and most powerful phase of folding, the Savic phase, began during the Burdigalian and culminated with the emergence of the whole Balearic domain. The present structures of the Balearic area were completed by Helvetian time. According to Colom and Escandell, the continuity and nature of sedimentation in the Sub-Betic–Balearic Geosyncline excludes any direct relationship with what they have termed the Pyrenean Geosyncline. These two geosynclines are considered as neighboring areas with different characteristics. The Balearic region, in both its sedimentational and tectonic history, is regarded as directly related to the Alpine and not the Pyrenean Zone. The Paleozoic of Menorca is seen as a fragment of the hinterland, perhaps related to the Paleozoic of Málaga but displaced to the north. The positions of the Sub-Betic–Balearic, Alpine, Apenine, and Sicilian Geosynclines tend to confirm the existence of an important Tyrrhenian Massif during the Paleogene, which has fragmented since the Oligocene. However, recent studies (L. M. Ríos, personal communication) indicate that the most intense folding in Mallorca was intra-Burdigalian, a tectonic pattern which is in accord with marine seismic refraction studies in suggesting that fragmentation may be somewhat later in time than had formerly been thought.

III. THE FLOOR OF THE WESTERN MEDITERRANEAN

The geomorphology of the floor of the western Mediterranean will be considered first, followed by a review of the continental platform and slope, and subsequently by the continental rise and deep ocean floor. The latter two will be considered with particular reference to the nature of the basement (here Hercynian), the materials of the Alpine cycle, and finally, the youngest sediments (see Stanley, 1972).

The last two decades have witnessed enormous activity and the growth of knowledge has proportionately advanced. In addition to the scientific stimulus of plate tectonics, economic interests (oil and mineral prospecting) and military needs have played a role. The means of research have been by drilling and geophysical techniques, and these have resulted in a large number of publications. The attempt will be made here to review the principal findings.

A. Morphology and Recent Sedimentation: Generalities

Several excellent maps of the Mediterranean have recently been published (Defense Mapping Agency, 1972), all of which convey more than any written description. The most important isobaths are the 200-m line marking the edge of the continental shelf, and the 2000-m isobath which marks, but not con-

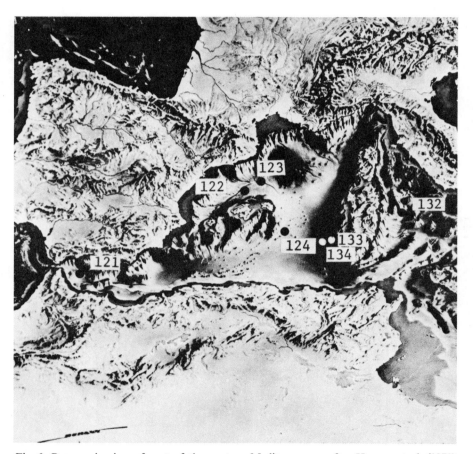

Fig. 6. Panoramic view of part of the western Mediterranean, after Heezen *et al.* (1970). The numbers refer to deep-sea drilling sites.

TABLE I

Results of the DSDP Deep Holes in the Western Mediterranean

No.	Locality	Water depth (m)	Penetration (m)	Formations recorded	
134	North of Sardinia	2864	364	Quaternary and Pliocene	325 m clastic and nannofossil muds, gypsum pebbles, detritus of phyllites. Lower Pliocene with open-sea fauna.
				Upper Miocene	Contains the M reflector, in halites, marine muds with foraminifera. It shows instability and sudden and rapid fluctuations from open-sea to evaporitic conditions.
				Basement	Nonigneous.
133	West of Sardinia	2563	192	Quaternary	Discordant.
				Miocene	Above, an evaporitic series that contains the M reflector. Below it a detrital series.
				Basement	Nonigneous.
124	Southeast of the Balearics	2726	422	Pleistocene and young Pliocene	Pelagic sediments with high rate of organic productivity (average biogenic accumulation 7 cm/year). Muds cut by intra-Pliocene unconformity.
				Miocene	At the bottom of the well an evaporitic series that contains the M reflector; the series extends several hundreds of meters below. The evaporitic formation is formed, from top to bottom, of Tortonian marls(?) with dwarf microfauna, laminated gypsum, and sapropels (60 m thick). Laminated anhydrite, diatomaceous, nodular marine anhydrite, laminated diatomaceous, dolomitic sapropels.
123	North of the Balearics	2290	398	Quaternary, Middle and Lower Pliocene	Well-rounded gravels and silty sandstones both in a channel excavated during the Pleistocene or Pliocene. Below the Quaternary gravels, nannomuds(?).

122	North of Mallorca	2146	162	Basement	Marine sediments. Erosion earlier than the Lower Pliocene removed the higher sediments. The lowest ones lie on a volcanic complex with hydrothermal veins which cut through andesitic ashes 80 m thick. They are probably pyroclastic layers accumulated on the flank of a volcano.
				Quaternary	Gravels and sands contained in an erosional channel excavated during the Pliocene and Pleistocene, overlying the Middle Pliocene.
				Lower Pliocene	Pelagic sedimentation after a period of subaerial exposure (Pontian).
				Lower Pliocene to uppermost Miocene	Marine sand of shallow facies, perhaps of the Pontian epoch, during which were excavated the canyons on the continental shelf around the Balearic Basin. Restricted marine circulation with prevalence of dwarf faunas.
				Basement	Andesitic volcanics. Vesicular or amygdaloid andesites associated with tuffs.
121	Alboran Sea	1163	867	Pleistocene	High rate of sedimentation; high level of planktonic production during the glacial epochs.
				Lower Pleistocene and Pliocene	Muds and marls, pelagic or turbiditic. Unconformity (erosional?) on top of the Lower Pliocene. The Alboran Sea must have existed at least since the Tortonian related to the orogenic deformation of the Betic mountains. The first sediments are already marine. At the bottom are rocks that were at first interpreted as representing marine crust of oceanic type, but which further studies revealed to be breccias of high-grade metamorphic rocks, as cordierite–biotite–feldspar hornfels. Therefore, there is no basalt in the basement but a series of crystalline metamorphic rocks: filadios (schists or phyllites?), gneisses and granodiorites rich in silica, biotite, and plagioclase. They are reminiscent of the rocks of Alpujarrides (ages 9–15 m.y.), according to verbal communications. However there is no certainty that the samples analyzed correspond to Well 121, and therefore both attributions are unsure.

sistently, the foot of the continental slope. The shelf is variable in width and its most conspicuous feature is the salient projecting toward the northeast, which carries the Balearic Islands and which is abruptly cut off with a steep slope into the deep ocean (Fig. 5). Another notable feature is the irregular relief around the Alboran Islands and adjacent areas, and the long (50 km), narrow (14 km) channel east of the Straits of Gibraltar. These features apart, the most conspicuous feature is the monotony of the sea floor of the Balearic Sea, the most extensive bathyal basin of the Mediterranean covered by a blanket of near horizontal sediments (Stanley, 1972). The depths range from 2900 to 2750 m over 240,000 km²; there is some minor isolated relief, little more than 50 m (Mauffret, 1969b). The relief has been identified with diapiric evaporitic cores (Hersey, 1965; Ryan et al., 1970a, b).

There is a considerable range in knowledge of the materials and their dispositions on the floor of the western Mediterranean. Those materials within direct reach either by dredging or drilling, or reached indirectly by geophysical means gravimetric, seismic, or echo-sounding, are fairly well known. They indicate a rather monotonous pattern of sedimentation. Knowledge of the deeper horizons, the result of the JOIDES deep-sea drilling campaigns (Ryan et al., 1970a, b), while important, is also sparse and localized.

The Spanish Mediterranean shelf begins to be better known as the result of exploration stimulated by the discovery of two oil-producing fields off the mouth of the Ebro. Some drill holes have penetrated to the rocks affected by the Alpine movements (Lower Cretaceous limestones).

Practically nothing is known from direct observations of the Hercynian, or older, basement in either the shelf or the deep basins, and very little more about the formations of the Alpine cycle or their structure, for they are seldom pierced in drilling. In consequence, interpretation is either purely hypothetical or obtained by palinspastic reconstructions that also rely heavily on geophysical observations. New observations (e.g., paleomagnetic) and new hypotheses (e.g., plate tectonics) in the area of concern here, have added few real facts and no unifying results. It may be that the association with the plate tectonics hypothesis obscures the many real problems, which are not really solved by the introduction of microplates. Whatever the end result, the value of the hypothesis lies in the flood of new data assembled.

The youngest sediments, those of the last 20,000 years, have been described by Vita-Finzi (1972), who shows the variability of the fluvial component over this period. This corresponds to the sea-level fluctuations, which show maxima between 10,000–5000 years and from 300 years to the present. In the southern Balearic Basin, Rupke and Stanley (1973) have distinguished two different but alternating types of sediment. Muds, a few centimeters to 50 cm thick, overlying turbidite sands or silts, and comparatively coarse sediments, varying from near zero thickness to 50 cm, forming silty plates or turbidity sandstones overlying

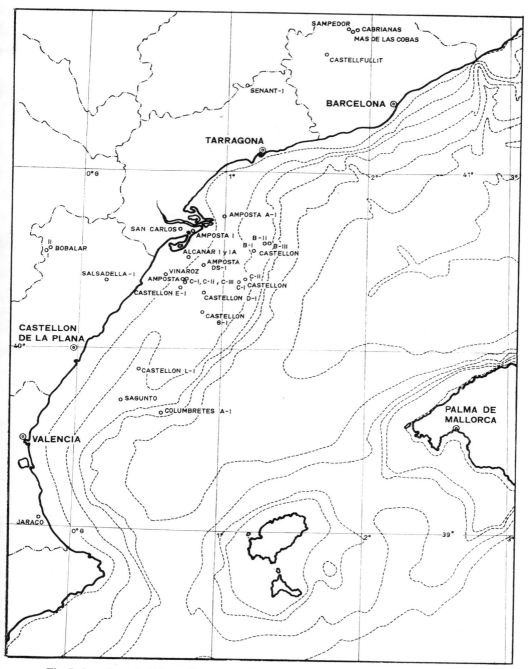

Fig. 7. Location of offshore oil exploration holes in the Iberian Mediterranean platform.

TABLE II. Summary of Results

Well locations are shown in Fig. 7. W⟨

Designation	Year	Water depth (m)	Bottom depth (m)	Quaternary	Pliocene	Miocene
Delta Ebro 1 (terrestrial)	1966	—	306.8	214		
Amposta 1 (terrestrial)	1965	—	1647.5	704	72	
Amposta A-1	1973	74	2603	576	1879	
Alcanar 1-A	1973	17	2038.5	818	639.5	492.5
Vinaroz 1	1970	31.1	2752.1	792.2	683	65
Barcelona C-1	1974	85				
Amposta DS-1	1973	64	3733	1399	1542	
Amposta B, B-1 to B-6	1971–1972	61		1351	1099	978 / 591
Amposta C-1	1970	62.2	2522.4	961.8		816
Amposta C-2	1970	62.2	2108	1034.8		781.5 / 799.5
Amposta C-3	1971–1972	62	3867	1108		711.5
Castellón E-1	1971	66.4	2032.4	1321.6		512.5
Castellón B-4	1974	112	2888	1519	1125	
Castellón B-2	1973	114.8	2740.8	609.2	1232	668
Castellón B-3	1973	717	2913	608	1352	733
Castellón B-1	1968	97.8	2833.8	628.2	974	996.5
Castellón D-1	1973	77.7	3049.7	1130.3	664	957
Castellón L-1	1973	84	4558.5	1021	115	2140
Sagunto	1972	130	4022	237.4		1419
Castellón C-1	1971	91	3129	897	1860.4	
Castellón C-2	1972	96	3005	892	1891	
Castellón G-1	1971	83	3054	902	1686	
Columbretes A-1	1974	126	4722.6	4723.6		

* *Explanation of table:* Quaternary–Upper Pliocene group known as Ebro Sandstones. Lower Pliocene–late Upper Miocene group known as Ebro Clays. Lower Upper Miocene–upper Middle Miocene group known as Castellón Sandstones. Lower Middle Miocene group known as Castellón Clays, but this term does not appear to be fixed. The late Upper Miocene is recorded in many wells as Tortonian. It is considered to be equivalent to the M reflector and in the Castellón L-1 well contains anhydrite. The lower Upper Miocene is combined with the middle as Helvetian. The Lower Miocene is Burdigalian. The Miocene transgression is found in most wells at the base of the Middle Miocene and is recognized in Amposta B-1, B-2, C-2; Castellón E-1, B-4, B-1, C-1. The Alcanar formation, more or less conglomeratic in type, forms the base of the transgression in Amposta A-1; Alcanar 1-A; Amposta BS-1; and Castellón B-3. Continental Oligocene is found in Castellón E-1. The top of the Cretaceous is recognized by karst character in wells Amposta B-1, B-2, B-3, B-4, B-5, B-6, C-1, C-2; and Castellón D-1. Undifferentiated Mesozoic reported from Castellón C-1. Upper Cretaceous and Turonian in Amposta A-1, and doubtfully in Castellón G-1 (with anhydrite). Lower Cretaceous, particularly Aptian, reported in Amposta A-1; Alcanar 1-A; Vinaroz 1; Amposta DS-1, B-1 to B-6, C-1, C-2; Castellón E-1, L-1; Sagunto 1; Columbretes. Jurassic in Vinaroz 1; Alcanar 1-A; Amposta DS-1, C-3. Liassic in Amposta DS-1. Undifferentiated Trias in Castellón B-2, B-4 (doubtful). Keuper in Amposta DS-1, C-3. Permo-Triassic (Buntsandstein facies) in Amposta DS-1, C-3; Castellón B-1, B-3, G-1. Undifferentiated Paleozoic in Castellón B-2. Carboniferous in Castellón B-1, C-1, C-2. Volcanic material in Castellón L-1 (Lower Cretaceous); Sagunto 1 (Miocene); and in great thickness

illing on the Iberian Mediterranean Platform*

grouped according to distance from the coast.

Oligocene	Mesozoic	Paleozoic	Observations
	92.8		Mesozoic undetermined
819	52.5		Mesozoic undetermined
	74		Ends in Turonian–Aptian
	71.5		Ends in Lower Cretaceous–Upper Jurassic
	1039		Ends in Cretaceous–Jurassic
			In progress; no information
	664		Passed through Lower Cretaceous and Jurassic and ended in Buntsandstein
			Lower Cretaceous
	622		Lower Cretaceous
	167		Lower Cretaceous
	1897.5		Cretaceous, Jurassic, Triassic, Permo-Triassic
65	192		Continental Oligocene; Lower Cretaceous
	20		?Triassic sands and red shales
	115.8		Triassic, Paleozoic
	85		Permo-Triassic
	137.3		Permo-Triassic, Carboniferous
	43		Lower Cretaceous and Upper Jurassic
	114.5		Lower Cretaceous with volcanics
	99		Aptian
	111.6	128	Mesozoic indeterminate; Lower Carboniferous
		30	Lower Carboniferous
	172	128	?Upper Cretaceous and Permo-Triassic
			No details known.

in Columbretes A-1. Anhydrite in Amposta DS-1 (Lower Lias, Keuper, Muschelkalk, and Bunter); Castellón L-1 (Upper Miocene), G-1 (?Upper Cretaceous). Average thickness of Miocene (from 12 wells) about 875 m with maxima 2140 m (Castellón L-1) and 1591 m (Amposta 1); minima 65 m (Vinaroz) and 512.5 m (Castellón E-1). Average thickness of Quarternary–Pliocene (from 13 wells) 1489 m with maxima 2374 m (Sagunto 1) and 1952 m (Castellón B-3); minima 962 m (Amposta C-1) and 1035 m (Amposta C-2). In examination of the distribution of pre-Miocene rocks, the Cretaceous to Miocene is systematically found in a band closer to the shore than the Triassic, with Paleozoic still further offshore. Although the number of wells is perhaps insufficient, and there is insufficiently extensive section of coast, there seems to be evidence for the existence of a depression parallel to the coast caused by a subsidence of the Paleozoic basement and its uplift further offshore, with the concomitant erosion and even total removal of the Mesozoic cover. There also appears to be a progressive increase in Neogene sedimentary thickness along the axis of the depression toward the Balearic Sea. This may be regarded as an indication of the thinning of the Paleozoic basement in this direction and the rise of lithospheric material under the deep sea. The intense volcanism of the Columbretes could be interpreted in this case as material injected into the rift produced by this tensional mechanism. The average thickness of the young sedimentary cover is 2418 m, ranging from maxima of 3793 m (Sagunto 1) and 3276 m (Castellón L-1) to minima of 1509 m (Vinaroz 1) and 1595 m (Amposta 1).

TABLE III

Correlation between Oil Exploration Wells and the DSDP Results

Correspondence between Terrestrial, Continental Platform and Sea-Floor Series

Formations	Onshore sequence	Continental platform Castellón E-1	Sea floor	
			DSDP 122 and 123	DSDP 124
Pliocene, uppermost Miocene	Nonexistent or continental	Sandstone or marine shales (Ebro), unconformity between Miocene and Pliocene (anhydrite in Castellón E-1)	Quaternary gravels and silty sands, Pliocene pelagic muds, erosional surface below the Lower Pliocene	Pelagic sediments, muds cut by intra-Pliocene unconformity
Upper Miocene	Arkoses and conglomerates with lenses of dominantly lacustrine shales to the NE and SW, diachronous and transitional to shallow marine facies, always ending in a continental facies	Sandstones and marine shales (of Castellón)	Marine sandstones (shallow-water facies)	Evaporitic facies
Middle and Lower Miocene	Bioclastic limestones above continental redbeds	Bioclastic limestones deposited in shallow water		
Oligocene	San Sadurní continental redbeds discordant on Mesozoic, Paleozoic, or granites.	Continental redbeds strongly discordant on lower Cretaceous		

the previous type. These rhythmic sediments have been deposited at the rate of about 8 cm/1000 years since the last glaciation.

The recent sedimentation in the Alboran Sea has been described (Milliman *et al.*, 1972) as consisting of about 200 m of Pliocene and Quaternary stratified sediments overlying a complex of basic volcanics. The floor lies at a depth of 100–120 m. The sediments are mostly organic carbonates, for the trough between the Alboran Islands and the African coast has acted as a sedimentary trap for coarse materials. The results of the cruises of the *R/V Akademik Vasilov* have been published by Emelyanov (1972), who considers the northern Mediterranean as a geosyncline, the southern part as a platform extending from the Lebanon to Tunisia, with the western Mediterranean intermediate in character, a conclusion based upon active, modern volcanism, organic and chemical sedimentation, and diagenetic redistribution. Texturally the terrigenous sediments in the western Mediterranean basin are either pelitic on aleuro-pelitic muds, the dominant fraction being of subcolloidal size (0.001 mm) typical of deep sea sediments at the transition from geosyncline to platform and from humid to arid climates. Illite, with more than 70% is the dominant clay mineral present.

Pyroclastic materials do not exceed 10% during the last few thousand years, but were much more important during the Pleistocene. The distribution of coarser material seems to be related to periods of shallow water and high hydrodynamic activity, even if in deeper water such material were carried down submarine canyons. The organic content is low, due to the paucity of phytoplankton and the rapid decomposition in warm oxygenated water, which results both in a low diatom population and a reduced silica content.

B. Neogene Sedimentation

The main contribution to the much discussed Neogene sedimentation has been from the JOIDES–DSDP program, the results of which were published by Nesteroff *et al.* (1972). The general succession can be summarized as follows:

Plio-Quaternary: Semipelagic oozes
Messinian: Evaporitic series
Tortonian Serravallian–Langhian: Blue marls

The evaporitic series (reflector Horizon M) is continuous over much of the Mediterranean. The series, clearly defined at top and bottom, consists of halite, anhydrite, gypsum, and dolomitic marl; it is of a facies similar to that found on the mainland, e.g., the Mediterranean coast of Spain. Figure 8 shows the western limit of the evaporitic zones. Nesteroff and colleagues (Nesteroff *et al.*, 1972; Nesteroff and Ryan, 1973) believe the logical (if geologically forced) interpretation to be that during the Messinian the Mediterranean was isolated by the closure of the Straits of Gibraltar and underwent extreme evaporation of an immense but shallow basin leading to the precipitation of salts. The Pliocene began with the invasion of marine water by the reopening of the Straits of Gibraltar in a fashion so sudden as to be almost catastrophic, returning the Mediterranean to normal marine conditions.

There have been six holes drilled in the western Mediterranean (121, 122, 123, 124, 133, 134), the detailed results of which are described by Ryan *et al.* (1970*a*, *b*) (Fig. 6; Table I). Two main points emerge from the examination of the borehole detail, the general and widespread character of the Messinian evaporitic episode and the brusque change from Pliocene to Quaternary. Stanley *et al.* (1972) summarize the sedimentation pattern in the Alboran Sea as follows:

(1) The sediments in the West Alboran Basin are almost exclusively semi-pelagic muds during the Upper Pleistocene and Holocene, with a few, thin, sandy horizons.

(2) The sedimentation rate in the Pleistocene and Upper Holocene was 20 cm/1000 years, rising to 40 cm during the Holocene on the basin flanks and to 80 cm/1000 years on the Alboran Basin plain during the Wurm.

The region to the east of the Alboran Sea, including the Balearic area, has been described by Leclaire (1972). Continuous profiling (Sparker, Flexotir) has given a clearer idea of the younger Quaternary sedimentation pattern (Alla *et al.*, 1972).

The reflector Horizon M is one outstanding feature of the Mediterranean Basin. It was first interpreted as possible basement, then as an evaporitic horizon of Triassic age (by analogy with the mainland), but only recognized as Messinian as a result of Cruise 13 of the *Glomar Challenger*. It provides the source for many salt domes which rise gently through the flat bottom sediments. These are particularly numerous in the region between the Balearic Islands and the French and Italian coasts, and they have been the subject of numerous studies (Alinat and Cousteau, 1962; Leenhardt, 1968; Watson and Johnson, 1968; Wong *et al.*, 1970; Mauffret *et al.*, 1969, 1970, 1971, etc.) Not all geologists accept the dessication hypothesis for the origin of the evaporites. Some believe they may be due to direct precipitation from deep-sea water, e.g., Schmalz (1969).

The discussion here has been restricted to only a small number of questions concerning the origin and sedimentation of the western Mediterranean. Much remains to be done.

C. The Balearic Sea and Gulf of Valencia

The Gulf of Valencia occupies the region between the Catalonian–Valencian marine platform and the Balearic Ridge. The bottom descends to 2000 m between Valencia and Ibiza. The sediments overlying the marine platform have been penetrated by DSDP holes 122, 123, and 124 (Fig 6; Table I). In Hole 124, gypsum was encountered and attributed to Horizon M, and in 122 and 123 andesitic ash and tuff were recorded. The sediments extend from the platform over the continental slope into deep water, where abundant salt domes and the presence of Horizon M are recorded (Alinat and Cousteau, 1962; Menard *et al.*, 1965; Hersey, 1965; Glangeaud, 1966; Leenhardt, 1968; Watson and Johnson, 1968; Wong *et al.*, 1970; Montadert *et al.*, 1970; and Mauffret *et al.*, 1969, 1970, 1971, 1973).

Horizon M, first attributed to the Keuper, was identified as the Messinian evaporitic horizon. It separates 1000 m of unconsolidated sediments above from about 1650 m of sediments below (Wong *et al.*, 1970). In the Balearic Sea, Finetti and Morelli (1972, 1973) find some 1200 m of Plio-Quaternary beds, and a subhorizontal sequence below the evaporites that is 4800 m thick and which they attribute to the Miocene.

In the absence of direct observation, the interpretation of the Balearic Sea, as of the western Mediterranean in general, is based largely upon geophysical surveys. (Pautot *et al.*, 1973; Auzende *et al.*, 1973; Allan and Morelli, 1971; Finetti and Morelli, 1972, 1973; Fahlquist, 1963; Fahlquist and Hersey, 1969).

Allan and Morelli (1971) concluded from the magnetic and gravimetric data that the Balearic Basin lies on material approximating oceanic crust, but were unable to establish any pattern of magnetic lineations. The seismic records of Finetti and Morelli appear to show a crustal thickness of 15.8 km, with Moho 4 km deeper than indicated by Fahlquist (1963) and Fahlquist and Hersey (1969). The NE–SW Flexotir reflection profile of Mauffret *et al.* (1973) from the Valencia–Columbretes coast toward Nice differs only in details from those of Finetti and Morelli. They show that in contrast to the coastal margin of France north of the Balearic Islands, there is an extensive shallow continental basement over which in restricted basins lies a thick infrasaline series without evaporites and a Plio-Quaternary sequence identical with that in the abyssal plain.

There seems little ground for an interpretation of the Gulf of Valencia–Balearic Ridge in terms of plate tectonics. The suggestion of Miocene fracture and rotation of the Balearic Ridge seems improbable, given the alignment of structural trends and the closer relationship of the Balearic Islands with Betic rather than the Iberian Ranges. The interpretation of the seismic profiles north of Mallorca shows a northward thinning of the 4 km of presumed Tertiary rocks below a Plio-Quaternary cover (Hinz, 1973). The central zone of the basin north of the islands has a crust not typically continental, but neither is it typically oceanic. Hinz (1973) proposed that the Balearic Islands were once closer to the Iberian Peninsula and constituted a link between the Betic Cordillera and the eastern Pyrenees.

An origin of the Balearic Basin by the rotation of the Corso-Sardinian block (see Moullade, Chapter 2, this volume), suggested by Carey (1958) and apparently supported by some paleomagnetic evidence (Nairn and Westphal, 1968), has found support from Allan and Morelli (1971) and Le Pichon *et al.* (1971) in terms of a plate model. Although clear, linear, magnetic anomalies are hard to discern, Bayer *et al.* (1973) did distinguish low-amplitude magnetic lineations and on this basis proposed the rotation of the Corso-Sardinian block plus the eastern Kabylie block, with a simultaneous offset along a transfer fracture of Sardinia and Corsica. The date of these movements is not given, but must be Aquitanian or even Oligocene because of the existence of the thick, infrasaline series. The new crust produced is presently being compressed by the African plate and it is possible that the original magnetic material is being reassimilated in the mantle.

Finetti and Morelli (1972, 1973) prefer to introduce a Mediterranean plate (western Mediterranean–Tyrrhenian Sea) which has a thin, continental crust and a general west-to-east motion. This triangular plate, its apex at the Straits of Gibraltar, is bordered by the Rif and Betic sedimentary troughs, its motion being the reaction to compression between the European and African plates. The Corso-Sardinian and Balearic blocks merely represent parts resisting the general post-Miocene subsidence, for they do not consider that the geophysical data sustain a crustal generation model. In this view the present Mediterranean was generally delineated before the salinity crisis, and they propose evaporite formation in shallow water.

D. The Alboran Sea and Islands

The Alboran Sea, lying between North Africa and Spain and extending as far west as the Straits of Gibraltar, is important for it is considered the source area of the Betic sheets. Following the 1967 cruise of the *Jean Charcot*, the complex submarine relief (Fig. 9) is much better known. Giermann *et al.* (1968) show that the relief is as much a function of volcanicity as of Mio-Plio-Quaternary tectonism. Most of the relief is of volcanic origin, the volcanos being located along Miocene to Quaternary faults. There is an important, fault-bounded, elevated zone, the Alboran threshold, elongated WSW–ENE, limited to the north by the western Alboran Basin and separated from the African coast to the south by the eastern Alboran Trough. The northern Alboran Basin extends to the Straits of Gibraltar and terminates to the east at the Alboran furrow, which in turn is the southeastern termination of the peninsular basement (Djibouti Bank).

A small (71,200 m²) island, Alboran Island, about 53 km off Cabo Tres Forcas on the African coast, rises from the northern edge of the Alboran threshold. This low, triangular-shaped island, is formed of alternating con-

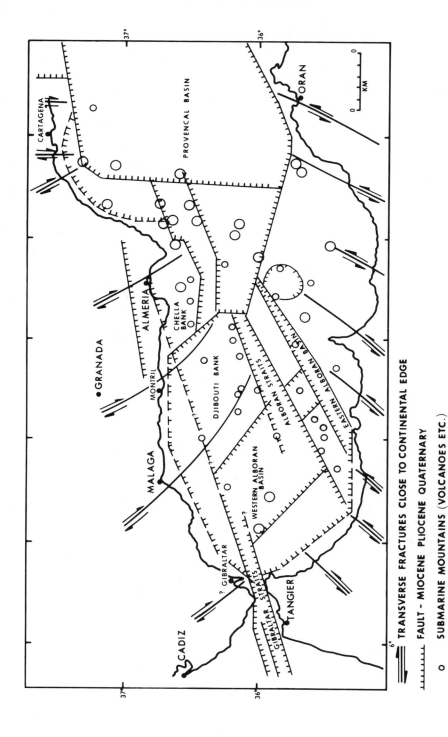

Fig. 9. The fracture pattern of the Alboran Sea (after Barrois *et al.*, in Giermann, 1968).

glomerate or breccia with thick tuff, capped by a thin layer of continental and marine Holocene. There two thin basalt flows (3–5 m thick). Within the conglomerate and breccia are several varieties of porphyritic rocks, (Gaibar-Puertas and Ruiz Lopez, 1969, 1970a, b). DSDP hole 121 (Fig. 6; Table I), drilled in the Alboran deep in the general vicinity, passed into basalts below Miocene

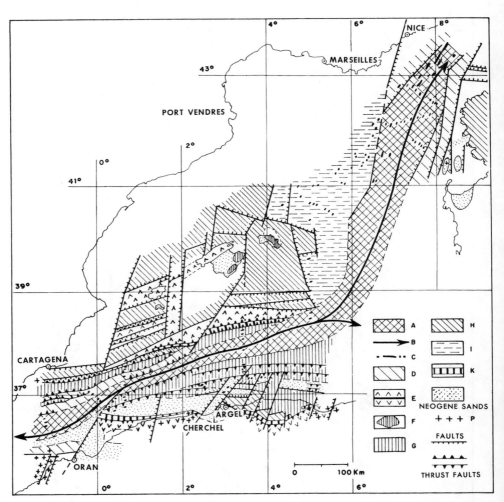

Fig. 10. The structure of the Alboran Sea (after Glangeaud, 1951, 1970). Key: A, Median zone; B, Axis of median zone; C, Diapirs; D, Zone of structural units vergence to north; E, Zone of structural units vergence to south, internal with respect to limestone chain; F, Zone of Menorca and Harta series, lateral transition from evaporites; G, Zone of structural units vergence to south, external with respect to limestone chain; H, Paleozoic autochthon; I, Epicontinental zone with gently folded Neogene; K, overthrust Paleozoic elements of Cape Génès and Chenona; P, Volcanic intrusions and extrusions.

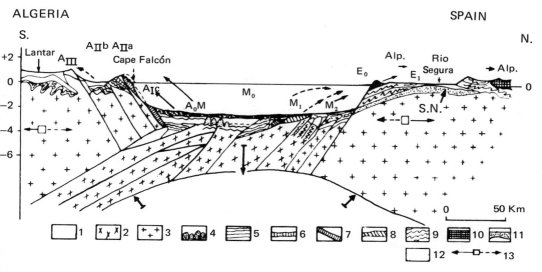

Fig. 11. Structural section between Cape Falcon and Benejuzar. Legend: (1) upper mantle; (2) Mesogean crust; (3) continental crust (Sierra Nevada and the African block); (4) domes and diapirs of Tertiary plastic sediments; (5) unmetamorphosed Paleozoic; (6) Mesozoic of the Oran unit; (7) Paleozoic and Liassic scales, terrestrial and submarine A_Ic; (8) hard rocks of the Spanish margin; (9) Mesogean mantle Alpujarrides; (10) Malaguides; (11) Plio-Quaternary; (12) cover of the Sierra Nevada (Nevada Filabrides) basement; (13) successive horizontal movement of the African and European blocks between Triassic and Middle Jurassic and between Cretaceous and Eocene. The length of the arrows is only relative; the displacements are more important. Symbols: A_{III}, Tessala Zone; A_{II}b, Sebkah of Oran Zone; A_Ic, Paleozoic and Liassic, mostly submerged; M_0, central Mesogean Zone; M_1, probable source of Malaguides; M_2, probable source of Alpujarrides; E_0, Sierra Nevada upon which lay the original Mesogean sediments; E_1, tectonic window through which the Sierra Nevada and Filabrides can be seen below their Mesogean cover.

sediments (Ryan *et al.*, 1970*a, b*; Olivet *et al.*, 1973*b*), proving the existence of the Alboran Sea as early as Tortonian. Its existence is linked to the deformation of the Betic Cordillera.

There are many interpretations of the structure and significance of the Alboran Sea. The idea of a mosaic structure of the western Mediterranean was developed by Glangeaud (1951, 1970; Glangeaud *et al.*, 1967*a,b*, 1970), summarized in Figs. 10 and 11, with a median zone extending from south of Cartagena to south of Nice, with a branch extending to the east from south of the Balearic Islands. This strip is characterized by positive Bouguer anomalies, suggesting a thinner crust than at continental margins. According to Bourcart (1960), the Mediterranean became progressively more restricted from Miocene time onward, ending with "Pliocene revolution," an inversion of relief with the rapid emergence of mountain chains, and a general deepening of a sea which had acquired its present configuration.

E. The Straits of Gibraltar

This westernmost outpost of the Mediterranean, at the western extremity of the Alboran Sea, and as the only opening to the oceans of the world has exerted a great influence in the post-Alpine history of that sea. The Straits are 12.5 km wide at the narrowest point with a maximum sill depth at 935 m (Reig Vilaplana, 1958) and a minimum of 320 m (Sonnenfeld, 1974a). The sill dips towards both the Atlantic and the Mediterranean, but has generally a complex relief. In nearly all hypotheses, the straits are due to fracturing, erosion, or both, of the Campo de Gibraltar Flysch (Fontboté et al., 1971), which lies in the tight arc formed by Betic and Rif chains. The similarity of this arc with the western Alps has been noted by Wills (1952), Klemme (1958), Carey (1955), Bogdanoff (1963), and Brinkmann (1966). Many regard the arc as the result of an orogenic event in a curved geosyncline, although Carey (1955) considered it a typical orocline and the idea of translation and rotation has interested many authors, such as Dietz and Holden (1970), Boccaletti and Guazzone (1974), and Le Pichon et al. (1972).

Former connection between Tethys and the Atlantic Ocean, which occurred along the Rif and Betic margins (Brinkmann, 1966), was broken by Miocene orogenesis and provoked the Messinian salinity crisis in some views. The present opening originated in the Upper Pliocene by fracture along the major conjugate system. There are many variants of this model since plate tectonics introduced the idea of a plate contact through the straits. Several interpretations are illustrated in Figs. 12 (Reig Vilaplana, 1958), 13 (Glangeaud et al., 1970), 14, and 15 (Araña and Vegas, 1974).

According to one group of theories the Gibraltar area formed the western end of Tethys until the end of the Triassic when, with the opening of the Atlantic, an important fracture separating Africa and Eurasia became active. Movements, commonest during the Jurassic and Cretaceous, were of a rotational character. The most important point for Dietz and Holden (1970) is that Gibraltar lay on a fracture line. A variant using the idea of microplates is that the Gibraltar Arc is the result of the vergence of two chains of opposed trends (Betic and Rif) (Andrieux et al., 1971). They accept the existence of an Alboran oceanic plate bounded by two branches of the Azores transform fault. On either side and underthrusting with relative movement of up to 100 km are sialic plates (or overthrusting of the Alboran microplate), resulting in the production of gravitational slides now forming the external sheets. The nappes of the internal zone (Nevada–Filabrides, Alpujarrides, Malaguides) are not then European or African margins but parts of the Alboran plate. The Gibraltar Arc is then the result of underthrusting at the western limit. The Alboran plate also moved to the east at a slower rate than the African and European plates. This hypothesis has the advantage, providing the

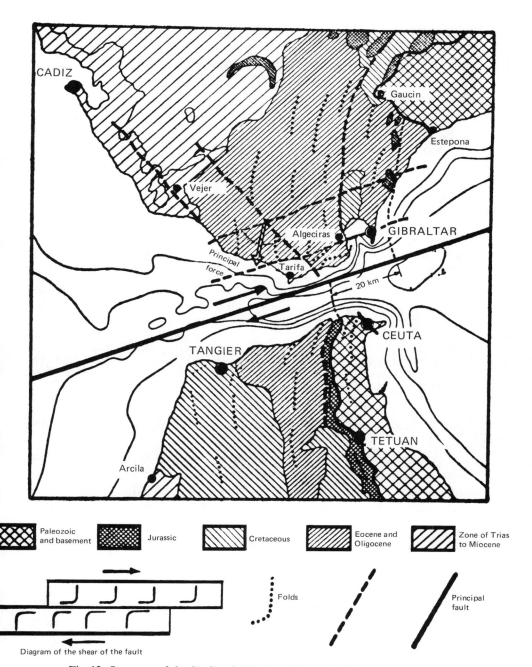

Fig. 12. Structure of the Straits of Gibraltar (after Reig Vilaplana, 1958).

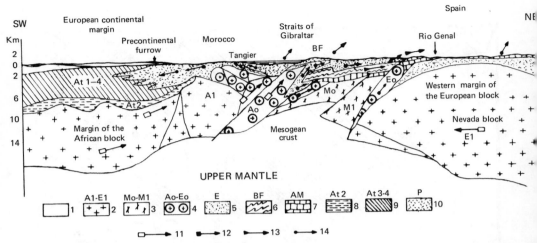

Fig. 13. Schematic crustal section across the Straits of Gibraltar between Tangier and Rio Genal (after Glangeaud): (1) upper mantle; (2) crust of the African and European blocks; (3) Mesogean crust; (4) "bisel" African front; (5) base of crystalline crust of the Sierra Nevada; (6) lower mantle (pre-Eocene); (7) Penbeticas and Rondaides; (8) Paleozoic and Mesozoic of pericontinental Atlantic margin; (9) Quaternary and Tertiary (mostly Oligo-Miocene) deposits of the pericontinental Atlantic margin; (10) parts of the Rif slide which grade laterally into the Oligo-Miocene of (9); (11) movements relative to the European block; (12) northeast movement of the frontal bulge (BF) (It is impossible to illustrate the shear motion of unit Al toward the southwest on this figure); (13) pre-Oligocene northwest motion; (14) post-Oligocene motion (northwest, west, and southeast) of the Numidian, which makes the "post-tectonic" Gibraltar Arc.

space reduction required without requiring the junction of the latter two plates.

Not all geologists can accept a fracture zone passing through the Straits of Gibraltar, given the many paleogeographic and tectonic similarities of the Rif and Betic orogens. According to Durand Delga (1961, 1973) and others the Mesogean development occurred within a single European–African plate.

F. The Origin and Structure of the Western Mediterranean

The historical development of the interpretation of the origin of the Mediterranean is complicated by the sheer volume of published material. The classical theory, exemplified by Klemme (1958), Bogdanoff (1963), or Brinkmann (1966), considers the progressive narrowing of the Tethys between the Angara (Eurasian) shield and Gondwana by successive tectonic cycles affecting marginal geosynclines. Modern hypotheses imply reorganization of former continental masses linked to rotations and of late complicated movement of plates.

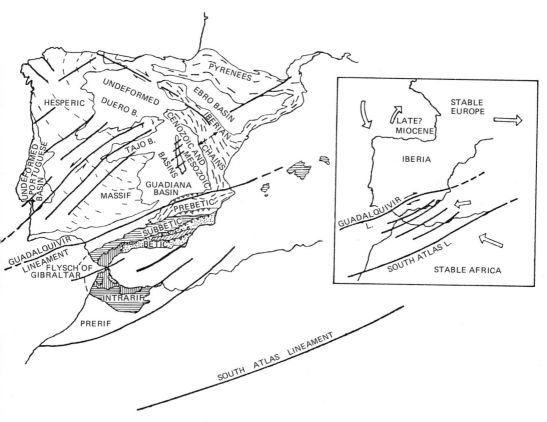

Fig. 14. Geotectonic sketch of the Iberian Peninsula and North Africa (after Araña and Vegas, 1974). The heavy lines in the left-hand figure indicate the principal tear faults; in the right-hand figure arrows show the relative movements of Africa and Europe at the end of the Cretaceous, including that along tear faults between the Guadalquivir and South Atlas faults.

It is a significant observation that the chains bordering the Mediterranean seem to be abruptly cut off, or disappear, at the margin of the platform without any trace of their prolongation in the featureless basins of the Mediterranean. This is true of the coast of Spain, with the termination of the Balearic salient where the Paleozoic basement forms a steep cliff. There may be several interpretations, continental displacement, migration, or the thinning of blocks with the disappearance of continental crust (i.e., "oceanization"). The latter view is presented by De Booy (1969) and De Roever (1969).

The attitude of De Booy stems from the observation that during the history of the western Mediterranean, prior to its present form, important segments of continental crust have disappeared, or at any rate have been concealed from direct observation, and some of these have already been referred to. If this

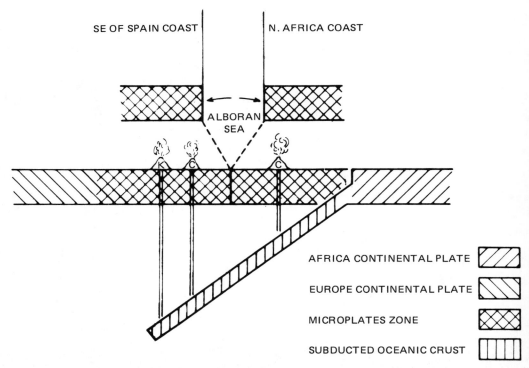

Fig. 15. Geometric relation between volcanism and subducted oceanic crust, before the opening of the Alboran Sea. C = calcalkaline rocks; K = potassic rocks (Araña and Vegas, 1974).

is true, and as structures cannot be traced out into the basins which appear to consist of oceanic crust, then it may be supposed that the continental crust was formerly more extensive and some has either separated and migrated, or disappeared. The analysis of sedimentary deposits and source areas shows that during the Upper Jurassic–Lower Cretaceous great areas of granitic crust were subjected to subaerial erosion and supplied sediment to the geosynclinal regions. These areas, (the Ebro Basin, etc.) disappeared between the Upper Cretaceous and, according to De Booy's sedimentological analysis, Middle Eocene. There was an independent, later phase of gravitational sliding during the Miocene. The "lost" areas, however, bear no direct relation to the present Mediterranean basins, whose origin is related to late Neogene–Quaternary events and linked with renewed loss of continental crust.

The alternative view is exemplified by the work of Carey (1955, 1958), in which the curvature of orogenic belts is explained as the result of the rotation of cratons rather than the passive result of the original shape of the geosyncline. One such orocline resulted from the rotation of the Iberian Peninsula around a point in the Bay of Biscay, a sinistral movement from the original near coincidence of the Spanish Cantabrian coast and the shelf off Aquitaine, resulting in a compressional closing of about 20° in the Pyrenean sector and an extensional opening of about 40° in the West Cantabrian sector. Later the rotation was corrected to 35°. His work has been superceded by the development of plate tectonic interpretations based upon either a simple plate scheme (e.g., Le Pichon, 1968) or one based upon many plates (Smith, 1971).

No general agreement has been reached concerning the rotation of the Iberian Peninsula; differences affect angle of rotation, location of the locus of rotation, and even the date of the rotation (Schwarz, 1963; Van Dongen, 1967; Van der Voo, 1967; Tarling and Tarling, 1971; and others). If only the present Mediterranean is considered, several interpretations are again possible. Aubouin (1971) considers that the Plio-Quaternary tectonic history of the western Mediterranean is dominated by fracturing, with one set a reactivation of older structures and another marking extension of the sea floor, thus replacing the earlier period of tectonic compression.

The floor of the western Mediterranean should be formed either of oceanic crust, or possibly by a thinned continental crust according to the available seismic and gravimetric data. Auboin raises the unanswered question of whether this existed prior to, or was the result of, the Alpine cycle, with, however, a tendency to regard it as new since the basins cut Alpine structures. Auboin, therefore, concludes that the Plio-Quaternary distension has ruptured the old continental crust and the neo-oceans are part of a new post-Alpine geosynclinal cycle.

ACKNOWLEDGMENTS

For reproduction of figures in this chapter, I gratefully acknowledge the permission of Dr. José Ignacio Izaguirre Rimmel, Director, Instituto Geologico y Mínero de España; R. J. Beaton, Technical Director of the Defense Mapping Agency, U.S. Department of Defense; Dr. C. Salle, l'Institut Français du Pétrole, le Centre National de l'Exploration Océanographique; Drs. Ramón Querol and José M. Fontboté; Dr. D. J. Stanley; and Dowden, Hutchinson and Ross, Inc.

REFERENCES

Alinat, J. and Cousteau, J. Y., 1962, Accidents de terrain en mer de Ligurie, *Océanogr. Géol. Géophys. Médit. Occid.*, Colloque National de C.N.R.S., Villefranche-sur-Mer, 1961, p. 121–123.

Alla, G., Dessolin, D., Got, H., Leenhardt, O., Rebuffatti, A., and Sabatier, R., 1969, Résultats préliminaires de la mission *François Blanc* en sondage sismique continu, *Vie et Milieu*, v. 20, p. 211–220.

Alla, G., Dessolin, D., Leenhardt, O., and Pierrot, S., 1972, Donneés du sondage sismique continu concernant la sédimentation Plio-Quaternaire en Méditerranée nord-occidental, in: *The Mediterranean Sea—A Natural Sedimentation Laboratory*, Stanley, D. J., ed., Stroudsburg: Dowden, Hutchinson and Ross, p. 471–487.

Allan, T. D. and Morelli, C., 1971, A geophysical study of the Mediterranean Sea, *Boll. Geofis., Teor. Appl.*, v. XIII, p. 99–142.

Alvarez, W., 1972, Rotation of the Corsica–Sardinia microplate, *Nature*, v. 235, p. 103–105.

Alvarez, W., 1973, The application of plate tectonics to the Mediterranean region, in: *Implications of Continental Drift to the Earth Sciences*, Tarling, D. S., and Runcorn, S. K., eds., New York: Academic Press, v. 2, p. 893–908.

Alvarez, W., Cocozza, T., and Wezel, F. C., 1974, Fragmentation of the Alpine Orogenic belt by microplate dispersal, *Nature*, v. 248 p. 309–314.

Andrieux, J. Y., and Mattauer, M., 1973, Précisions sur un modèle explicatif de Gibraltar, *Bull. Soc. Géol. France*, v. XV, p. 115–119.

Andrieux, J., Fontboté, J. M., and Mattauer, M., 1971, Sur un modèle explicatif de l'Arc de Gibraltar, *Earth Planet. Sci. Lett.*, v. 12, p. 191–198.

Araña, V., and Vegas, R., 1974, Plate tectonics and volcanism in the Gibraltar Arc, *Tectonophysics*, v. 24, p. 197–212.

Arcyana, X. Y., 1975, Un premier bilan du project Famous, *La Recherche*, no. 53, p. 167–170.

Auboin, J., 1971, Réflexions sur la tectonique de faille plioquaternaire, *Geol. Rundschau*, v. 60, p. 833–848.

Auzende, J. M., and Olivet, J. L., 1973, The origin of the Western Mediterranean Basin, *J. Geol. Soc. London*, v. 129, p. 607–620.

Auzende, J. M., Bonnin, J., and Olivet, J. L., 1971, Hypotheses on the origin of the Western Mediterranean Basin, *Dept. Sci. Centre Océanol. Bretagne*, Contrib. 110.

Auzende, J. M., Olivet, J. L., and Bonnin, J. L., 1972, Une structure compressive au nord de l'Algérie? *Deep Sea Res.*, v. 19, p. 149–155.

Auzende, J. M., Bonnin, J., and Olivet, J. L., 1973a, The origin of the Western Mediterranean Basin, *J. Geol. Soc. London*, v. 129, p. 607–620.

Auzende, J. M., Olivet, J. L., and Pautot, G., 1973b, balearic Islands: Southern prolongation, in: *Initial Rept. Deep Sea Drilling Project*, 48.3, *Structural Framework of the Western Mediterranean*, Ryan, W. B. F., *et al.*, eds., Washington, D.C.: U.S. Govt. Printing Office, v. XIII, p. 1441–1447.

Auzende, J. M., Rehault, J. P., Pastouret, L., Szep, G., and Olivet, J. L., 1975, Les bassins sédimentaires de la mer d'Alboran *Bull. Soc. Géol. France*, v. XVII, p. 98–107.

Azema, J., Bourrouilh, R., Champetier, Y., Fourcade, E., and Rangheard, Y., 1974, Rapports stratigraphiques, paléogéographiques et structuraux, entre la Chaîne ibérique, les Cordillères bétiques et les Baléares, *Bull. Soc. Géol. France*, v. XVI, p. 140–161.

Bandy, O. L., 1972, Messinian deep-water evaporite deposition and the Miocene–Pliocene boundary, *24th Int. Geol. Congr.*

Bayer, R., Le Mouel, J. L., and Le Pichon, X., 1973, Magnetic anomaly pattern in the Western Mediterranean, *Earth Planet. Sci. Lett.*, v. 19, p. 168–176.

Bemmelen, R. W. Van, 1969, Origin of the Western Mediterranean Sea, in: *Symposium on the Problem of Oceanization in the Western Mediterranean*, *Verh. K. Ned. Geol. Mijnb. Genoot.*, v. 26, p. 13–52.

Benson, R. M., 1972, Ostracodes as indicators of threshold depth in the Mediterranean during the Pliocene, in: *The Mediterranean Sea—A Natural Sedimentation Laboratory*, Stanley, D. J., ed., Stroudsburg: Dowden, Hutchinson and Ross, Part. 2, p. 63–73.

Benson, R. H., and Sylvester-Bradley, P. C., 1971, Deep sea ostracodes and the transformation of ocean to sea in the Tethys, *Bull. Centre Rech. Pau, S.N.P.A.*, Suppl. 5, p. 53–91.

Berry, M. J., and Knopoff, L., 1967, Structure of the Upper Mantle under the Western Mediterranean Basin, *J. Geophys. Res.*, v. 72, p. 3613–3626.

Berthois, L., 1968, Sur la présence d'affleurements calcaires organogènes miopliocènes au large de Gibraltar: Ses conséquences tectoniques, *C. R. Acad. Sci. Paris.* v. 267, p. 561–571.

Blumenthal, M., 1930, Beiträge zur Geologie der Betischen Cordilleren beiderseits des Rio Guadalhorce, *Eclogae Geol. Helv.*, v. 23, p. 41–293.

Boccaletti, M., and Guazzone, G., 1974, Remnant arcs and marginal basins in the Cainozoic development of the Mediterranean, *Nature*, v. 252, p. 18–21.

Boer, J. de, 1965, Paleomagnetic movements in the Tethys, *J. Geophys. Res.*, v. 70, p. 937–945.

Bogdanoff, A., 1963, Problèmes de structure et d'histoire de la plateforme de l'Europe orientale, *Bull. Soc. Géol. France*, v. IV p. 898–911.

Booy, T. De, 1968, Mobility of the earth's crust: A comparison between the present and past *Tectonophysics*, v. 6, p. 177–206.

Booy, T. De, 1969, Repeated disppearance of continental crust during the geological development of the Western Mediterranean Area, in: *Symposium on the Problems of Oceanization in the Western Mediterranean*, Verh. K. Ned. Geol. Mijnb. Genoot., p. 79–103.

Bosellini, A., and Hsü, K., 1973, Mediterranean plate tectonics and Triassic palaeogeography, *Nature*, v. 244, p. 144–146.

Bourcart, J., 1960, La Méditerranée et la révolution du Pliocène, *Livre Mém. P. Fallot, Soc. Géol. France*, v. 1, p. 103–116.

Bourrouilh, R., 1974, *Stratigraphie, Sédimentologie et Tectonique de l'Ile de Minorque et du Nord-Est de Majorque (Baléares)*, Thesis, Univ. Paris-VI (Dept. Géol. Struct.), Trav. Lab. Géol. Médit. Associée au C.N.R.S., v. 2, 822 p.

Bourrouilh, R., and Magné, J., 1963, A propos de dépôts du Pliocène supérieur et du Quaternaire sur la côte nord de l'île de Minorque, *Bull. Soc. Géol. France*, v. V, p. 298–302.

Brinkmann, R., 1954, *Abriss der Geologie: Historische Geologie*, Stuttgart: Ferd. Enke, 359 p. [Spanish Edition, *Compendio de Geológia Histórica*, Labor Ed., 337 p.].

Brune, J. N., 1969, Surface waves and crustal structure in: *The Earth's Crust and Upper Mantle*, Belousov, V. V., and Hart, P. J., eds., Geophys. Monogr. 13, Amer. Geophys. Union.

Burollet P. F., and Byramjee, R., 1974, Réflexions sur la tectonique globale. Exemples africains et méditerranéens, *Notes Mém. Compagnie Française du Pétrole*, no. 11, p. 71–120.

Carey, S. W., 1955, The orocline concept in geotectonics, *Papers and Proc. Roy. Soc. Tasmania*, v. 89, p. 255–288.

Carey, S. W., 1958, A tectonic approach to continental drift, in: *Continental Drift—A Symposium*, Carey, S. W., ed., Hobart: Univ. of Tasmania, p. 177–355.

Cita, M. B., 1972, Il significato della transgressione pliocenica alla luce delle nuove scoperte nel Mediterraneo, *Riv. Ital. Paleontol. Stratigr.* v. 78, p. 527–594.

Cita, M. B., and Ryan, W. F. B., 1973, Time scale and general synthesis, in: *Initial Rep. Deep Sea Drilling Project*, Ryan, W.F.B., *et al.*, eds., Washington, D.C.: U.S. Govt. Printing Office, v. 13, p. 1404–1447.

Colom, G., and Escandell, B., 1962, L'évolution du géosynclinal Baléare, *Livre Mém. P. Fallot, Soc. Géol. France*, v. 1, p. 125–136.

Cornet, C., 1965, Evolution tectonique et morphologique de la Provence depuis l'Oligocène, *Mém. Soc. Géol. France*, v. 103, p. 252.

Defense Mapping Agency Hydrographic Center, 1972, Mediterranean Sea. Map scale 1:2,849,300, Washington, D.C.: 17th ed.

Dewey, J. F., Pittman, W. C., III, Ryan, W. B. F., and Bonnin, J., 1973, Plate tectonics and the evolution of the Alpine system, *Bull. Geol. Soc. Amer.*, v. 84, p. 3137–3180.

Didon, J., 1960, Le flysch gaditan au nord et au nord-est d'Algésiras (province de Cadiz, Espagne), *Bull. Soc. Géol. France*, v. II, p. 352–361.

Didon, J., 1962, Les unités ultrabétiques de la zone du flysch gaditan au nord et nord-est d'Algésiras, *Livre Mém. P. Fallot, Soc. Géol. France*, v. 1, p. 265–272.

Didon, J., 1973, Accidents transverses et coulissages longitudinaux dextrés dans la partie nord de l'arc de Gibraltar (Cordillères bétiques occidentales, Espagne), *Bull. Soc. Géol. France*, v. XV, p. 121–129.

Didon, J., and Durand Delga, M., 1973, Colloque de Gibraltar de l'action thématique programmée de l'I.N.A.G., (Géodynamique de la Méditerranée Occidentale), *Bull. Soc. Géol. France*, v. XV, p. 160–189.

Didon, J., Durand Delga, M., and Kornprobst, J., 1973, Homologies géologiques entre les deux rives du detroit de Gibraltar, *Bull. Soc. Géol. France*, v. XV, p. 77–106.

Dietz, R. S., and Holden, J. C., 1970, Reconstruction of Pangaea: Breakup and dispersion of continents, Permian to present, *J. Geophys. Res.*, v. 26, p. 4939–4956.

Dongen, P. G., Van, 1967, The rotation of Spain: Paleomagnetic evidence from the eastern Pyrenees, *Palaeogeogr. Palaeoclimatol. Palaeoecol.*, v. 3, p. 417–432.

Drooger, C. W., ed., 1973, *Messinian Events in the Mediterranean*, Amsterdam-North-Holland Publ. Co., 272 p.

Durand Delga, M., 1961, Au sujet du sillon méso-méditérranéen du flysch au Crétacé et au Nummulitique, *C. R. Soc. Géol. France*, fasc. 2, p. 45–47.

Durand Delga, M., 1973, Hypothèses sur la genèse de la courbure de Gibraltar, *Bull. Soc. Géol. France*, v. XV, p. 119–121.

Durand Delga, M., and Mattauer, M., 1960, Sur l'origine ultra-rifaine des certaines nappes du Rif Septentrional, *C. R. Soc. Géol. France*, fasc. 2, p. 22–25.

Eggeler, G. C., and Bodenhausen, J. W. A., 1964, Distinct phases of alpine overthrusting in the eastern part of the Betic zone of Spain, *Verh. K. Ned. Geol. Mijnb. Genoot.*, v. 43, p. 316–320.

Eggeler, G. C., and Simon, O. J., 1969, Sur la tectonique de la zone Bétique (Cordillères Betiques, Espagne), *K. Ned. Akad. Wet.*, v. 25, p. 1–90.

Emelyanov, E. M., 1972, Principal types of recent bottom sediments in the Mediterranean Sea: Their mineralogy and geochemistry, in: *The Mediterranean Sea—A Natural Sedimentation Laboratory*, Stanley, D. J., ed., Stroudsburg: Dowden, Hutchinson and Ross, p. 355–387.

Emelyanov, E. M., and Shimkus, K. M., 1972, Suspended matter in the Mediterranean Sea, in: *The Mediterranean Sea—A Natural Sedimentation Laboratory*, Stanley, D. J., ed., Stroudsburg: Dowden, Hutchinson and Ross, p. 417–439.

Escandell, B., and Colom, G., 1960, Sur l'existence de diverses phases de plissements alpins dans l'île de Majorque (Baléares), *Bull. Soc. Géol. France*, v. II, p. 267–272.

Fahlquist, D. A., 1963, *Seismic Refraction Studies in the Western Mediterranean*, Ph.D. Thesis, M.I.T., Cambridge, Mass.

Fahlquist, D. A., and Hersey, J. B., 1969, Seismic refraction measurements in the Western Mediterranean Sea, *Bull. Inst. Océanogr. Monaco*, v. 67, p. 52.

Fairbridge, R. W., 1969, The changing level of the sea, *Sci. Amer.*, v. 1, p. 43.

Fairbridge, R. W., 1971, Quaternary sedimentation in the Mediterranean region controlled by tectonics, paleoclimates and sea level, in: *The Mediterranean Sea—A Natural Sedimentation Laboratory*, Stanley, D.J., ed., Stroudsburg: Dowden, Hutchinson and Ross, p. 99–113.

Fallot, P., 1948, Les Cordillères Bétiques, *Est. Geol.*, no. 4, C.S.I.C. Madrid, p. 83–172.

Fallot, P., 1954, Comparaison entre Cordillères Bétiques et Alpes Orientales, *Bol. R. S., Esp. Hist. Nat.*, 80th anniversary volume for Prof. E. Hernandez-Pacheco, p. 259–279.

Finetti, I., and Morelli, C., 1972, Wide scale digital seismic exploration of the Mediterranean Sea, *Boll. Geofis. Teor. Appl.*, v. XIV, p. 291–342.

Finetti, I., and Morelli, C., 1973, Geophysical exploration of the Mediterranean Sea, *Boll. Geofis. Teor. Appl.*, v. XV, p. 263–340.

Fontboté, J. M., 1957, Tectoniques superposées dans la Sierra Nevada, Cordillères Bétiques, *C. R. Acad. Sci. Paris*, v. 245, p. 1324–1326.

Fontboté, J. M., 1970a, Sobre la historia Preorogenica de las Cordilleras Beticas, *Cuad. Geol. Univ. Granada*, no. 1, p. 71–78.

Fontboté, J. M., 1970b, La cobertura sedimentaria de la isla de Alboran (Almeria), *Bol. Geol. Min.*, v. LXXXI–IV, p. 39–62.

Fontboté, J. M., 1973, El campo de pesantez y la estructura del Estrecho de Gibraltar, *Inst. Geogr. Cat. Proyecto Geodinamico*, v. 85.

Fontboté, J. M., Estevez, A., Navarro-Vila, F., Orozco, M., and Sainz de Galdeano C., 1972, Mem. Expl. de la hoja 4-12 (num. 87), Algeciras. Mapa Geol. Nac. a la escala 1:200,000, *Inst. Geol. Min. España*.

Gaibar-Puertas, C., and Ruiz Lopez, J., 1969, Estudio geologico de la isla de Alboran (Almeria). I. Las rocas eruptivas, *Acta Geol. Hisp.*, v. 4, p. 72–80.

Gaibar-Puertas, C., and Ruiz Lopez, J., 1970a, Las anomalias de la pesantez en la isla de Alboran (Almeria), *Rev. Geofis*, no. 4, p. 267–292.

Gaibar-Puertas, C., and Ruiz Lopez, J., 1970b, Las anomalias magneticas de la isla de Alboran (Almeria), *Bol. Geol. Min.*, v. LXXXI–IV, p. 372–393.

Gass, I. G., 1968, Is the Troodos massif of Cyprus a fragment of Mesozoic ocean floor? *Nature*, v. 220, p. 39–42.

Giermann, G., 1961, Erlauterung zur bathymetrischen Karte der Strasse von Gibraltar, *Bull. Inst. Océanogr. Monaco*, v. 58, p. 28.

Giermann, G., Pfannenstiel, M., and Wimmenauer, W., 1968, Relations entre morphologie, tectonique et volcanisme en mer d'Alboran (Méditerranée occidentale). Résultats preliminaires de la campagne *Jean Charcot* 1967, *C. R. Soc. Géol. France*, fasc. 4, p. 116–118.

Gilbert Smith, A., 1971, Alpine deformation and the Oceanic areas of the Tethys, Mediterranean and Atlantic, *Bull. Geol. Soc. Amer.*, v. 82, p. 2039–2070.

Girdler, R. W., 1965, Continental drift and the rotation of Spain, *Nature*, v. 207, p. 396–398.

Glangeaud, L., 1951, Interprétation tectono-physique des caractères, *Bull. Soc. Géol. France*, p. 753–762.

Glangeaud, L., 1956, Sur les structures scalaires, *XXe Semaine de Synthèse*, Centre Int. de Synthèse, p. 228–231.

Glangeaud, L., 1962, Paléogéographie dynamique de la Méditerranée et de ses bordures. Le rôle des phases ponto-plio-quaternaires, *Océanogr. Géol. Geophys. Médit. Occid., Centre Nat. Recherche Sci.*, p. 125–165.

Glangeaud, L., 1963, Les transferts d'échelle en géologie et geophysique. Application à la Méditerranée occidentale et aux chaînes peripacifiques, *Bull. Soc. Géol. France*, v. IV, p. 912–961.

Glangeaud, L., 1966, Les grands ensembles structuraux de la Méditerranée occidentale d'après les données de Géomède 1, *Acad. Sci. Paris*, v. 262, p. 2405–2408.

Glangeand, L., 1970, Les structures mégamétriques de la Méditerranée occidentale, *C. R. Acad. Sci. Paris*, v. 270, p. 3184–3190.

Glangeaud, L., and Olivet, P. N., 1970, Structures mégamétriques de la Méditerranée. Evolution de la Mésogée de Gibraltar à l'Italie, *C. R. Acad. Sci. Paris*, v. 271, p. 1161–1166.

Glangeaud, L., Alinat, J., Garate, C. A., Leenhardt, O., and Pautot, G., 1967a, Les phénomènes ponto-plio-quaternaires dans la Méditerranée occidentale d'après le Geomede I, *C. R. Acad. Sci. Paris*, v. 264, p. 208–211.

Glangeaud, L., Bobier, C., and Bellaiche, G., 1967b, Evolution néotectonique de la Mer d'Alboran et ses conséquences paléogéographiques, *C. R. Acad. Sci. Paris*, v. 265, p. 1672–1675.

Glangeaud, L., Bobier, C., and Szep, B., 1970, Les structures mégamétriques de la Méditerranée: la mer d'Alboran et l'arc de Gibraltar, *C. R. Acad. Sci. Paris*, Sér. D, v. 271, p. 473–481.

Haarman, E., 1935, *Um das Geologische Weltbild*, Stuttgart: Ferd. Enke Ed., 108 p.

Heezen, B. C., 1959, Géologie sous-marine et déplacement des continents, en la topographie et la géologie des profondeurs océaniques, *Colloque Int.*, *Centre Nat. Recherche Sci.*, no. LXXXIII, p. 295–304.

Heezen, B. C., 1960, Geologic mapping of submerged continental margins, *Bull. Amer. Assoc. Petrol. Geol.*, v. 44, p. 1250.

Heezen, B. C., Tharp, M., and Ryan, W. B. F., 1970, Panorama of the Mediterranean Sea, *Geotimes*, v. 15, p. 12–13.

Hersey, J. B., 1965, Sedimentary basins of the Mediterranean Sea, in: *Submarine Geology and Geophysics*, Whittard, W. F., and Bradshaw, R., eds., London: Butterworths, p. 75–91.

Hess, H. H., 1955, Serpentines, orogeny and epeirogeny, *Geol. Soc. Amer.*, Spec. Paper 62.

Hess, H. H., 1962, History of the ocean basins, in: *Petrologic Studies*, Buddington Mem. Vol., Geol. Soc. Amer., p. 599–620.

Hess, H. H., 1965, Mid-oceanic ridges and tectonics of the sea floor, in: *Submarine Geology and Geophysics*, Whittard, W. F., and Bradshaw, R., eds., London: Butterworths, p. 313–334.

Hilten, D. Van, and Zijderveld, J. D. A., 1966, The magnetism of the Permian porphyries near Lugano (Northern Italy–Switzerland), *Tectonophysics*, v. 3, p. 429–446.

Hill, M. N., ed., 1962, *The Sea: Ideas and Observations on Progress in the Study of the Seas, Vol. 1: Physical Oceanography*, New York: Wiley, 863 p.

Hill, M. N., 1963, *The Sea: Ideas and Observations on Progress in the Study of the Seas, Vol. 3: The Earth Beneath the Sea, History*, New York: Wiley, p. 963.

Hinz, K., 1972, Zum Diapirismus in westlichen Mittelmeer, *Geol. Jahrb.*, v. 90, p. 389–396.

Hinz, K., 1973, Crustal Structure of the Balearic Sea, *Tectonophysics*, v. 20, p. 295–302.

Hofman, B. J., 1952, The gravity field of the West Mediterranean area, *Verh. K. Ned. Geol. Mijnb. Genoot.*, v. 8, p. 297–305.

Holmes, A., 1945, *Principles of Physical Geology*, London: Nelson and Sons, 532 p.

Hsü, K. J., 1965, Franciscan rocks of Santa Lucia Range, California, and the argille scagliose of the Apennines, Italy: A comparison in style of deformation, *Geol. Soc. Amer.* (Abstracts), p. 210.

Hsü, K., 1971, Origin of the Alps and Western Mediterranean, *Nature*, v. 233, p. 44–48.

Hsü, K., 1971, History of the Mediterranean Basins, *Abstracts with Programs*, Geol. Soc. Amer., v. 3, p. 607.

Hsü, K., 1972, When the Mediterranean dried up, *Sci. Amer.*, v. 277, p. 27–36.

Hsü, K. J., Cita, M. B., and Ryan, W. B. F., 1973a, The origin of the Mediterranean evaporites, in: *Initial Rep. Deep Sea Drilling Project*, Ryan, W. B. F., *et al.*, eds., Washington, D.C.: U.S. Govt. Printing Office, v. 13, p. 1–74.

Hsü, K. J., Ryan, W. B. F., and Cita, M. B., 1973b, Late Miocene desiccation of the Mediterranean, *Nature*, v. 242, p. 240–244.

Huang, T., 1972, Sedimentological evidence for current reversal at the Straits of Gibraltar, *J. Tech. Soc.*, v. 6, no. 4, p. 25–33.

Huang, T. C., and Stanley, D. J., 1972, Western Alboran Sea: Sediment dispersal, ponding and reversal currents, in: *The Mediterranean Sea—A Natural Sedimentation Laboratory*, Stanley, D. J., ed., Stroudsburg: Dowden, Hutchinson and Ross, Part 9, p. 521–559.

Institut Français du Pétrole et Centre National de Exploration Océanographique, 1974, Carte Géologique et Structurale des Bassins du Domaine Méditerranéen, 1:2,500,000, Mercator, Paris: Proj. Ed. Technip.

Inst. Geol. y Min. de España, 1929, Hoja 522, Tortosa–Tarragona, a la escala 1:50,000.

Inst. Geol. y Min. de España, 1930, Hoja 420, San Baudilio del Llobregat, a escala 1:50,000.

Inst. Geol. y Min. de España, 1931, Hoja 498, Hospitalet, a escala 1:50,000.

Inst. Geol. y Min. de España, 1932, Hoja 448, Gava, a escala 1:50,000.

Inst. Geol. y Min. de España, 1933, Hoja 473, Tarragona, a escala 1:50,000.

Inst. Geol. y Min. de España, 1934, Hoja 446, Valls, a escala 1:50,000.

Inst. Geol. y Min. de España, 1941, Hoja 297, Estartit, a escala 1:50,000.

Inst. Geol. y Min. de España, 1952, Hoja 472, Reus, a escala 1:50,000.

Inst. Geol. y Min. de España, 1952, Hoja 447, Villanueva y Geltru, a escala 1:50,000.

Inst. Geol. y Min. de España, 1953, Hoja 419, Villafranca del Panades, a escala 1:50,000.

Inst. Geol. y Min. de España, 1953, Hoja 747, Sueca, a escala 1:50,000.

Inst. Geol. y Min. de España, 1956, Hoja 418, Montblanch, a escala 1:50,000.

Inst. Geol. y Min. de España, 1957, Hoja 770, Alcira, a escala 1:50,000.

Inst. Geol. y Min. de España, 1958, Hoja 593, Cuevas de Vinroma, a escala 1:50,000.

Inst. Geol. y Min. de España, 1959, Hoja 668, Sagunto, a escala 1:50,000.

Inst. Geol. y Min. de España, 1961, Hoja 795, Jativa, a escala 1:50,000.

Inst. Geol. y Min. de España, 1963, Hoja 594, Alcala de Chisvert, a escala 1:50,000.

Inst. Geol. y Min. de España, 1971, Hoja 25, Figueras, a escala 1:200,000 (sintesis).

Inst. Geol. y Min. de España, 1971, Hoja 42, Tarragona, a escala 1:200,000 (sintesis).

Inst. Geol. y Min. de España, 1971, Hoja 41, Tortosa, a escala 1:200,000 (sintesis).

Inst. Geol. y Min. de España, 1971, Hoja 35, Barcelona, a escala 1:200,000 (sintesis).

Inst. Geol. y Min. de España, 1972, Hoja 48, Vinaroz, a escala 1:200,000 (sintesis).

Inst. Geol. y Min. de España, 1972, Hoja 56, Valencia, a escala 1:200,000 (sintesis).

Inst. Geol. y Min. de España, 1972, Hoja 55, Liria, a escala 1:200,000 (sintesis).

Inst. Geol. y Min. de España, 1973, Hoja 447, Villanueva y Geltru, a escala 1:50,000.

Inst. Geol. y Min. de España, 1973, Hoja 64, Alcoy, a escala 1:200,000 (sintesis).

Inst. Geol. y Min. de España, 1973, Hoja 63, Albacete–Onteniente, a escala 1:200,000 (sintesis).

Inst. Geol. y Min. de España, 1974, Mapa Tectónico de la Península Ibérica y Baleares, a escala 1:1,000,000.

Inst. Geol. y Min. de España, 1974, Mapa Tectónico de la Península Ibérica y de las Baleares (Contribución al Mapa Tectónico de Europa), 113 p.

Inst. Geol. y Min. de España, 1973, Hoja de Vinaroz (571–571 bis), a escala 1:50,000.

Inst. Geol. y Min. de España, 1973, Hoja de Ulldecona (546), a escala 1:50,000.

Inst. Geol. y Min. de España, 1973, Hojas de Tarragona, Reus, Valls, Montblanch, Villafranca y Villanueva, a escala 1:50,000.

Inst. Geol. y Min. de España, 1973, Hoja de Villafames (616), a escala 1:50,000.

Inst. Geol. y Min. de España, 1973, Hoja de Faro de Oropesa (617), a escala 1:50,000.

Inst. Geol. y Min. de España, 1973, Hoja de Albocacer (570), a escala 1:50,000.

Inst. Geol. y Min. de España, 1973, Hoja de Alcanar (547), a escala 1:50,000.

Inst. Geol. y Min. de España, 1973, Hoja de Burjasot (696), a escala 1:50,000.

Inst. Geol. y Min. de España, 1973, Hoja de Moncofar (669), a escala 1:50,000.

Inst. Geol. y Min. de España, 1973, Hoja de Sagunto (668), a escala 1:50,000.

Inst. Geol. y Min. de España, 1974, Hoja de Castellón de la Plana (641), a escala 1:50,000.

Inst. Geol. y Min. de España, 1974, Hoja de Segorbe (640), a escala 1:50,000.

Inst. Geol. y Min. de España, Hoja de Islas Columbretes (642), a escala 1:50,000 (in press).

Inst. Geol. y Min. de España, Mem. provinciál sintética de Gerona sobre los mapas y cortes a escala 1:50,000, elaborada entre los años 1968–1971 (unpublished).

Instituto Hidrográfico de la Marina, Cartas de navegación de los fondos de la costa a escala 1:10,000, obtenidos por el buque planero, *Juan de la Cosa*, Cadiz.

Jong, K. de, 1973, Mountain building in the Mediterranean Region, in: *Gravity and Tectonics*, New York: J. Wiley and Sons, Van Bemmelen Volume, Part 2, p. 125–139.

Kelling, G., and Stanley, D. J., 1972, Sedimentation in the vicinity of the Strait of Gibraltar, in: *The Mediterranean Sea—A Natural Sedimentation Laboratory*, Stanley, D. J., ed., Stroudsburg: Dowden, Hutchinson and Ross, Part 9, p. 489–519.

Klemme, H. D., 1958, Regional geology of circum-Mediterranean region, *Bull. Amer. Assoc. Petrol. Geol.*, v. 42, p. 477–512. [Spanish translation: Ríos, J. M., 1958, Geológia regional del país circum-mediterráneo, *N. y C. Inst. Geol. Min. España*, no. 51, p. 63–129.]

Kober, L., 1931, *Das Alpine Europa*, Berlin: Borntraeger, p. 310.

Lacombe, H., 1961, Contribution à l'étude du régime du Détroit de Gibraltar, *Cah. Océanogr.*, v. 13, no. 2, p. 73–107.

Lacombe, H., and Tchernia, P., 1972, Caractères hydrologiques et circulation des eaux en Méditerranée, in: *The Mediterranean Sea—A Natural Sedimentation Laboratory*, Stanley, D. J., ed., Stroudsburg: Dowden, Hutchinson and Ross, p. 25–36.

Leclaire, L., 1972, Aspects of Late Quarternary sedimentation on the Algiers–Balearic Basin, in: *The Mediterranean Sea—A Natural Sedimentation Laboratory*, Stanley, D. J., ed., Stroudsburg: Dowden, Hutchinson and Ross, p. 561–582.

Leenhardt, O., 1968, Le problème des dômes à la Méditérranée occidentale. Etude géographique d'une colline abyssale, la Structure A, *Bull. Soc. Géol. France*, v. X, p. 497–509.

Leenhardt, O., 1969, Analyse en sondage sismique continu, *Inst. Hydrogr. Rev.*, v. 46, p. 51–79.

Leenhardt, O., 1970, Sondages sismiques continus en Méditerranée occidentale, *Mém. Inst. Océanogr. Monaco*, no. 1, 120 p.

Leenhardt, O., Rebuffati, A., Sabatier, R., and Bruneton, A., 1969, Profil sismique dans le bassin nord-Baléares, *C. R. Soc. Géol. France*, v. 7, p. 249–251.

Leenhardt, O., Rebuffatti, A., and Sancho, J., 1970, Carte du Plioquaternaire entre Ibiza et le Cap San Antonio (Méditerranée Occidentale), *Rev. Inst. Franç. Petrole*, v. XXV, p. 165–173.

Le Pichon, X., 1968, Sea floor spreading and continental drift, *J. Geophys. Res.*, v. 73, p. 3661–3697.

Le Pichon, X., Pautot, G., Auzende, J. M., and Olivet, J. L., 1971, La Méditérranée occidentale depuis l'Oligocène, schema d'évolution, *Earth Planet. Sci. Lett.*, v. 13, p. 145–152.

Le Pichon, X., Pautot, G., and Weill, J. P., 1972, A model of the opening of the Alboran Sea, *Nature Phys. Sci.*, v. 263, p. 83–85.

Le Pichon, X., Francheteau, J., and Bonnin, J., 1973, *Plate Tectonics, Developments in Geotectonics 6*, Amsterdam: Elsevier, 300 p.

Lloyd, R. M., and Hsü, K. J., 1972, Preliminary isotopic investigations of samples from Deep Sea Drilling Cruise to the Mediterranean, 1971, in: *The Mediterranean Sea—A Natural Sedimentation Laboratory*, Stanley, D. J., ed., Stroudsburg: Dowden, Hutchinson and Ross, p. 681–686.

Mauffret, A., 1969a, Sur la rotation d'Espagne, *Earth Planet. Sci. Lett.*, v. 7, p. 315–320.

Mauffret, A., 1969b, Les dômes et les structures "anticlinales" de la Méditerranée occidentale au nord-est des Baléares, *Rev. Inst. Franç. Petrole*, v. 24, p. 953–960.

Mauffret, A., 1970, Structure des fonds marins au tour des Baléares, *Cah. Océanogr.*, v. 22, p. 33–42.

Mauffret, A., and Sancho, J., 1970, Etude de la marge continentale au nord de Majorque, Baléares, *Rev. Inst. Franç. Pétrole*, v. 25, p. 714–730.

Mauffret, A., Auzende, J. M., Olivet, J. L., and Pautot, G., 1971, Upper Miocene salt layer in the Western Mediterranean Basin, *Nature Phys. Sci.*, v. 230.

Mauffret, A., Auzende, J. M., Olivet, J. L., and Pautot, G., 1972, Le bloc continental Baléares (Espagne). Extension et évolution, *Marine Geol.*, v. 12, p. 289–300.

Mauffret, A., Fail, J. P., Montadert, L., Sancho, J., and Winnock, E., 1973, Northwestern Mediterranean sedimentary basin from seismic reflection profile, *Bull. Amer. Assoc. Petrol. Geol.*, v. 57, p. 2245–2262.

Maxwell, A. E., 1971, *The Sea: Ideas and Observations on Progress in the Study of the Sea*, v. 4, Pts. 1–3, New York: Wiley.

Miller, A. R., 1972, Speculations concerning bottom circulation in the Mediterranean Sea, in: *The Mediterranean Sea—A Natural Sedimentation Laboratory*, Stanley, D. J., ed., Stroudsburg: Dowden, Hutchinson and Ross, p. 37–42.

Milliman, J. D., Weiler, Y., and Stanley, D. J., 1972, Morphology and Carbonate Sedimentation on Shallow Banks in the Alboran Sea, in: *The Mediterranean Sea—A Natural Sedimentation Laboratory*, Stanley, D. J., ed., Stroudsburg: Dowden, Hutchinson and Ross, p. 241–259.

Montadert, L., Sancho, S., Fail, J. P., Debysser, J. P., and Winnock, E., 1970, De l'âge tertiaire de la série salifère responsable des structures diapiriques en Méditérranée occidentale, *C. R. Acad. Sci. Paris.*, v. 271, p. 812–815.

Montenat, C., 1970, Sur l'importance des mouvements orogéniques récents dans le sud-est de l'Espagne (provinces d'Alicante et de Murcia), *C. R. Acad. Sci. Paris*, v. 270, p. 663–677.

Montenat, C., Bizon, G., and Bizon, J. J., 1975, Remarques sur le Néogène du forage JOIDES 121 en Mer d'Alboran (Méditérranée occidentale), *Bull. Soc. Géol. France*, v. XVII, p. 45–51.

Nairn, A. E. M., and Westphal, M., 1968, Possible implications of the Paleomagnetic study of late Paleozoic igneous rocks of northwestern Corsica, *Palaeogr. Palaeoclimatol. Palaeoecol.*, v. 5, p. 179–204.

Nairn, A. E. M., and Westphal, M., 1972, La crise de salinité messinienne en Méditérranée: enseignement des forages JOIDES et du Bassin de Sicile, in: *Symposium sur la Geodynamique de la Méditerranée*, 33rd Congr. Assemblée Comm. Int. Expl. Sci., Athens.

Nesteroff, W. D., and Ryan, W. B. F., 1973, Séries stratigraphiques et implications tectoniques du forage JOIDES 121 en mer d'Alboran, *Bull. Soc. Géol. France*, v. XV, p. 113–115.

Nesteroff, W. D., Ryan, W. B. F., Hsü, K. J., Pautot, G., Wezel, F. C., Lort, J. M., Cita, M. B., Mayne, W., Stradner, H., and Dumitrica, P., 1972, Evolution de la sédimentation pendant le Néogène en Méditérranée d'après les forages JOIDES–DSPD, in: *The Mediterranean Sea—A Natural Sedimentation Laboratory*, Stanley, D. J., ed., Stroudsburg: Dowden, Hutchinson and Ross, p. 47–62.

North, R. G., 1974, Seismic slip rates in the Mediterranean and Middle East, *Nature*, v. 252, p. 560–563.

Obrador, G. S., 1973, Estudio estratigráfico y sedimentológico de materiales miocénicos de la isla de Menorca, *Acta Geol. Hisp.*, v. V, p. 21–23.

Obrador, A., Mercadal, B., and Rosell, J., 1971, Geology of Menorca, in: *10th Int. Field Inst. Guidebook Spain*, Freeman, T., and Simancas, R., eds., Washington, D. C.: Amer. Geol. Inst., p. 139–148.

Olivet, J. L., Pautot, G., and Auzende, J. M., 1969, Uplift and subsidence in and around the Western Mediterranean since the Oligocene: A review, in: *Symposium on the Problems of Oceanization in the Western Mediterranean*, *Verh. K. Ned. Geol. Mijnb. Genoot.*, v. 26, p. 53–77.

Olivet, J. L., Auzende, J. M., and Bonnin, J., 1973a, Structure et évolution tectonique du bassin d'Alboran, *Bull. Soc. Géol. France*, v. XV, p. 108–113.

Olivet, J. L., Pautot, G., and Auzende, J. M., 1973b, Alboran Sea, in: *Initial Rept. Deep Sea Drilling Project*, 48-1. *Structural framework of selected regions of the Western Mediterranean*, Ryan, W. B. F., *et al.*, eds., Washington, D.C.: U.S. Govt. Printing Office, v. XIII, p. 1417–1430.

Olivet, J. L., Pautot, G., and Auzende, J. M., 1974, Tectonique éocène dans les Cordillères Bétiques; vers une nouvelle conception de la paléogéographie en Méditérranée occidentale, *Bull. Soc. Géol. France*, v. XVI, p. 58–71.

Parga, J. R., 1969, Spätvariszische Bruchsysteme in Hesperischen Massiv, *Geol. Rundschau*, v. 59, p. 323–336.

Parga, J. R., 1972, Et si la Méditerranée avait été à sec? *Science, Progrès, Découverte*, v. 3.443, p. 42–47.

Pautot, G., Auzende, J. M., Olivet, J. L., and Mauffret, A., 1973, Valencia basin structural framework, in: *Initial Rep. Deep Sea Drilling Project*, Ryan, W. B. F., *et al.*, eds., Washington, D.C.: U.S. Govt. Printing Office, v. XIII, p. 1430–1431.

Payo, G., 1967, Crustal structure of the Mediterranean Sea by surface waves. Part I: Group velocity, *Bull. Seism. Soc. Amer.*, v. 56, p. 151–172.

Payo, G., 1975, Estructura, sismicidad y tectónica del Mar Mediterráneo, *Inst. Geogr. Cat.*, 39 p.

Petit, G., and Laubier, L., 1962, Les canyons de la Côte Catalane. Aperçu de nos connaissances et programme de recherches, in: *Océanogr. Géol. Géophys. Médit. Occid.*, Centre de la Recherche Scientifique, p. 89–95.

Pfannenstiel, M., 1963, Das Relief des Mittelmeerbodens, seine geotektonische Deutung an Hand bathymetrischen Karten, 1:750,000, *Geol. Rundschau*, v. 53.

Pitman, C. W., and Talwani, M., 1972, Sea floor spreading and the North Atlantic, *Bull. Geol. Soc. Amer.*, v. 83, p. 619–643.

Querol, R., 1969, Petroleum exploration in Spain, in: *The Exploration for Petroleum in Europe and North Africa*, A Symposium, The Institute of Petroleum, Hepple, P., ed., p. 49–72.

Rangheard, Y., 1970a Principales données stratigraphiques et tectoniques des îles d'Ibiza et de Formentera (Baléares); situation paléogéographiques et structurales de ces îles dans les Cordillères Bétiques, *C. R. Acad. Sci. Paris.*, v. 270, p. 1227–1230.

Rangheard, Y., 1970b, Tectonique d'Ibiza et de la Sierra de Majorque, *Ann. Soc. Geol. Nord.*, v. XC, 4 p.

Rangheard, Y., 1971, Etude géologique des îles d'Ibiza et de Formentera (Baléares), *Mem. Inst. Geol. Min. España*, v. 82, 340 p.

Raoult, J. F., 1973, Compte rendu de la Réunion Extraordinaire de la Soc. Géol. France, l'Arc de Gibraltar (1972), *Bull. Soc. Géol. France*, v. XV, p. 129–160.

Reig Vilaplana, F., 1958, La estructura del Estrecho de Gibraltar y las posibilidades de las obras del cruce del mismo, *C.S.I.C.*, Madrid, 48 p.

Ríos, J. M., 1961, A geological itinerary through the Spanish Pyrenees, *Proc. Geol. Assoc.*, v. 72, p. 360–371.

Ríos, J. M., 1963, Materiales salinos del suelo español, *Mem. Inst. Geol. Min. España*, v. 64, 166 p.

Ríos, J. M., 1968, Saline deposits of Spain, *Geol. Soc. Amer.*, Spec. Paper, p. 59–74.

Ríos, J. M., 1971, Mar Mediterráneo, E. N. Adaro de Inv. Míneras y Escuela Técnica Superior de Ingenieros de Minas, 238 p.

Ríos, J. M., 1975a, El mar Mediterráneo occidental y sus costas Ibéricas. Las areas tierras firmes, *Vol. Homeriaje D. Manuel Lora Tamayo, Acad. Cienc, E. F. y N. Madrid*, p. 147–192.

Ríos, J. M., 1975b, El mar Mediterráneo occidental y sus costas Ibéricas. Las areas marinas, *Acad. Cienc, E. F. y N. Madrid*, v. LXIX, rev. 12, p. 285–473.

Ríos, J. M., and Almela, A., 1953, Analogies entre les séries stratigraphiques de la Sierra

de Ricote (Espagne) et de l'Apennin septentrional, *Bull. Soc. Géol. France*, v. III, p. 767–773.

Ritsema, A. R., 1969, Seismic data of the West Mediterranean and the problem of oceanization, in: *Symposium on the Problems of Oceanization in the Western Mediterranean*, *Verh. K. Ned. Geol. Mijnb. Genoot.*, v. XXVI p. 105–120.

Roederer, D. H., 1973, Subduction and orogeny, *J. Geophys. Res.*, v. 78, p. 5005–5024.

Roever, W. P. de, 1969, Genesis of the Western Mediterranean Sea: enigmatic oceanization disruption of continental crust or also upheaval above sea level and later subsidence of oceanic floor? *Verh. K. Ned. Geol. Mijnb. Genoot.*, v. XXVI, p. 9–11.

Ruggieri, G., 1967, The Miocene and later evolution of the Mediterranean Sea, in: *Aspects of Tethyan Biogeography*, Adams, C. G., and Ager, D. V., eds., London: Syst. Assoc. Publ. no. 7, p. 283–290.

Rupke, N. A., and Stanley, D. J., 1973, Distinguishing tubiditic and hemipelagic muds in the Balearic Basin, Western Mediterranean, *Abstracts with Programs, Geol. Soc. Amer.*, no. 5, p. 789.

Ryan, W. B. F., 1969, *The Floor of the Mediterranean Sea*, Ph.D. Thesis, Columbia Univ., 236 p.

Ryan, W. B. F., 1971, Can an ocean dry up? Results of deep-sea drilling in Mediterranean, *Bull. Amer. Assoc. Petrol. Geol.*, v. 55(2), p. 362.

Ryan, W. B. F., and Allan, T. D., 1970, The tectonics and geology of Mediterranean Sea, in: *The Sea*, Maxwell, A. E. ed., New York: Wiley, v. 4, p. 387–492.

Ryan, W. B. F., Hsü, K. J., Nesteroff, W. D., Wezel, F. C., Lort, J. M., Cita, M. B., Maync, W., Stradner, H., and Dumitrica, P., 1970*a*, Deep Sea Drilling Project: Leg 13, *Geotimes*, Amer. Geol. Inst., v. 15 (10), p. 12–13.

Ryan, W. B. F., Hsü, K. J., Nesteroff, W. D., Pautot, G., Wezel, F. C., Lort, J. M., Cita, M. B., Maync, W., Stradner, M., and Dumitrica, P., 1970*b*, Summary of Deep Sea Drilling Project Leg XIII, Scripps Institution of Oceanography, University of California, San Diego.

Schmalz, R. F., 1969, Deep water evaporite deposition. A genetic model, *Bull. Amer. Assoc. Petrol. Geol.*, v. 53, p. 798–823.

Schreiber, C. B., 1974, Vanished evaporites, revisited, *Sedimentology*, v. 21 (2), p. 329–331.

Schuiling, R. D., 1969, A geothermal model of oceanization in: *Symposium on the Problems of Oceanization in the Western Mediterranean, Verh. K. Ned. Geol. Mijnb. Genoot.* p. 143–148.

Schwarz, E. J., 1963, A palaeomagnetic investigation of Permo-Triassic red beds and andesites from the Spanish Pyrenees, *J. Geophys. Res.*, v. 68, p. 3265–3271.

Serra Raventos, J., and Got, H., 1974, Resultados preliminares de la campaña marina realizada en el precontinente catalán entre los canones de la Fonera y Blanes, *Acta Geol. Hisp.*, v. IX, p. 73–80.

Smith, A. G., 1971, Alpine deformation and the oceanic areas of the Tethys, Mediterranean and Atlantic, *Bull. Geol. Soc. Amer.*, v. 82, p. 2039–2070.

Smith, A. G., 1973, The so called tethyan ophiolites, in: *Implications of Continental Drift to the Earth Sciences*, Tarling, D. S., and Runcorn, S. K., eds., New York: Academic Press, v. 2, p. 977–986.

Smith, P. J., 1974, Ophiolites and oceanic lithosphere, *Nature*, v. 250, p. 99–100.

Société Géologique de France, 1973, *Bull. Soc. Géol. France*, v. XV, p. 77–189.

Sole Sabaris, L., and Llopis Llado, N., 1939, Terminación septentrional de la Cadena Costera Catalana, *Asoc. Esp. Geol. Medit. Occid., Lab. Geol. Univ. Barcelona*, v. VI, 87 p.

Sonnenfeld, P., 1974*a*, The Upper Miocene evaporite basins in the Mediterranean Region, *Geol. Rundschau*, v. 63, p. 1133–1172.

Sonnenfeld, P., 1974b, A Mediterranean catastrophe? *Geol. Mag.*, v. 101, p. 19–80.

Sonnenfeld, P., 1975, The Significance of Upper Miocene (Messinian) evaporites in the Mediterranean Sea, *J. Geol.*, v. 83, p. 287–311.

Stanley, D. J., ed., 1972, *The Mediterranean Sea—A Natural Sedimentation Laboratory*, Stroudsburg: Dowden, Hutchinson and Ross, 765 p.

Stanley, D. J., and Mutti, E., 1968, Sedimentological evidence of an emerged land mass in the Ligurian Seas during the Paleogene, *Nature*, v. 218, p. 32–36.

Stanley, D. J., and Terchien, M., 1971, Multiple origin of hemipelagic mud fill in Mediterranean Basin, *Bull. Amer. Assoc. Petrol. Geol.*, v. 55, p. 365.

Stanley, D. J., Cita, M. B., Flemming, N. C., Kelling, G., Lloyd, R. M., Milliman, J. D., Pierce, W. J., Ryan, W. B. F., and Weiler, Y., 1972, Guide for future sediment related research on the Mediterranean Sea, in: *The Mediterranean Sea—A Natural Sedimentation Laboratory*, Stanley, D. J., ed., Stroudsburg: Dowden, Hutchinson and Ross, p. 723–741.

Stanley, D. J., Got, H., Leenhardt, O., and Weiler, Y., 1974, Subsidence of the Western Mediterranean Basin in Pliocene–Quaternary time: Further evidence, *Geology*, v. 2, p. 345–351.

Storetvedt, K. M., 1974, Genesis of West Mediterranean basins, *Earth Planet. Sci. Lett.*, p. 22–28.

Suess, E., 1908, *Das Antlitz der Erde*, Vienna: E. Temsky, 778 p.

Tarling, D., and Tarling, M., 1971, *Continental Drift. A Study of the Earth's Moving Surface*, New York: Doubleday.

Termier, P., 1911, Les problèmes de la géologie tectonique dans la Méditerranée occidentale, *Rev. Gen. Sci. Pures Appl.*, 80 p.

Tex, E. den, 1969, Summary of the discussion on the Symposium on the Problems of Oceanization in the Western Mediterranean, *Verh. K. Ned. Geol. Mijnb. Genoot.*, p. 164–165.

Todd, R., 1958, Foraminifera from the Western Mediterranean deep sea cores, *Reports of the Swedish Deep Sea Expedition*, 1947–1948, v. 3, p. 169–211.

Udias, A., and Lopez Arroyo, A., 1972, Plate tectonics and the Azores–Gibraltar Region, *Nature Phys. Sci.*, v. 237, p. 67–69.

Udias, A., and Lopez Arroyo, A., 1975, Wrench (transcurrent) fault system of the southwestern Iberian Peninsula, paleogeographic and morphostructural implications, *Geol. Rundschau*, v. 64, p. 266–278.

Vita-Finzi, C., 1972, Supply of fluvial sediment to the Mediterranean during the last 20,000 years, in: *The Mediterranean Sea, a Natural Sedimentation Laboratory*, Stanley, D. J., ed., Stroudsburg: Dowden, Hutchinson and Ross, p. 43–62.

Vogt, P. R., Higgs, R. H., and Johnson, G. L., 1971, Hypotheses on the origin of the Mediterranean Basin, magnetic data, *J. Geophys. Res.*, v. 76, p. 3207–3228.

Vogt, P. R., Higgs, R. H., and Johnson, G. L., 1972, Hypotheses on the origin of the Mediterranean Basin, magnetic data, *J. Geophys. Res.*, v. 77, p. 391–393.

Volk, H. R., 1966, Geologische Grunde für die Existenz Sialischen Krustenmaterials im Mittelmeer Ostlich von Vera (SE. Spanien), zur Zeit des jungeren Pliozäns, *Proc. K. Ned. Akad. Wet.*, v. 69, p. 446–451.

Voo, R. Van der, 1967, The rotation of Spain: paleomagnetic evidence from the Spanish Meseta, *Palaeogeogr. Palaeoclimatol. Palaeoecol.*, v. 3, p. 393–416.

Voo, R. Van der, 1969, Paleomagnetic evidence for the rotation of the Iberian Peninsula, *Tectonophysics*, v. 7, p. 5–56.

Voo, R. Van der, and Zijderveld, J. D. A., 1969, Paleomagnetism in the western Mediterranean area, in: *Symposium on the Problems of Oceanization in the Western Mediterranean*, *Verh. K. Ned. Geol. Mijnb. Genoot.*, v. XXVI, p. 121–139.

Vuagnat, M., 1963, Remarques sur la trilogie diabase–gabbro–serpentinite dans le Bassin de la Méditerranée, *Geol. Rundschau*, v. 53, p. 336–358.

Watkins, N. D., and Richardson, A., 1970, Rotation of the Iberian Peninsula, *Science*, v. 107, p. 209.

Watson, J. A., and Johnson, G. L., 1968, Mediterranean diapiric structures, *Bull. Amer. Assoc. Petrol. Geol.*, v. 52, p. 2247–2249.

Weiler, Y., and Stanley, D. J., 1973, Sedimentation on the Balearic Rise, a foundered block in the Western Mediterranean, *Bull. Amer. Assoc. Petrol. Geol.*, v. 57(4), p. 811–812.

Westphal, M., 1967, *Etude Paléomagnétique des Formations Volcaniques Primaires de Corse. Rapports avec la Tectonique du Domaine Ligurien*, Ph. D. Thesis, Strassbourg.

Wills, L. M., 1952, *A Palaeogeographical Atlas of the British Isles and Adjacent Parts of Europe*, London: Blackie and Sons.

Wong, H. K., and Zarudzki, E. F. K., 1969, Thickness of unconsolidated sediments in the Eastern Mediterranean Sea, *Bull. Geol. Soc. Amer.*, v. 80, p. 2611–2614.

Wong, H. K., Zarudzki, E. F. K., Knott, S. T., and Hays, E. E., 1970, Newly discovered group of diapiric structures in Western Mediterranean, *Bull. Amer. Assoc. Petrol. Geol.*, v. 54, p. 2200–2204.

Woodside, J., and Bowin, C., 1970, Gravity anomalies and inferred crustal structure in the Eastern Mediterranean Sea, *Bull. Geol. Soc. Amer.*, v. 81, p. 1107–1122.

Zijderveld, J. D. A., De Jong, K. A., and Voo, R. Van der, 1970, Rotation of Sardinia: paleomagnetic evidence from Permian rocks, *Nature*, v. 226, p. 933–934.

Chapter 2

THE LIGURIAN SEA AND THE ADJACENT AREAS

Michel Moullade*

Centre de Recherche Micropaléontologique Jean Cuvillier
Laboratoire de Géologie Structurale
Université de Nice
Nice, France

I. INTRODUCTION

The Ligurian Sea forms the most northerly element of the western Mediterranean Basin (Fig. 1). More exactly, it appears as the northeastern extension of a vast abyssal depression, sometimes called the Balearic or the Algero-Provençal Basin, and which we propose to call the Algero-Balearic-Provençal Basin. Largely open toward the southwest, the Ligurian Sea is almost everywhere else delimited by continental or island regions: to the northwest, the north, and east by the southeastern coast of France (Provençal coast, Côte d'Azur *s.s.*) and by the Italian Riviera and Tuscany coast; to the southeast by the islands of the Tuscan Archipelago (principally Elba), which permit open communication by deeps which may exceed 400 m with another Mediterranean basin, the Tyrrhenian Sea. To the south the Ligurian Sea is bounded by the northern and western coasts of Corsica.

Within the framework of this article, therefore, the definition of the Ligurian Sea is somewhat extensive. In effect, its limit with the Tyrrhenian Sea is localized along the line of the Tuscan Archipelago, that is an eastern limit at

---

* With the collaboration of F. Irr, University of Nice, for the Neogene and Quaternary.

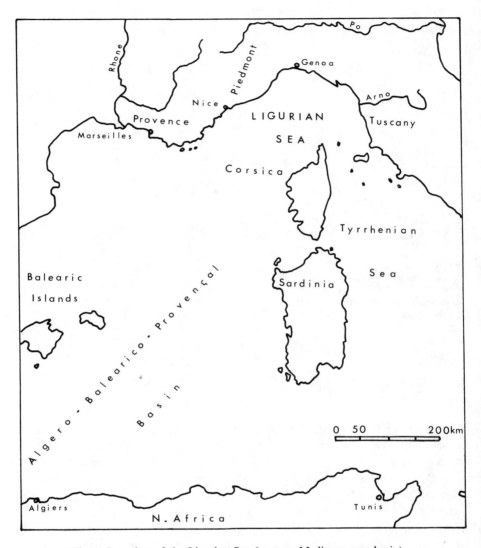

Fig. 1. Location of the Ligurian Sea (western Mediterranean basin).

10o30′ E. Further, the margins of the Provençal coast (which give a western limit at about 6° E, roughly the meridian of Toulon) and the western coast of Corsica, which may be extended even to include the northwestern part of Sardinia which belongs to the same continental block, are included. We are thus led to include within the domain of the Ligurian Sea, in the sense described above, a section which is poorly defined and which could also be equally well treated as part of the Balearic Basin.

Around most of its periphery the Ligurian Sea is bordered by mountainous regions, usually with strong relief. The principal units are the Maures–Esterel Massifs and the southern termination of the western Franco-Italian Alps, forming the northern rim; the northern Apennines to the east; and the Corsican Massif to the south. In contrast to the western section of the Algero-Balearic-Provençal Basin, where the sedimentation is nourished by two important rivers, the Rhône and Ebro, the Ligurian region receives from its bordering zone only relatively modest sedimentation from torrents or smaller rivers, such as the Gapeau west of the Maures Massif, the Argens between the Maures Massif and Esterel and the Siagne to the east of the latter massif, the Loup and the Var to the west of Nice, and the Paillon, the stream which enters the sea at Nice. Along the Italian coast from Ventimiglia to Piombino there are many streams, e.g., the Roya, Nervia, Armea, Argentina–Taggia, Impero, Arroscia–Centa, the rivers of Genoa (Leira, Polcevera, Bisagno, Sturla), Entella, Magra, Serchio, Arno, and Cecina. In Corsica the Fango, Liamone, Gravone, Taravo, and the Ortolo deposit their sediments within the Ligurian domain.

In this article a summary of the principal geological and geophysical information available for the marine domain of the Ligurian Sea and its surrounding continental region is presented. Such an analysis includes a summary of the bathymetry and the morphology of the sea floor, the principal geophysical results (submarine and terrestrial), as well as the survey of the structure and geological evolution of the emergent marginal areas. On the basis of the geological and geophysical information available we will then attempt to choose a model interpreting the creation and evolution of this sector of the Mediterranean, the object of so much study and research.

II. BATHYMETRY AND SUBMARINE MORPHOLOGY

A. Sources of Information

The bathymetry of the Ligurian Sea (Fig. 2) can be obtained from such general or regional publications as those of the Oceanographic Museum of Monaco (Bourcart, 1958b, 1960a,b; Alinat et al., 1969), the Mediterranean map of the U.S. Naval Oceanographic Office (Carter et al., 1972), the maps of Debrazzi and Segre (1959), Angrisano and Segre (1969), Rehault et al. (1974), and of the Geophysical Laboratory of Trieste (Allan and Morelli, 1971). Additional information is available from the files of the Geodynamic Station at Villefranche-sur-Mer.

Physiographic maps and discussion of the morphology of the sea floor are frequently found accompanying the sources cited above. Carter et al. (1972) have thus published an excellent, relatively detailed physiographic map of the Tyrrhenian Basin, including the eastern part of the Ligurian Sea. In contrast,

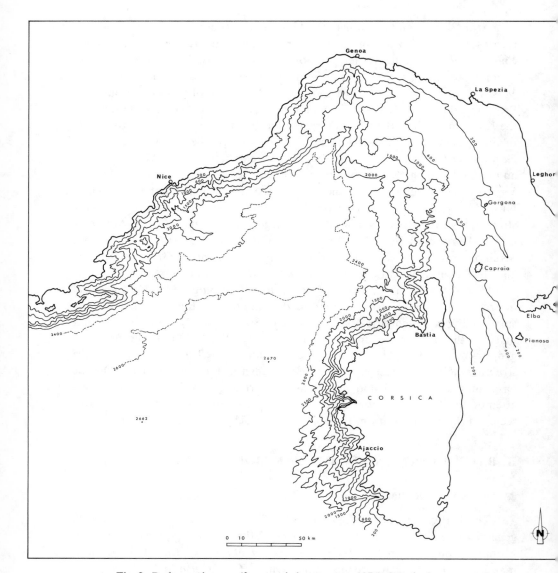

Fig. 2. Bathymetric map (from varied sources to 1974). Depths in meters.

the map they have produced for the Algero-Balearic-Provençal Basin—the major part of the region of interest here—is much less satisfactory because it is on much too small a scale. The same is true for the maps, both bathymetric and physiographic, produced on the scale of the western Mediterranean as a whole by Ryan *et al.* (1969). Finally, many data on the submarine morphology of the Ligurian Basin have been synthesized in recent publications by Rehault *et al.* (1974) and Gennesseaux and Rehault (1975), and for that part which concerns the continental shelf by Fierro *et al.* (1973) and Fanucci *et al.* (1974*a*).

B. The Continental Shelf

The continental shelf of the Ligurian Sea is characterized by its narrowness, to the point that locally it may even disappear. Where it is found it shows a regular and very gentle slope. From Toulon to Genoa it has an average width of 3 to 10 km, and similar values are found over most of the western coast of Corsica. It is, for example, a few kilometers wide immediately west of Cape Corse, a dozen or so kilometers wide from the Gulf of Saint-Florent to Calvi, once again less than 10 km from Calvi to the Gulf of Valinco. From this point there is, in contrast, a sudden and appreciable enlargement (30 km on average) of the shelf off the extreme southwest of Corsica and northwestern margin of Sardinia. The continental shelf progressively widens between Genoa and Piombino–Isle of Elba via Leghorn along the Italian coast, where from La Spezia it is linked to a coastal plain and may be as much as 50 to 60 km wide (even 65 km in the latitude of the Isle of Capraia).

The edge of the continental shelf is generally clearly defined and occurs at variable depths never less than 100 m. It lies between 100 to 120 m from Nice to Genoa, and from 100 to 180 m between La Spezia and the Isle of Elba. In a recent study Fierro *et al.* (1973) provided a detailed description of the morphology and constitution of the continental platform in the northern sector of the Ligurian Sea, from Nice to Genoa. They showed, for example, that it was almost completely absent off headlands and even in certain bays (the Baie des Anges at Nice). The continental shelf in the whole of this region is characterized by being extremely narrow, and it can be divided into three zones from the shoreline seaward: (1) a littoral fringe corresponding to the zone between the 0–10-m isobaths, with a floor generally rocky and irregularly covered by the products of littoral erosion and alluvium from coastal torrents (sandy and gravelly material); (2) an internal platform (between the 10–70-m isobaths) cut in the sedimentary or crystalline substratum and covered by marine plants in sheltered areas or by bioclastic sands; and (3) an external zone at depths from 70–120 m, constructed ("progradation") from Plio-Quaternary deposits (blue pericontinental clays) with a form ultimately dependent on the phases of regression and transgression of the Quaternary.

C. The Continental Slope and Continental Rise

The close spacing of the isobaths (Fig. 2) is evidence of the steep slope characteristic of the continental slope off Provence and off the Riviera from Toulon to Genoa, with an attenuation from Imperia, and again along the western margin of Corsica. The angle of the continental slope is much less between Genoa and Cape Corse, a region where, in addition, it appears to possess an intermediate terrace called "Terrasse de La Spezia," which has a submeridional orientation (between 9°30′ and 9°50′ E) and lies at an average depth of 500 m. Toward the south this terrace rises slightly until it reaches depths scarcely greater than 400 m, forming the "Corso-Tuscan" sill between Cape Corse to the west and one of the islands of the Tuscany Archipelago, the Isle of Capraia, to the east. Across the sill and to the south there begins a morphologically different region, the Tuscany Trench, which has the same alignment as the "Terrasse de La Spezia" and forms the northernmost physiographic element of the Tyrrhenian Sea, between the gentle eastern continental slope of Corsica and the continental slope off the coast of Tuscany.

On both sides of the Ligurian Abyssal Plain the continental slope is found to be dissected by deep canyons which may be incised to a depth of about 400 m. These canyons, which are structurally controlled, are very commonly located along the submarine prolongation of a terrestrial valley, and therefore originate on the continental platform. Where the canyons cut the shelf the edge of the continental platform is more clearly marked. According to Gennesseaux (1962b) the canyons appear to play an active role in the transport of coarse sediment to the abyssal plain. The two most remarkable canyons have their heads situated just south of Genoa on the continental platform, where they appear as the prolongation of the "Genoa rivers," and they can be followed from there in a south-southwest direction to a depth of 2500 m on the abyssal plain. In addition to these two spectacular elements, other canyons which cut the continental slope off the French coast and off the western Corsican coast have northwest–southeast or, more rarely, east–west and, in southwestern Corsica, northeast–southwest orientations.

In addition to the canyons and, locally, the terraces, the continental slope may sometimes be interrupted by shoals such as, for example, that of Mejean lying to the south of the Esterel.

The continental rise is not always present in the Ligurian Sea. It is not observed at the foot of certain island slopes, such as those west of Corsica and Sardinia, where the junction with the abyssal plain is abrupt. On the other hand it is well developed with a relatively weak slope between Toulon and Genoa. Off the Franco-Italian Maritime Alps, according to Rehault et al. (1974), it has a stepped appearance.

D. Abyssal Plain

This structure has a general northeast–southwest orientation, and is often referred to as the Ligurian Trench. It is delimited approximately by the 2400–2500-m isobath and nowhere has depths much greater than 2700 m. The slope is, in general, gentle and toward the southwest. It is a zone of sedimentary accumulation, generally flat-lying on the whole. The floor of the abyssal plain, however, is slightly asymmetrical in the sense that the deeper part (here at depths scarcely greater than 2600 m) is closer to the French coast of the Maritime Alps than it is to the coast of Tuscany or the northwest margin of Corsica. On the other hand, further to the southwest where the abyssal plain becomes larger and merges into the Balearic Basin, this asymmetry reverses, and henceforth the greater depths are closer to the Corsican margin (cf. bathymetric maps of the U.S. Naval Oceanographic Office, Carter *et al.*, 1972).

The bathymetric measurements have, in addition, demonstrated the existence of knolls of low relief (about 100 m) and small size (a few kilometers in diameter) within the abyssal plain. They appear to be dispersed, but may often be grouped or aligned as, for example, in the sector situated to the southwest of the Gulf of Genoa, off Imperia, where they may have a common border with the continental rise. The significance of this type of relief will be discussed later in Section IV, which is devoted to the geophysical results. There are a few rare shoals, too (cf. Rehault *et al.*, 1974), mostly located along the eastern margin of the Ligurian Abyssal Plain.

III. STRUCTURAL FRAMEWORK (CONTINENTAL AND ISLAND BORDERS)

Lying at the point of intersection of three Tertiary orogenic belts (Pyreneo-Provençal, Alpine, Apennine) and with the history of varying sedimentation during the Neogene and Quaternary further complicated by tectonic and volcanic events, the continental and island borders of the Ligurian Sea are characterized by a complex, fragmented structure (Fig. 3).

In order to present an analysis of this structure, the exposed crystalline basement and the overlying folded sedimentary sequence, which most often form the mountainous massifs, will be treated first. These contrast with the intervening basins, which are often of relatively small dimensions and generally filled with Neogene and Quaternary sediments. To this discontinuous sequence, including the coastal basins, can be added the islands of the Tuscan Archipelago —Capraia, Elba (in part), Pianosa, Monte Cristo, Giglio, etc.—which result from Neogene to Recent magmatism, with the exception of the island of Pianosa, constructed entirely of Neogene sediments.

A. The Crystalline Massifs and Their Sedimentary Envelopes—The Folded Sedimentary Massifs

Following the littoral of the Ligurian Sea from west to east, beginning at Toulon and returning toward the west along its southern margin, five distinct

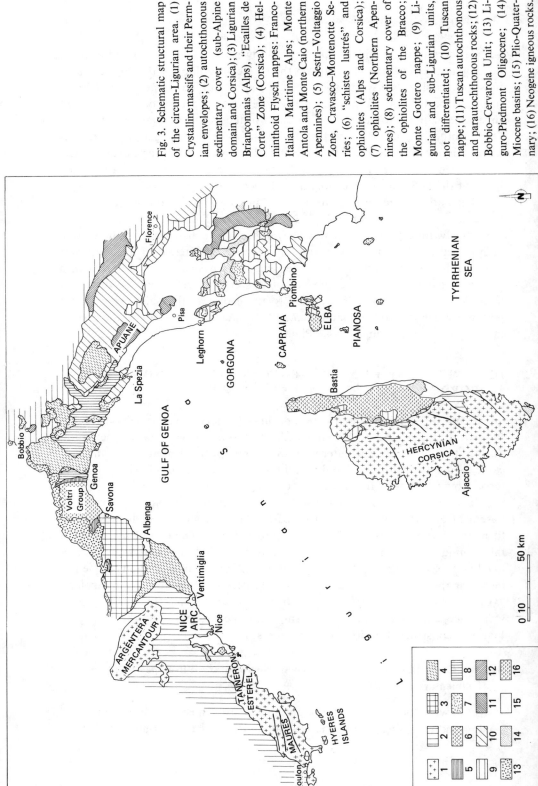

Fig. 3. Schematic structural map of the circum-Ligurian area. (1) Crystalline massifs and their Permian envelopes; (2) autochthonous sedimentary cover (sub-Alpine domain and Corsica); (3) Ligurian Briançonnais (Alps), "Ecailles de Corte" Zone (Corsica); (4) Helminthoid Flysch nappes: Franco-Italian Maritime Alps; Monte Antola and Monte Caio (northern Apennines); (5) Sestri–Voltaggio Zone, Cravasco–Montenotte Series; (6) "schistes lustrés" and ophiolites (Alps and Corsica); (7) ophiolites (Northern Apennines); (8) sedimentary cover of the ophiolites of the Bracco; Monte Gottero nappe; (9) Ligurian and sub-Ligurian units, not differentiated; (10) Tuscan nappe; (11) Tuscan autochthonous and parautochthonous rocks; (12) Bobbio–Cervarola Unit; (13) Liguro-Piedmont Oligocene; (14) Plio-Quaternary; (15) Neogene igneous rocks.

sectors can be recognized on the basis of structural form. These are: (1) from Toulon to Cannes ("Provence Cristalline"), a zone essentially formed by crystalline, volcanic, and sedimentary Paleozoic rocks, which, according to marine data from farther to the west (Alla *et al.*, 1973) may be considered as the eastern extension of the Pyrenean basement; (2) from Cannes to Genoa (more exactly to Sestri Ponente, a dozen kilometers west of Genoa) there are a number of subalpine sedimentary massifs (external alpine zone) terminating at the coast, which then cuts alpine units belonging to progressively more internal zones; (3) from Genoa to Piombino–Isle of Elba (in part), the coast cuts in a more or less orderly fashion the internal zones of the Apennines (Ligurian Apennine), taking into account their structural disposition, and then the internal part of the external zones (Tuscany Apennine); (4) northern and eastern Corsica, a structurally complex area formed of units which are essentially related to the internal alpine zones, although some of them, according to certain authors (Nardi, 1968*a*), may be considered as being of Apennine type; and (5) western Corsica, which corresponds essentially to a crystalline Hercynian massif whose geological and structural affinities with the Maures–Esterel Massif have been underlined by Nairn and Westphal (1968).

1. *"Provence Cristalline"*

The structural analysis of this region (Fig. 4) is based on the excellent studies of Lutaud (1924) and of Aubouin and Mennessier (1960–1963), completed by several more recent works (particularly Aubouin, 1974).

The framework of this region comprises three Hercynian massifs, which are geographically distinct and of differing importance. From southwest to northeast these are: the massifs of the Toulon region, represented in fact by relatively poor outcrops; the Maures Massif, which is the most extensive and includes the outcrops of the Hyères Islands (Porquerolles, Port-Cros, and the Ile du Levant) a dozen or so kilometers beyond the Bay of Hyères; and the Tanneron Massif west of Cannes.

To these may be added their envelope of Permian volcano-sedimentary rocks in synclines separating the Toulon Massifs from the Maures Massif and the latter from the Tanneron Massif. These Permian "depressions" border the Maures to the west and north and contain volcanic flows which form the backbone of the Esterel between the Maures and Tanneron.

The Massifs of the Toulon Region. The outcrops consist of schists ("phyllades"), forming the massif of Cape Sicié and the islands of Embiez southwest of Toulon and the hills immediately east of the town. Part of these schists are clearly detached and thrust as a slice over a substratum most often Permo-Triassic in age, but which may be formed of autochthonous schists.

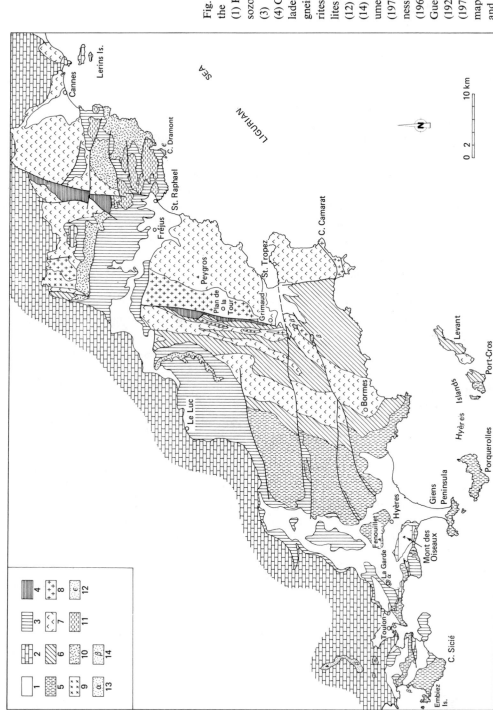

Fig. 4. Structural map of the "Provence Cristalline." (1) Plio-Quaternary; (2) Mesozoic sedimentary cover; (3) sedimentary Permian; (4) Carboniferous; (5) "Phyllades"; (6) micaschists; (7) gneiss; (8) granites; (9) diorites; (10) Permian rhyolites; (11) Permian dolerites; (12) esterellite; (13) andesite; (14) basalts. Maps and documents consulted: Aubouin (1974), Aubouin and Mennessier (1960–1963), Bordet (1966), Boucarut (1971), Guérard (1957), Lutaud (1924), Maluski and Allegre (1970), B.R.G.M. geological maps (1/200,000; 1/80,000; and 1/50,000).

The Maures Massif. This is the central element and geographically the most extensive within the "Provence Cristalline." A series of successive bands aligned in a NNE–SSW to N–S direction parallel the metamorphic gradient, which increases from west to east: The Hyères "phyllades" form the western band and crop out at the extremity of the Giens Peninsula and in the island of Porquerolles. They are the easterly prolongation of the schists of the Toulon region. A few kilometers west of Hyères at Mont Fenouillet, schists with graptolites of Silurian age have been described (Schoeller, 1938; Gueirard *et al.*, 1970). Toward the east the schists pass progressively to the following formations:

(1) Micaschists, called Micaschists I (Maluski, 1968) or the Upper Micaschists, within which are found gneissic facies at several horizons.

(2) The Bormes gneiss, which forms the axial region of the western Maures.

(3) Micaschists II or Lower Micaschists. Along the eastern margin of this band of micaschists are gneiss and granite intercalations with a general N–S trend. There is also a narrow, elongate basin, the basin of Plan de la Tour, in which a discordant Carboniferous sequence was deposited (sandstones and shales with plants, intercalated rhyolites, all of Stephanian age).

(4) The Saint-Tropez and the Peygros Massif gneiss (migmatitic and anatexic), forming the eastern Maures which encloses a granitic body oriented NNE–SSW (the Plan de la Tour granite) in the west and in the southeast an ENE–WSW granite (the Cape Camarat granite).

The radiometric age dates (e.g., Maluski and Allegre, 1970; Roubault *et al.*, 1970*a*, *b*; etc.) suggest a Lower Paleozoic age for the crystalline rocks of the Maures and Tanneron Massifs. The gneisses appear to be a little older than the granites (Plan de la Tour, etc.) but the ages obtained must be treated with some caution, for it would seem that they most often date the metamorphism of the region.

The major structural features of the Maures Massif are the result of several tectonic phases and are still the object of diverging interpretations. It seems that the following can be considered as dominant: (1) the local anticlinal bearing of the Bormes Gneiss band; (2) the Grimaud Fault, a zone of vertical NNE–SSW to N–S faulting with mylonites, as a result of which the Micaschists II and the sedimentary fill in the basin of Plan de la Tour were thrown into contact with the gneiss and granites of the eastern Maures; (3) a series of ENE–WSW faults well marked in the western Maures where they pinch out the Permian. An identical structural trend has been observed in the massifs of the Toulon area. In the Maures these faults are generally reversed and the significance of this thrusting is still under discussion; and (4) a general updoming of the basement with an uncertain orientation, but apparently NE–SW.

It also leads to the uplift witnessed indirectly in the North Varois doming (of the beds covering the northwest of the Maures Massif) and the presence at outcrops of parts of the deeper, more metamorphosed basement in the eastern Maures.

The Tanneron Massif. This massif is formed essentially of gneiss and granite, with the same N–S to NNE–SSW orientation as in the Maures Massif. Petrographic and structural analogies between the various elements indicate that the Tanneron Massif is only the northern prolongation of the Maures Massif, from which it is separated by the Bas–Argens Syncline, with its fill of Permian sediments, and by the Esterel Massif. From west to east the following units can be recognized: micaschists of the Gardanne Zone; a narrow band of gneiss [these two units being separated by a N–S fault (with mylonite) which is none other than the prolongation of the Grimaud Fault of the Maures]; then the Prignonet diorite; Rouet granite (the northern prolongation of the Plan de la Tour granite); gneiss which corresponds to the Peygros–Saint Tropez gneiss; the Carboniferous basin of the Reyran which is narrow, elongate, and limited by N–S faults; and finally the eastern Tanneron gneiss which crops out at the coast at Cannes.

The Permian Depression and the Esterel Massif. Lying discordantly on the Carboniferous and the crystalline Hercynian basement, a thick detrital series (conglomerates, arkoses, sandstones, and red pelites) of Permian age is seen almost everywhere around the massifs previously described. Since they are more easily removed by erosion, these beds outcrop in a vast depression with a NE–SW trend extending from Toulon to Saint-Raphael. Intercalated within this Permian series to the northeast are many thick volcanic flows, dolerites, pyromerides, and especially rhyolites. While disseminated in the Luc Cuvette west of the eastern Maures, the volcanic material predominates to the northeast of the Maures, at the contact with the Tanneron Massif, where it is responsible for the orographic relief and constitutes the famous Esterel sites, in particular with the celebrated amarinthine rhyolite in the eastern littoral region.

Structurally the most distinctive feature of the Tanneron and Esterel Massifs lies in the existence of important N–S to NNE–SSW faults, which in the volcanic massif cut in addition a dense system of E–W trending faults.

At a few points within the zone of crystalline rocks of Provence, a number of outcrops of the overlying Mesozoic occur, particularly in the region of Toulon where the Triassic (sandstones, conglomerates, limestones and dolomites, marls and gypsiferous marls) contains two distinct levels of décollement. Although the Toulon Massifs and particularly the Maures Massif are bordered by a thick Permian sequence, it should be noted that toward the north it is the Triassic which rests directly with discordance on the crystalline rocks of the Tanneron Massif. Some remains of a Jurassic cover, which rest on a Triassic substratum, are found near the coastal region southwest of Hyères, where they form the core of the Mont des Oiseaux synclinal zone between the Toulon Massifs and the Maures Massif.

In addition to the Paleozoic volcanism (Carboniferous and particularly Permian), there are dispersed outcrops within or just outside the region of the crystalline rocks of Provence, which are evidence of Tertiary igneous activity (see Section IIIB).

2. The Sub-Alpine Massifs

Structure. The sub-Alpine massifs extend along the Mediterranean coast from Cannes to Ventimiglia (Fig. 5). In contrast with the preceding sector, composed almost exclusively of crystalline rocks and a Late Paleozoic volcano-sedimentary sequence, this area is formed of folded sediments, primarily Mesozoic in age, within which it is not easy to make structural subdivisions. Very schematically, the following zones may be distinguished:

(1) A marginal zone more or less developed (up to 20 km) and always along the coast where, in addition to Neogene deposits (which will be studied later), Mesozoic outcrops predominate. These are moderately folded and may be considered with Gèze (1960–1963) as the natural northern sedimentary margin of the Maures–Esterel–Tanneron Massifs and of their submerged eastern extension.

(2) A tectonically complex hinterland where overturning, stretching, and thrusting of the folds of the Mesozoic and Lower Tertiary cover predominates. This moved by décollement over a Triassic substratum which includes plastic members. Only here can one speak strictly of the sub-Alpine chains (the Castellane Arc to the west of the River Var, the arcs of Nice and Roya to the east). Locally, the extreme southern margin of the Nice Arc and the Roya Arc may extend farther to the south and come to margin the littoral where the first zone disappears under the sea.

In reality the passage between the hinterland—strictly, the Sub-Alpine Zone—and the autochthonous zone which acts as a foreland, is scarcely, if at all, marked by any facies change. Even in the structural form, if it is well defined to the west (clear evidence of thrusting of the "Baous" from the extreme southeast of the Castellane Arc over the autochthonous zone of Vence), it is less clear farther to the east, at Nice and Monte Carlo, for example, where a marked decrease in the intensity of thrusting in the frontal zone is apparent.

The major structural trends (folds and faults) have a prevailing north–south orientation in the autochthonous zone, particularly between the massif of Tanneron and the Pliocene basin of the Var. The subsidence of the latter basin was itself doubtless controlled by north–south faults. On the other hand, thrust faults and folds have an east–west orientation in the Castellane Arc and

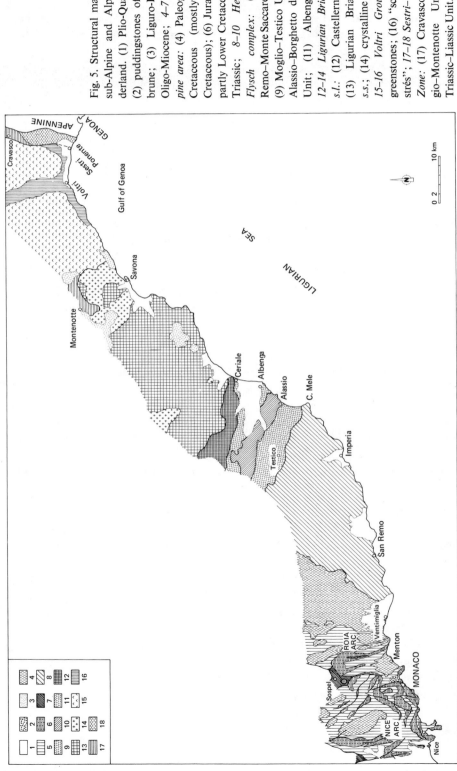

Fig. 5. Structural map of the sub-Alpine and Alpine borderland. (1) Plio-Quaternary; (2) puddingstones of Roquebrune; (3) Liguro-Piedmont Oligo-Miocene; *4–7 Sub-Alpine area:* (4) Paleogene; (5) Cretaceous (mostly Upper Cretaceous); (6) Jurassic (and partly Lower Cretaceous); (7) Triassic; *8–10 Helminthoid Flysch complex:* (8) San Remo–Monte Saccarello Unit; (9) Moglio–Testico Unit; (10) Alassio–Borghetto d'Arroscia Unit; (11) Albenga Unit; *12–14 Ligurian Briançonnais s.l.:* (12) Castellermo Unit; (13) Ligurian Briançonnais *s.s.;* (14) crystalline massifs; *15–16 Voltri Group:* (15) greenstones; (16) "schistes lustrés"; *17–18 Sestri–Voltaggio Zone:* (17) Cravasco–Voltaggio–Montenotte Unit; (18) Triassic–Liassic Unit.

trend toward NNW–SSE in the Nice Arc and the Roya Arc. According to several authors (Gèze, 1960–1963; Lanteaume, 1968; Perez, 1975a, b) an important NNE–SSW strike-slip faulting zone, called the Breil–Sospel–Monaco Fault, acts as a division between the Nice Arc and the Roya Arc. This fault cuts the sedimentary cover as deep as the Triassic and could even affect the crystalline basement, which outcrops further to the north (Argentera Massif).

Lithology. The complete lithological succession of the Triassic can only be determined in the region of Grasse–Vence, where it is of the germanic type (sandstones, dolomites and limestones, marls and clays, gypsum, carneoles, dolomites, and finally a Rhaetic marly limestone). In the Nice Arc, where décollement and pseudodiapirism involve Middle and Upper Triassic, the latter are present in the form of chaotic masses of dolomites, carneoles, gypsum, and shales.

The Jurassic is essentially calcaro-dolomitic and forms a tectonically more rigid mass, reacting by fracturing. The Lower Cretaceous is nearly absent in the Grasse–Vence area, and is apparently reduced in thickness (condensation, gaps) in the Nice Arc, with a facies which is often sandy and glauconitic, and thus plays no tectonic role. It is thus in contrast to the Upper Cretaceous, a relatively thick, marly limestone sequence which behaves as a pliable mass and is affected by disharmonic folding.

The Paleocene is not represented, and the discordantly overlying Eocene (Middle and Upper) is represented by three successive facies: sandy nummulitic limestones, marls with planktonic foraminifera, and finally a thick sequence of flysch sandstones. These facies are not isochronous, for each formation becomes younger the farther it is traced toward the west, to the point that in the Castellane Arc the top of the marls and particularly the uppermost sandy member, the Grès d'Annot Formation, already lies within the Oligocene. These coarse detrital Paleogene successions form the heart of large synclines in the Nice Arc and the Roya Arc and they are equally found in the core of the semisyncline (cut by the Mediterranean Sea) in the coastal area of Menton

The most recent deposits involved in the sub-Alpine folding are Oligocene. At Roquebrune–Cape Martin (between Monaco and Menton) the sub-Alpine folds are discordantly overlain by a marine conglomeratic formation, the puddingstones of Roquebrune. These rocks have been assigned to the Miocene (probably Middle Miocene) by Anglada *et al.* (1967).

3. *The Internal Alpine Zones*

From Bordighera just to the east of Ventimiglia, to Sestri Ponente a few kilometers to the west of Genoa, the Ligurian coast cuts across, more or less at right angles, a series of allochthonous units belonging to the alpine internal zones (Fig. 5). The direction of thrusting appears to be directed from the northeast toward the southwest, or from the east to the west, but frequently one can observe superposed an inverse sense of tectonic movement, i.e., retrothrusting or underthrusting.

In following the Mediterranean Ligurian coast the main structural units can be distinguished:

The Helminthoid Flysch Zone (s.l.). This zone outcrops from Bordighera to beyond Alassio, some 3 km to the southwest of Albenga. It is subdivided into three units or "festons," separated from each other by major tectonic contacts:

(a) The San Remo–Mte. Saccarello Unit or Helminthoid flysch *s.s.*, consisting of a "complexe de base" of red and green pelitic beds with manganiferous horizons, of Vraconian to Turonian age, passing progressively into a Senonian flysch, which is initially dominantly detrital but becomes progressively more calcaro-argillaceous. The San Remo Unit has the structural form of wide, sinuous folds which are easily distinguished at the level of the basal complex. It forms, in fact, an enormous triangular klippe each side of which measures at least 40 km, and which along its southwestern margin thrusts directly over autochthonous flyschoid sandstones (the formation Grès d'Annot–Grès de Menton) at the top of the sedimentary sequence in the external domain. It must be noted, however, that toward the northwest, that is with increasing distance from the coast, the San Remo Unit no longer rests directly on the external zone but upon allochthonous tectonic units belonging to the margin of the internal domain—the parautochthonous unit of the Col de Tende, and then the zone of the "Unités dilacérées," which is related to the Subbriançonnais. This clearly demonstrates the complete tectonic independence of the Helminthoid Flysch Zone.

(b) The Moglio–Testico Unit, considered by some authors (Lanteaume, 1968) as a more internal "feston" of the Helminthoid Flysch group and by others (Haccard *et al.*, 1972) as a distinct unit belonging to the ophiolitic series, which may be found in a more internal part of the Alps and in the northern Apennines. According to the first interpretation these beds are of Upper Cretaceous–Paleocene age, rather than Lower Cretaceous as inferred in the second interpretation.

(c) The Alassio–Borghetto d'Arroscia Unit, which lithologically is very close to that of the San Remo Series. Along its internal margin it thrusts over the following units:

The Albenga Unit (sensu Boni and Vanossi, 1972). This unit equals the Arnasco Unit plus the Sandy, Calcareous Shale Unit in the sense of Lanteaume (1968). Like the preceding units, it forms a narrow band, here with a predominately east–west elongation, and terminates on the littoral at Ceriale. The sandy argillaceous and calcareous beds of which it is formed are assigned to the Paleocene–Eocene.

The Ligurian Briançonnais. This can be followed along the coast from Ceriale to just beyond Savone. It corresponds to the prolongation in the southern Alps of the Briançonnais Zone, which form the frame of the Penninic domain along the whole of the western Alps. In the Ligurian region, where it follows a general WNW–ESE orientation, its development is important particularly in the coastal area. The detailed structure is very complex as the result of tectonic cleavage with numerous second order units whose geometry and interrelationships are not always clear. Along the southern margin, in the zone near the coast, the Castellermo Unit borders on the Albenga Unit. It is composed of calcareous and dolomitic material attributed to the Upper Triassic and Jurassic. Tectonically the highest level, this Castellermo Unit is considered as the most internal unit of the Briançonnais Zone. Nearly all the other Briançonnais units contain well developed Triassic members and a reduced Jurassic sequence, as well as dissociated elements of the overlying Upper Cretaceous–Eocene cover. Paleozoic (Permo-Carboniferous) rocks, essentially in a metamorphic and crystalline form, are also observed within certain of these units. Among these the crystalline massif of the Savona region (granites, gneiss, amphibolites), which crops out in the coastal area along the northeastern margin of the Ligurian Briançonnais, is interpreted either classically as belonging to the Briançonnais Zone (Boni and Vanossi, 1972), or as a basement "Insubric" klippe (basement of the Austro-Alpine–Apennine type) (Haccard *et al.*, 1972), resting on the Briançonnais and consequently strongly allochthonous.

Voltri Group. The crystalline rocks of Savona are followed by an ensemble of polymetamorphic rocks belonging to the Piedmont Zone. They are represented by the "schistes lustrés" (sericitic–calcareous schists), associated with a considerable volume of greenstones (ophiolites) in the form of folded nappes capped by a thick slice of peridotites and gabbros. A Jurassic (probably Middle–Upper) to Lower Cretaceous age is assigned to this ophiolitic sequence.

Sestri–Voltaggio Zone. This zone forms a narrow band about 3 km wide, oriented north–south, and extends from Sestri Ponente at the coast to Voltaggio inland, where it disappears under discordant Oligocene deposits. It is in faulted contact (the Sestri–Voltaggio Fault) to the west, with the eastern margin of the metamorphic complex of the Voltri Group, and to the east supports the first Apennine units. The stratigraphic and structural interpretation of this zone is still the object of controversy. It is formed of Mesozoic rocks within which Haccard (1971) distinguished at least two structural units: (1) a Triassic–Liassic unit of dolomites and reef limestones (Upper Triassic–Rhaetic), followed by Liassic shales and limestones; (2) the Cravasco–Voltaggio Unit, in which may be recognized a ?Liassic sequence of shaly limestones and platy, bedded limestones, followed by radiolarites and diabases associated with gabbros of ?Upper Jurassic age.

The paleogeographic origin of these two units, the second particularly, has recently been located (Haccard *et al.*, 1972) in a more internal position than the Piedmont Zone to which, however, they appear to be most closely related.

4. *The Northern Apennine (Ligurian and Tuscany)*

In following the Ligurian coast to Genoa the various Alpine units have been traversed progressively from the external to the internal zones, but with the Apennine units directed in the opposite sense a continuation of the traverse reveals, in general, the internal zones progressing toward the external. In detail, however, as a result of intense deformation, break-up, and interpenetration of different units, a more clear exposition can be given by first presenting schematically the geometry as a whole and referring to coastal locations where these units can be demonstrated (Fig. 6).

Internal Domain (Ligurides). The Ligurides are made up of a complex of allochthonous units; most of them tend to cover the external units of the Apennines. They crop out principally along the margin of the Ligurian Sea from Genoa–Sestri Ponente to Monterosso 15 km northwest of La Spezia. They reappear sporadically south of La Spezia as far as southern Tuscany. Schematically the Ligurides form two major structural ensembles which show marked lithological analogies with certain Alpine units already described. There is an ophiolitic ensemble with associated sediments, which includes the Bracco, Busalla, Val Lavagna, and Monte Gottero units, and an ensemble belonging to the Helminthoid Flysch category, which along the Ligurian margin is represented by the Monte Antola Unit, but, farther to the southeast, by several other units such as the Monte Caio and the Monte Cassio Units, which crop out in the hinterland.

From Sestri Ponente the coastline cuts successively: (1) Immediately before Genoa, a relatively narrow band overturned to the west onto elements of the Sestri–Voltaggio (*s.s.*) Zone. According to Haccard (1971) two structural units can be recognized: (a) the ophiolitic unit of Monte Figonia, made up of serpentines, diabases, and radiolarites, followed by limestones and shales. To this ensemble, by analogy with other ophiolitic units in the Apennines, may be assigned a Jurassic–Lower Cretaceous age; and (b) the strongly folded flysch Busalla Unit. It has been compared by Haccard (1971, 1975) with the Moglio–Testico Unit of the Franco-Italian Maritime Alps and with the Ligurian Val Lavagna shales. The recent discovery of an Albian ammonite in the upper beds of the Busalla Flysch (Haccard and Thieuloy, 1973) permits the unit to be assigned to the top of the Lower Cretaceous. (2) From Genoa to Chiavari, tectonically overlying [or stratigraphically discordant upon(?), cf. Haccard and Thieuloy, 1973] the Busalla Unit, is a thick plate of Helminthoid

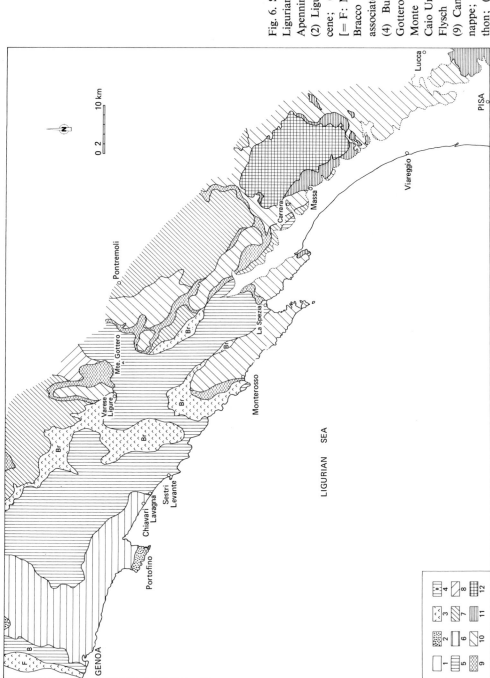

Fig. 6. Structural map of the Ligurian part of the northern Apennine. (1) Plio-Quaternary; (2) Liguro-Piedmont Oligo-Miocene; (3) Ophiolitic complex [= F: Monte Figonia Unit; Br: Bracco Unit (greenstones and associated sedimentary cover)]; (4) Busalla Unit; (5) Monte Gottero–Val Lavagna units; (6) Monte Monte Antola Unit; (7) Monte Caio Unit; (8) other Helminthoid Flysch units, undifferentiated; (9) Canetolo Unit; (10) Tuscan nappe; (11) Tuscan parautochthon; (12) Tuscan autochthon.

Flysch, the Monte Antola Unit. It includes a "complexe de base" of variegated shales sometimes enclosing blocks from the ophiolitic units. (3) From Chiavari to Monterosso the coastline cuts obliquely across another dominantly ophiolitic ensemble. The structural interpretation varies from author to author, but it is possible to distinguish: (a) the Bracco Unit, formed almost exclusively of rocks of the ophiolitic suite (lherzolitic serpentines, gabbros, and diabases) of Jurassic age and which structurally appears today as a kind of NNW–SSE-oriented axis within the Ligurides. The greenstones are associated with a sedimentary cover, principally of Upper Jurassic radiolarites, Calpionellid limestones (Tithonian–Berriasian), Palombini shales and limestones (Neocomian–Barremian), and a complex of shales several hundred meters in thickness, the Val Lavagna shales, assigned to the Aptian–Albian; and (b) the Monte Gottero Unit. Overlying the Val Lavagna shales, with which they seem locally to possess a stratigraphic relationship (lateral passage to the northwest?) lies a thick flysch sequence, the Monte Gottero sandstones. This unit rests tectonically upon different folded members of the ophiolitic suite and its western margin is thrust in turn by the Monte Antola nappe. Up to the present time these sandstones have not been dated with certainty. Some authors considered this flysch to be related to the Apenninic units belonging to the group of Upper Cretaceous Helminthoid Flysch. On the other hand, Haccard (in Haccard *et al.*, 1972; Haccard and Thieuloy, 1973) leans to equivalence in facies (and eventually in age) between the supra-ophiolitic detrital flyschoid ensemble which embraces the Busalla Flysch, Val Lavagna shales, and Monte Gottero sandstones with the Moglio–Testico Unit of the Franco-Italian Maritime Alps, and considers that all these are elements of a single Lower–Middle Cretaceous sedimentary complex, subsequently disrupted. This interpretation slightly differs from that of Elter (1975*b*), according to whom the Ligurides can be paleogeographically subdivided into an internal part, which includes the Val Lavagna shales–Monte Gottero sandstones basin, and an external part, where the Helminthoid Flyschs have been deposited. They are separated by an ophiolitic ridge of which remains can be found in the form of the Bracco Unit.

Canetolo Sub-Ligurian Unit (*or Clays and Limestones Complex*). This unit is very broken up and irregularly developed. Often it does not form a true nappe, and sometimes appears to have given rise to olistostromes which fed the sedimentary basin lying in the western part of the external Apennine domain. Tectonically it is inserted between the Ligurides and the most internal elements of the external domain, i.e., the Toscanides. The distribution of outcrops is sporadic; along the margin of the Ligurian Sea a few elements are to be seen in the area near La Spezia, again to the north of Pisa (east of Viareggio), and finally along the southern Tuscany coast in the vicinity of Leghorn and to the north east of Piombino.

Where the section is most complete (found only in the more continental sector of the Apennines) it consists of an argillaceous series (Canetolo Formation *s.s.*) which encloses beds of limestone and sandstone and lenses of a calcaro-calcarenitic flysch (Groppo del Vescovo limestones), and at various levels, but particularly toward the top, horizons of coarsely detrital material formed mostly of andesitic volcanic elements (Petrignacola Formation). The majority of authors assign a Lower to Middle Eocene age (or even younger as far as the uppermost members are concerned) to the material forming the Clays and Limestones Complex.

External Domain (In Part). In the northern Apennines the external domain is made up of the following structural zones (from the most external to the most internal): (1) the Ombrian Zone; (2) the Cervarola–Bobbio Unit essentially formed of Upper Oligocene to Middle Miocene sandstones. Neither of these zones is geographically involved in the structures of the present Ligurian borderland; and (3) the Tuscan Zone (Toscanides) which the Ligurian Sea (*s.l.*) cuts between Monterosso and Piombino–Isle of Elba, and which is subdivided into: (a) an external Tuscan Zone (or Tuscan autochthon) seen only in tectonic windows (Apuanes north of Viareggio, Monte Pisano east of Pisa, Montagnola Senese in southern Tuscany) through the internal Tuscan Zone. The most complete exposure of the series is seen in the metamorphic and strongly tectonized Apuanes Massif, where essentially it is composed of porphyroids, quartzites and conglomerates ("Verrucano") of Permo-Triassic age overlain by a Mesozoic–Eo-Tertiary cover (Upper Triassic dolomites, Jurassic marbles, limestones with flints, radiolarites, crinoidal limestones and shales; Upper Cretaceous–Eocene platy limestones, chloricitic and sericitic shales, and Oligocene "pseudomacigno"); and (b) an internal Tuscan zone, which consists of (i) a metamorphic basement whose margin overthrusts the western part of the Tuscan autochthon (Apuanes) in the form of a series of slices of Paleozoic schists and of Verrucano (= Massa Scales Zone) associated with parautochthonous elements torn from the Apuanes. The whole is referred to as the Tuscan parautochthon; and (ii) the Mesozoic and Paleogene cover. By décollement over the Upper Triassic evaporitic horizon it forms the principal part of the Tuscan nappe. More or less the same facies, but unmetamorphosed, are found here as in the Tuscan autochthon, with a series a little more complete north of the Arno than to the south of this river in southern Tuscany. In any case the series ends with a thick flyschoid or "Macigno" member of Upper Oligocene–Lower Miocene age.

From La Spezia to Pisa, as a result of local uplift of the northern Apennines, the Ligurian maritime borderland (excluding coastal and alluvial Recent deposits) mostly corresponds to outcrops of the Toscanides. Then south of the Arno a certain sinking both facilitates a significant Neogene and Quaternary sedimentation and reduces on the surface the importance of the Tuscan deeper elements (the series, in any event, is less developed than it is to the north; see above). As a corollary the Ligurian superstructures reappear (ophiolitic series, flysch). A final parameter must be added, the Neogene–Quaternary magmatic activity subsequent to the last phase of tectogenesis, which is especially active further to the south (Tyrrhenian façade of the Apennines), but of which some signs already occur in SW Tuscany and which also created the greater part of the islands of the Tuscan Archipelago.

In summary, from this structural analysis of the maritime Ligurian side of the northern Apennines, the following major characteristics can be restated:

(1) An intense tectonic dislocation of the cover.

(2) Probably a significant shortening; the autochthonous part of the external domain is exposed far to the west, close to the Ligurian coast in a window SE of La Spezia.

(3) The heterogeneity and variability of the major structural trends, as a result of a polyphase tectonic activity. The most obvious constraint (because it is the last) seems to be oriented SW–NE (Liguria, NW Tuscany) to W–E (SW Tuscany).

5. The Isle of Elba (In Part)

The Isle of Elba lies in line with the southwest extension of the Piombino Peninsula and forms the largest of the islands of the Tuscan Archipelago, lying between Tuscany in the east and Corsica in the west. As in the case of the other islands, products of acidic magmatic activity during the later part of the Tertiary and Quaternary are present, in this case in the form of a granodioritic body dated at 7 m.y. approximately, i.e., close to the Miocene–Pliocene boundary. This forms the western part of the island (Monte Capanne), but the major part is made up of elements related to the Tuscan domain (Permo-Triassic Verrucano, Triassic and Jurassic limestones) and Ligurian domain (greenstones, Calpionellid limestones, Palombini shales and Helminthoid Flysch). In other words, geologically the island is no more than a westerly extension of southern Tuscany. In a section from Corsica to Tuscany, if one excludes the interpretation of the upper Corsican units according to Nardi (1968a), Elba represents the most westerly limit of the Apennines, given that farther to the north on approximately the same meridian and at the latitude of Leghorn, the little island of Gorgona is formed of "schistes lustrés" and ophiolites analogous to those of the Voltri Group of the internal Alpine zone.

6. *Alpine Corsica*

Less than 50 km west of Elba lies the promontory of Cape Corse, the farthest projection of Corsica into the Ligurian Sea (Fig. 7). The northeastern third of the island, structurally exceedingly complex and of whose significance

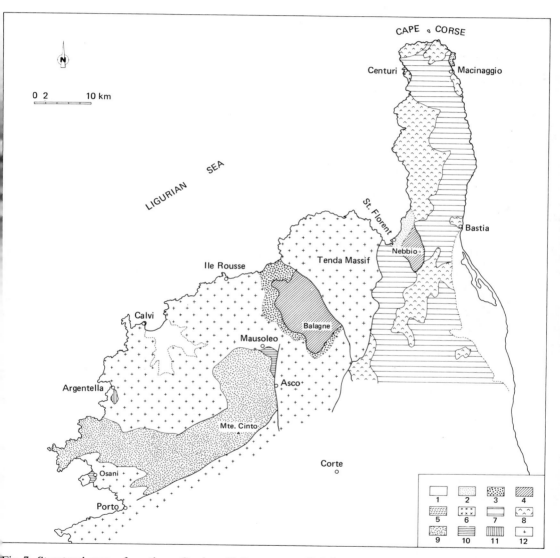

Fig. 7. Structural map of northern Corsica. (1) Quaternary; (2) Miocene; (3) autochthonous Eocene; (4) units of the Nebbio and Balagne Basins; (5) Macinaggio Flysch; (6) gneiss of Centuri; (7) "schistes lustrés"; (8) ophiolites; (9) Permian rhyolites and ignimbrites; (10) Upper Carboniferous; (11) Argentella–Tour Margine Series (Devonian and Lower Carboniferous); (12) crystalline basement.

there are still diverse interpretations, is often called Alpine Corsica because of the probable affinity of the units of which it is composed. This contrasts with the rest of the island, called Hercynian Corsica, made up of Paleozoic crystalline and metamorphic rocks and volcano-sedimentary rocks.

Cape Corse, as well as the greater part of the region extending to the south, is formed of epimetamorphic Mesozoic rocks with calc-schists, sericito-schists, greenstones, etc., which are similar to the ophiolitic complex of the Voltri Group of the western Alpine Piedmont. The assignment of the Corsican "schistes lustrés" to this corresponding Alpine zone appears to be accepted by all authors (Amaudric du Chaffaut *et al.*, 1972). They form the elements of a nappe which to the west rests upon units whose nature varies according to both the section examined and the appearance of the contact (vertical or directed sometimes to the east, sometimes to the west) which separates them.

The interpretation of two small zones of outcrop found toward the tip of Cape Corse has been much debated. To the northeast (Macinaggio), tectonically overlying the "schistes lustrés" and separated from them by a sole of carbonates (Trias? Lias?), is a Senonian flysch composed of shaly sandstones with black lydites which Nardi (1968a) correlated with the Helminthoid Flysch of the Ligurian units in the northern Apennines and the Franco-Italian Maritime Alps. To the northwest (Centuri) are gneisses considered by Haccard *et al.* (1972) as an "Insubric" klippe (Apennine–Austro-Alpine basement) and as an element of the basement of the "schistes lustrés" by Durand Delga (1972 in observations on a note by Grandjacquet *et al.*, 1972b). On the other hand the latter author requires a less internal origin (non-Apenninic) for the flysch of Macinaggio.

Similarly, south of Saint-Florent in the Nebbio Basin the "schistes lustrés" are tectonically overlain by various units of which the highest (Mortola Unit, in particular) contain members (spilites, radiolarites, fine-grained limestones, red and green conglomerates and flysch) which also resemble the Ligurian series.

In following the northern coast of Corsica from east to west the "schistes lustrés" can be seen in tectonic contact with the basement of the Tenda Massif, which is formed of crushed, schistose granite (the "protogine" of some authors). This massif is apparently considered as allochthon or at least as parautochthon by some (Nardi, 1968a, b; Nardi *et al.*, 1971; Haccard *et al.*, 1972), but the westerly thrusting implied has been cast in doubt by Durand Delga (1972 in observations on a note by Grandjacquet *et al.*, 1972b).

The Tenda Massif is only an outlier of the Hercynian basement of Corsica, for immediately to the west a vertical fault brings it into contact with a sedimentary ensemble, La Balagne, which rests upon the basement and separates the Tenda Massif from Hercynian Corsica. The section through La Balagne shows the following structural superposition: (1) autochthonous Paleozoic basement (granites, gneiss, micaschists, and porphyroids) of Hercynian Corsica; and (2) autochthonous Tertiary complex beginning with a basal con-glomerate of granite pebbles trangressing over the basement, followed by sandy limestones. These members are of Lutetian (nummulites) age and are topped by a Middle–Upper Eocene(?) shaly–sandy flysch which encloses slices and klippes of sedimentary material interpreted (Durand Delga, 1974) as elements torn from the original cover of the Tenda Massif or from the demantling of the front of the Ligurian units (Nardi *et al.*, 1971); (3) a composite nappe, containing greenstones (gabbros and diabases), radiolarites, Calpionellid limestones, flysch with flint beds (lydites), breccias with fragments of diabases, granite and limestone (= Toccone breccias dated as Middle Cretaceous). Nardi (1968*a*, *b*) and Haccard *et al.* (1972) also equate the Balagne nappe to the Ligurian units, the Toccone breccias and their envelope being compared to the "complexe de base" of the Helminthoid Flysch (Haccard *et al.*, 1972); (4) the final mem-ber, the Lutetian Annunciata sandy flysch, is the subject of diverse interpre-tations. According to Durand Delga (1974) it represents the last member crowning the allochthonous series of La Balagne. Nardi (1968*b*) considers it associated with the autochthonous Eocene flysch containing blocks or olisto-lithes of the Balagne Ligurian unit in the strictest sense, while according to Haccard *et al.* (1972) it forms a member of the allochthonous sequence but corresponds to the Corsican (flyschoid here) equivalent of the Clays and Lime-stones Complex of the Apennines.

From this rapid survey of the northern part of Alpine Corsica, despite the varying interpretations, it is possible to discern one element similar to the internal Alpine units of the Franco-Italian Maritime Alps (ophiolitic formation of the "schistes lustrés") thrusting toward the west over an autochthonous basement whose sedimentary cover, here restricted to the later Eocene members although a more complete sequence is seen to the south, seems to correspond to that observed in the external sub-Alpine region (e.g., Grès d'Annot).*

South of La Balagne additional structural units are emplaced, such as the zone of "écailles de Corte" interpreted as an extension either of the Ligurian Briançonnais (Nardi) or of the "ultradauphinois" (Durand Delga), as a kind of parautochthon, or of the external part of the Piedmont zone (Amaudric du Chaffaut, 1975), completing a traverse of an Alpine type. Finally several authors (see also Mattauer and Proust, 1975*a*, *b*) believe they have identified in some nappes in Alpine Corsica (Balagne, upper member of Nebbio, Maci-naggio) elements of Ligurian origin and even (Centuri) basement of the In-subric type.

---

* Nevertheless, the recent works of Amaudric du Chaffaut *et al.* (1969) and Amaudric du Chaffaut and Lemoine (1974) demonstrate the existence at various points of a Mesozoic series better referred to as "Briançonnais" transgressive over the basement and apparently autochthonous. The external sub-Alpine "Dauphinois" of the western Alps does not appear to have any strict equivalent in Corsica.

The main features of the northern part of Alpine Corsica reside, according to Durand Delga (1974) in major NNW–SSE to N–S structures, from the west to the east, are: Balagne syncline; Tenda anticline; Saint-Florent syncline (syncline of the "schistes lustrés" wrapping around the Nebbio nappe); Cape Corse anticline; and major submeridional faults, such as that which limits Corsica to the east and brings down the Miocene of Aleria with respect to the "schistes lustrés".

Finally the abnormal contact separating Alpine Corsica from Hercynian Corsica, considered (excluding subsequent events) as the result of the thrusting of Alpine units over Hercynian Corsica and its cover, is interpreted by Maluski *et al.* (1973) rather as a zone of sinistral strike-slip faults (today oriented NNW–SSE), operative throughout the Alpine tectogenesis.

7. *Hercynian Corsica*

Facing the Ligurian Sea, Hercynian or crystalline Corsica (Fig. 7) begins a dozen kilometers east of Ile Rousse, where it succeeds a zone with an Eocene cover, the tectonic sole of the Balagne nappe. It has, in fact, already been observed further to the east, between the Gulf of Saint-Florent and La Balagne in the form of the Tenda Massif.

The greater part of the Hercynian is represented by metamorphic rocks (gneiss and schists) of probable Lower Paleozoic age, and especially by granite and granitoid intrusives (granodiorites and monzonitic granites dating the end of the Lower Carboniferous, and Permian alkaline granites). In some places volcano-sedimentary series are also present: sandstones, conglomerates, and keratophyres of Devonian to Lower Carboniferous age to the north of Porto (Argentella–Tour Margine series); conglomerates with granite pebbles, coaly sandstones with plant remains, tuffs and andesitic flows of Upper Carboniferous age in the coastal region near Osani, north of the Gulf of Porto, and in the north-central part of the island northeast of Monte Cinto (Asco–Mausoleo); rhyolites and ignimbrites of Permian age covering a large area of the north-western part of the island, and forming Monte Cinto itself.

Such material has obvious petrographic and chronologic affinities with that previously described in the Maures–Esterel–Tanneron Massifs (see also Nairn and Westphal, 1968). Although from a structural point of view the majority of authors rather compare the Hercynian basement of Corsica with crystalline external massifs of the Alpine chain (e.g., Argentera–Mercantour Massif), it must also be pointed out that Amaudric du Chaffaut and Lemoine (1974) compared the Permian cover with its volcanics not only with the Esterel but also with the Permian of the Ligurian Briançonnais.

The directions of petrographic differentiation as well as the major structural features (e.g., faults) affecting the Hercynian show a dominant SW–NE to WSW–ENE orientation. There are also a number of N–S structures, such as the thrusting of the Asco gneiss over the Permian of Cinto, whose trend parallels those already described in Alpine Corsica (N–S folds, fault west of Tenda, Ecailles de Corte, etc.). It should be noted that similar dominant directions have been found along the northern margin of the Ligurian Sea ("Provence Cristalline," sub-Alpine and Alpine domains).

B. The Coastal Sedimentary Basins

The location of the Neogene and Quaternary sedimentary basins (Fig. 8) which border on the Ligurian Sea is closely related to the structural units of the Alps, Apennines, and Corsica. Two groups may be distinguished:

(1) The basins along the northern shore from Toulon to Genoa. They are impressed in differing structural zones from the "Provence Cristalline" to the region of the internal domain of the Apennines. In the whole of this sector the form of the Ligurian coastline was acquired as early as the Miocene, the basins thus being confined to the present coastal region. The same is true of the Corsican Neogene and Quaternary basins.

(2) The basins of the maritime Tuscany. The main paleogeographic features, predating the present geography, were not acquired until after the Pliocene. The basins in this zone, in fact, are only the western part of a complex of basins linking the Ligurian and Adriatic shores across Tuscany, Marche, and Romagne.

1. The Basins of the Northwestern Littoral (between Toulon and Genoa)

The Miocene Marine Basins of the French Maritime Alps. Essentially the whole of this zone was emergent during the Oligocene. It seems indeed unlikely that the conglomerates with molasse intercalations, sands and clays with a marine fauna (undated and possibly reworked) from Roquefort-les-Pins (Ginsburg, 1970) may represent an Oligocene sedimentary cycle. However, recent investigations (Bellaiche *et al.*, 1976) have indicated evidence of marine end-Oligocene material in the bottom of some western Ligurian canyons.

During the Miocene two gulfs formed the larger Vence Basin lying west of the Var, the smaller restricted to east of Monaco (Roquebrune–Cape Martin Basin), containing a quite different sedimentary suite.

The Vence basin was overthrust by the sub-Alpine Castellane Arc and now is exposed as a narrow band bordering the arc some 12 km from the present coast. It is filled (Ginsburg, 1960) with shallow water marine sediments (basin margin sediments), beginning with a Burdigalian molasse with, locally, a basal conglomerate and algal limestone horizons. Toward the Middle Miocene

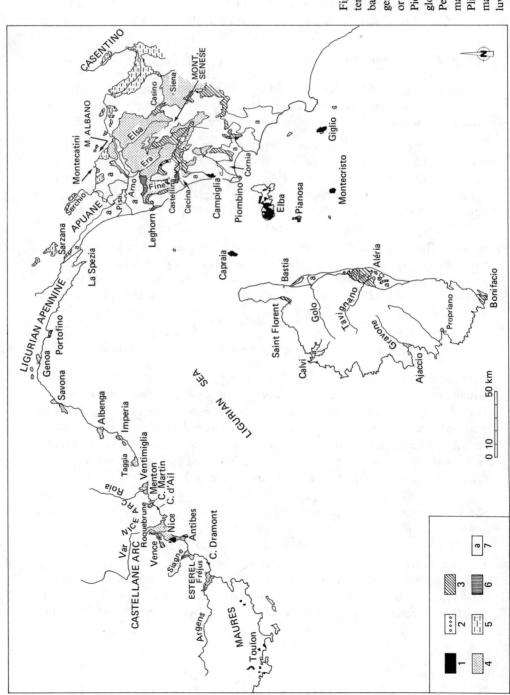

Fig. 8. Neogene and Quaternary; coastal sedimentary basins, magmatism. (1) Neogene and Quaternary plutonic or volcanic rocks; (2) Liguro-Piedmont Oligo-Miocene conglomerates from the Portofino Peninsula; (3) Miocene; (4) marine Pliocene; (5) lacustrine Pliocene and Quaternary; (6) marine uaternary; (7) alluvial coastal plains.

these give way to more argillaceous sediments still with algal limestones along the margin. The Upper Miocene coarse molasse and conglomerates are transgressive and discordant. They are not younger than Tortonian. The Miocene sequence would thus be incomplete at the bottom and top. In reality, the molasse called "Burdigalian" and the "Helvetian" marls locally have yielded an Aquitanian microfauna (Gohau and Veslin, 1960), and this age is consistent with the age of the synsedimentary volcanics in the basal beds of the series (Irr, 1973). The Miocene transgression then dates from the Aquitanian, as in the peri-Alpine basin in the Rhône Valley. No regressive Messinian deposits are known.

The Roquebrune–Cape Martin Basin is located at the front of the sub-Alpine Nice Arc and borders the Menton syncline to the west. It contains conglomerates with marly horizons at the base and top. As in the Vence Basin, the sequence begins in the Aquitanian (Le Calvez and Vernet, 1966) and continues until the Tortonian (Anglada *et al.*, 1967). No regressive Messinian is known.

West of these basins, to the north of the Esterel and Tanneron, the Miocene is represented only by continental marls and breccias with intercalated lacustrine limestone.

The Marine Pliocene Basins of the Franco-Italian Littoral. The marine incursion of the Lower Pliocene into the lower part of the hydrographic system, developed at the end of the Miocene, resulted in a number of small basins: from west to east the basins of Argens, Siagne, Var, Roya, Taggia, Imperia, Albenga, Savona, and Genoa.

The Var Basin (Irr, 1971*b*, 1975), the most important of them, lies in a trough which is assumed to be tectonic between the Castellane and Nice Arcs. Until the beginning of the Upper Pliocene this basin behaved as a submarine alluvial fan; following a progressive southward tilting, coarse detrital elements were laid down as puddingstones in the nearshore region at the outlet of the river, passing outward to muds. Along the margins of the basin, remote from the detrital influx, sands and organo-detrital limestones accumulated.

Of the basins west of the Var, the Siagne Basin resulted from a superimposed drainage on the eastern margin of the Tanneron Massif, and the Argens Basin lies in the Permian depression between the Esterel and Maures Massifs. In these basins, tectonically the least disturbed, conglomerates thin and disappear westward; the succession is incomplete, representing only the Lower Pliocene (Irr, 1975).

East of the Var, as far as the Albenga Basin, puddingstones are still well developed (Boillot, 1957; Clochiatti, 1968), but disappear farther to the east. Synsedimentary tectonic movements, detectable in the Var and Roya Basins as early as the Lower Pliocene, appear later farther to the east (Middle Pliocene) and attenuate near Genoa (Lorenz, 1971; Irr, 1975). The succession is again incomplete and does not extend beyond the Middle Pliocene.

Tertiary volcanism. Evidence of volcanism is found at several localities, which form two distinct groups: (1) East of Frejus they belong to the Ligurian province of calc-alkaline affinities represented by quartz-bearing microdiorites in the Esterel (Dramont esterellite east of Saint-Raphael), and andesitic rocks (marine conglomerates, aeolian ashes and associated volcano-sedimentary deposits) in the Biot–Antibes–Vence and Cap d'Ail (SW of Monaco) regions. The esterellite is assigned to the Oligocene (Baubron, 1974), the andesitic material to the end-Oligocene–basal Miocene (Bellon and Brousse, 1971). (2) West of Toulon and in the Maures, the volcanism is different and belongs to the Provençal domain. It consists of a continuous suite from alkali-olivine basalts to olivine-tholeites (Coulon, 1967a, b). Within the area, however, there are traces of Ligurian-type volcanism, e.g., some andesites east of Toulon. The Provençal-type volcanism began during the Miocene, reaching its peak at the Miocene–Pliocene boundary, and at some localities (NW of Toulon) continued into the Pliocene (Baubron, 1974).

Quaternary Deposits. Between Nice and Ventimiglia, where the best evidence of the Quaternary is found, in addition to the many prehistoric sites (De Lumley 1963, 1969), there are terraces at 8, 22, 24, 34, 84, 108, 114, and 117 m; they represent the maxima of interglacial transgressions, and there are littoral sediments at intermediate altitudes (11, 17, 23, 42, 56, 93 m), reflecting minor fluctuations (Iaworsky, 1973). In the lower Var Valley fluvial terraces are correlated with the principal marine terraces (Iaworsky, 1971).

West of the Var the marine terraces become fewer and lower. Along the margin of the Esterel only two are well defined, at 10–15 m and 45–65 m, while along the margin of the Maures there are none. Such a variation in altitude and number of the marine terraces from Ventimiglia to the Maures is due to the unequal amplitude of the Quaternary movements in different sectors.

At the present time the Pliocene beds are largely eroded in the river valleys and in the lower reaches they are covered by Recent or present alluvium.

2. The Portofino Conglomerate

There are no Neogene deposits along the Ligurian coast between Genoa and the basins of maritime Tuscany. However, east of Genoa deltaic conglomerates form the Portofino Peninsula, the most southerly occurrence of Liguro-Piedmont Oligo-Miocene sediments which farther to the north are discordant over all the internal Apennine and Alpine units (Lorenz, 1969). Some elements of these conglomerates were derived from the south from a zone in the region of the present Gulf of Genoa, which was emergent during the Oligocene (see Section VA).

3. The Basins of Maritime Tuscany

North of the Arno two large troughs developed, to the west the Sarzana graben or coastal graben, which sank between La Spezia and the Apuanes Massif, and to the east of the Apuanes Massif the Serchio Valley graben.

South of the Arno there is a succession of at least five basins, from west to east: Fine, Era, Elsa, Arno Valley, and Casentino. South of this group in the coastal region two further basins, Cecina Valley and Upper Cornia Valley, can be distinguished. Finally, inland the Siena Basin lies along the prolongation of the Elsa Valley Basin.

Neogene Basins South of the Arno Valley (Central and Southern Tuscany). The Tortonian compressive phase was immediately followed by a distensional period during which several basins were formed by down-faulting. The faults delimiting the basins were active on several occasions, directly controlling the sedimentation pattern during the greater part of the Upper Neogene (Trevisan, 1952; Mazzanti, 1961; Giannini and Tongiorgi, 1958). Movements were particularly marked during the Messinian, around the Miocene–Pliocene boundary, and between the Lower and Middle Pliocene. In the central part of the basins sedimentation was concordant, continuous, and thick; only along the step-faulted margins do various members thin, transgress one over the other, and even show discordances.

The earliest foundering during the Messinian was accompanied in every basin by lacustrine sedimentation, beginning with coarse detrital deposits, essentially conglomeratic beds, rapidly followed by more varied and fine-grained sediments, bituminous marls, molasse, sands, clays, and lignite horizons, ending in lacustrine clays. Such an ensemble forms the Lignite Group.

In all the basins located west of a structural line formed by Montagnola Senese, which continues southeastward of the Apuanes–Mte. Pisano alignment, following a Messinian transgression, marine beds of the Evaporite Group were deposited. These consist at first of various beds, including locally a basal conglomerate and/or marls with sandy or gypsiferous intercalations, organo-detritic limestone, clays and marls with intercalations of tripoli, followed everywhere by an evaporitic regime with clays, gypsum, and halite.

In contrast, east of the Mte. Pisano–Montagnola Senese line in the Elsa Valley and Siena Basin, lacustrine conditions continue and predominate (Signorini, 1966; Ghelardoni *et al.*, 1968) except at Casino (Siena), where there is evidence of some evaporites (Merla and Bortolotti, 1967).

In maritime Tuscany the thickest and most complete Messinian sections are found in the Upper Cornia Valley (Lazzaroto *et al.*, 1964; Lazzaroto and

Mazzanti, 1965) and in the Cecina Valley Basins, where the sequence reaches a thickness of 1000 m. The Fine Valley Basin in the northwest is the only basin in coastal Tuscany which remained in direct communication with the open sea during the Messinian. At the base the detrital sequence in the Mounts of Leghorn consists of a sandy conglomerate with intercalations of gray clay, giving evidence of a brackish water (and not lacustrine) environment (Giannini, 1960). The Messinian transgression is signaled by the occurrence of marine conglomerates followed by organo-detrital limestones. The Upper Messinian deposits are not evaporitic but consist of conglomeratic and regressive sands (Giannini, 1960, 1962). However, during the Upper Miocene the Mounts of Leghorn and those of Castellina Marittima (formed of allochthonous Ligurian elements) are presumed to have remained emergent (Mazzanti *et al.*, 1963; Barsotti *et al.*, 1974).

The sea reached its greatest extent in the entire Tuscany region during the Lower Pliocene, extending to Elsa Valley and the vicinity of Siena. The most constant facies is marine clays with some beds of conglomerate and sand at the base of the series and around the basin margins. The Pliocene deposits (Mazzanti *et al.*, 1963; Squarci and Taffi, 1963; Ghelardoni *et al.*, 1968) exceed 1000 m in the Era, Cecina, and Elsa Valleys, although not extending above the Middle Pliocene (Dallan and Salvatorini, 1967); finally the regression was marked by sands.

In maritime Tuscany the Pliocene is restricted to the Fine, Cecina, and Cornia Valleys. The Mounts of Leghorn, Castellina Marittima, Campiglia Marittima, and Piombino appear to have emerged as islands at this time.

The Plio-Pleistocene Basins North of the Arno. In the two grabens (Sarzana, Serchio Valley and its southern prolongation Montecatini–Cerbaie) which border the Apuanes–Mte. Pisano structural line, Messinian deposits are only known in the extreme southern part (below Pisa and between Mte. Pisano and Mte. Albano) (Giglia, 1974). The basin fill consists mainly of Plio-Pleistocene lacustrine and fluvio-lacustrine materials (conglomerates, sands, clays, lignite intercalations) (Federici, 1973; Masini, 1933; Nardi, 1961; Bortolotti, 1964; Trevisan *et al.*, 1971; Calistri, 1974).

The Coastal Quaternary Deposits. The lower reaches of the Arno, Cecina, and Cornia Rivers form wide alluvial plains which extend almost continuously from Carrara to the Piombino Peninsula, with a break at the foot of the Mounts of Leghorn.

Following the regression at the end of the Middle Pliocene, a final important phase of foundering principally affecting the lower Cecina and Arno Valleys is reflected in a marine Calabrian succession of fine conglomerates, clays, and sands with *Arctica islandica* (Barsotti *et al.*, 1974). These beds can be seen clearly transgressing along the margin of the Cecina plain. In the lower

Arno Valley the beds sometimes lie concordantly, sometimes discordantly, upon Pliocene sediments (Struffi and Sommi, 1960).

During the Quaternary (Barsotti *et al.*, 1974) eustatic movements in the vicinity of Leghorn resulted in ?10 m, 20 m, 40 m, and 120 m marine terraces; they are associated with a variety of marine sediments, sands, clays, "panchina" (i.e., calcarenites) deposited during interglacial periods, and continental or fluvio-lacustrine detrital sands, conglomerates, aeolian clays during glacial periods. Epeirogenic movements have raised the highest terrace a few tens of meters (Barsotti *et al.*, 1974). After the Versilian transgression the Recent deposits consist of aeolian, marshy, and fluvio-lacustrine sediments.

4. *The Tuscan Archipelago*

Pianosa Island. In the interpretation of the Recent history of the Corsican Channel and the paleogeographical relationship of Corsica to Tuscany, Pianosa forms an important intermediate link. It is composed of a series of northwest-dipping (15 to 20°) clays, conglomerates, and calcarenite intercalations of Burdigalian age (Dallan, 1964). Upon these Miocene beds rests a discordant series of Pliocene organo-detrital limestones. Geophysical methods provide some information on the Neogene deposits in the Corsican Channel (see Section VB).

The relation of Pianosa to Tuscany is uncertain, although Gabin (1972) suggested that the –80 m surface which links the two may consist of the Pianosa Pliocene limestones dropped by faults trending NNE–SSW.

The Islands Composed of Igneous Material—Capraia, Elba (in part), Monte Cristo, Giglio, etc. For information on Tuscan magmatism see Boccaletti and Manetti (this volume) and Bortolotti and Passerini (1970).

5. *The Corsican Neogene and Quaternary Basins*

Northern Corsica. Along the northern coast a Miocene basin at Saint-Florent, and the Quaternary alluvial plain of Calvi, can be recognized; east of Cape Corse occurs the northern part of the Eastern Plain.

The Miocene Saint-Florent Basin has been affected by Miocene and Plio-Quaternary tectonic movements with the uplift of the eastern and submergence of the western part of the basin. The sediments reflect marginal deposits in a basin open to the northwest (Orszag-Sperber, 1975), and consist of sandy, bioclastic limestones followed by reef limestones with intercalations of encrusting algae. According to Orszag-Sperber (oral communication), only part of the Middle Miocene, and perhaps part of the Lower Miocene, may be present. No Pliocene beds are known along the north Corsican littoral.

In the Calvi alluvial plain, above the present alluvium of the lower plain,

Quaternary alluvial terraces at 15 to 20 m, 30 to 35 m, and 55 to 60 m are known.

The Eastern Plain borders the Tyrrhenian margin and at its northern tip (Bastia Plain) is formed of Quaternary materials. Only further to the south do the sub-Quaternary sediments appear to provide the most complete Neogene section in Corsica.

Upon a substratum formed by the "schistes lustrés," the lowest horizon consists of Upper Burdigalian–Langhian conglomerates and marine marls. There is a break in the marine series, indicated by the deposition of sandy beds with paleosols, giving evidence of a regression in the lower part of the Middle Miocene (Orszag-Sperber and Freytet, 1973; Orszag-Sperber, 1975). A return to marine conditions was marked in the Middle Miocene (zone N 9, with *Praeorbulina*) by the deposition of sands alternating with organodetrital limestones at the base and with reef limestones at the top. The Upper Miocene is represented by marls with a Tortonian–basal Messinian microfauna (Magné *et al.*, 1975a).

As along the Franco-Italian Maritime Alps littoral the end of the Miocene was marked by a phase of erosion, and the Lower Pliocene transgression invaded the lower reaches of the rivers. In the Eastern Plain a regression followed the deposition of the Tortonian–Lower Messinian beds prior to the Lower Pliocene; it is recorded by the conglomerates with brackish water marl intercalations of the Aléria Formation. These detrital sediments may be Messinian in age and could thus be equivalent to part of the evaporitic horizons (Magné *et al.*, 1975a). Only Lower Pliocene is known in the Eastern Plain, in a classic blue marl facies (Magné *et al.*, 1975a, b).

The northeastern part of the plain is formed of Quaternary deposits. The Golo, Bevinco, and Fiume Alto Rivers, flowing from the zone of the "schistes lustrés," spread the alluvium over the plain where a thickness of 60 m is known (Conchon, 1975); however the thickness probably reaches to 100–150 m (results of the Compagnie Générale de Géophysique, 1972). The effects of Quaternary climatic variations are recorded in the deposits; for while in the valleys phases of alluviation alternated with periods of erosion, in the plain seven alluvial sheets separated by weathering horizons are found (Ottman, 1958; Conchon, 1975). The thickness and mode of superposition of the deposits were controlled by movements along a flexure. Until recent times this flexure was responsible for the downwarping of the plain, while Cape Corse and the "schistes lustrés," which prolong it toward the south, were uplifted (Conchon, 1975).

Western Corsica. Magné *et al.* (1975b) attributed a Lower Pliocene age to marls which rest upon a granitic base in the Gulf of Ajaccio and of Propriano (Ottman, 1958).

Southern Corsica. There is a Miocene basin located in the surroundings of Bonifacio, in southernmost Corsica, filled with sandy bioclastic and reef deposits. The lower part of the series is contemporaneous with beds containing *Praeorbulina* from the Eastern Corsican Plain (Orszag-Sperber, 1975). A southwestern entension of the Miocene basin of Bonifacio is known offshore, where sandy limestones form the possible substratum of the continental shelf between Corsica and Sardinia (Gennesseaux, 1972; Gennesseaux *et al.* 1974, 1975).

IV. GEOPHYSICAL DATA

Geophysical methods provide information concerning two regions not accessible to direct observation—the deeper zones under the continental borderlands whose superstructure has been studied geologically, and the submarine realm.

In order not to burden the text with an analysis of the numerous works which have appeared during the last decade in particular, the attempt will be made here to present a synthesis which will comprise a geological interpretation of the results provided by a variety of geophysical methods.

A. Seismicity

Within the Mediterranean as a whole the distribution of earthquakes and seismic mechanisms have been used to delimit plates and the constraints prevailing in the crust and mantle. In this sense the Ligurian domain appears removed from the highly active zones (Ritsema 1969; Mackenzie 1970). On a more restricted scale, however, the works of Recq (1972, 1973, 1974), Bossolasco and Eva (1965, 1971), and Bossolasco *et al.* (1972, 1973*b, c, d*) provide some interesting details concerning the deep structure of the Ligurian Sea and surrounding area.

They demonstrate the stability and independence of the Maures Massif, suggesting an attachment to a domain other than Alpine (rather Pyrenean, cf. Section IIIA). The other crystalline massifs (e.g., the Argentera Massif) and the Ligurian Briançonnais appear to be aseismic. The existence of an intermediate basaltic layer below the Esterel, an idea already proposed on quite different grounds by Boucarut (1971) is also suggested.

The Mohorovicic discontinuity rises in the Gulf of Genoa with respect to the depth measurement in the adjacent continental area (with the sole exception of the Ivrea area, further to the north). This was interpreted by Recq (1973) as evidence for possible oceanization. A preferential alignment or NNE–SSW seismic axis stretches from Genoa, marked by epicenters in the Ligurian Sea. There is some seismic activity on land west of this line in the Ligurian–Piedmont area, but none to the east. According to Bossolasco *et al.* (1973)

this seismic axis might represent a system of westerly dipping faults and some form of trough in the central part of the Ligurian Sea.

Seismic data thus indicate the existence of an unstable zone centered on the Franco-Italian Maritime Alps and the Lower Var Valley, separating the stable Maures to the west from the equally stable eastern part of the Ligurian Sea. Still further to the east the Apennine represents a new area with constraints (Ritsema, 1969). For more precise information, however, it is necessary to use seismic exploration techniques (refraction and reflection).

B. Seismic Exploration

Since the first work of Ewing *et al.* (1955, 1959) many seismic profiles have been shot in the western Mediterranean in general and in the Ligurian and circum-Ligurian domain in particular. Refraction shooting has provided information on deep structure which may be correlated with earthquake data, while reflection work has been used to provide more detailed information on the superficial layers.

1. *Seismic Refraction Data on Deep Structure*

Characteristics of the Crust and Upper Mantle. The earliest research showed that the abyssal zone of the Ligurian Sea rested upon a thin crust in common with the rest of the deep Algero-Balearic-Provençal Basin. The average depth of the Mohorovicic discontinuity is approximately 12 km, contrasting with 30 km under the bordering continental areas. Below the water layer and the upper sedimentary sequence, Fahlquist (1963) demonstrated the existence of a layer 3 to 4 km thick, which thickens towards Corsica and the Italian coastline and whose seismic characteristics have led some authors (Muraour, 1970) to interpret it as a granitic or metamorphic layer, i.e., of continental type. This layer does not exist in the southwest beyond the Ligurian Basin in the Balearic–Corsican–Sardinian Basin whose seismic characteristics are otherwise the same. The crust here has been compared (Muraour, 1970) to the "subcontinental" type of Drake and Kosminskaya (1969), known in island arcs or under the mid-oceanic ridges.

A further notable feature is the rise to relatively shallow depths (about 50 km instead of 100 km) of the top of the low velocity layer and of the whole low velocity zone, in fact, in the axial section of the Algero-Balearic-Provençal Basin (Berry and Knopoff, 1967). This elevated zone of the low velocity layer has a NNE–SSW orientation between Corsica and Provence. Ritsema (1969, 1970) concludes from the thin crust over an abnormal upper mantle that in the Ligurian Sea region the crust is transitional between continental and oceanic.

Map of the Depth of the Mohorovicic Discontinuity. As a result of the numerous seismic refraction surveys in the western Alpine–Mediterranean sector (Closs and Labrouste, 1963; Aubrat *et al.*, 1967; Choudhury *et al.*, 1971; Recq, 1967; Labrouste *et al.*, 1968; Giese *et al.*, 1970; Giese and Morelli, 1973*a, b*; etc.) and in Provence (Recq 1970, 1972, 1973; Bellaiche *et al.*, 1971; Recq and Bellaiche, 1972; Bellaiche and Recq, 1973) there has been produced a map of the depth of the Moho for southeastern France (Labrouste

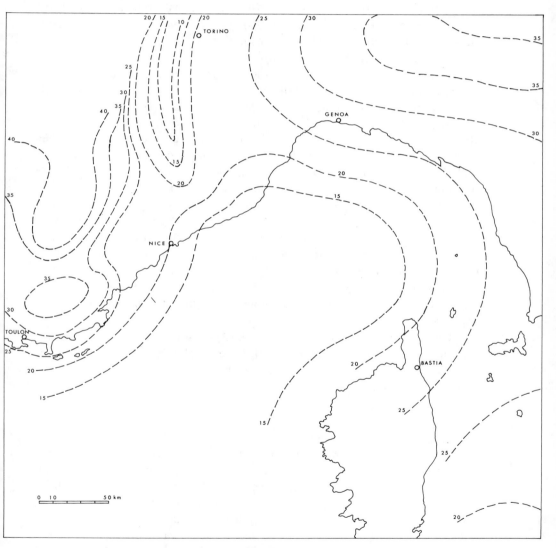

Fig. 9. Contour map of the depth of the Mohorovicic discontinuity (depths in kilometers).

et al., 1968; Recq, 1974*a*, *b*), Provence (Recq, 1970), and Italy (Giese and Morelli, 1973*a*, *b*). The attempt has been made here to extract for the Ligurian domain a map (Fig. 9) which shows the development of the deep structure in this area.

From the continental borderland (southwestern edge of the Argentera Massif, Maures–Tanneron–Esterel Massifs, northern Apennines, Corsican–Sardinian Massif) there is apparent a rapid thinning of the continental crust in the direction of the Ligurian Sea, where it is reduced to a thickness of 11 km (cf. Section IVA). Among the sectors with a thick crust, some correspond to a zone of Permian (north of the Maures) or Mesozoic (north of Castellane) subsidence. The gradient of depth reduction of the Moho is at first parallel to the littoral off the Maures, but then becomes perpendicular to it east of Cannes where, in Recq's (1974*a*, *b*) interpretation the isobaths appear to be along the continuation of those underlying the Ivrea–Verbano Zone, farther to the north, which, although continental, is also characterized by a rapid rise of the mantle. The Moho is thus at progressively shallower depths from Nice to Imperia (cf. Recq, 1972*b*) to a minimum of 12 km. This terrestrial zone of thin crust comparable to the Ivrea Zone, which may not be totally continental in character, underlies the outcrops of the Voltri group, where seismic refraction has also indicated an updoming of the basement with its top to 0.3–0.5 km below the surface (Recq, 1972*b*). Thus, these geophysical results may be translated into geological terms, in the trans-Alpine and Apennine sections of Giese *et al.* (1970) and Haccard *et al.* (1972), which imply to the former a type of thrusting of the Apennine upper mantle northwestward over a much broken Alpine crust, or for the latter authors the thrusting of an Insubric basement (= Apennine–Austro-Alpine) over the European (= Western Alpine) basement along a sole formed by a series of disrupted ophiolitic units, peridotites, and gabbros.

On a smaller scale the refraction results give some structural information (of faults affecting the deeper basement, for example), but for a more detailed interpretation it is preferable to combine these results with reflection seismic and magnetic data.

2. Seismic Reflection

Structural Data. The interpretation of seismic reflection records permits the recognition or definition of an acoustic basement (bedrock) of variable lithological nature, sedimentary or otherwise, which forms a contrast with the "transparent," less consolidated superficial layers. The large volume of data accumulated between 1966 and 1973 (by the Geodynamic Research Station of Villefranche-sur-Mer, the Oceanographic Museum of Monaco, the French Petroleum Institute, and the C.N.E.X.O.) has been synthesized by Rehault *et al.* (1974*a*) in the form of isochrons to the acoustic basement in the Ligurian region. In some cases this basement may correspond to the continental basement (Auzende *et al.*, 1973).

The morphology which this map reveals reflects the major topographic trends of the sea floor. This indicates, once again, the steep slope along the margins of Liguria and northwestern Corsica, indicating the rapid downwarping in a series of steps and the gentler but more irregular slope of the eastern zone between the Gulf of Genoa and Cape Corse, which can only be interpreted (Section IVD) with reference to magnetic results and dredge samples (Bellaiche *et al.*, 1974*a*). South of the "Provence Cristalline" (Bellaiche *et al.*, 1971; Recq and Bellaiche, 1972; Recq, 1974*b*) and west of the Corsican–Sardinian block (Auzende *et al.*, 1973; Ryan *et al.*, 1973*b*) the rapid plunge and disappearance of the reflector representing the acoustic basement corresponds approximately with the tracing of the continental slope and the limits of the seaward extension of a continental type basement (Maures, "Hercynian" Corsica). The latter is confirmed by dredge samples.

The tectonic control implied by this sinking of the margins of the Ligurian Basin suggests the existence of faults more or less parallel to the shoreline, with an E–W orientation off Provence, then NE–SW off the Côte d'Azur, Riviera, and northwest Corsica. A third structural trend is defined by faults which appear to be followed by the principal submarine canyons, and of which the most important, the Cape Mele Fault, begins a little to the east of Imperia. Seismic reflection shows that this fault downthrows the southwestern segment with respect to the northeastern.

Many studies (Menard *et al.* 1965; Hersey, 1965; etc.) following those of Fahlquist (1963) have defined a number of domelike structures, more or less coalescent, subsequently interpreted as salt domes in the abyssal part of the Ligurian Basin. It is even possible to map the various evaporitic types seismically (Schreiber *et al.*, 1972), thus to the south off Toulon, Bellaiche and Recq showed that halite was more restricted within the basin than gypsum and anhydrite, the order of evaporites reflecting their solubility. It should also be noted that the appearance in the record of the reflector announcing the presence of evaporites is often where either by faulting or flexuring the continental crust also disappears (Recq and Bellaiche, 1972). The geophysical data thus indicates that the contours of the abyssal zone correspond both to the region where the crust is no longer dominantly continental and (approximately) to the sedimentary basin fill which contains an important evaporitic member, locally flowing to give diapirs.

Stratigraphic Data. The detailed studies by Montadert *et al.* (1970), Mauffret *et al.* (1973), and Rehault *et al.* (1974*a*) of the seismic reflection data in the Balearic–Provençal Basin, and more especially in the Ligurian Sea region, show that the sedimentary sequence may be subdivided into four principal members (from bottom to top):

(1) Infrasalt series found particularly in the central part of the basin. The lower part appears as an infilling covering an irregular basement, often not distinguishable seismically. The series may reach a thickness of a few thousand meters and appears to be tectonically undeformed or only slightly so. The results from Leg 13 DSDP (Ryan *et al.*, 1973*a*) confirmed the Messinian age assigned to the overlying evaporites and suggests a pre-Upper Miocene date for this series, but the age of the lowest members remains open.

(2) Salt series, formed of halite of variable thickness in the Ligurian Sea, where the top of this series lies at about 4 km below the sea level and where it has given rise to numerous domes and diapirs. Its overall distribution corresponds to the limits of the central basin, except in the vicinity of the Gulf of Genoa where the salt extends a short way over the margin.

(3) Upper evaporitic series which is a few hundred meters in thickness and appears to be well bedded in view of the many reflecting horizons found. As a result of the examination of the Leg 13 DSDP cores (Ryan *et al.*, 1970, 1973*a*) this upper series is regarded as formed of interbedded marine and evaporitic sediments, end-Miocene in age. In the Ligurian Basin its distribution exceeds that of the underlying salt series to the point that it is exposed on the flanks of the continental slope some 15 km off Savone and Imperia. It appears from some dredge samples (Bellaiche *et al.*, 1974*a*; Rehault *et al.*, 1974*b*) that the series may pass laterally into a more detrital facies, marking a continental influx probably related to a general Pontian emergent phase known on land.

(4) Plio-Quaternary series separated from the underlying upper evaporitic series by a well-marked reflecting horizon H or M of various authors (Alinat *et al.*, 1966; Mauffret, 1968; Montadert *et al.*, 1970; Ryan, 1969). This reflector has been considered as evidence of a disconformity*; a further slight disconformity is also observed within the Pliocene. This series appears, seismically, to be well stratified and may be up to 1000 m thick. Sampled at outcrop, in regions of strong erosion (i.e., steep slopes or canyon walls) these Pliocene beds are shown to consist of indurated gray-blue clays with bands of coarser material. Locally (Stoechades Canyon north of Mejean Shoal) sandy clays of the Lower Pliocene, which crop out at a depth of 2000 m, contain a microfauna characteristic of deposition in a shallow water environment (Bellaiche, 1972*b*), indicating a significant subsidence of the margin of the basin in post-Lower Pliocene times.

---

* It has recently been questioned whether this reflector is truly an isochron (Biscaye *et al.*, 1972; Bellaiche *et al.*, 1974*a*), for samples taken in order to date it sometimes give a Messinian age in some places (Glaçon and Rehault, 1973), but a Lower Pliocene age at other points.

The seismic reflection survey in the Corsican Channel between Corsica and the Isle of Elba indicates a somewhat different sedimentation history in this region (belonging in fact to the northern extremity of the Tyrrhenian Basin) from that of the Ligurian Basin in the strictest sense (Gabin, 1972). At least 4000 m of Mio-Plio-Quaternary sediments indicate subsidence since Miocene times, interrupted during the Pontian by a brief phase of emergence which linked Corsica and Tuscany via the Tuscan Archipelago (Mio-Pliocene discordance, pre-Pliocene erosion). The evaporitic members are not recorded here. There is evidence of tectonic activity, in the form of vertical movements at the end of the Miocene and in Plio-Quaternary times.

C. Gravimetry

The Bouguer anomaly maps (e.g., de Bruyn, 1955) indicate a gravity high in the Algero-Balearic-Provençal Basin extending from the Ligurian trough in a SSW direction, a pattern which has some similarities with the zone in which the rise of the low velocity layer was determined by deep seismic soundings by Berry and Knopoff (1967). It was these gravity results, combined with the first seismic refraction data, which led Glangeaud (1962; Glangeaud et al., 1966) to introduce the term "antéclise ligurien" for the postulated existence of a zone of crustal thinning of a particular oceanic type in the center of the Ligurian Sea. Collette (1969) and Van Bemmelen (1969) noted that these positive gravity anomalies of the western Mediterranean contrast with the negative values characteristic of the present continents and may be interpreted in various ways: "... a great many models of density distribution in the crust and the upper mantle would satisfy the gravity measurements at the surface..." They agree, however, that the gravimetric results are compatible with the model derived by Berry and Knopoff. According to Van Bemmelen (1969) "... the general gravimetric situation indicates that the young Cenozoic differential vertical movements in the western Mediterranean area have not yet achieved an isostatic equilibrium. The basin areas still have some overweight, tending to subsidence; whereas in the surrounding area mantle–crust columns of too low weight tend to rise..."

On a smaller scale Coron and Guillaume (1966) and Guillaume (1969) interpreted a positive gravity anomaly whose maximum lies near Imperia as an indication of the rise of denser material at that point, perhaps related to a southern extension of the Ivrea Zone (cf. seismic refraction data, Recq, 1972b). Blanc et al. (1970) also confirmed the rise of denser material under the Voltri group from the study of gravity anomalies and the importance even in the crust of the fault separating the latter from the Ligurian Briançonnais in the region of Savone. Van Bemmelen (1969) had already remarked that the strong positive anomalies in the vicinity of Genoa could be related to the outcrop on land of mafic and ultramafic rocks of the ophiolitic complex. The gravity results thus merely serve to confirm the results obtained seismically and structurally without unfortunately adding any precision.

D. Magnetism

In the western Mediterranean and its borders, the principal aeromagnetic data are reproduced on a map of the U.S. Naval Oceanographic Office (1958) interpreted by Vogt *et al.* (1971), and in maps published by the B.R.G.M.

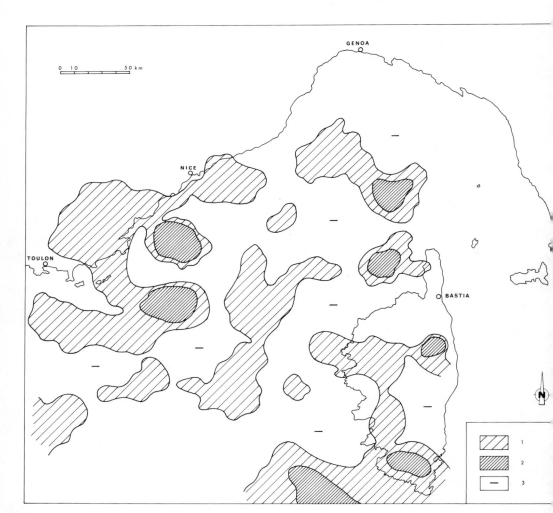

Fig. 10. Simplified map of the magnetic anomalies. (1) Weak positive anomalies (0 to 100 γ, usually less than 50 (2) strong positive anomalies ($\geq$ 100 γ); (3) negative anomalies ($\geq$ $-$ 50 γ).

(France) from flights carried out in 1966 (northern sector) and 1971 (southern sector), interpreted by Le Moüel and Le Borgne (1970), Le Borgne *et al.* (1971), Auzende *et al.* (1973), and Bayer *et al.* (1973). The difficulty is that, as a result of different interpretative methods, contradictory results have been obtained from the same basic data.

Two types of magnetic anomaly (Fig. 10) may be recognized in a first analysis of the Ligurian area (in a broad sense). The most intense, with a clear circular pattern and magnitudes of +100 to 150 gammas, occur around the periphery of the basin, southeast of the coast of Provence, in the Gulf of Genoa and off Corsica and Sardinia. The majority of authors agree in referring these to a form of andesitic volcanism (locally confirmed by sampling in the Gulf of Genoa, Bellaiche *et al.*, 1974*a*). It has been suggested by analogy with events on shore (Bellon and Brousse, 1971) that they date from the Oligocene–Miocene boundary. These anomalies, then, are discerned within a layer of sialic-type crust, which seems borne out by the fact that in Corsica certain of them may be traced without change from sea to onshore (cf. Storetvedt, 1973). The strong positive anomalies paralleling 10°E longitude from the southeastern angle of the Ligurian Sea to the northernmost limits of the Tyrrhenian Sea are also almost certainly related to volcanism (Tuscan magmatism) and are evidence of the existence at depth of abundant magmatic rocks aligned along the axis of which the crest is marked by the islands of Elba and Pianosa (Le Borgne, 1969, in Gabin, 1972).

It is the second type of anomaly restricted more to the center of the Ligurian Basin which has been the source of divergent interpretation. Some authors (Storetvedt, 1973) only accord a tectonic (vertical movement) or volcanic significance to all the anomalies measured, and consider that they are not to be related to a crust resulting from sea-floor spreading. On the basis of the magnetic data then available, Le Borgne *et al.* (1971), Vogt *et al.* (1971), and Auzende *et al.* (1973), without ever questioning the oceanic nature of the crust of the western Mediterranean and more particularly the abyssal part of the northern sector, could not recognize magnetic lineations of the type described by Vine and Matthews (1963) in the major oceanic basins. Bayer *et al.* (1973), however, were able to discern in the Ligurian Sea a pattern of low amplitude lineations with a symmetric NE–SW orientation and offset by fractures, by the use of new aeromagnetic measurements and a more sophisticated calculation technique. According to these authors the anomalies, which have not been dated, are comparable with, but less clear than, the oceanic lineations of Vine and Matthews, and are thus evidence of the operation in the Ligurian Sea of a mechanism comparable to that assumed in the Atlantic, for example. This interpretation has also been refuted by Storetvedt (1973) on grounds which hardly seem scientific, and questioned by Gonnard *et al.* (1975), using new data obtained by seismic reflection methods.

The definite result from the magnetic data is thus the confirmation of Oligo-Miocene volcanism restricted to the margins of the Ligurian Basin. There is, further, the possibility (still disputed) of oceanic processes in the said basin analogous to sea-floor spreading.

E. Paleomagnetism

Within the Ligurian domain the first paleomagnetic work was carried out on Permian volcanic material from Corsica (Ashworth and Nairn, 1965; Westphal, 1967; Nairn and Westphal, 1967a, 1968). On the basis of the declination values obtained, which differed appreciably from those of stable Europe, as well as other arguments detailed later, the authors concluded that an anticlockwise post-Permian rotation of Corsica was possible, an interpretation consistent with hypotheses previously proposed by Argand (1924) and Carey (1955, 1958).

Further research in Sardinia, considered as belonging to the same continental block as Corsica, provided Zijderveld et al. (1970a) with results similar to those found by Nairn and Westphal in Corsica. De Jong et al. (1973), studying the Upper Oligocene–Lower Miocene volcanics of Sardinia, obtained declinations which implied a certain rotation after that period. At first these results appeared to be confirmed by Bobier and Coulon (1970), but their examination of post-Tortonian (probably Pliocene) basalts suggested rotation was completed by that time. The latter conclusion was supported by Alvarez et al. (1973).

The validity of certain paleomagnetic results has, however, been questioned. Based on the work of Henry (1971) on the Barrot dome (southeastern France), Recq (1974b) questioned the value of paleomagnetic techniques applied to rocks subjected to tectonic effects, as is evidently the case in the circum-Ligurian region. Following new measurements on Corsican rocks (Storetvedt and Petersen) Storetvedt (1973) questioned the results of all his predecessors and argued against any rotation of Corsica and Sardinia. Recently, Bobier and his co-workers (in Coulon et al., 1973, 1974) restudied the paleomagnetism of Oligo-Miocene and Plio-Quaternary volcanics in Sardinia using new radiometric dates and concluded that if there was movement of the Corsican–Sardinian block, it had to have taken place prior to the emplacement of the Upper Oligocene calc-alkali volcanism, and that the directions of magnetization which led them to their first conclusions (in 1970) may be no more than the emplacement of flows during a phase of transition of the geomagnetic field.

De Jong et al. (1973), however, reached quite different conclusions, claiming that if "this change in declination could be the result of a long lasting late Oligocene–early Miocene excursion of the geomagnetic dipole field,

... such an event has not been demonstrated for this period." Based on their earlier results augmented by new measurements on rocks from an area adjoining that studied by Bobier *et al.*, de Jong *et al.* observed that "the oldest rocks show the most westerly declinations..., younger rocks a NNW declination and the youngest rocks (...) possess a direction close to due north..." The authors conclude that there was a progressive anticlockwise rotation of Sardinia of about 50°, this movement cannot predate the Upper Oligocene–Lower Miocene and was completed immediately prior to Middle Miocene times. The results recently obtained by Manzoni (1974, 1975) from new measurements on rocks from southern Sardinia appear to be consistent with the conclusions of de Jong *et al.* (1973).

At the present state of knowledge it seems, however, that paleomagnetic arguments for a post-Oligocene rotation of the Corsican–Sardinian block cannot be accepted without some reservations.

V. COMPLEMENTARY DATA

The study of the structural framework of the Ligurian Sea (Section III) and the summary of the geophysical results (Section IV) provide a geological context in which to consider the evidence leading to a better understanding of its creation and evolution. Prior to this, however, it is necessary to be cognizant of certain other facts and arguments from a variety of other fields, sedimentology, paleogeography, paleobiogeography, etc. which complement the information previously acquired and are an essential prerequisite to attempts at interpretation.

A. Sedimentological and Paleogeographical Data

For the many authors who have studied the peri-Mediterranean Alpine chains, it was necessary that there should exist in the region of the present western Mediterranean an emergent landmass to form the principal source area for important terrigenous detritus which fed the basins of sedimentation (flysch in particular) of the future orogens (cf. Flandrin, 1948; Kuenen *et al.*, 1957; etc.).

In the Ligurian domain, of particular concern here, there are many paleogeographic indications favoring the existence to the south of the "Alpine Sea" during the Mesozoic and Paleogene of an emergent landmass of which the Maures–Esterel–Tanneron Massifs are a relict. The sedimentological studies of Stanley and Mutti (1968) on the sub-Alpine (Annot sandstones) and Apennine (Ranzano sandstones) detrital beds of end-Eocene and Oligocene age, based on mineralogical composition, show that channels and paleo-current directions are important in this respect. Further, the studies of Lorenz (1969; in Haccard *et al.*, 1972), sometimes using quite old observations (Issel, 1892) have demonstrated that until the end of the Oligocene a landmass occupied the position of the present Gulf of Genoa, the sea lying further to the north, as evidenced by outcrops in the Ligurian–Piedmont Basin.

From this then, the sedimentological and paleogeographical evidence leads to the conclusion of a relatively young age for the origin of the Ligurian Sea (post-Oligocene), in agreement with the reflection seismic data on sediment thickness below the Messinian salt in the abyssal zone.

Nevertheless, the observations of Cornet (1965) lead one to think that there existed in Provence even in the Upper Miocene (Tortonian), a southerly source of continental material, that is in the opposite direction to that of the present time. Although this is in general conformity with ideas introduced earlier, it is more than a little difficult from the point of view of chronology in the sense that there is paleogeographic evidence of a Tethyan marine transgression from Lower to Middle Miocene times (in the vicinity of Marseilles, the Vence–Monaco region, etc., the sea progressively occupying the sector which has been previously demonstrated continental at least until Oligocene times). It is, however, possible that the reversal of the hydrographic pattern may have been delayed by the initial elevation of the margins.

B. Results Based upon Geological and Morphological Similarities

Following Nairn and Westphal (1968) and using the ideas of Argand (1924) and Carey (1955), other authors (e.g., Alvarez, 1972a; Ryan *et al.*, 1973b; etc.) have made use of contours, not of the present coast but of certain isobaths (e.g., 1000 m), or better the limits of the abyssal zone (regarded as the line of division between the continental and oceanic crust) in reassembling the western margin of Corsica–Sardinia against the Mediterranean margin of southern France. These reconstructions obviously imply a subsequent drift. The limited value of such reconstructions based upon submarine morphology is remarkably enhanced if it is considered how, as a result, structures and geological features now some hundreds of kilometers apart (cf. Section III) are aligned. Among these the following may be noted: Permo-Carboniferous volcanic and sedimentary sequences of northwestern Corsica and the Esterel; the structural orientations in Liguria and northwestern Corsica (including submarine canyons); agreement of tectonic alignments (E–W in particular) in the "Provence Cristalline" and the Corsican–Sardinian microcontinent once the latter is placed in its supposed former position; Lower Paleozoic of Sardinia and the Montagne Noire; Urgonian facies and bauxites of Sardinia and western Provence; magnetic anomalies (and associated andesitic volcanism) of the Ligurian Sea margins.

The geological history of the two zones thus juxtaposed then begins to diverge after Oligocene times (Chabrier and Mascle, 1974, 1975).

C. Paleobiogeographical Data

In 1962 Dubourdieu drew attention to the faunal similarities between the Lower Paleozoic assemblages of Sardinia and the Montagne Noire, two regions now more than 500 km apart. Dieni and Massari (1965) also underlined the microfaunal similarities of the Lower Cretaceous of Provence' and Sardinia, a point more recently developed by Cherchi and Schroeder (1973), Chabrier and Fourcade (1975a), who concluded that particular endemic associations of Valanginian and Barremian microfossils were confined to southeastern France, western Sardinia, and the northeast of Spain.

There is, thus, a convergence of geological and paleobiological evidence implying a common evolution of the continental borders of the present Ligurian Sea up to Oligocene–Miocene times, all seemingly consistent with the paleomagnetic results (see Section IVE), although the latter are still subject to question.

VI. PRE-MIOCENE GEOLOGICAL HISTORY OF THE CIRCUM-LIGURIAN AREA

In the western part of the circum-Ligurian domain, the feeble and local discordance of the Triassic on earlier deposits may be interpreted as reflecting the last, Palatine phase of the Hercynian orogeny. The underlying Permian consists of a continental or lacustrine volcano-detrital sequence deposited in subsiding basins, nourished by a relief formed during earlier tectonic phases, and constantly rejuvenated by a dominantly rhyolitic volcanism.

According to the classical interpretation, the Triassic paleogeography (Fig. 11) of middle Europe was influenced by the effect of the Vindelician Ridge, which followed the trend indicated approximately by the Alpine external crystalline massifs and the Corsican–Sardinian block. The greater part of the European platform, to which belong the Provençal and sub-Alpine regions of the circum-Ligurian area, was invaded in mid-Triassic times by a shallow epicontinental sea. The fluctuation in level of this sea, becoming regressive during the Upper Triassic, resulted in the formation of evaporitic members. To the east and south of this hypothetical ridge lay the "Alpine" sea, a northwestern extension of the Tethys. Here accumulated a shallow water, carbonate sequence in which calcareous algae and even evaporitic sediments (e.g., Tuscan domain) are found. It would thus seem that at this time the region of the future Alpino-Apennine geosyncline was still poorly defined.

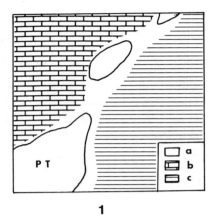

1

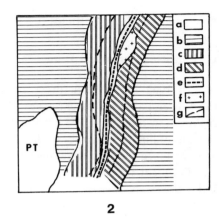

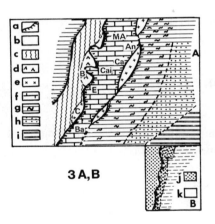

2

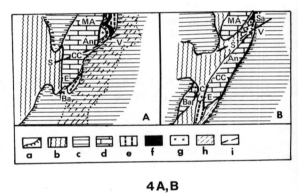

3 A,B

4 A,B

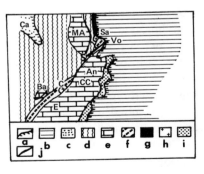

5

The presence within certain of the Alpine and Apennine internal nappes of ophiolitic rocks assigned to the middle and lower part of the Upper Jurassic led some authors to suppose that an oceanic zone linked to the Tethys formed between a "European" and an "Apennino-Austro-Alpine" block during the Jurassic. According to this model, following a period of distension during the Triassic–Liassic, a paleo-ridge formed separating a western or Piedmont Basin from an eastern or Ligurian* Basin. Upon an oceanic crust of peridotites and gabbros, and associated with ophiolites, were deposited sediments, which in their present state are represented by "schistes lustrés" in the Piedmont Basin and by a sequence of radiolarites, *Calpionella* limestones and Palombini shales

---

* It would be preferable to use a more appropriate term for this Mesozoic basin, which does not present any geological relation with the present Ligurian Sea.

Fig. 11. Paleogeographical maps. These maps have been constructed by taking into account the influence of the oceanic structures involving the continental drift–sea-floor spreading hypothesis, i.e., in this particular case the Mesozoic Western Tethysian hiatus and the Upper Tertiary Ligurian sphenochasm. (1) *Triassic:* a. emerged areas (PT: "Proto-Tyrrhenide" massif, including the Corsico-Sardinian block, the Maures–Esterel massifs, etc.); b. "Germanic" facies; c. Mediterranean facies ("Alpine" sea). (2) *Upper Jurassic–Lower Cretaceous* (after Haccard et al., 1972, simplified and modified): a. emerged areas; b. dominantly platform deposits; c. Piedmont Basin; d. Ligurian Basin; e. ophiolitic ridge; f. sedimentation area of the future Cravasco–Voltaggio–Busalla, etc., units; g. limits of the area with sediments deposited upon an oceanic crust: to the west, area with a "European" continental crust; to the east, area with an "Appennine–Austro-Alpine" continental crust. (3) *End-Cretaceous–early Tertiary* (after Haccard et al., 1972, simplified): A. Paleogeographical map—a. limit of the emerged areas (barbs on the marine side); b. external part of the Alpine basin; c. Piedmont Zone (emerged); d. Bracco–Val Lavagna–Mt. Gottero Units (emerged, cf. Bracco Ridge sense of Elter et al., 1966); e. "Insubric" ("Apennine–Austro-Alpine") basement, allochthonous to the west (klippes of Savona, Centuri) and forming the emerged insubric ridge to the east; f. Helminthoid Flysch Basin; g, h. Tuscan domain—g. "scisti policromi," h. "Scaglia"; i. platform facies. B. Relationships (thrusting) between the "European"—j—and the "Apennine–Austro-Alpine"—k—basement resulting from the Austrian uppermost Albian orogenic phase. An = Antola, B = Bracco, Ba = Balagne, C = Centuri, Ca = Monte Cassio, Cai = Monte Caio, E = Elba, MA = Maritime Alps, Sa = Savona. (4) *Eocene* (after Haccard et al., 1972, simplified and modified): A. Lower Eocene; B. Middle–Upper Eocene: a. limit of the emerged areas (barbs on the marine side); b. emerged parts of the external Alpine and Briançonnais domains; c. areas of marine sedimentation; d. Helminthoid Flysch Units; e. "schistes lustrés" (Piedmont Zone) unit; f. Cravasco–Voltaggio, etc., units; g. insubric basement; h. area of sedimentation (in A) or of outcropping structural unit (in B) of the "Clays and Limestones" Complex; i. Sestri–Voltaggio strike-fault. Letters: as in 3, and CC = Caio–Cassio, SV = Sestri–Voltaggio. (5) *Oligocene* (after Haccard et al., 1972, simplified and modified): a. limit of the emerged areas (barbs on the marine side); b. area of marine sedimentation; c. area of lagunal–lacustrine (Castellane–Digne Basin) sedimentation; d. external Alpine and Briançonnais domains; e. Helminthoid Flysch Units; f. "schistes lustrés" unit; g. Cravasco–Voltaggio, etc., units; h. insubric basement; i. NW Apennine Upper Eocene; j. Sestri–Voltaggio strike-fault. Letters as in 3 and 4 plus Ca = Castellane, Vo = Voltri.

in the Ligurian Basin. Following Haccard *et al.* (1972), to the latter must also be added the greater part of the material forming the units of Cravasco–Voltaggio–Busalla (see Section IIIA3 and IIIA4), as well as the Val Lavagna shales and the Monte Gottero sandstones. The major problem is, however, the precise location within the Ligurian Basin of the source of each of these units subjected to several and strong phases of tectonism.

It must be noted that only the internal part of the two basins has a crust of oceanic type; the external zones rest on a sialic crust forming the eastern margin of the European craton for the Piedmont Basin and the western margin of the Apennino-Austro-Alpine craton for the Ligurian Basin. This arrangement is inescapable, for the series mentioned are not always uniquely associated with greenstones but also with rocks derived from a continental crust.

According to the view of Haccard *et al.* (1972), the Austrian phase (end-Albian), well known in the eastern Alps, had also an important tectonic effect in the Alpino-Apennine circum-Ligurian region. Following a period of distension and spreading from the Jurassic into Lower Cretaceous, it corresponds to a shrinking of the basement of the cratons, perhaps related to movement of the African plate with respect to the European. It resulted in the crushing of the Liguro-Piedmont oceanic hiatus and in the expulsion of the sediments in the form of nappes. It is even possible that this movement, which could also initiate the metamorphism of the Piedmont "schistes lustrés," continued with the overthrusting of the European plate by the Apennino-Austro-Alpine plate. Haccard *et al.* base their interpretation primarily upon the occurrence of reworked elements and blocks of the ophiolitic series in the "complexes de base" (Albo-Cenomano-Turonian) of the Senonian Helminthoid Flysch of the Apennines and Corsica in particular. These Middle and Upper Cretaceous units are further discordant and detached from the substratum, and are thus tectonically independent. The initial thrusting of the cratons was probably renewed during subsequent phases, and is also suggested now by outcrops of the upper mantle in the Ivrea Zone, by gravimetric anomalies associated with this zone and with the Voltri group, and by seismic transverse sections across the Alpine–Apennine region (cf. Section IVB, C).

On both sides of the Liguro-Piedmont Basin, during the Jurassic and Lower Cretaceous, platform deposits formed unaffected by the Austrian orogenic phase, except in the Provençal region (see also Moullade and Porthault, 1970).

Attractive though the interpretation of Haccard *et al.* (1972) may be, it is not universally accepted. According to Elter (1975*b*) the Liguro-Piedmont oceanic opening could not have been created by an expansion process, for no material attributable to new oceanic crust can be found and the final closure would only occur following the Eocene orogenic phase.

The tectonic movements at the end of the Albian radically altered the paleogeographic framework of the Ligurian–Piedmont region. During the Upper Cretaceous the central part of the region was represented by a basin, where upon a tectonized substratum the future Helminthoid Flysch was deposited. The basin was limited to the west by a more or less emergent ridge (Bracco ridge of Elter *et al.*, 1966) formed of ophiolitic material. A second ridge of sialic material from the Apennino-Austro-Alpine basement formed the eastern margin; it was called the "Insubric ridge" and thus separates the Helminthoid Flysch from the Tuscan region farther to the east where "Scisti polichromi" and "Scaglia" were being deposited.

It is not easy to define the paleogeography of the Upper Cretaceous in the region west of the Bracco Ridge. Elter (1975*b*), who assigns the greater part of the Val Lavagna shales and Monte Gottero sandstones to the Upper Cretaceous, regards these sediments as having been deposited immediately west of the ridge in a zone he calls "internal Ligurian domain." These units, however, were associated by Haccard *et al.* (1972) with the preliminary tectonized (during the Middle Cretaceous) ophiolitic units, that is, the Piedmont Zone (being emergent following the Austrian phase), which is located by them west of the Bracco Ridge. Still farther to the west is found the external Alpine basin in which were deposited limestones and calcareous shales, which will subsequently belong to the Briançonnais and autochthonous domains.

At the end of the Cretaceous, or rather within the Paleocene, Haccard *et al.* (1972) consider that another slight tectonic movement resulted in the emersion and then displacement westward in an Alpine direction of the sediments deposited in the Helminthoid Flysch basin. This material moved by décollement at the horizon of the "complexe de base." According to these authors, the space vacated by the flysch was subsequently filled by the "clays and limestones" of the future Canetolo Unit.

The external Alpine domain was also affected by a period of emersion and even of "préstructuration" at the end of the Cretaceous or early Tertiary, prior to the deposition of the detrital and calcareous beds of the Paleogene (Campredon, 1972). Numerous authors, in particular Stanley and Mutti (1968), have demonstrated that the source of the Paleogene detrital material must be sought to the south (the Maures-"Tyrrhenide" Massif). The thermoluminescence studies of Ivaldi (1971, 1973), however, suggest a complementary easterly source from the Helminthoid Flysch material; in consequence, this must have been tectonized and already in process of erosion as early as Upper Eocene–Lower Oligocene times.

It is during Middle and Upper Eocene times that Haccard *et al.* (1972) locate a series of tectonic events; an updoming and torsion in the Savona region (with the initiation of the Sestri–Voltaggio strike-slip), resulting in cleavage under gravity in the Helminthoid Flysch, one part continuing to glide following the Alpine way, the other going back into the Apennine domain; tectogenesis of part of the Pennine domain (Piedmont and internal Briançonnais) as well as the "clays and limestones"; and finally the tectogenesis of the sub-Alpine domain and external Briançonnais zone. All the tectonic contacts between the various Ligurian and Pennine units (including the Sestri–Voltaggio Fault) are sealed by Stampian deposits laid down in a northeastern marine basin (called "Oligo-Miocene Liguro-Piedmont Basin"), which is fed from the unmantling of the internal zones of the newly formed Alpino-Apennine chain. The slow transgression of Oligocene age developed from east to west toward the Apennines and Gulf of Genoa, at that time occupied by an emergent zone (cf. Section VA). In contrast in the extreme northwest of the region, in the sub-Alpine domain, a regression of the "Alpine Sea" toward the north took place during the Oligocene.

It can thus be unequivocally stated that *up to the beginning of the Oligocene, the geological history of the region leaves no place for a marine structure which could be compared with the present Ligurian Sea or even which could be regarded as ancestral to it.* The history is one of the evolution of the different phases of the tectogenesis of the Alps and the internal Apennine domain (active sedimentation continued in the external Apennine), and whatever model is used to interpret the origin of the Ligurian Sea must be consistent with the fact that during the Upper Eocene–Oligocene the region was a continental one.

The Lower Miocene was a period of renewed tectonic activity affecting the structure of the Alps–Apennines, witness the coarse sandy-conglomerate deposits in the Liguro-Piedmont Basin (in Aquitanian) and even locally (vicinity of Roquebrune near Monaco), in the sub-Alpine domain where a NE–SW-oriented constraint is exerted. There are also numerous indications during the Aquitanian–Burdigalian of marine ingressions from the south (in the regions of Marseilles, Vence, Monaco, Finale Ligure); thus, this sea invades the Oligocene continent and anticipates the present Mediterranean. Even if the exact mechanisms leading to the formation of the Ligurian Sea are not yet well known, it can, nonetheless, be stated that from the study of the marginal regions it must have occurred toward the end of the Oligocene or at the beginning of the Miocene.

This fact may be compared with the age of the andesitic volcanism of the Ligurian province, also end-Oligocene–Lower Miocene. Deduced from actual outcrops, dredged marine material, and certain magnetic anomalies, the position of most of the vents can be located with some confidence along the large ENE–WSW faults running parallel to the coast and which affect or control the Ligurian–Provençal and Corsican continental margin (Fig. 12). Given the chronology, it is obviously tempting to relate the faulting and perhaps associated volcanism with the initial phenomena in the differentiation of the Ligurian Sea.

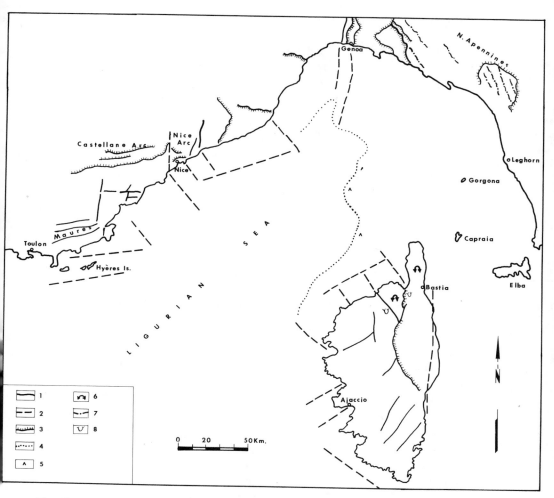

Fig. 12. The most significant structural trends of the Ligurian and circum-Ligurian areas. (1) Observed faults; (2) inferred faults; (3) thrust faults; (4) limit of the Upper Miocene salt basin, which corresponds more or less with the presumed extension of the continental margin; (5) Cenozoic volcanics; (6) anticlinal structure; (7) anticlinal axis; (8) synclinal structure.

VII. HYPOTHESES ON THE GENESIS AND EVOLUTION OF THE LIGURIAN SEA

The preceding data are in chronologic agreement with most of the paleomagnetic observations (Section IVE). Yet, if it seems that there are no major disagreements concerning the time of formation of the Ligurian Sea, there is no unanimity concerning the process.

A. The Origin of the Ligurian Sea

Several authors (e.g., Van Bemmelen, 1969, 1972a, b; de Roever, 1969; Ritsema, 1970) have already discussed the details of the various oceanization models of the western Mediterranean. Here will be emphasized only the consequences of these models insofar as they apply to the Ligurian Basin.

1. *Mediterranean-Type Oceanization*

Van Bemmelen referred to corrosion and subcrustal erosion of the continental crust by upward motions in the upper mantle (upper mantle diapirism), followed by the sinking of the newly formed, more dense material. Such a model has the advantage of linking the process of orogenesis (in itself also the result of upper mantle diapirism) and oceanization. Depending upon vertical motion, it does not necessarily require horizontal displacements of blocks or plates; however, taking the results of paleomagnetism into account, Van Bemmelen does not completely exclude translation, such as the rotation of the Corso-Sardinian block. He reduces the importance of such phenomena as drift, considering them merely as superimposed to the oceanization process.

2. *"Atlantic"-Type Oceanization*

The majority of authors seeking an explanation for the formation of the western Mediterranean basin are compelled to resort to horizontal displacement. Most of them favor Atlantic-type oceanization commonly involving crustal extension (rifting–sea-floor spreading–continental drift), previously invoked (section VI) for the formation of the Mesozoic Tethys. In the Mediterranean this leads to the concept of microplates, with the Mediterranean basin more or less the result of the relative motions of the European and African plates related to the different spreading rates in the North and South Atlantic (Dewey et al., 1973). Applying this model to the Ligurian sector and taking into account the interpretation of the margin using bathymetric, geological, and geophysical data, a variety of reassembly models of the Corso-Sardinian block (either as a unit or decoupled), supposed or calculated, have been presented (Nairn and Westphal 1968; Boccaletti and Guazzone, 1970; Hsü, 1971; Smith, 1971; Alvarez, 1972a, 1974; Ryan et al., 1973b; Westphal et al., 1973, 1975; Chabrier and Mascle, 1974, 1975; Mattauer, 1974; Arthaud and Matte, 1975; Faure-Muret and Choubert 1975; etc.). In the process of displacement or translation the importance of certain structural directions, e.g., strike-slip faults, has been emphasized by Muraour (1970), Auzende et al. (1973), Bayer et al. (1973), Rehault et al. (1974a), Gennesseaux et al. (1974), and Arthaud and Matte (1975). In addition Alvarez (1973) used the evidence of the eastward migration of Tuscan magmatism to infer a subduction zone at the leading edge of the Corso-Sardinian block, that is, in the Tyrrhenian Sea to the east.

In all reconstructions accent has been placed, with minor variations, on the geological affinities of Hercynian Corsica with the Maures–Tanneron–Esterel Massifs, and of Sardinia with Provence, the Montagne Noire, and the Balearic Islands (Fig. 13).

Objections have been made concerning the validity of this superficially attractive model. They concern the paleomagnetic (cf. Section IVE), magnetic (Storetvedt, 1973; Gonnard et al., 1975; Montadert and Letouzey, 1975), and geological (Bourrouilh, 1975) data upon which it is based. The paleobio-geographic data, while certainly consistent with the proximity of the Corso-Sardinian block with the French Mediterranean coast, do not necessarily imply it, for they cannot provide a means of gauging the original dimensions of the basin. In other words, while there are many arguments and presumptions in favor of such a model, there does not yet exist the concrete proof which would permit its unqualified acceptance. It should also be noted that it is more difficult to explain the possible existence of a residual sialic layer, such as that geophysically supposed below the Ligurian Sea, by such a model than by that proposed by Van Bemmelen.

The review of circum-Ligurian geology (Section III) clearly indicates that the configuration of the sea transects all earlier structures, which appears consistent with the idea that the distensional oceanization process is unrelated to the orogenic process (Aubouin, 1975). Some authors have, however, sought to link the opening of the Ligurian sphenochasm, responsible for the basin of the same name (with the rotation of the Corso-Sardinian block), with some of the final stages of Apennine tectogenesis (e.g., Haccard et al., 1972, Radicati di Brozolo and Giglia, 1973; Giglia 1974). This association provides another way of considering the chronology of the formation of the Ligurian Sea. The metamorphism of the Apuanes sector of the Apennines being interpreted as the result of a tectonic overburden by the emplacement of Tuscan and Ligurian nappes on the Tuscan autochthon, means that the dating of the Apuanes metamorphism can provide an idea of the age of the tectonic process, and consequently of the rotation of Corso-Sardinia which would be the cause. Radicati di Brozolo and Giglia (1973) arrive at a date of 16 m.y. for the onset of the phenomenon. Relying on Cornet's (1965) geomorphological data, the significance of which has already been discussed, as well as on arguments based upon the age of the Tuscan magmatism, Alvarez (1972a) independently arrived at a date of 11.5 m.y. (which he later increased a little (1973) to take into account the results of Radicati di Brozolo and Giglia). On the other hand, it is unanimously agreed that the rotation was completed before the formation of the Messinian evaporitic basin, which, allowing time for the deposition of the evaporites, gives an age of about 9 m.y.

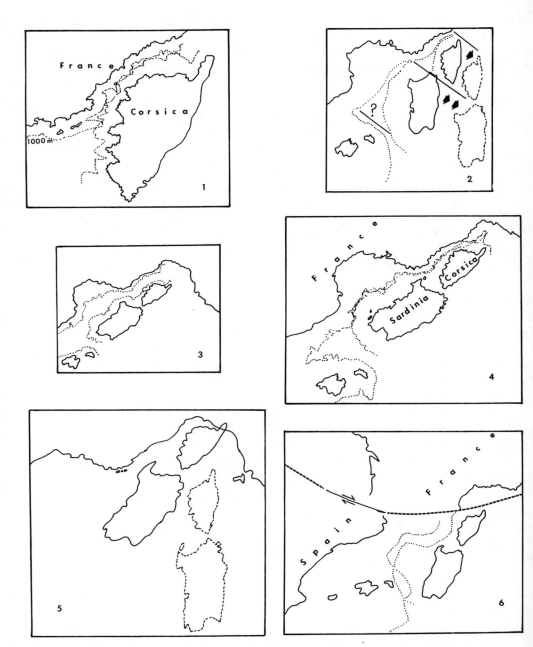

Fig. 13. Some hypotheses on the presumed initial location of the Corso-Sardinian block. (1) Nairn and Westphal, 1968; (2) Auzende *et al.*, 1973; (3) Alvarez, 1972*a*; (4) Westphal *et al.*, 1973; (5) Chabrier and Mascle, 1975; (6) Arthaud and Matte, 1975.

These deductions are relatively compatible with the paleomagnetic results (de Jong *et al.*, 1973) and the arguments derived from regional geology (e.g., end-Oligocene–basal-Miocene age of the andesitic volcanism in the Ligurian region, perhaps linked to rifting; and the Aquitanian age for the first Neogene transgressions known in the circum-Ligurian area), although the latter would rather suggest an older date (from the end of the Oligocene) for the initiation of the Ligurian rift. Since a certain time must elapse between the time of rifting and the drift process eventually responsible for the Apennine tectogenesis, there is no fundamental disagreement between these conclusions and those of Radicati di Brozolo and Giglia (1973).

3. *"Pacific"-Type Oceanization*

The nature of the crust of the Ligurian Sea (see Section IVB1) and the configuration of the intra- and circum-Ligurian continental margins has led certain authors (e.g., Boccaletti and Guazzone, 1971, 1974*a, b*) to invoke a third type of model. Their interpretation of a Pacific-type of back-arc marginal basin on the scale of the Mediterranean has the Corso-Sardinian block representing a remnant arc whose migration during the Neogene resulted in the rear in the opening of the Ligurian Basin.

At the present state of knowledge it is difficult to make a choice among the three models proposed, and it hardly seems profitable to introduce yet more to attempt an explanation of the various basins in the western Mediterranean. The role of hypothesis and intellectual speculation is still too great, and to reduce it, it is essential to accumulate new observations (for example, by better seismic penetration and by deep drilling into the Ligurian sea floor, which would give the age and nature of the infra-evaporitic series) and to improve the credibility of other results (e.g., paleomagnetic). Unless these steps are taken, further progress seems unlikely.

B. Evolution of the Ligurian Sea

If the origin of the Ligurian Sea is still uncertain, there is, at least, a much clearer knowledge of its history from the Neogene onward. It may be deduced from geological observations in the present coastal basins (Section IIIB), from marine geophysics (Section IVB), and from dredging and coring of the bottom.

Evidence of an Aquitanian–Burdigalian transgression in the northeastern part of the western Mediterranean has already been noted. It seems logical to suppose that this transgression, known on land, could have followed or accompanied the initial Ligurian oceanization process. In the present state of knowledge, these observations, and the interpretation which follows from them, cannot be confirmed in the marine domain, for the only means of investigation (indirect and distant) available consists of the upper beds of the Messinian evaporitic sequence drilled during D.S.D.P. Leg 13, west of Sardinia, or else from the petroleum exploration bores in the Golfe du Lion (Cravatte *et al.*, 1974), where marine Miocene and perhaps even Oligocene sediments are found to lie on a continental-type crust linked to the margin.* For the same reason there is no direct information concerning the sedimentation and geological events in the Ligurian Sea for the rest of the Miocene, at least up to the Messinian. The sedimentation in the coastal basins was of shallow water type, with molasse, sands, and reefal formations, whose nature obviously reflects their location on the margin of a continent tectonized at the end of the Eocene and/or beginning of the Oligocene. Locally, evidence of further compressive tectonism can be observed, as, for example, during the Tortonian in the northern Apennines.

The Ligurian region was obviously involved in the important regression and desiccation phase which affected the Mediterranean at the end of the Miocene. The magnitude of the evaporitic basin, known from geophysical exploration, implies that by Messinian times the Ligurian Sea was about the same size it is now (cf. earlier comment on ending of the Corso-Sardinian rotation). To explain the relatively large volume of evaporites (with intercalated marine beds), intermittent replenishment essentially from the Atlantic has been invoked. The thickness of the evaporites also implies an appreciable subsidence of the central part of the basin. It has indeed already been shown (Section IIIB) that locally on land (e.g., Tuscany) tensional tectonics were established during the Messinian; however, over the greater part of the circum-Ligurian area, the end of the Miocene marked a phase of emersion with continental deposits, erosion, and even in certain regions renewal of tectogenesis.

Authors have questioned the depth of the salt depositional basin, and except for certain surprising models (Le Pichon *et al.*, 1971; Hsü, 1971; Cita 1974) it has been generally deduced (Nesteroff *et al.*, 1972; Fredj, 1974) that the evaporites formed in relatively shallow water and in a shallow basin. Given the characteristics of the Pliocene transgression both in the coastal region and offshore, it can be concluded that the sea rapidly invaded a basin that finally dried at the end of the Miocene, but one in the process of rapid deepening.

The passage from Miocene to Pliocene was, in effect, marked by a radical change in regime. Both in the coastal region and offshore the contact is sharp with erosion and a lacuna of greater or lesser importance. In the center of the basin the facies of the deposits was pelagic, as at many points along the border conglomerates at the base of the series are the exception. During the Pliocene and Quaternary the most constant facies is represented by bluish marls, detrital beds being more abundant and coarser in the marginal basins.

* See, however, recent results of Bellaiche *et al.*, 1976.

The best-established tectonic events of the Pliocene and Quaternary involve vertical motion. The play of numerous E–W or NE–SW faults leads to the foundering of the continental margin and the central part of the basin, accompanied by halokinetic phenomena in the central graben and locally an uplift of the marginal zone (e.g., Arc of Nice). Data on land indicates that movements were particularly marked at the Mio-Pliocene boundary, between the Lower and Middle Pliocene, and at the Plio-Quaternary confines.

Dating from Middle Pliocene, a marine regression set in and nowhere is a continuous passage seen from the Pliocene into the Quaternary. These observations may be correlated with the evidence of a hardground, reflecting a break in sedimentation during part of the Upper Pliocene and Lower Pleistocene, seen in D.S.D.P. Leg 13, Site 134 (Ryan et al., 1973a).

The climatic and glacio-eustatic variations during the Quaternary resulted in numerous oscillations of sea level, witness the marine terraces and alluvial deposits in the circum-Ligurian littoral. Their present altitudes also reflect subsequent local deformation and movements.

If it has been possible to determine the structure of the Ligurian sector and its surroundings, and to reconstruct from the currently available geological and geophysical data its history both prior to and subsequent to the creation of this particular western Mediterranean basin, it has also been remarked that a better appreciation of the process of oceanization demands continued research. The key to the problem may lie in certain technical improvements in seismic reflection techniques and in deep-water drilling, which would remove the indeterminacy which exists concerning the nature and age of the infra-Messinian substratum.

ACKNOWLEDGMENTS

The author wishes to express his gratitude to Dr. F. Irr (University of Nice, France) for his help on the section concerning the coastal basins and the Neogene history of the Ligurian Sea, and to Mrs. F. Orszag for her information concerning the Corsican Neogene. He also wishes to thank Drs. A. E. M. Nairn and F. G. Stehli for a critical review of the manuscript, and the former for its translation.

REFERENCES

Abbate, E., and Sagri, M., 1970, The eugeosynclinal sequences, in: *Development of the Northern Apennines Geosyncline*, Sestini, G., ed., *Sediment. Geol.*, v. 4, p. 251–340.

Abbate, E., Bortolotti, V., Passerini, P., and Sagri, M., 1970a, Introduction to the geology of the Northern Apennines, in: *Development of the Northern Apennines Geosyncline*, Sestini, G., ed., *Sediment. Geol.*, v. 4, p. 207–249.

Abbate, E., Bortolotti, V., Passerini, P., and Sagri, M., 1970*b*, The geosyncline concept and the Northern Apennines, in: *Development of the Northern Apennines Geosyncline*, Sestini, G., ed., *Sediment. Geol.*, v. 4, p. 625–636.

Abbate, E., Bortolotti, V., Passerini, P., and Sagri, M., 1970*c*, The Northern Apennines geosyncline and continental drift, in: *Development of the Northern Apennines Geosyncline*, Sestini, G., ed., *Sediment. Geol.*, v. 4, p. 637–642.

Alinat, J., Giermann, G., and Leenhardt, O., 1966, Reconnaissance sismique des accidents de terrain en mer Ligure, *C.R. Acad. Sci. Paris*, sér. B, v. 262, p. 1311–1314.

Alinat, J., Cousteau, J. Y., Giermann, G., Leenhardt, O., Perrien, C., and Pierrot, S., 1969, Lever de la carte bathymétrique de la mer ligure, *Bull. Inst. Océanogr. Monaco*, v. 68 (1395), 12 p.

Alla, G., 1970, Etude sismique de la plaine abyssale au Sud de Toulon, *Rev. Inst. Franç Pétrole*, v. 25, p. 291–304.

Alla, G., Dessolin, D., Leenhardt, O., Pierrot, S., 1972, Données du sondage sismique continu concernant la sédimentation Plio-Quaternaire en Méditerranée nord-occidentale, in: *The Mediterranean Sea—A Natural Sedimentation Laboratory*, Standley, D. J., ed., Stroudsburg Pa.: Dowden, Hutchinson and Ross, p. 471–487.

Alla, G., Byramjee, R., Didier, J., Durand, J., Louis, J., Montadert, L., Mugniot, J. F., and Valery, P., 1973, Structure géologique de la marge continentale du Golfe du Lion, 23rd CIESM Congr., Athens (1972), *Bull. Geol. Soc. Greece*, v. X, p. 1–2.

Allan, T. D., and Morelli, C., 1971, A geophysical study of the Mediterranean Sea, *Boll. Geofis. Teor. Appl.*, v. 13, p. 99–142.

Alvarez, W., 1972*a*, Rotation of the Corsica–Sardinia microplate, *Nature*, v. 235, p. 103–105.

Alvarez, W., 1972*b*, Uncoupled convection and subcrustal current ripples in the Western Mediterranean, in: *The H. H. Hess Volume—Studies in Earth and Space Science*, Shagam, R., ed., *Mem. Geol. Soc. Amer.*, no. 132, p. 119–132.

Alvarez, W., 1973, The application of plate tectonics to the Mediterranean region, in: *Implications of Continental Drift*, Tarling, D. H., and Runcorn, S. K., eds., v. 2, New York: Academic Press; p. 893–908.

Alvarez, W., 1974, Sardinia and Corsica: one microplate or two? *Paleog. Terziario Sardo nell'ambito Medit. Occid., Rend. Sem. Fac. Sci. Univ. Cagliari* (1973), Suppl. v. XIII, p. 1–4.

Alvarez, W., 1976, A former continuation of the Alps, *Bull. Geol. Soc. Amer.*, v. 87, p. 891–896.

Alvarez, W., and Cocozza, T., 1974, The tectonics of central-eastern Sardinia and the possible continuation of the Alpine chain to the South of Corsica, *Paleog. Terziario Sardo nell'ambito Medit. Occid., Rend. Sem. Fac. Sci. Univ. Cagliari*, (1973) Suppl. v. XIII, p. 5–34.

Alvarez, W., Franks, S. G., and Nairn, A. E. M., 1973, Palaeomagnetism of Plio-Pleistocene basalts from north-west Sardinia, *Nature*, v. 243, p. 10–11.

Amaudric du Chaffaut, S., 1975, L'unité de Corte: un témoin de "Piémontais externe" en Corse? *Bull. Soc. Géol. France*, (7), v. XVII, p. 739–745.

Amaudric du Chaffaut, S., and Lemoine, M., 1974, Découverte d'une série jurassico-crétacée d'affinités briançonnaises transgressive sur la marge interne de la Corse granitique, *C.R. Acad. Sci. Paris*, sér. D, v. 278, p. 1317–1320.

Amaudric du Chaffaut, S., Boulanger, D., and Jauzein, A., 1969, Présence de formations transgressives maestrichtiennes, paléocènes et yprésiennes sur la côte sud-orientale de la Corse, *C.R. Acad. Sci. Paris*, sér. D, v. 268, p. 1706–1709.

Amaudric du Chaffaut, S., Caron, J. M., Delcey, R., and Lemoine, M., 1972, Données nouvelles sur la stratigraphie des schistes lustrés de Corse: la série de l'Inzecca. Comparaisons avec les Alpes occidentales et l'Apennin ligure, *C.R. Acad. Sci. Paris*, sér. D, v. 275, p. 2611–2614.

Anglada, R., Follacci, J. P., and Meneroud, J. P., 1967, Sur la présence du Miocène marin en bordure sud de l'arc de Nice, dans la région de Roquebrune–Cap-Martin (Alpes-Maritimes), *Bull. Soc. Géol. France*, (7), v. IX, p. 526–529.

Angrisano, G., and Segré, A., 1969, Carta batimetrica del Mediterraneo Nord-Occidentale, Map 1501 F.C. 1030/2, Genoa: Ist. Idrogr. Marina.

Argand, E., 1916, Sur l'arc des Alpes occidentales, *Eclogae Geol. Helv.*, v. 14, p. 145–191.

Argand, E., 1924, La tectonique de l'Asie, *Proc. XIIIth Int. Geol. Congr.*, Bruxelles (1922), p. 171–372.

Arthaud, F., and Matte, P., 1970, Contribution à l'étude des tectoniques superposées dans la chaîne hercynienne: étude microtectonique des séries métamorphiques du massif des Maures (Var), *C.R. Acad. Sci. Paris.*, sér. D, v. 270, p. 18–21.

Arthaud, F., and Matte, P., 1975, Les décrochements tardi-hercyniens du Sud-Ouest de l'Europe. Géométrie et essai de reconstitution des conditions de la déformation, *Tectonophysics*, v. 25, p. 139–171.

Ashworth, T. P., and Nairn, A. E. M., 1965, An anomalous Permian pole from Corsica, *Palaeogeogr. Palaeoclimatol. Palaeoecol.*, v. 1, p. 119–125.

Aubouin, J., 1973, Paléotectonique, tectonique, tarditectonique et néotectonique en Méditerranée moyenne: à la recherche d'un guide pour la comparaison des données de la géophysique et de la géologie, 23rd CIESM Congr., Athens (1972), *Bull. Geol. Soc. Greece*, v. X, p. 3–10.

Aubouin, J., 1974, La Provence, in:*Géologie de la France*, Debelmas, J., ed., Paris: Doin, v. 2, p. 346–386.

Aubouin, J., 1975, De la Méditerranée aux Caraïbes; de quelques problèmes structuraux, *3rd Earth Sci. Annu. Meeting*, Montpellier (Abstracts), p. 17.

Aubouin, J., and Mennessier, G., 1960–1963, Essai sur la structure de la Provence, *Livre Mém. Prof. P. Fallot, Mém. h.s. Soc. Géol. France*, v. II, p. 45–98.

Auzende, J. M., Bonnin, J., Olivet, J. L., Pautot, G., and Mauffret, A., 1971, Upper Miocene salt layer in the western Mediterranean basin, *Nature Phys. Sci.*, v. 230, p. 82–84.

Auzende, J. M., Bonnin J., and Olivet J. L., 1973, The origin of the western Mediterranean basin, *J. Geol. Soc. London*, v. 129, p. 607–620.

Barsotti, G., Federici, P. R., Giannelli, L., Mazzanti, R., and Salvatorini, G., 1974, Studio del Quaternario livornese, con particolare riferimento alla stratigrafia ed alle faune della formazione del bacino di Carenaggio della Torre del Fanale, *Mem. Soc. Geol. Ital.*, v. XIII, p. 425–495.

Baubron, J. C., 1974, Etude de l'évolution magmatique des formations calco-alcalines tertiaires de Provence et Haute-Provence par la géochimie du rubidium et du strontium, *Rapport B.R.G.M.*, v. 74 SGN 003 LAB, 37 p.

Baud, J. P., Boyon, M. J., and Rollet, M., 1974, Observations sur les roches vertes en Haute Balagne. Conséquences structurales, *2nd Earth Sci. Ann. Meeting*, Pont-à-Mousson (Nancy) (Abstracts), p. 35.

Bayer, R., 1974, *Anomalies Magnétiques et Évolution Tectonique de la Méditerranée Occidentale*, Thesis (3rd cycle), Univ. Paris.

Bayer, R., Le Mouel, J. L., and Le Pichon, X., 1973, Magnetic anomaly pattern in the western Mediterranean, *Earth Planet. Sci. Lett*, v. 19, p. 168–176.

Bellaiche, G., 1969, *Etude Géodynamique de la Marge Continentale au Large du Massif des Maures et de la Plaine Abyssale Ligure*, Thesis, Univ. Paris.

Bellaiche, G., 1970, Géologie sous-marine de la marge continentale au large du massif des Maures (Var, France) et de la plaine abyssale ligure, *Rev. Géogr. Phys. Géol. Dynam.*, v. 12, p. 403–440.

Bellaiche, G., 1972a, Les dépôts quaternaires immergés du golfe de Fréjus (Var), France, in: *The Mediterranean Sea—A Natural Sedimentation Laboratory*, Stanley, D. J., ed., Stroudsburg, Pa: Dowden, Hutchinson and Ross, p. 171–176.

Bellaiche, G., 1972*b*, Prélèvement par 2000 m de profondeur d'un réflecteur acoustique d'âge pliocène inférieur à faciès peu profond (canyon de Stoechades, Méditerranée nord-occidentale). Nouvelles données sur le creusement des canyons sous-marins et l'amplitude des mouvements verticaux ponto-plio-quaternaires, *C.R. Acad. Sci. Paris*, sér. D, v. 275, p. 321–322.

Bellaiche, G., 1973, Morphologie et sédimentologie de la terminaison des canyons sous-marins au large de la région toulonnaise, *Rev. Inst. Franç. Pétrole*, v. XXVIII, p. 251–158.

Bellaiche, G., and Mascle, J., 1970, Premiers résultats par sismique continue de l'ennoyage du massif des Maures sous la plaine abyssale, *C.R. Somm. Soc. Géol. France*, v. 3, p. 88.

Bellaiche, G., and Mauffret, A., 1971, Résultats des dragages réalisés sur un des dômes situés au large de Toulon (Dôme T1 Sud-Toulon), *C.R. Somm. Soc. Géol. France*, v. 3, p. 187–188.

Bellaiche, G., and Recq, M., 1973, About off-shore seismological experiments south of Provence, in relation to "Glomar Challenger" deep sea drillings (JOIDES-DSDP, leg 13), *Marine Geol.*, v. 15, p. 49–52.

Bellaiche, G., Vergnaud-Grazzini, C., and Glangeaud, L., 1969, Les épisodes de la transgression flandrienne dans le golfe de Fréjus, *C.R. Acad. Sci. Paris*, sér. D, v. 268, p. 2765–2770.

Bellaiche, G., Recq, M., and Rehault, J. P., 1973, Nouvelles données sur la structure du haut-fond du Méjean obtenues par la "sismique réfraction", *C.R. Acad. Sci. Paris*, sér. D, v. 276, p. 1529–1532.

Bellaiche, G., Genesseaux, M., Mauffret, A., and Rehault J. P., 1974*a*, Prélèvements systématiques et caractérisation des réflecteurs acoustiques. Nouvelle étape dans la compréhension de la géologie de la Méditerranée occidentale, *Marine Geol.*, v. 16, p. 47–56.

Bellaiche, G., Mauffret, A., and Rehault, J. P., 1974*b*, Nature, âge et origine des affleurements rocheux prélevés par 2400 m de fond sur le sommet d'un dôme au large de la Provence, *C.R. Acad. Sci. Paris*, sér. D, v. 278, p. 41–43.

Bellaiche, G., Irr, F., and Labarbarie, M., 1976, Découverte de sédiments marins fini-oligocènes-aquitaniens au large du massif des Maures (canyon des Stoechades), *C.R. Acad. Sci. Paris*, sér. D, v. 283, p. 319–322.

Bellini, A., 1963, Petrogenesi e significato stratigrafico dei porfiroidi—cosiddetti besimauditi—dell'areale savonese delle Alpi Liguri, *Atti Ist. Geol., Univ. Genova*, v. 1, p. 275–320.

Bellini, A., 1964, Nuove osservazioni petrogenetiche e geologiche sul cristallino delle Alpi Liguri e del Savonese in particolare, *Atti Ist. Geol., Univ. Genova*, v. 2, p. 99–192.

Bellini, A., 1966, Sulla geologia della Corsica alpina. Nota preliminare, *Atti Ist. Geol., Univ. Genova*, v. 2, p. 465–474.

Bellon, H., and Brousse, R., 1971, L'âge oligo-miocène du volcanisme ligure, *C.R. Acad. Sci. Paris*, sér. D, v. 272, p. 3109–3111.

Berry, M. J., and Knopoff, L., 1967, Structure of the upper mantle under the western Mediterranean Basin, *J. Geophys. Res.*, v. 72, p. 3613–3626.

Bezzi, A., and Piccardo, G. B., 1971, Structural features of the Ligurian ophiolites: petrologic evidence for the "oceanic" floor of the Northern Apennines geosyncline. A contribution to the problem of the alpine-type gabbro-peridotite associations, *Mem. Soc. Geol. Ital.*, v. 10, p. 53–63.

Biju-Duval, B., and Letouzey, J., 1974, Comments about the new "Carte géologique et structurale des bassins tertiaires du domaine méditerranéen", *24th CIESM Congr.*, Monaco (Abstracts).

Biju-Duval, B., Letouzey, J., Montadert, L., Courrier, P., Mugniot, J. F., and Sancho, J., 1974, Geology of the Mediterranean Sea basins, *Rapport Inst. Franç. Pétrole*, v. 18464, 38 p.

Biscaye, P. E., Ryan, W. B. F., and Wezel, F. C., 1972, Age and nature of the pan-Mediterranean subbotom reflector M., in: *The Mediterranean Sea—A Natural Sedimentation Laboratory*, Stanley, D. J., ed., Stroudsburg, Pa.: Dowden, Hutchinson and Ross, p. 83–90.

Blanc, J. J., 1968, Sedimentary geology of the Mediterranean Sea, *Oceanogr. Mar. Biol. Annu. Rev.*, v. 6, p. 377–454.

Blanc, J. J., 1972, Observations sur la sédimentation bioclastique en quelques points de la marge continentale de la Méditerranée, in: *The Mediterranean Sea—A Natural Sedimentation Laboratory*, Stanley, D. J., ed., Stroudsburg, Pa.: Dowden, Hutchinson and Ross, p. 225–240.

Blanc, M., Coron, S., Guillaume, A., and Guillaume, S., 1970, Nouvelles mesures de pesanteur en Ligurie. Conséquences géologiques, *C.R. Acad. Sci. Paris*, sér. D, v. 270, p. 1766–1769.

Blanc-Vernet, L., 1972, Données micropaléontologiques et paléoclimatiques d'après les sédiments profonds de la Méditerranée, in: *The Mediterranean Sea—A Natural Sedimentation Laboratory*, Stanley, D. J., ed., Stroudsburg, Pa.: Dowden, Hutchinson and Ross, p. 115–127.

Bloch, J. P., 1971, Trias briançonnais et Trias ligure: précisions stratigraphiques, *C.R. Acad. Sci. Paris*, sér. D, v. 272, p. 519–522.

Bobier, C., 1974a, La signification de l'aimantation rémanente des laves de la "série des ignimbrites inférieures." Conséquence pour l'étude de la rotation du bloc corso-sarde durant le Tertiaire, *Paleog. Terziario Sardo nell'ambito Medit. Occid., Rend. Sem. Fac. Sci. Univ. Cagliari* (1973), Suppl. v. XIII, p. 35–56.

Bobier, C., 1974b, Les mouvements du microcontinent Sarde durant le Néogène d'après les données paléomagnétiques, *2nd Earth Sci. Annu. Meeting*, Pont-à-Mousson (Nancy) (Abstracts), p. 56.

Bobier, C., and Coulon, C., 1970, Résultats préliminaires d'une étude paléomagnétique des formations volcaniques tertiaires du Logudoro (Sardaigne septentrionale), *C.R. Acad. Sci. Paris*, sér. D, v. 270, p. 1434–1437.

Boccaletti, M., and Guazzone, G., 1970, La migrazione terziaria dei bacini toscani e la rotazione dell'Appennino settentrionale in una "zona di torsione" per deriva continentale, *Mem. Soc. Geol. Ital.*, v. 9, p. 177–195.

Boccaletti, M., and Guazzone, G., 1971, Gli archi appenninici, il mar ligure ed il Tirreno nel quadro della tettonica dei bacini marginali retro-arco, *Mem. Soc. Geol. Ital.*, v. 11, p. 201–216.

Boccaletti, M., and Guazzone, G., 1974a, Il microcontinente sardo come un arco residuo di un sistema arco-fossa miocenico, *Paleog. Terziario Sardo nell'ambito Medit. Occid., Rend. Sem. Fac. Sci. Univ. Cagliari* (1973), Suppl. v. XIII, p. 57–68.

Boccaletti, M., and Guazzone, G., 1974b, Remnant arcs and marginal basins in the Cainozoic development of the Mediterranean, *Nature*, v. 252, p. 18–21.

Boccaletti, M., Elter, P., and Guazzone, G., 1971a, Plate tectonic models for the development of the western Alps and northern Apennines, *Nature, Phys. Sci.*, v. 234, p. 108–111.

Boccaletti, M., Elter, P., and Guazzone, G., 1971b, Polarità strutturali delle Alpi e dell'Appennino settentrionale in rapporto all'inversione di una zona di subduzione nordtirrenica, *Mem. Soc. Geol. Ital.*, v. 10, p. 371–378.

Bodelle, J., and Campredon, R., 1969, L'Eocène des Alpes-Maritimes et des Basses Alpes, *Mem. B.R.G.M.*, v. 69, p. 409–415.

Boillot, G., 1957, Conditions de dépôt des terrains pliocènes dans la région de Vintimille, *Rev. Géogr. Phys. Géol. Dynam.*, v. 1, p. 229–243.

Boiteau, A., 1971a, Le massif gneissique de Loano: témoin méridional extrême de la nappe des gneiss ligures, *C.R. Acad. Sci. Paris*, v. 272, p. 370–373.

Boiteau, A., 1971*b*, Un exemple de la tectonique des Alpes Ligures: la région du Monte Carmo (Italie), *Géol. Alpine*, v. 47, p. 117–132.

Boni, A., and Vanossi, M., 1972, Carta geologica dei terreni compresi tra il Brianzonese Ligure *s.l.* ed il flysch ad Elmintoidi *s.s.*, *Atti Ist. Geol.*, *Univ. Pavia*, v. 23.

Bonifay, E., 1965, Remarques sur le Pleistocène marin des Alpes-Maritimes, *Bull. Soc. Géol. France*, (7), v. VII, p. 946–956.

Bordet, P., 1951, *Etude Géologique et Pétrographique de l'Estérel*, Thesis Univ. Paris, *Mém. Expl. Carte Géol. Dét. France*, p. 1–207.

Bordet, P., 1966, *L'Estérel et le Massif du Tanneron* (*Guide Géologique*), Paris: Hermann, 114 p.

Bortolotti, V., 1964, Geologia dell'alta Garfagnana tra Poggio, Dalli e Gramolazzo, *Boll. Soc. Geol. Ital.*, v. 83.

Bortolotti, V., and Passerini, P., 1970, Magmatic activity, in *Development of the Northern Apennines Geosyncline*, Sestini, G., ed., *Sediment. Geol.*, v. 4, p. 599–624.

Bortolotti, V., Passerini, P., Sagri, M., and Sestini, G., 1970, The miogeosynclinal sequences, in: *Development of the Northern Apennines Geosyncline*, Sestini, G., ed., *Sediment. Geol.*, v. 4, p. 341–444.

Bosellini, A., 1973, Modello geodinamico e paleotettonico delle Alpi Meridionali durante il Giurassico–Cretacico. Sue possibili applicazioni agli Appennini, in: *Moderne Vedute sulla Geologia dell'Appennino*, Accad. Naz. Lincei, Roma (1972), v. 183, p. 163–213.

Bossolasco, M., and Eva, C., 1965, Il terremoto del 19 Luglio 1963 con epicentro nel Mar Ligure, *Geofis. Meteorol.*, v. XIV, p. 6–18.

Bossolasco, M., and Eva, C., 1971, The seismicity in Maritime Alps and adjacent areas, *XII*ᵉ *Assemblée Gén. Comm. Seism. Europe, Luxembourg, 1970, Observat. Roy. Belgique, Sér. Géophys.*, v. 101, p. 104–110.

Bossolasco, M., Cicconi, G., Eva, C., and Pasquale, V., 1972, La rete sismica dell'Istituto Geofisico di Genova e primi risultati sulla sismo-tettonica delle Alpi Marittime ed Occidentali, e del Mar Ligure, *Riv. Ital. Geofis.*, v. XXI, p. 229–247.

Bossolasco, M., Dagnino, I., Elena, A., and Bozzo, E., 1973*a*, Ricerche geofisiche sul gruppo di Voltri. I. Caratteristiche geomagnetiche, *Riv. Ital. Geofis.*, v. XXII, p. 75–84.

Bossolasco, M., Cicconi, G., and Eva, C., 1973*b*, Spectral characteristics of microseisms recorded near the coast, *Riv. Ital. Geofis.*, v. XXII, p. 302–308.

Bossolasco, M., Eva, C., and Pasquale, V., 1973*c*, Il terremoto del 25 Ottobre 1972 in Lunigiana, *Riv. Ital. Geofis.*, vol. XXII, p. 314–322.

Bossolasco, M., Eva, C., and Pasquale, V., 1973*d*, On the seismicity of the Ligurian Sea, 23rd CIESM Congr., Athens (1972), *Bull. Geol. Soc. Greece*, v. 1, p. 19–22.

Boucarut, M., 1971, *Etude Volcanologique et Pétrographique de l'Estérel*, Thesis, Univ. Nice, 487 p.

Boucarut, M., 1974, Arguments en faveur de l'existence d'un réseau de fractures crustales d'âge alpin en Provence. Leurs relations avec le volcanisme de même âge, *2nd Earth Sci. Annu. Meeting*, Pont-à-Mousson (Nancy) (Abstracts), p. 72.

Bourcart, J., 1958*a*, Morphologie du précontinent des Pyrénées à la Sardaigne, *La Topographie et la Géologie des Profondeurs Océaniques*, Coll. C.N.R.S., Villefranche-sur-Mer, p. 33–50.

Bourcart, J., 1958*b*, Précontinent entre Antibes et Gênes, 1/200,000ème. Précontinent entre Marseille et Antibes, 1/200,000ème, Topogr. maps, *Musée Océanogr. Monaco*.

Bourcart, J., 1959, Le plateau continental de la Méditerranée occidentale, *C.R. Acad. Sci. Paris*, v. 249, p. 1380–1382.

Bourcart, J., 1960*a*, Carte topographique du fond de la Méditerranée occidentale, *Bull. Inst. Océanogr. Monaco*, v. 1163, 20 p.

Bourcart, J., 1960*b*, Précontinent sous-marin corse de Porto aux bouches de Bonifacio, 1/200,000ème. Précontinent sous-marin corse du nord de Porto au Cap Corse, Topogr. maps, *Musée Océanogr. Monaco.*

Bourcart, J., 1960–1963, La Méditerranée et la révolution du Pliocène, *Livre Mém. Prof. P. Fallot, Mém. h.s. Soc. Géol. France*, v. I, p. 103–116.

Bourcart, J., 1963, Morphologie de la Méditerranée occidentale, *Geol. Rundschau*, v. 53, p. 1–18.

Bourgeois, R., Cacan, C., Monnet M., and Tofani, R., 1975, *Les Fonds marins du Golfe Juan (Alpes-Maritimes), Un Modèle Sédimentologique ou un Cas Particulier? Perspectives d'Aménagement*, Thesis (3rd Cycle), Univ. Nice, 294 p.

Bourrouilh, R., 1973, La structure des iles Baléares et de leur promontoire (analyse optique par laser de documents géologiques structuraux), 23rd CIESM Congr., Athens (1972), *Bull. Geol. Soc. Greece*, v. X, p. 23–24.

Bourrouilh, R., 1974, La terminaison nord-orientale des Cordillères bétiques en Méditerranée occidentale, *2nd Earth Sci. Annu. Meeting*, Pont-à-Mousson (Nancy) (Abstracts), p. 80.

Bourrouilh, R., 1975, Le Paléozoïque supérieur des Baléares et du S.W. de la Sardaigne: une première comparaison, *3rd Earth Sci. Annu. Meeting*, Montpellier (Abstracts), p. 72.

Brandi, G. P., Dallan, L., Lazzarotto, A., Mazzanti, R., Squarci, P., Taffi, L., and Trevisan, L., 1968, Note illustrative della carta geologica d'Italia alla scala 1/100,000, foglio 119, Massa Marittima, *Serv. Geol. d'Italia*, Roma, 65 p.

Breslau, L. R., and Edgerton, H. E., 1972, The Gulf of la Spezia, Italy: A case history of seismic-sedimentologic correlation, in: *The Mediterranean Sea—A Natural Sedimentation Laboratory*, Stanley, D. J., ed., Stroudsburg, Pa.: Dowden, Hutchinson and Ross, p. 177–188.

Burollet, P. F., and Byramjee, R., 1974, Evolution géodynamique néogène de la Méditerranée occidentale, *C.R. Acad. Sci. Paris*, sér. D., v. 278, p. 1321–1324.

Calistri, M., 1974, Studi di geomorfologia e neotettonica: II. Il Pliocene fluviolacustre della conca di Barga, *Mem. Soc. Geol. Ital.*, v. 13.

Campredon, R., 1972, *Les Formations Paléogènes des Alpes-Maritimes Franco-Italiennes*, Thesis, Univ. Nice, 520 p.

Campredon, R., and Boucarut, M., 1975, *Alpes-Maritimes, Maures, Estérel*, Guides Géol. Régionaux, Paris: Masson and Co.

Campredon, R., and Porthault, B., 1971, Sur la présence de Maestrichtien dans l'Autochtone des Alpes-Maritimes franco-italiennes, *C.R. Acad. Sci. Paris*, v. 272, p. 1580–1582.

Caputo, M., Panza, G. F., and Postpichl, D., 1970, Deep structure of the Mediterranean Basin, *J. Geophys. Res.*, v. 75, p. 4919–4923.

Carey, S. W., 1955, The orocline concept on geotectonics, *Papers Proc. Roy. Soc. Tasmania*, v. 89, p. 255–288.

Carey, S. W., 1958, The tectonic approach to continental drift in: *Continental Drift: A Symposium*, Carey, S.W., ed., Hobart: Univ. Tasmania, p. 177–354.

Carmigiani, L., and Giglia, G., 1975, Aperçu sur la géologie des Apuanes, *Bull. Soc. Géol. France*, (7), v. XVII, p. 963–978.

Carter, T. G., Flanagan, J. P., Jones, C. R., Marchant, F. L., Murchison, R. R., Rebman, J. A., Sylvester, J. C., and Whitney, J. C., 1972, A new bathymetric chart and physiography of the Mediterranean Sea, in: *The Mediterranean Sea—A Natural Sedimentation Laboratory*, Stanley, D. J., ed., Stroudsburg, Pa.: Dowden, Hutchinson and Ross, p. 1–23.

Cerro, A., Gianotti, R., and Vanossi, M., 1969*a*, Nuovi dati sulla distribuzione del Carbonifero nella Liguria Occidentale, *Accad. Naz. Lincei, Rend. Sci. Fis. Math. Nat.*, v. 46, p. 449–456.

Cerro, A., Gianotti, R., and Vanossi, M., 1969b, Osservazioni preliminari sulle rocce verdi del Permo-Carbonifero brianzonese ligure, *Accad. Naz. Lincei, Rend. Sci. Fis. Math. Nat.*, v. 46, p. 737–744.

Cerro, A., Gianotti, R., and Vanossi, M., 1969c, Osservazioni sulla posizione strutturale dei massicci cristallini della Liguria Occidentale (F. Albenga–Savona), *Accad. Naz. Lincei, Rend. Sci. Fis. Math. Nat.*, v. 47, p. 337–347.

Cerro, A., Gianotti, R., and Vanossi, M., 1970, Sull'inizio dell'attività vulcanica acida nel tardo Paleozoico brianzonese ligure, *Accad. Naz. Lincei, Rend. Sci. Fis. Math. Nat.*, v. 48, p. 102–107.

Chabrier, G., 1970, Tectonique de socle d'âge alpin en Sardaigne centro-orientale, *C.R. Acad. Sci. Paris*, ser. D., v. 271, p. 1252–1255.

Chabrier, G., and Fourcade, E., 1975a, Sur le Crétacé du Nord-Ouest de la Sardaigne (présence de Valanginien à faciès pyrénéo-provençal), *C.R. Acad. Sci. Paris*, sér. D, v. 280, p. 563–566.

Chabrier, G., and Fourcade, E., 1975b, Sur le Jurassique du Nord-Ouest de la Sardaigne, *C.R. Acad. Sci. Paris*, sér. D, v. 281, p. 493–496.

Chabrier, G., and Mascle, G., 1974, Les rapports de la Provence et du domaine sarde, *C.R. Acad. Sci. Paris*, sér. D, v. 278, p. 2881–2884.

Chabrier, G., and Mascle, G., 1975, Comparaison des évolutions géologiques de la Provence et de la Sardaigne (à partir d'exemples de la région toulonnaise et de la Nurra sarde), *Rev. Géogr. Phys. Géol. Dynam.*, v. XVII, p. 121–135.

Cherchi, A., Maxia, C., and Ulzega, A., 1974, Evoluzione paleogeografica del Terziario della Sardegna, *Paleog. Terziario Sardo nell'ambito Medit. Occid., Rend. Sem. Fac. Sci. Univ. Cagliari* (1973), Suppl. v. XIII, p. 73–89.

Chessex, R., Delaloye, M., and Bordet, P., 1967, Ages "plomb total" déterminés sur des zircons des massifs des Maures et de l'Esterel, *C.R. Séances Soc. Phys. Hist. Nat. Genève*, v. 2, p. 97–106.

Choudury, M., Giese, P., and de Visintini, G., 1971, Crustal structure of the Alps: some general features from explosion seismology, *Boll. Geofis. Teor. Appl.*, v. XIII, p. 211–240.

Chumakov, I. S., 1973, Geological history of the Mediterranean at the end of the Miocene. The beginning of the Pliocene according to new data, in: *Init. Rep. Deep Sea Drilling Project*, v. XIII, Ryan, W. B. F., *et al.*, eds., Washington D.C.: U.S. Govt. Printing Office, p. 1241–1242.

Ciani, A., Gantar, C., and Morelli, C., 1960, Rilievo gravimetrico sullo zoccolo epicontinentale dei mari italiani, *Boll. Geofis. Teor. Appl.*, v. 2, p. 289–386.

Cita, M. B., 1973, Pliocene biostratigraphy and chronostratigraphy, in: *Init. Rep. Deep Sea Drilling Project*, v. XIII, Ryan, W. B. F., *et al.*, eds., Washington D.C.: U.S. Govt. Printing Office, p. 1343–1364.

Cita, M. B., 1974, I pozzi profondi perforati nel 1970 nel quadro paleogeografico e geodinamico del Mediterraneo occidentale, *Paleog. Terziario Sardo nell'ambito Medit. Occid., Rend. Sem. Fac. Sci. Univ. Cagliari* (1973), Suppl. v. XIII, p. 91–143.

Cita, M. B., and Ryan, W. B. F., 1973, Time scale and general synthesis, in: *Init. Rep. Deep Sea Drilling Project*, v. XIII, (Ryan, W. B. F., *et al.*, eds.), Washington D.C.: U.S. Govt. Printing Office, p. 1405–1415.

Clauson, G., 1973, The eustatic hypothesis and the pre-Pliocene cutting of the Rhône Valley, in: *Init. Rep. Deep Sea Drilling Project*, v. XIII, Ryan, W. B. F., *et al.*, eds., Washington D.C.: U.S. Govt. Printing Office, p. 1251–1256.

Clocchiatti, R., 1968, *Contribution à l'Etude Sédimentologique du Bassin Pliocène d'Albenga (Italie)*, Thesis (3rd Cycle), Univ. Paris.

Cocozza, T., Jacobacci, A., Nardi, R., and Salvadori, I., 1974, Schema stratigrafico–strutturale del massiccio sardo-corso e minerogenesi della Sardegna, *Mem. Soc. Geol. Ital.*, v. 13, p. 85–186.

Collette, B. J., 1969, Mediterranean oceanization. A comment, *Verhandl. Kon. Ned. Geol. Mijn. Gen.* v. 26, p. 139–142.

Compagnie Générale de Géophysique, 1972, Reconnaissance géophysique par sondages électriques dans la plaine de Marana, commune de Biguglia, Bastia: Direction Dep. Agriculture.

Conchon, O., 1970, Géologie du Quaternaire. Précisions sur la chronologie des formations fluviatiles de Corse orientale, *C.R. Acad. Sci. Paris*, sér. D., v. 270, p. 283–286.

Conchon, O., 1975, *Les formations Quaternaires de Type Continental en Corse Orientale*, Thesis, Univ. Paris, v. I, 514 p., v. II, 243 p.

Conti, S., and Andri, E., 1966, Sulla geologia dei Monti Livornesi e suoi riferimenti nel quadro più generale dell'Appennino Settentrionale, *Atti Ist. Geol., Univ. Genova*, v. 4, p. 265–456.

Cornet, C., 1965, Evolution tectonique et morphologique de la Provence depuis l'Oligocène, *Mém. Soc. Géol. France*, n.s., v. XLIV, no. 103, 252 p.

Cornet, C., 1968, Le graben médian (zone A) de la Méditerranée occidentale pourrait être pontien, *C.R. Somm. Soc. Géol. France*, v. 5, p. 149–150.

Cornet, C., 1969, La Provence de l'Oligocène à nos jours, *Rev. Géogr. Phys. Géol. Dynam.*, v. XI, p. 101–122.

Coron, S., and Guillaume, A., 1966, Nouvelles mesures de pesanteur dans les Alpes-Maritimes, *C.R. Acad. Sci. Paris*, sér. D., v. 264, p. 13–16.

Coulon, C., 1967a, Le volcanisme tertiaire de la région toulonnaise (Var), *Bull. Soc. Géol. France*, v. IX, p. 691–700.

Coulon, C., 1967b; *Le Volcanisme Basique de la Basse-Provence Occidentale (Région Toulonnaise et Massif des Maures)*, Thesis (3rd Cycle), Univ. Paris, 117 p.

Coulon, C., 1971, La genèse du massif rhyolitique du Mt. Traessu (Sardaigne septentrionale): évolution de son dynamisme volcanique, *Boll. Soc. Geol. Ital.*, v. 90, p. 73–90.

Coulon, C., 1974, Données géochronologiques, géochimiques et paléomagnétiques sur le volcanisme cénozoïque calco-alcalin de la Sardaigne nord-occidentale. Le problème de la dérive de la Sardaigne, *Paleog. Terziario Sardo nell'ambito Medit. Occid., Rend. Sem. Fac. Sci. Univ. Cagliari* (1973), Suppl. v. XIII, p. 163–169.

Coulon, C., Bobier, C., and Demant, A., 1973, Contribution du paléomagnétisme à l'étude des séries volcaniques cénozoïques et quaternaires du Logudoro et du Bosano (Sardaigne nord-occidentale). Le problème de la dérive de la Sardaigne, 23rd CIESM Congr., Athens (1972), *Bull. Geol. Soc. Greece*, v. X, p. 37–38.

Coulon, C., Demant, A., and Bobier, C., 1974, Contribution du paléomagnétisme à l'étude des séries volcaniques cénozoïques et quaternaires de Sardaigne nord-occidentale, *Tectonophysics*, v. 22, p. 59–82.

Cravatte, J., Dufaure, P., Prim, M., and Rouaix, S., 1974, Les sondages du Golfe du Lion: stratigraphie, sédimentologie, *Notes Mém. C.F.P.*, v. 11, p. 209–274.

Dallan, L., 1964, I foraminiferi miocenici dell'Isola di Pianosa, *Boll. Soc. Geol. Ital.*, v. 83, p. 167–182.

Dallan, L., and Salvatorini, G., 1967, Biostratigrafia del Pliocene della Toscana Marittima: *Giorn. Geol.*, Bologna, ser. 2, v. XXXV, p. 333–339.

Dallan-Nardi, L., and Nardi, R., 1974, Schema stratigrafico e strutturale dell'Appennino settentrionale, *Mem. Accad. Lunig. Sci. "G. Capellini,"* v. XLII, p. 1–212.

Dallan-Nardi, L., Elter, P., and Nardi, R., 1971, Considerazioni sull'arco dell'Appennino settentrionale e sulla "linea" Ancona-Anzio, *Boll. Soc. Geol. Ital.*, v. 90, p. 203–211.

Debelmas, J., 1972, A propos de quelques hypothèses récentes sur la genèse de l'arc Alpino-Apenninique, *Rev. Géogr. Phys. Géol. Dynam.*, v. XIV, p. 229–243.

Debelmas, J., 1975*a*, Les Alpes et la théorie des plaques, *Rev. Géogr. Phys. Geol. Dynam.*, v. 17, p. 195–208.

Debelmas, J., 1975*b*, Liaisons paléogéographiques entre Alpes et Apennins, *Bull. Soc. Géol. France*, (7), v. XVII, p. 1002–1012.

Debelmas, J., and Lemoine, M., 1970, The Western Alps: Paleogeography and structure, *Earth Sci. Rev.*, v. 6, p. 221–256.

De Boer, J., 1965, Paleomagnetic indications of megatectonic movements in the Tethys, *J. Geophys. Res.*, v. 70, p. 931–944.

De Booy, T., 1969, Repeated disappearance of continental crust during the geological development of the Western Mediterranean area, *Verhandl. Kon. Ned. Geol. Mijn. Gen.*, v. XXVI, p. 80–103.

Debrazzi, E., and Segre, A. G., 1959, Carta batimetrica del Mediterraneo; carta no. 1250, Mare Ligure e Tirreno settentrionale, Genoa: Ist. Idrogr. Marina.

De Bruyn, 1955, Isogam maps of Europe and North Africa, *Geophys. Prospecting*, v. 3, p. 1–14.

Decandia, F. A., and Elter, P., 1972, La "zona" ofiolitifera del Bracco nel settore compreso Fra Levanto e la Val Graveglia (Appennino ligure), *Mem. Soc. Geol. Ital.*, v. 11, p. 503–530.

De Jong, K. A., Manzoni, M., Stavenga, T., Van der Voo, R., Van Dijk, F., and Zijderveld, J. D. A., 1973, Paleomagnetic evidence for rotation of Sardinia during the Early Miocene, *Nature*, v. 243, p. 281–283.

Delcey R., 1974, Données sur deux nouvelles séries lithostratigraphiques de la zone des schistes lustrés de la Corse nord-orientale, *C.R. Acad. Sci. Paris*, sér. D, v. 279, p. 1693–1696.

Delcey, R., Limasset, J. C., and Routhier, P., 1964, Les bassins sédimentaires du nord de la Corse: essai de synthèse stratigraphique et aperçu tectonique, *Bull. Soc. Géol. France*, (7), v. 6, p. 324–333.

Delcey, R., Meunier, A., and Routhier, P., 1965, Stratigraphie, épaisseur et style tectonique des schistes lustrés de la Corse, *C.R. Acad. Sci. Paris*, v. 260, p. 6644–6647.

Delteil, J., Fenet, B., Guardia, P., Laval, F., and Polvêche, J., 1974, Les observations tectoniques dans l'Algérie nord-occidentale et l'origine du bassin nord-africain, *24th CIESM Congr.*, Monaco (Abstracts).

De Lumley, H., 1963, Les niveaux quaternaires marins des Alpes-Maritimes. Corrélations avec les industries préhistoriques, *Bull. Soc. Géol. France*, (7), v. 5, p. 562–579.

De Lumley, H., 1969, Les civilisations préhistoriques en France. Corrélations avec la chronologie du Quaternaire, *VIIIth Int. Congr. I.N.Q.U.A.*, Paris (1969), p. 151–169.

Demant, A., 1972, *Contribution à l'Etude du Volcanisme Tertiaire de la Sardaigne Nord-Occidentale. Le Cycle Andésitique Terminal (Région Mara-Romana)*, Thesis (3rd Cycle), Univ. Provence, Marseilles, 120 p.

Deriu, M., 1962, Stratigrafia, cronologia e caratteri petrochemici delle vulcaniti "oligoceniche" in Sardegna, *Mem. Soc. Geol. Ital.*, v. 3, p. 675–706.

De Roever, W. P., 1969, Genesis of the Western Mediterranean Sea: enigmatic oceanization, disruption of continental crust, or also upheaval above sea-level and later subsidence of oceanic floor? *Verhandl. Kon. Ned. Geol. Mijn. Gen.*, v. XXVI, p. 9–11.

Dewey, J. F., Pitman III, W. C., Ryan W. B. F., and Bonnin, J., 1973, Plate tectonics and the evolution of the alpine system, *Bull. Geol. Soc. Amer.*, v. 84, p. 3137–3180.

Drooger, C. W., ed., 1974, *Messinian Events in the Mediterranean*, Amsterdam: Geodynam. Sci. Rep., v. 7, 272 p.

Dubourdieu, A., 1962, Le déplacement de l'Europe occidentale: *C.R. Acad. Sci. Paris*, v. 254, p. 510–512.

Duplaix, S., 1972, Les minéraux lourds de sables de plages et de canyons sous-marins de la

Méditerranée française, in: *The Mediterranean Sea—A Natural Sedimentation Laboratory*, Stanley, D. J., ed., Stroudsburg, Pa.: Dowden, Hutchinson and Ross, p. 293–303.

Duplaix, S., and Gennesseaux, M., 1966, Preuves minéralogiques de manifestations volcaniques dès l'Eocène inférieur dans les Alpes-Maritimes, *C.R. Acad. Sci. Paris*, sér. D, v. 262, p. 2424–2426.

Duplaix, S., and Gennesseaux, M., 1968, Les minéraux lourds des formations tertiaires et quaternaires de la région niçoise (Alpes-Maritimes), *Rev. Géogr. Phys. Géol. Dynam.*, vol. 10, p. 353–374.

Durand Delga, M., 1974, La Corse, in: *Géologie de la France*, Debelmas, J., ed., Paris: Doin, v. 2, p. 465–478.

Edgerton, H. E., Giermann, G., and Leenhardt, O., 1967, Etude structurale de la baie de Monaco en sondage séismique continu, *Bull. Inst. Océanogr. Monaco*, v. 67 (1377), 6 p.

Elter, P., 1973, Lineamenti tettonici ed evolutivi dell'Appennino settentrionale, in: *Moderne Vedute sulla Geologia dell'Appennino, Accad. Naz. Lincei*, Roma (1972), v. 183, p. 97–118.

Elter, P., 1975a, Introduction à la géologie de l'Apennin septentrional, *Bull. Soc. Géol. France*, (7), v. XVII, p. 956–962.

Elter, P., 1975b, L'ensemble ligure, *Bull. Soc. Géol. France*, (7), v. XVII, p. 984–997.

Elter, P., and Pertusati, P., 1973, Considerazioni sul limite Alpi-Appennino e sulle sue relazioni con l'arco delle Alpi occidentali, *Mem. Soc. Geol. Ital.*, v. 12, p. 359–375.

Elter, G., Elter, P., Sturani, C., and Weidmann, M., 1966, Sur la prolongation du domaine ligure de l'Apennin dans le Montferrat et les Alpes et sur l'origine de la Nappe de la Simme *s.l.* des Préalpes romandes et chablaisiennes, *Arch. Sci.*, v. 19, p. 279–377.

Ewing, J., and Ewing, M., 1959, Seismic refraction measurements in the Atlantic Ocean basins, in the Mediterranean Sea, on the mid-Atlantic ridge and in the Norwegian Sea, *Bull. Geol. Soc. Amer.*, v. 70, p. 291–318.

Ewing, M., and Press, F., 1955, Seismic measurements in ocean basins, *Marine Res.*, v. 14, p. 417–422.

Fahlquist, D. A., 1963, *Seismic Measurements in the Western Mediterranean Sea*, Thesis, M.I.T., Cambridge, Mass., 173 p.

Fahlquist, D. A., and Hersey, J. B., 1969, Seismic refraction measurements in the western Mediterranean Sea, *Bull. Inst. Oceanogr. Monaco*, v. 67, 52 p.

Fanucci, F., Fierro, G., and Rehault, J., 1974a, Evoluzione quaternaria della piattaforma continentale ligure, *Reun. Spec. Soc. Geol. Ital.*, 26 X 1974, Parma, 19 p.

Fanucci, F., Fierro, G., Gennesseaux, M., Rehault, J. P., and Tabbo, S., 1974b, Indagine sismica sulla piattaforma litorale del Savonese (Mar Ligure), *Boll. Soc. Geol. Ital.*, v. 93, p. 421–435.

Fanucci, F., Fierro, G., Rehault, J. P., and Terranova, R., 1974c, Le plateau continental de la mer Ligure de Portofino à la Spezia: étude structurale et évolution plioquaternaire, *C.R. Acad. Sci. Paris*, sér. D, v. 279, p. 1151–1154.

Faure-Muret, A., and Choubert, G., 1975, Proposition d'un nouveau modèle tectonique pour la Méditerranée occidentale, *C.R. Acad. Sci. Paris*, sér. D, v. 280, p. 1947–1950.

Federici, P. R., 1973, La tettonica recente dell'Appennino: I. Il bacino villafranchiano di Sarzana e il suo significato nel quadro dei movimenti distensivi a nord-ovest delle Alpi Apuane, *Boll. Soc. Geol. Ital.*, v. 92, p. 287–301.

Feugueur, L., and Le Calvez, Y., 1962, Mise en évidence de mouvements tectoniques post-miocènes dans les travaux de percement du tunnel ferroviaire de Monaco, *C.R. Acad. Sci. Paris*, v. 254, p. 3113–3115.

Fierro, G., Piacentino, G. B., and Ricciardi, G., 1972, Répartition des sédiments marins près de Livourne (Mer Ligure), in: *The Mediterranean Sea—A Natural Sedimentation Laboratory*, Stanley, D. J., ed., Stroudsburg Pa.: Dowden, Hutchinson and Ross, p. 321–332.

Fierro, G., Gennesseaux, M., and Rehault, J. P., 1973, Caractères structuraux et sédimentaires du plateau continental de Nice à Gênes (Méditerranée nord-occidentale), *Bull. B.R.G.M.*, (2), Sect. IV, v. 4, p. 193–208.

Finetti, I., and Morelli, C., 1973*a*, Wide scale digital seismic exploration of the Mediterranean Sea, 23rd CIESM Congr., Athens (1972), *Bull. Geol. Soc. Greece*, v. X, p. 65–66.

Finetti, I., and Morelli, C., 1973*b*, Geophysical exploration of the Mediterranean Sea, *Boll. Geofis. Teor. Appl.*, v. 15, p. 263–342.

Finetti, I., and Morelli, C., 1974, Esplorazione geofisica dell'area mediterranea circostante il blocco sardo-corso, *Paleog. Terziario Sardo nell'ambito Medit. Occid., Rend. Sem. Fac. Sci. Univ. Cagliari* (1973), Suppl. v. XIII, p. 213–238.

Finetti, I., Morelli, C., and Zarudsky, E., 1970, Continuous seismic profiling in the Tyrrhenian Sea, *Boll. Geofis. Teor. Appl.*, v. 12, p. 311–345.

Fredj, G., 1974, *Essai de Stockage et d'Exploitation des Données en Ecologie Marine. Application à l'Etude Biogéographique du Benthos Méditerranéen*, Thesis, Univ. Nice, p. 1–176, A1-A164.

Gabin, R., 1970, *Etude des Structures Géologiques Sous-Marines de la Mer Nord-Tyrrhénienne et du Canal de Corse*, Thesis, Univ. Paris, 120 p.

Gabin, R., 1972, Résultats d'une étude de sismique réflexion dans le canal de Corse, et de sondeur de vase dans le bassin toscan, *Marine Geol.*, v. 13, p. 267–286.

Galli, M., Bezzi, A., Piccardo, P. G., Cortesogno, L., and Pedemonte, G. M., 1972, Le ofioliti dell'Appennino ligure, un fragmento di crosta-mantello "oceanici" dell'antica Tetide, *Mem. Soc. Geol. Ital.*, v. 11, p. 467–502.

Gelati, R., and Pasquare, G., 1970, Interpretazione geologica del limite Alpi-Appennini in Liguria, *Riv. Ital. Paleontol.*, v. 76, p. 513–578.

Gennesseaux, M., 1962*a*, Travaux du Laboratoire de Géologie sous-marine concernant les grands carottages effectués sur le précontinent de la région niçoise, *Océanogr. Géol. et Géophys. de la Méditerranée Occid., Coll. Int. C.N.R.S.*, Villefranche, p. 177–181.

Gennesseaux, M., 1962*b*, Les canyons de la baie des Anges, leur remplissage sédimentaire et leur rôle dans la sédimentation profonde, *C.R. Acad. Sci. Paris*, v. 254, p. 2409–2411.

Gennesseaux, M., 1963, Structure et morphologie de la pente continentale de la région niçoise, *Rapport C.I.E.S.M.*, v. 17, p. 991–998.

Gennesseaux, M., 1966, Prospection photographique des canyons sous-marins du Var et du Paillon (Alpes-Maritimes) au moyen de la troika, *Rev. Géogr. Phys. Géol. Dynam.*, v. 8, p. 3–38.

Gennesseaux, M., 1972, La structure du plateau occidental des Bouches de Bonifacio (Corse), *C.R. Acad. Sci. Paris*, sér. D, v. 275, p. 2295–2297.

Gennesseaux, M., 1974, Relations structurales entre la Corse cristalline et sa marge sous-marine occidentale, *2nd Earth Sci. Annu. Meeting*, Pont-à-Mousson (Nancy) (Abstracts), p. 184.

Gennesseaux, M., and Glaçon, G., 1972, Essai de stratigraphie du Pliocène sous-marin en Méditerranée nord-occidentale, *C.R. Acad. Sci. Paris*, sér. D., v. 275, p. 1863–1866.

Gennesseaux, M., and Rehault, J. P., 1975, La marge continentale corse, *Bull. Soc. Géol. France*, (7), v. XVII, p. 505–518.

Gennesseaux, M., and Thommeret, Y., 1968, Datation par le radiocarbone de quelques sédiments sous-marins de la région niçoise, *Rev. Géogr. Phys. Géol. Dynam.*, v. 10, p. 375–382.

Gennesseaux, M., Auzende, J. M., Olivet, J. L., and Bayer, R., 1974, Les orientations structurales et magnétiques sous-marines au Sud de la Corse et la dérive corso-sarde, *C.R. Acad. Sci. Paris*, sér. D, v. 278, p. 2003–2006.

Gennesseaux, M., Glaçon, G., Rehault, J. P., and Fierro, G., 1975, Les affleurements sédimentaires néogènes dans la vallée sous-marine d'Asinara (Sardaigne septentrionale), *3rd Earth Sci. Annu. Meeting*, Montpellier (Abstracts), p. 163.

Gerber, J. P., De Lumley, H., Miskovsky, J. C., and Renault-Miskovsky, J., 1974, Paléoclimatologie du Midi méditerranéen pendant le Quaternaire d'après les études stratigraphiques, sédimentologiques, géochimiques, paléontologiques et palynologiques des remplissages de grottes et abris sous-roches, *2nd Earth Sci. Annu. Meeting*, Pont-à-Mousson (Nancy) (Abstracts), p. 185.

Gèze, B., 1960–1963, Caractères structuraux de l'arc de Nice, *Livre Mém. Prof. P. Fallot, Mém. h.s. Soc. Géol. France*, v. II, p. 289–300.

Ghelardoni, R., Giannini, E., and Nardi, R., 1968, Ricostruzione paleogeografica dei bacini neogenici e quaternari nella bassa valle dell'Arno sulla base dei sondaggi e dei rilievi sismici, *Mem. Soc. Geol. Ital.*, v. 7, p. 91–106.

Gianmarino, S., Nosengo, S., and Vannucci, G., 1959, Risultanze geologico-paleontologiche sul conglomerato di Portofino (Liguria orientale), *Atti Ist. Geol., Univ. Genova*, v. 7, p. 305–363.

Giannini, E., 1955, Osservazioni sulla geologia del Bacino della Fine e delle colline fra Rosignano Marittimo e la valle della Cecina, *Boll. Soc. Geol. Ital.*, v. 74, p. 219–296.

Giannini, E., 1960, Studio di alcune sezioni stratigrafiche nel Miocene superiore delle Colline Livornesi ed osservazioni sui caratteri e sui limiti del Messiniano in Toscana, *Giorn. Geol.*, v. 28, p. 35–58.

Giannini, E., 1962, Geologia del bacino della Fine (province di Pisa e Livorno), *Boll. Soc. Geol. Ital.*, v. 81, p. 101–224.

Giannini, E., and Tongiorgi, M., 1958, Osservazioni sulla tettonica neogenica della Toscana Marittima, *Boll. Soc. Geol. Ital.*, v. 77, p. 147–170.

Giannini, E., and Tongiorgi, M., 1962, Les phases tectoniques néogènes de l'orogenèse alpine dans l'Apennin septentrional, *Bull. Soc. Géol. France*, (7), v. 4, p. 682–690.

Giermann, G., 1969, Le précontinent entre le cap Ferrat et le cap Martin, *Bull. Inst. Océanogr. Monaco*, v. 68 (1392).

Giese, P., and Morelli, C., 1973a, La struttura della crosta terrestre in Italia, in: *Moderne Vedute sulla Geologia dell'Appennino*, Accad. Naz. Lincei, Roma (1972), v. 183, p. 317–366.

Giese, P., and Morelli, C., 1973b, Crustal structure of Italy. Some general features from explosion seismology, 23rd CIESM Congr., Athens (1972), *Bull. Geol. Soc. Greece*, v. X, p. 94–98.

Giese, P., Günther, K., and Reutter, K. J., 1970, Vergleichende geologische und geophysikalische Betrachtungen der Westalpen und des Nordapennins, *Z. Deutsch. Geol. Ges.*, v. 120, p. 151–195.

Giglia, G., 1974, L'insieme Corsica–Sardegna e i suoi rapporti con l'Appennino Settentrionale: rassegna di dati cronologici e strutturali, *Paleog. Terziario Sardo nell'ambito Medit. Occid., Rend. Sem. Fac. Sci. Univ. Cagliari* (1973), Suppl. v. XIII, p. 245–275.

Giglia, G., and Radicati di Brozolo, F., 1970, K/Ar age of metamorphism in the Apuane Alps (Northern Tuscany), *Boll. Soc. Geol. Ital.*, v. 89, p. 485–497.

Ginsburg, L., 1960, Etude géologique de la bordure subalpine, à l'Ouest de la basse vallée du Var, *Bull. Serv. Carte Géol. France*, v. 57, p. 1–38.

Ginsburg, L., 1970, Carte géologique détaillée de la France, Grasse-Cannes, 1/50,000.

Giraud, J. P., 1975, *Une Intrusion Calco-Alcaline Type: L'Estérellite; Position dans le Contexte Volcanologique Tertiaire Régional*, Thesis (3rd Cycle), Univ. Nice, 184 p.

Glaçon, G., and Rehault, J. P., 1973, Le Messinien marin de la pente continentale ligure (−1750 m) (Italie), *C.R. Acad. Sci. Paris*, sér. D, v. 277, p. 625–628.

Glangeaud, L., 1962, Paléogéographie dynamique de la Méditerranée et de ses bordures.

Le rôle des phases ponto-plio-quaternaires, *Océanogr. Géol. et Géophys. de la Méditerranée Occid., Coll. Int. C.N.R.S.*, Villefranche, p. 125–165.

Glangeaud, L., 1966, Les grands ensembles structuraux de la Méditerranée occidentale d'après les données de Géomède 1, *C.R. Acad. Sci. Paris*, v. 262, p. 2405–2408.

Glangeaud, L., 1967, Epirogenèses ponto-plio-quaternaires de la marge continentale francoitalienne du Rhône à Gênes, *Bull. Soc. Géol. France*, v. IX, p. 426–449.

Glangeaud, L., 1968, Les méthodes de la Géodynamique et leurs applications aux structures de la Méditerranée occidentale, *Rev. Géogr. Phys. Géol. Dynam.*, v. X, p. 83–136.

Glangeaud, L., and Olive, P., 1970, Les structures mégamétriques de la Méditerranée. Evolution de la Mésogée de Gibraltar à l'Italie, *C.R. Acad. Sci. Paris*, sér. D, v. 271, p. 1161–1166.

Glangeaud, L., and Rehault, J. P., 1968, Evolution ponto-plio-quaternaire du golfe de Gênes, *C.R. Acad. Sci. Paris*, sér. D, v. 266, p. 60–63.

Glangeaud, L., Agaraté, C., Bellaiche, G., and Pautot, G., 1965a, Morphotectonique de la terminaison sous-marine orientale des Maures et de l'Estérel, *C.R. Acad. Sci. Paris*, sér. D, v. 261, p. 4795–4798.

Glangeaud, L., Schlich, R., Pautot, G., Bellaiche, G., Patriat, P., and Ronfard, M., 1965b, Morphologie, tectonophysique et évolution géodynamique de la bordure sous-marine des Maures et de l'Estérel. Relations avec les régions voisines, *Bull. Soc. Géol. France*, (7), v. VII, p. 998–1009.

Glangeaud, L., Alinat, J., Polvêche, J., Guillaume, A., and Leenhardt, O., 1966, Grandes structures de la mer ligure, leur évolution et leurs relations avec les chaînes continentales, *Bull. Soc. Géol. France*, (7), v. VIII, p. 921–937.

Glangeaud, L., Alinat, J., Agaraté, C., Leenhardt, O., and Pautot, G., 1967, Les phénomènes ponto-plio-quaternaires dans la Méditerranée occidentale d'après les données de Géomède 1, *C.R. Acad. Sci. Paris*, v. 264, p. 208–211.

Gohau, G., and Veslin, J., 1960, A propos de l'âge de la série du Miocène de Vence, *Bull. Soc. Géol. France*, (7), v. 2, p. 764–767.

Gonnard, R., Letouzey, J., Biju-Duval, B., and Montadert, L., 1975, Apports de la sismique réflection aux problèmes du volcanisme en Méditerranée et à l'interprétation des données magnétiques, *3rd Earth Sci. Annu. Meeting*, Montpellier (Abstracts), p. 171.

Grandjacquet, C., and Haccard, D., 1972, Tectoniques superposées et orientation des accidents principaux dans les Alpes méridionales et l'Apennin, *C.R. Acad. Sci. Paris*, sér. D, v. 274, p. 2845–2847.

Grandjacquet, C., Haccard, D., and Lorenz, C., 1972a, Sur l'importance de la phase tectonique aquitanienne dans l'Apennin et les Alpes occidentales, *C.R. Acad. Sci. Paris*, sér. D, v. 275, p. 807–810.

Grandjacquet, C., Haccard, D., and Lorenz, C., 1972b, Essai de tableau synthétique des principaux évènements affectant les domaines alpin et apennin à partir du Trias, *C.R. Somm. Soc. Géol. France*, v. 4, p. 158–163.

Gueirard, S., 1957, Description pétrographique et zonéographique des schistes cristallins des Maures (Var), *Trav. Lab. Géol. Fac. Sci. Marseille*, v. 6, p. 71–274.

Guillaume, A., 1969, Contribution à l'étude géologique des Alpes liguro-piémontaises, *Doc. Lab. Géol. Fac. Sci. Lyon*, no. 30.

Günther, K., 1973, Ergebnisse meeresgeologischer und geophysikalischer Untersuchungen in der nördlichen Tyrrhenis und im Ligurischen Meer und ihre Konsequenzen für die Deutung der Orogenese des Nordapennins, *N. Jb. Geol. Paläont. Abh.*, v. 142, p. 191–264, 265–296.

Haccard, D., 1971, Nouvelles données stratigraphiques sur la zone de Sestri–Voltaggio (Genovesato, Italie), *C.R. Acad. Sci. Paris*, sér. D, v. 272, p. 1063–1066.

Haccard, D., 1975, Les grands ensembles structuraux de la Ligurie entre Nice et Sestri–Levante, *Bull. Soc. Geol. France*, (7), v. XVII, p. 919–921.

Haccard, D., and Lemoine, M., 1970, Sur la stratigraphie et les analogies des formations sédimentaires associées aux ophiolites dans la zone piémontaise des Alpes ligures (zones de Sestri–Voltaggio et de Montenotte) et des Alpes cottiennes (zone du Gondran, Queyras, Haute-Ubaye), *C.R. Somm. Soc. Géol. France*, v. 6, p. 209.

Haccard, D., Lorenz, C., and Grandjacquet, C., 1972, Essai sur l'évolution tectogénétique de la liaison Alpes-Apennins (de la Ligurie à la Calabre), *Mem. Soc. Geol. Ital.*, v. 11, p. 309–341.

Hersey, J. B., 1965, Sedimentary basins of the Mediterranean Sea, in: *Submarine Geology and Geophysics*, Whittard, W. F., and Bradshaw, R., eds., Proc. 17th Symp. Colston Res. Soc., London: Butterworths, p. 75–91.

Hirn, A., 1975, Structure profonde de la Corse et du bassin provençal, *3rd Earth Sci. Annu. Meeting*, Montpellier (Abstracts), p. 191.

Hsü, K. J., 1971, Origin of the Alps and western Mediterranean, *Nature*, v. 233, p. 44–48.

Hsü, K. J., 1972, When the Mediterranean dried up, *Sci. Amer.*, v. 227, p. 27–36.

Hsü, K. J., and Ryan, W. B. F., 1973, Summary of the evidence for extensional and compressional tectonics in the Mediterranean, in: *Init. Rep. Deep Sea Drilling Project*, v. XIII, Ryan, W. B. F., *et al.*, eds., Washington, D.C.: U.S. Govt. Print. Office, p. 1011–1019.

Hsü, K. J., Ryan, W. B. F., and Cita, M. B., 1973a, Late Miocene dessication of the Mediterranean, *Nature*, v. 242, p. 240–244.

Hsü, K. J., Cita, M. B., and Ryan, W. B. F., 1973b, The origin of the Mediterranean evaporites, in: *Init. Rep. Deep Sea Drilling Project*, v. XIII, Ryan, W. B. F., *et al.*, eds., Washington D.C.: U.S. Govt. Print. Office, p. 1203–1231.

Hsü, K. J., Ryan, W. B. F., Cocozza, T., and Magnier, P., 1973c, Comparative petrography of three suites of basement rocks from the Western Mediterranean, in: *Init. Rep. Deep Sea Drilling Project*, v. XIII, Ryan, W. B. F., *et al.*, eds., Washington D.C.: U.S. Govt. Print. Office, p. 775–780.

Iaworsky, G., 1971, La présence de six cycles sédimentaires associés à six terrasses de l'estuaire du Var dans la stratigraphie des limons rouges à Nice, *C.R. Acad. Sci. Paris*, sér. D, v. 273, p. 1775–1778.

Iaworsky, G., 1973, L'âge des niveaux marins interglaciaires et interstadiaires dans les Alpes-Maritimes et la stratigraphie de leur couverture continentale, *Le Quaternaire. Géodynamique, stratigraphie et Environnement. Travaux français récents (9th Int. Congr. I.N.Q.U.A.)*, p. 160–163.

Iaworsky, G., 1974, Sur la part de la tectonique et de l'eustatisme dans les discordances entre les dépôts sédimentaires à la limite du Pliocène et du Quaternaire dans les Alpes-Maritimes, *24th CIESM Congr.*, Monaco (Abstracts).

Iaworsky, G., and Le Calvez, Y., 1974, Découverte du Calabrien froid transgressif sur du Pliocène supérieur dans les A.M. à Nice, *C.R. Acad. Sci. Paris*, sér. D, v. 278, p. 2007–2010.

Irr, F., 1971a, Sur l'intérêt stratigraphique des Foraminifères du Pliocène des Alpes-Maritimes, *C.R. Acad. Sci. Paris*, sér. D, v. 272, p. 2281–2284.

Irr, F., 1971b, Livret-guide de l'excursion dans la région de Nice (Alpes Maritimes, France), *Vth Int. Congr. Medit. Neogene*, 43 p.

Irr, F., 1973, Sur le caractère synsédimentaire des phénomènes volcaniques dans la molasse du bassin de Vence (Alpes Maritimes), *C.R. Acad. Sci. Paris*, sér. D., v. 277, p. 1275–1277.

Irr, F., 1975, Evolution de la bordure du bassin méditerranéen nord-occidental au Pliocène: nouvelles données biostratigraphiques sur le littoral franco-ligure et leurs implications tectoniques, *Bull. Soc. Géol. France*, (7), v. XVII, p. 945–955.

Irr, F., and Dardeau, G., 1976, Mise en evidence d'une série fini-miocène dans l'Arc de Nice, *C.R. Acad. Sci. Paris*, sér. D, v. 283, p. 749–752.

Issel, A., 1892, *Liguria Geologica e Preistorica*, Genoa: 3 v., 816 p.

Ivaldi, J.-P., 1971, Le phénomène de thermoluminescence appliqué á l'étude du flysch "Grès d'Annot" (France). Conséquences paléogéographiques, *Rev. Géogr. Phys. Géol. Dynam.*, (2), v. XIII, p. 521–526.

Ivaldi, J.-P., 1973, *Contribution de la Thermoluminescence à l'Etude des Séries "Grès d'Annot." Origines du Matériel Détritique. Conséquences Paléogéographiques*, Thesis (3rd Cycle), Univ. Nice, 163 p.

Klemme, H. D., 1958, Regional geology of circum-Mediterranean region, *Bull. Amer. Assoc. Petrol. Geol.*, v. 42, p. 477–512.

Kremer, Y., 1974a, *Littoral et Précontinent de Menton. Morphologie, Sédimentologie et Structure*, Thesis (3rd Cycle), Univ. Nice, 160 p.

Kremer, Y., 1974b, Morphologie et structure de la plate-forme continentale de Menton en sondage sismique continu, *24th CIESM Congr.*, Monaco (Abstracts).

Kuenen, P. H., 1959, Age d'un bassin méditerranéen, *La Topographie et la Géologie des Profondeurs Océaniques, Coll. Int. C.N.R.S.*, Nice-Villefranche, p. 157–162.

Kuenen, P. H., Faure-Muret, A., Lanteaume, M., and Fallot, P., 1957, Observations sur les flyschs des Alpes-Maritimes françaises et italiennes, *Bull. Soc. Géol. France*, (6), v. VII, p. 11–26.

Kuhn, R., and Hsü, K. J., 1974, Bromine content of Mediterranean halite, *Geology*, v. 2, p. 213–216.

Labrouste, Y. H., Baltenberger, P., Perrier, G., and Recq, M., 1968, Courbes d'égale profondeur de la discontinuité de Mohorovicic dans le Sud-Est de la France, *C.R. Acad. Sci. Paris*, sér. D, v. 266, p. 663–665.

Lacazedieu, A., and Parsy, A., 1970, Etude géologique de la Balagne sédimentaire (Corse), *D.E.A.*, Univ. Paris, p. 140.

Lanteaume, M., 1962, Considérations paléogéographiques sur la patrie supposée des nappes de flysch à Helminthoïdes des Alpes et des Apennins, *Bull. Soc. Géol. France*, (7), v. 4, p. 627–643.

Lanteaume, M., 1968, Contribution à l'étude géologique des Alpes-maritimes franco-italiennes, *Mém. Serv. Carte Géol. dét. France*, Paris: Imp. Natl., p. 1–405.

Lanteaume, M., and Haccard, D., 1961, Stratigraphie et variations de faciès des formations constitutives de la nappe du flysch à Helminthoïdes des Alpes–Maritimes franco-italiennes, *Boll. Soc. Geol. Ital.*, v. 80, p. 3–15.

Laubscher, H. P., 1971a, Das Alpen-Dinariden-Problem und die Palinspastik der südlichen Tethys, *Geol. Rundschau*, v. 60, p. 813–833.

Laubscher, H. P., 1971b, The large scale kinematics of the Western Alps and the Northern Apennines and its Palinspastic implications, *Amer. J. Sci.* v. 271, p. 193–226.

Laubscher, H. P., 1975, Plate boundaries and microplates in alpine history, *Amer. J. Sci.*, v. 275, p. 865–876.

Lauro, C., 1973, Magmatismo terziario e recente, plutonico e vulcanico, della Italia peninsulare e moderne vedute sulla geologia dell'Appennino, in: *Moderne Vedute sulla Geologia dell'Appennino, Accad. Naz. Lincei*, Roma (1972), v. 183, p. 251–271.

Lazzarotto, A., 1967, Geologia della zona compresa fra l'alta valle del fiume Cornia ed il torrente Pavone (Prov. di Pisa e Grosseto), Mem. Soc. Geol. Ital., v. 6, p. 151–197.

Lazzarotto, A., and Mazzanti, R., 1965, Sulle caratteristiche di alcune strutture tettoniche frequenti nelle formazioni neoautoctone delle alte valli dei fiumi Cecina, Cornia e Milia, *Boll. Soc. Geol. Ital.*, v. 84, p. 177–196.

Lazzarotto, A., Mazzanti, R., and Salvatorini, G., 1964, Stratigrafia neogenica toscana:

Esame geologico et micropaleontologico di alcune sezioni del complesso neoautoctono delle valli di Cornia e di Milia (Province di Pisa e Grosseto), *Boll. Soc. Geol. Ital.*, v. 83, p. 401–460.

Le Borgne, E., Le Mouël, J. L., and Le Pichon, X., 1971, Aeromagnetic survey of South Western Europe, *Earth Planet. Sci. Lett.*, v. 12, p. 287–299.

Le Calvez, Y., and Vernet, J., 1966, Présence de l'Aquitanien fossilifère dans les formations volcaniques du cap d'Aîl, *C.R. Acad. Sci. Paris*, v. 262, p. 841–842.

Leenhardt, O., 1968, Le problème des dômes de sel de la Méditerranée occidentale: étude géophysique d'une colline abyssale, la structure A, *Bull. Soc. Géol. France*, (7), v. X, p. 497–509.

Leenhardt, O., 1970, Sondages sismiques continus en Méditerranée occidentale (enregistrement, analyses, interprétation), *Mém. Inst. Océanogr. Monaco*, v. 1, 120 p.

Leenhardt, O., Pierrot, S., Rebuffatti, A., and Sabatier, R., 1970, Sub-Sea Floor structure south of France, *Nature*, v. 226, p. 930–932.

Le Mouël, J., and Le Borgne, E., 1970, Les anomalies magnétiques du Sud-Est de la France et de la Méditerranée occidentale, *C.R. Acad. Sci. Paris*, sér. D, v. 271, p. 1348–1350.

Le Pichon, X., Pautot, G., Auzende, J. M., and Olivet, J. L., 1971, La Méditerranée occidentale depuis l'Oligocène: schéma d'évolution, *Earth Planet. Sci. Lett.*, v. 13, p. 145–152.

Lesquer, A., 1975, Les anomalies gravimétriques du Nord de la Corse, *C.R. Acad. Sci. Paris*, sér. D., v. 280, p. 2297–2300.

Lloyd, R. M., and Hsü, K. J., 1972, Preliminary isotopic investigations of samples from Deep Sea Drilling Cruise to the Mediterranean, in: *The Mediterranean Sea—A Natural Sedimentation Laboratory*, Stanley, D. J., ed., Stroudsburg, Pa.: Dowden, Hutchinson and Ross, p. 681–686.

Lorenz, C., 1969, Contribution à l'étude stratigraphique de l'Oligocène et du Miocène inférieur des confins liguro-piémontais (Italie), *Atti Ist. Geol., Univ. Genova*, vol. VI, p. 255–888.

Lorenz, C., 1971, Observations sur la stratigraphie du Pliocène ligure; la phase tectonique du Pliocène moyen, *C.R. Somm. Soc. Géol. France*, v. 8, p. 441–445.

Lowrie, W., and Alvarez, W., 1974a, Rotation of the Italian peninsula, *Nature*, v. 251, p. 285–288.

Lowrie, W., and Alvarez, W., 1974b, Paleomagnetism of the scaglia rossa limestone in the Northern Apennines, *24th CIESM Congr.*, Monaco (Abstracts).

Lowrie, W., and Alvarez, W., 1975, Paleomagnetic evidence for rotation of the Italian Peninsula, *J. Geophys. Res.*, v. 80, p. 1579–1592.

Lutaud L., 1924, *Etude Tectonique et Morphologie de la Provence Cristalline*, Thesis, *Rev. Geogr.*, v. 12, 270 p.

MacKenzie, D. P., 1970, Plate tectonics of the Mediterranean region, *Nature*, v. 226, p. 239–243.

Magné, J., Orszag-Sperber, F., and Pilot, M. D., 1975a, La formation d'Aléria: le problème de la limite Miocène-Pliocène en plaine orientale corse, *C.R. Acad. Sci. Paris*, sér. D, v. 280, p. 247–250.

Magné, J., Orszag-Sperber, F., and Pilot, M. D., 1975b, Nouvelles données sur le Pliocène corse: le problème de la limite Miocène-Pliocène, *VIth Congr. Medit. Neogene Strat.*, Bratislava (in press).

Malaroda, R., 1970, Carta geologica del Massicio dell'Argentera alla scala 1/50,000, *Mem. Soc. Geol. Ital.*, v. 9, p. 557–663.

Maluski, H., 1968, *Etude Tectonique, Microtectonique et Géochronologique de la Zone Méridionale du Massif des Maures (Var)*, Thesis (3rd cycle), Univ. Montpellier.

Maluski, I., 1971, Etude 87Rb/87Sr des minéraux des gneiss des Bormes (Maures, France), *C.R. Acad. Sci. Paris*, sér. D, v. 273, p. 1470–1473.

Maluski, H., and Allegre, C., 1970, Problème de la datation par le couple 87Rb/87Sr des socles gneissiques: exemple des gneiss de Bormes (massif hercynien des Maures), *C.R. Acad. Sci. Paris*, sér. D, v. 270, p. 18–21.

Maluski, H., Mattauer, M., and Matte, P., 1973, Sur la présence de décrochements alpins en Corse, *C.R. Acad. Sci. Paris*, sér. D, v. 276, p. 709–712.

Manzoni, M., 1974, Un'interpretazione dei dati paleomagnetici del Terziario della Sardegna ed alcuni nuovi resultati, *Paleog. Terziario Sardo nell'ambito Medit. Occid., Rend. Sem. Fac. Sci. Univ. Cagliari* (1973) Suppl. v. XIII, p. 283–295.

Manzoni, M., 1975, Paleomagnetic data from Tertiary volcanics of the Campidano and associated grabens, Sardinia, *Earth Planet. Sci. Lett.*, v. 27, p. 275–282.

Manzoni, M., de Jong, K. A., Carobene, L., and Pasini, G., 1972, Paleomagnetism of some basalts from Sardinia, *Giorn. Geol.*, Bologna, v. 38, p. 5–9.

Marchand, J. P., and Ceccaldi, X., 1966, Etude par sondages séismiques des prolongements vers l'Est et l'Ouest des structures de la Baie des Anges, *Rec. Trav. Stat. Marit. Endoume*.

Mascle, J. R., 1968, *Contribution à l'Etude de la Marge Continentale et de la Plaine Abyssale au Large de Toulon*, Thesis (3rd Cycle), Univ. Paris.

Mascle, G., and Mascle, J. R., 1972, Aspects of some evaporitic structures in western Mediterranean Sea, *Bull. Amer. Assoc. Petrol. Geol.* v., 56, p. 2260–2267.

Mascle, J., and Pautot, G., 1967, Nouvelles données sur la géologie sous-marine du canyon de Toulon, *C.R. Somm. Soc. Géol. France*, v. 8, p. 361–363.

Masini, R., 1933, I due laghi pliocenici di Barga e di Castelnuovo di Garfagnana e i loro rapporti con le direttrici di frattura e le aree sismiche, *Riv. Accad. Lucchese Sci. Lettere Arti*, v. 4, p. 152–175.

Mattauer, M., 1968, Le problème des déplacements de la Péninsule ibérique par rapport au bloc européen et ses conséquences possibles sur la genèse des fonds océaniques de la Méditerranée occidentale, *Sur les Méthodes de Sismique et les Cartes de Géologie Sous-Marine, Coll. C.N.R.S.*, Villefranche.

Mattauer, M., 1974, Une nouvelle hypothèse sur la position de la microplaque corso-sarde avant sa rotation d'âge cénozoïque, *Paleog. Terziario Sardo nell'ambito Medit. Occid., Rend. Sem. Fac. Sci. Univ. Cagliari* (1973), Suppl. v. XIII, p. 297–300.

Mattauer, M., and Proust, F., 1975a, Arguments microtectoniques en faveur de l'origine ultra des nappes de Balagne et de Saint-Florent (Corse), *3rd Earth Sci. Ann. Meeting*, Montpellier (Abstracts).

Mattauer, M., and Proust, F., 1975b, Données nouvelles sur l'évolution structurale de la Corse alpine, *C.R. Acad. Sci. Paris*, sér. D, v. 281, p. 1681–1684.

Mauffret, A., 1968, *Etude des Profils Sismiques Obtenus au cours de la Campagne Geomède 1 au Large des Baléares et en Mer Ligure*, Thesis (3rd Cycle), Univ. Paris.

Mauffret, A., Fail, J. P., Montadert, L., Sancho, J., and Winnock, E., 1973, Northwestern Mediterranean sedimentary basin from seismic reflection profiles, *Bull. Amer. Assoc. Petrol. Geol.*, v. 57, p. 2245–2262.

Mazzanti, R., 1961, Geologia della zona di Mentaione tra le valli dell'Era e dell'Elsa (Toscana), *Boll. Soc. Geol. Ital.*, v. 80, p. 1–90.

Mazzanti, R., 1966, Geologia della zona di Pomarance-Larderello (Provincia di Pisa), *Mem. Soc. Geol. Ital.*, v. V, p. 105–138.

Mazzanti, R., Squarci, P., and Taffi, L., 1963, Geologia della zona di Montecatini Val di Cecina in provincia di Pisa, *Boll. Soc. Geol. Ital.*, v. 82, p. 1–68.

Mechler, P., and Rocard, Y., 1966, Relief de la couche de Mohorovicic sous la Provence, *C.R. Acad. Sci. Paris*, sér. D, v. 262, p. 241–244.

Menard, H. W., 1967, Transitional types of crust under small ocean basins, *J. Geophys. Res.*, v. 72, p. 3061–3073.

Merla, G., and Bortolotti, V., 1967, Note illustrative della carta geologica d'Italia, foglio 113, Castelfiorentino, *Serv. Geol. Italia*, Roma, 62 p.

Montadert, L., and Letouzey, J., 1975, Structure de la marge méditerranéenne d'après les études de détail effectuées pour le leg 42 du Glomar Challenger, *Bull. Soc. Géol. France*, (7), v. XVII (Abstracts), p. 519–520.

Montadert, L., Sancho, J., Fail, J. P., Debyser, J., and Winnock, E., 1970, De l'âge tertiaire de la série salifère responsable des structures diapiriques en Méditerranée occidentale (Nord-Est des Baléares), *C.R. Acad. Sci. Paris*, sér. D, v. 271, p. 812–815.

Morelli, C., 1973, La gravimetria dell'area italiana, in: *Moderne Vedute sulla Geologia dell'Appennino, Accad. Naz. Lincei*, Roma (1972), v. 183, p. 297–315.

Moullade, M., and Porthault, B., 1970, Sur l'âge précis et la signification des grès et conglomérats crétacés de la vallée du Toulourenc (Vaucluse). Répercussions de la phase orogénique "autrichienne" dans le Sud-Est de la France, *Géol. Alpine*, v. 46, p. 141–150.

Muraour, P., 1970, Considérations sur la genèse de la Méditerranée occidentale et du Golfe de Gascogne (Atlantique), *Tectonophysics*, v. 10, p. 663–677.

Muraour, P., and Genesseaux, M., 1965, Quelques remarques à la suite d'une étude de sismique réfraction sur la pente continentale niçoise, *C.R. Acad. Sci. Paris*, sér. D, v. 260, p. 227–230.

Muraour, P., and Groubert, E., 1963, Etude sismique par réfraction sur le précontinent italien entre La Spezia et l'ile de Capraia, *Bull. Inst. Océanogr. Monaco*, v. 61, 16 p.

Muraour, P., Ducrot, J., and Marchand, J. P., 1965a, Etude seismique sur la pente continentale provençale (région golfe Juan, baie des Anges, baie de Beaulieu), *C.R. Somm. Soc. Géol. France*, v. 7, p. 228–229.

Muraour, P., Ceccaldi, X., Ducrot, J., and Marchand, J. P., 1965b, Sur la structure profonde entre la Provence et la Corse, *C.R. Somm. Soc. Géol. France*, v. 9, p. 314–315.

Muraour, P., Ducrot, J., Gennesseaux, M., Groubert, E., and Marchand, J. P., 1965c, Etude sismique par réfraction de la pente continentale niçoise, *Bull. Inst. Océanogr. Monaco*, v. 65, 44 p.

Muraour, P., Ceccaldi, X., Ducrot, J., and Marchand, J. P., 1966, Remarques sur la structure profonde du précontinent de la région de Calvi (Corse), *C.R. Acad. Sci. Paris*, sér. D, v. 262, p. 17–19.

Muraour, P., Ducrot, J., Marchand, J. P., and Ceccaldi, X., 1969, Contribution à l'étude de la structure profonde entre la Provence et la Corse, *Rec. Trav. Stat. Marit. Endoume*, v. 46, p. 269–284.

Nairn, A. E. M., and Westphal, M., 1967a, Sur l'aimantation de rhyolites permiennes en Corse, *C.R. Acad. Sci. Paris*, sér. D, v. 265, p. 319–321.

Nairn, A. E. M., and Westphal, M., 1967b, A second virtual pole from Corsica, the Ota Gabbrodiorite, *Palaeogeogr. Palaeoclimatol. Palaeoecol.*, v. 3, p. 277–286.

Nairn, A. E. M., and Westphal, M., 1968, Possible implications of the paleomagnetic study of late Palaeozoic igneous rocks of Northwestern Corsica, *Palaeogeogr. Palaeoclimatol. Palaeoecol.*, v. 5, p. 179–204.

Nairn, A. E. M., Roche, A., Westphal, M., and Zijderveld, J. D. A., 1967, Sur l'aimantation de deux coulées volcaniques tertiaires en Provence, *C.R. Somm., Soc. Géol. France*, p. 360–361.

Nardi, R., 1968a, Le unità alloctone della Corsica e loro correlazione con le unità delle Alpi e dell'Appennino, *Mem. Soc. Geol. Ital.*, v. 7, p. 323–344.

Nardi, R., 1968b, Contributo alla geologia della Balagne (Corsica nord-occidentale), *Mem. Soc. Geol. Ital.*, v. 7, p. 471–489.

Nardi, R., Puccinelli, A., and Verani, M., 1971, Sezioni geologiche interpretative nella Balagne sedimentaria (Corsica nord-occidentale), *Mem. Soc. Geol. Ital.*, v. 10, p. 191–202.

Nesteroff, W., 1965, Recherches sur les sédiments marins actuels de la région d'Antibes, *Ann. Inst. Océanogr. Monaco*, v. 43, 136 p.

Nesteroff, W., 1971, Histoire sédimentaire du domaine méditerranéen et alpin depuis le Burdigalien, *C.R. Somm. Soc. Géol. France*, v. 8, p. 418–420.

Nesteroff, W., 1973a, Mineralogy, Petrography, distribution and origin of the Messinian Mediterranean evaporites, in: *Init. Rep. Deep Sea Drilling Project*, v. XIII, (Ryan, W. B. F., *et al.*, eds.), Washington, D.C.: U.S. Govt. Printing Office, p. 673–694.

Nesteroff, W., 1973b, The Sedimentary history of the Mediterranean area during the Neogene, in: *Init. Rep. Deep Sea Drilling Project*, Ryan, W. B. F., *et al.*, eds., v. XIII, Washington, D.C.: U.S. Govt. Printing Office, p. 1257–1261.

Nesteroff, W., 1973c, La crise de salinité messinienne en Méditerranée: enseignements des forages Joides et du bassin de Sicile, 23rd CIESM Congr., Athens (1972), *Bull. Geol. Soc. Greece*, vol. X, p. 154–155.

Nesteroff, W. D., Ryan, W. B. F., Hsü, K. J., Pautot, G., Wezel, F. C., Lort, J. M., Cita, M. B., Maync, W., Stradner, H., and Dumitrica, P., 1972, Evolution de la sédimentation pendant le Néogène en Méditerranée d'après les forages JOIDES–DSDP, in: *The Mediterranean Sea—A Natural Sedimentation Laboratory*, Stanley, D. J., ed., Stroudsburg, Pa.: Dowden, Hutchinson and Ross, p. 47–62.

Ogniben, L., 1973, Conclusioni sullo stato attuale delle conoscenze nella geologia dell'Appennino, in: *Moderne Vedute sulla Geologia dell'Appennino, Accad. Naz. Lincei*, Roma (1972), v. 183, p. 367–445.

Ohnenstetter, M., Ohnenstetter, D., and Rocci, G., 1975, Essai de reconstitution du puzzle ophiolitique corse, *C.R. Acad. Sci. Paris*, sér. D, v. 280, p. 395–398.

Olivet, J. L., Auzende, J. M., Mascle, J. R., Monti, S., Pastouret, L., and Pautot, G., 1971, Description géologique de la bordure provençale. Résultats de la campagne de flexo-électro-forage en Méditerranée nord-occidentale, *C.N.E.X.O.*, v. 2, p. 375–394.

Orsini, J. B., 1968, *Etude Pétrographique et Structurale du Massif du Tanneron (Var) (Partie Occidentale et Centrale)*, Thesis (3rd Cycle), Univ. Grenoble, 109 p.

Orsini, J. B., Bruneton, S., Hermitte, D., and Pezeril, G., 1975, Vue d'ensemble sur les granitoïdes calco-alcalins du bloc corso-sarde, *3rd Earth Sci. Annu. Meeting*, Montpellier (Abstracts), p. 280.

Orszag-Sperber, F., 1971, Mise en évidence de repères stratigraphiques et structuraux dans le Miocène de la plaine orientale corse, *C.R. Acad. Sci. Paris*, sér. D, v. 272, p. 8–11.

Orszag-Sperber, F., 1974, Existence d'une phase de régression dans le Langhien de la plaine orientale corse: son extension probable en Méditerranée occidentale, *Paleog. Terziario Sardo nell'ambito Medit. Occid.*, Rend. Sem. Fac. Univ. Cagliari (1973), Suppl. v. XIII, p. 301–304.

Orszag-Sperber, F., 1975, Importance régionale des deux phases régressives enregistrées dans le Néogène de la Méditerranée occidentale, *IXth Int. Sediment. Congr.*, Nice, v. 5, p. 325–328.

Orszag-Sperber, F., and Freytet P., 1973, Sur l'intercalation de paléosols dans le Miocène marin de la plaine orientale corse. Conséquences paléogéographiques, *C.R. Acad. Sci. Paris*, sér. D, v. 276, p. 253–255.

Ottmann, F., 1958, Les formations pliocènes et quaternaires sur le littoral corse, *Mém. Soc. Géol. France*, n.s., v. XXXVII, fasc. 4, no. 84, 176 p.

Pannekoek, A. J., 1969, Uplift and subsidence in and around the western Mediterranean since the Oligocene: a review, *Verhandl Kon. Ned. Geol. Mijn. Gen.*, v. XXVI, p. 53–77.

Pautot, G., 1968, Etude géodynamique de la pente continentale au large de Cannes et Antibes, *Bull. Soc. Géol. France*, (7), v. X, p. 253–260.

Pautot, G., 1969, *Etude Géodynamique de la Marge Continentale au Large de l'Estérel*, Thesis, Univ. Paris.

Pautot, G., 1970, La marge continentale au large de l'Estérel (France) et les mouvements verticaux pliocènes, *Marine Geophys. Res.*, v. 1, p. 61–84.

Pautot, G., 1972, Histoire sédimentaire de la région au large de la Côte d'Azur, in: *The Mediterranean Sea—A Natural Sedimentation Laboratory*, Stanley, D. J., ed., Stroudsburg Pa.: Dowden, Hutchinson and Ross, p. 583–613.

Pautot, G., and Bellaiche, G., 1968, La marge continentale au large des Maures et de l'Estérel (France), *Coll. Int. C.N.R.S.*, Villefranche.

Perez, J. L., 1975a, *Etude Structurale de la Zone Limite entre l'Arc de Nice et l'Arc de la Roya (Alpes Maritimes)*, Thesis (3rd Cycle), Univ. Nice, 141 p.

Perez, J. L., 1975b, Observations structurales sur la zone limite entre l'Arc de Nice et l'Arc de la Roya, *Bull. Soc. Géol. France*, (7), v. XVII, p. 930–938.

Plesi, G., 1975, La nappe de Canetolo, *Bull. Soc. Géol. France*, 7, v. XVII, p. 979–983.

Radicati di Brozolo, F., and Giglia, G., 1973, Further data on the Corsica–Sardinia rotation, *Nature*, v. 241, p. 389–391.

Raggi, G., Squarci, P., and Taffi, L., 1966, Considerazioni stratigrafico-tettoniche sul flysch dell'isola d'Elba, *Boll. Soc. Geol. Ital.*, v. 84, p. 1–14.

Recq, M., 1967, Structure de la croûte terrestre en Provence, d'après les expériences du Revest et du lac Nègre, *C.R. Acad. Sci. Paris*, sér. D, v. 264, p. 1588–1591.

Recq, M., 1970, Courbes d'égale profondeur de la discontinuité de Mohorovicic en Provence, *C.R. Acad. Sci. Paris*, sér. D, v. 270, p. 11–13.

Recq, M., 1972a, La structure de la croûte profonde sous le massif de l'Estérel, *Boll. Geofis. Teor. Appl.*, v. XIV, p. 253–268.

Recq, M., 1972b, Profils de réfraction en Ligurie, *Pure Appl. Geophys.*, v. 101, p. 155–161.

Recq, M., 1972c, Sur la stabilité récente du massif des Maures et son indépendance par rapport aux régions environnantes, *C.R. Acad. Sci. Paris*, sér. D, v. 275, p. 333–336.

Recq, M., 1973, Contribution à l'étude de la structure profonde de la croûte terrestre dans la région de Nice, *Boll. Geofis. Teor. Appl.*, v. XV, p. 161–180.

Recq, M., 1974a, *Contributions à l'Etude de la Zone de Transition entre la Structure Continentale et la Structure Océanique de la Croûte Terrestre entre Toulon et Genes*, Thesis, Univ. Paris, 140 p.

Recq, M., 1974b, Contribution à l'étude de l'évolution des marges continentales du Golfe de Gênes, *Tectonophysics*, v. 22, p. 363–375.

Recq, M., and Bellaiche, G., 1972, La structure de la croûte terrestre au Sud des Maures, *Boll. Geofis. Teor. Appl.*, v. XIV, p. 128–149.

Rehault, J. P., 1968, *Contributions à l'Etude de la Marge Continentale au Large d'Imperia et de la Plaine Abyssale Ligure*, Thesis (3rd Cycle), Univ. Paris, 115 p.

Rehault, J. P., Olivet, J. L., and Auzende, J. M., 1974a, Le bassin nord-occidental méditerranéen: structure et évolution, *Bull. Soc. Géol. France*, (7), v. XVI, p. 281–294.

Rehault, J. P., Recq, M., Gennesseaux, M., Estève, J. P., and Bellaiche, G., 1974b, Nouvelles données de sismique-réfraction obtenues dans la région de Nice, *24th CIESM Congr.*, Monaco (Abstracts).

Reutter, K. J., 1968, Die tektonischen Einheiten des Nordapennins, *Eclogae Geol. Helv.*, v. 61, p. 184–224.

Richter, M., 1960, Beziehungen zwischen ligurischen Alpen und Nordapennin, *Geol. Rundschau*, v. 50, p. 529–537.

Richter, M., 1962a, Alpen, Apennin und Dinariden, *Neues Jb. Geol. Paläont., Mh.*, v. 9, p. 466–480.

Richter, M., 1962b, Bemerkungen zur Geologie der Insel Elba, *Neues Jb. Geol. Paläont., Mh.*, v. 10, p. 495–505.

Richter, M., 1963, Der Bauplan des Apennins, *Neues Jb. Geol. Paläont., Mh.*, v. 9, p. 509–518.

Ritsema, A. R., 1969, Seismic data of the west Mediterranean and the problem of oceanization, *Verhandl. Kon. Ned. Geol. Mijn. Gen.*, v. 26, p. 105–120.

Ritsema, A. R., 1970, On the origin of the western Mediterranean Sea basins, *Tectonophysics*, v. 10, p. 609–623.

Rothé, J. P., 1968, Seismicité de l'Atlantique oriental et de la Méditerranée occidentale, *Coll. Int. C.N.R.S.*, Villefranche.

Roubault, M., Bordet, P., Leutwein, F., Sonet, J., and Zimmermann, J. L., 1970a, Ages absolus des formations cristallophylliennes des Massifs des Maures et du Tanneron, *C.R. Acad. Sci. Paris*, v. 271, p. 1067–1070.

Roubault, M., Bordet, P., Leutwein, F., Sonet, J., and Zimmermann, J. L., 1970b, Ages absolus des formations volcaniques du Tanneron–Estérel (Var), *C.R. Acad. Sci. Paris*, sér. D, v. 271, p. 1157–1160.

Routhier, P., 1968, Sur les relations entre Corse "hercynienne" et Corse alpine; âge des schistes lustrés; inventaire des acquisitions et des problèmes, des progrès et des régressions, *Bull. Soc. Géol. France*, (7), v. X, p. 13–35.

Royant, G., Rioult, M., and Lanteaume, M., 1970, Horizon stromatolithique à la base du Crétacé supérieur dans le Briançonnais ligure, *Bull. Soc. Géol. France*, v. 12, p. 372–374.

Ruggieri, G., 1967, The Miocene and later evolution of the Mediterranean Sea, Aspect of Tethyan Biogeography, Adams, P. J., and Ager, D. V., eds., *Syst. Assoc. Publ.*, v. 7, p. 283.

Ryan, W. B. F., 1969, *The Floor of the Mediterranean Sea*, Thesis, Columbia Univ., 196 p.

Ryan, W. B. F., and Hsü, K. J., 1970, Deep Sea Drilling Project, Leg 13, *Geotimes*, v. 15, p. 12–15.

Ryan, W. B. F., Stanley, D. J., Hersey, J. B., Fahlquist, D. A., and Allan, T. D., 1970, The tectonics and geology of the Mediterranean Sea, in: *The Sea*, v. 4, pt. 2, Maxwell, A.E., ed., New York: Wiley–Interscience, p. 387–492.

Ryan, W. B. F., Hsü, K. J., Cita, M. B., Dumitrica, P., Lort, J., Maync, W., Nesteroff, W. D., Pautot, G., Stradner, H., and Wezel, F. C., eds., 1973a, *Init. Rep. Deep Sea Drilling Project*, v. XIII, Washington D.C.: U.S. Govt. Printing Office, 1447 p.

Ryan, W. B. F., Hsü, K. J., Cita, M. B., Dumitrica, P., Lort, J., Maync, W., Nesteroff, W. D., Pautot, G., Stradner, H., and Wezel, F. C., 1973b, Boundary of Sardinia slope with Balearic abyssal plain. Sites 133 and 134, in: *Init. Rep. Deep Sea Drilling Project*, v. XIII, Ryan, W. B. F., *et al.*, eds., Washington D. C.: U.S. Govt. Printing Office, p. 465–514.

Scholle, P., 1970, The Sestri–Voltaggio line: a transform fault induced tectonic boundary between the Alps and the Apennines, *Amer. J. Sci.*, v. 269, p. 343–359.

Segre, A., 1960, Carta batimetrica del Mediterraneo centrale; Mari Ligure e Tirreno settentrionale. Carta no. 1250, Genoa: Ist. Idrogr. Marina.

Seligmann, F., 1973, Ophiolites and continental fragments in the northern Apennines, 23rd CIESM Congr., Athens (1972), *Bull. Geol. Soc. Greece*, v. X, p. 180–184.

Selli, R., 1973, Caratteri geologici dei Mari italiani, in: *Moderne Vedute sulla Geologia dell'Appennino*, Accad. Naz. Lincei, Roma (1972), v. 183, p. 273–280.

Sestini, G., 1970a, Sedimentation of the late geosynclinal stage, in: *Development of the Northern Apennines Geosyncline*, Sestini, G., ed., *Sediment. Geol.*, v. 4, p. 445–479.

Sestini, G., 1970b, Postgeosynclinal deposition, in: *Development of the Northern Apennines Geosyncline*, Sestini, G., ed., *Sediment. Geol.*, v. 4, p. 481–520.

Signorini, R., 1966, I terrini neogenici del Foglia "Sierra", *Boll. Soc. Geol. Ital.*, v. 85, p. 639–654.

Smith, A. G., 1971, Alpine deformation and the oceanic areas of the Tethys, Mediterranean and Atlantic, *Bull. Geol. Soc. Amer.*, v. 82, p. 2039–2070.

Soffel, H., 1972, Anticlockwise rotation of Italy between the Eocene and Miocene: Paleomagnetic evidence from the Colli Euganei, Italy, *Earth Planet. Sci. Lett.*, v. 17, p. 207–210.

Squarci, P., and Taffi, L., 1963, Geologia della zona di Chianni–Laiotico–Orciatico (provincia di Pisa), *Boll. Soc. Geol. Ital.*, v. 83, p. 219–290.

Stanley, D. J., and Bouma, A. H., 1964, Methodology and paleogeographic interpretation of flysch formations: a summary of studies in the Maritime Alps, in: *Turbidites*, Bouma, A. H., and Brouwer, A., eds., *Developments in Sedimentology*, v. 3, Amsterdam: Elsevier, p. 34–64.

Stanley, D. J., and Mutti, E., 1968, Sedimentological evidence for an emerged land mass in the Ligurian Sea during the Paleogene, *Nature*, v. 218, p. 32–36.

Stanley, D. J., Cita, M. B., Flemming, N. C., Kelling, G., Lloyd, R. M., Milliman, J. D., Pierce, J. W., Ryan, W. B. F., and Weiler, Y., 1972, Guidelines for future sediment-related research in the Mediterranean Sea, in: *The Mediterranean Sea—A Natural Sedimentation Laboratory*, Stanley, D. J., ed., Stroudsburg Pa.: Dowden, Hutchinson and Ross, p. 723–741.

Stanley, D. J., Got, H., Leenhardt, O., and Weiler, Y., 1974, Subsidence of the Western Mediterranean Basin in Pliocene–Quaternary time, *Geology*, v. 2, p. 345–350.

Storetvedt, K. M., 1973, Genesis of west Mediterranean basins, *Earth Planet. Sci. Lett.*, v. 21, p. 22–28.

Struffi, G., and Sommi, M., 1960, Il limite Pliocene–Quaternario lungo il margine settentrionale delle Colline Livornesi, *Boll. Soc. Geol. Ital.*, v. 79, no. 2.

Sturani, C., 1973, Considerazioni sui rapporti tra Appennino settentrionale et Alpi occidentali, in: *Moderne Vedute sulla Geologia dell'Appennino*, Accad. Naz. Lincei, (Roma 1972), v. 183, p. 119–145.

Tempier, C., 1973, Les faciès calcaires du Jurassique provençal, *Trav. Labor. Sci. Terre, St Jérôme, Marseilles*, no. 4, 361 p.

Trevisan, L., 1952, Sul complesso sedimentario del Miocene superiore e del Pliocene della Val di Cecina e sui movimenti tettonici tardivi in rapporto ai giacimenti di lignite e di salgemma, *Boll. Soc. Geol. Ital.*, v. 70, p. 65–78.

Trevisan, L., Brandi, G. P., Dallan, L., Nardi, R., Raggi, G., Rau, A., Squarci, P., Taffi, L., and Tongiorgi, M., 1971, Note illustrative della Carta Geologica d'Italia, foglio 105, Lucca, *Serv. Geol. Italia*.

Van Bemmelen, R. W., 1969, Origin of the Western Mediterranean sea, *Verhandl. Kon. Ned. Geol. Mijn. Gen.*, v. XXVI, p. 13–52.

Van Bemmelen, R. W., 1972a, Geodynamic models. An evolution and a synthesis, in: *Developments in Geotectonics*, v. 2, Amsterdam: Elsevier, 267 p.

Van Bemmelen, R. W., 1972b, Driving forces of Mediterranean orogeny, *Geol. en Mijnb.*, v. 51, p. 548–573.

Van den Berg, J., Klootwijk, C. T., and Wonders, T., 1975, Implications for the rotational movement of Italy from current palaeomagnetic research in the Umbrian sequence, Northern Apennines, *Progress in Geodynamics*, Amsterdam: Royal Neth. Acad. Arts and Sci., p. 165–175.

Van der Voo, R., and Zijderveld, J. D. A., 1969, Paleomagnetism in the Western Mediterranean area, *Verhandl. Kon. Ned. Geol. Mijn. Gen.*, v. XXVI, p. 121–138.

Vanossi, M., 1970, Contributi alla conoscenza delle unità stratigrafico–strutturali del Brianzonese Ligure, s.l., IV. Messa a punto generale, *Ist. Geol., Univ. Pavia*, v. 21, p. 109–114.

Vernet, J., 1962, Contribution à l'étude du Pliocène niçois, *Trav. Lab. Geol. Fac. Sci. Grenoble*, v. 38, p. 249–274.

Vernet, J., 1968, Tectonique et problème de tectorogenèse antépliocène de l'Arc de Nice dans sa marge externe frontale, *Rev. Géogr. Phys. Géol. Dynam.* v. 10, p. 49–64.

Veslin, J., 1958, Contribution à l'étude de la stratigraphie du Miocène de Vence, *D.E.S.*, Univ. Paris.

Vogt, P. R., Higgs, R. H., and Johnson, E. L., 1971, Hypotheses on the origin of the Mediterranean Basin: magnetic data, *J. Geophys. Res.*, v. 76, p. 3207–3228.

Watson, J. A., and Johnson, G. L., 1968, Mediterranean diapiric structures, *Bull. Amer. Assoc. Petrol. Geol.*, v. 52, p. 2247–2249.

Watson, J. A., and Johnson, G. L., 1969, The marine geophysical survey in the Mediterranean, *Int. Hydrogr. Rev.*, v. 46, p. 81–107.

Westphal, M., 1967, *Etude Paléomagnétique de Formations Volcaniques Primaires de Corse. Rapports avec la Tectonique du Domaine Ligurien*, Thesis, Univ. Strasbourg.

Westphal, M., Bardon, C., Bossert, A., and Hamzeh, R., 1973, A computer fit of Corsica and Sardinia against southern France, *Earth Planet. Sci. Lett.*, v. 18, p. 137–140.

Westphal, M., Orsini, J., and Vellutini, P., 1975, Le microcontinent corso-sarde, sa position initiale: données paléomagnétiques et raccords géologiques, *3rd Earth Sci. Annu. Meeting*, Montpellier (Abstracts), p. 388.

Wezel, F. C., 1970, Interpretazione dinamica della "eugeosynclinale meso-mediterranea," *Riv. Min. Siciliana*, v. 21, p. 124–126 and 187–198.

Wong, H. K., Zarudzki, E. F. K., Knott, S. T., and Hays, E. E., 1970, Newly discovered group of diapiric structures in Western Mediterranean, *Bull. Amer. Assoc. Petrol. Geol.*, v. 54, p. 2200–2204.

Wunderlich, H. G., 1964, Gebirgsbildung im Alpen–Nordapennin-Orogen, *Tectonophysics*, v. 1, p. 73–84.

Zacher, W., 1974, Rotation and paleotectonic evolution of Corsica, *24th CIESM Congr.*, Monaco, (Abstracts).

Zijderveld, J. D. A., and Van der Voo, R., 1970, Paleomagnetism in the Mediterranean area, in: *Implications of Continental Drift to the Earth Sciences*, v. 1, Tarling, D. H., and Runcorn S. K., eds., London, New York: Academic Press, p. 133–161.

Zijderveld, J. D. A., De Jong, K. A., and Van der Voo, R., 1970a, Rotation of Sardinia, palaeomagnetic evidence from Permian rocks, *Nature*, v. 226, p. 933–934.

Zijderveld, J. D. A., Hazeu, G. J. A., Nardin, M., and Van der Voo, R., 1970b, Shear in the Tethys and the Permian Paleomagnetism in the Southern Alps, including new results, *Tectonophysics*, v. 10, p. 639–662.

Chapter 3

THE TYRRHENIAN SEA AND ADJOINING REGIONS

M. Boccaletti

Istituto di Geologia dell'Università
Florence, Italy
and Centro di Geologia dell'Appennino, Florence, Italy

and

P. Manetti

Istituto di Mineralogia, Petrografia e Geochimica dell'Università
Florence, Italy

I. INTRODUCTION

The Tyrrhenian Sea, in the central Mediterranean area, is bounded on the northern and eastern sides by the Italian Peninsula, and on the western and southern sides by the Ligurian Sea, Corsica, Sardinia, and Sicily. It covers a triangular area of 231,000 km^2 (Carter *et al.*, 1972) and reaches a maximum depth of 3620 m (Selli, 1970a). The Tyrrhenian Sea is widely open toward the western Mediterranean. Structurally the Tyrrhenian Sea may be divided into two sections: a smaller, northern area lying behind the northern Apennines,[*] and a larger, southern basin backing on the Sicilian–Calabrian Arc and the southern Apennines.

---

[*] This northern section is often regarded as the southeastern part of the Ligurian Sea, and is discussed in detail by Moullade in the preceding chapter.

II. PHYSIOGRAPHY

A. The Southern Tyrrhenian Sea

The morphology of the southern Tyrrhenian Sea is complex. Selli (1970a) recognized seven different physiographic units (Fig. 1), namely: 1) continental shelf; 2) upper continental slope, cut by deep canyons; 3) peri-Tyrrhenian basins; 4) peri-Tyrrhenian seamounts, which form the outer rim of the peri-Tyrrhenian basins; 5) lower continental slopes; 6) Bathyal Plain; and 7) central seamounts.

The continental shelf, relatively wide in the north, is nearly lacking along the Calabrian coast. The average depth of the shelf break is 128 m. The continental slope is divided by the peri-Tyrrhenian basins and seamounts into an upper slope with an average dip of 3°29′ and a lower slope with an average dip of 3°11′. At the base of the slope, the Bathyal Plain slopes very gently (average dip 0°30′–0°40′) and has an average depth of 3184 m.

Seamounts occur in both the peri-Tyrrhenian area and in the Bathyal Plain (Segre, 1955). In the Bathyal Plain, the three main seamounts are submarine volcanoes, with a general NW–SE trend. The largest of them, Marsili Seamount, is 50 km long and 2900 m high. The largest submarine volcano in the peri-Tyrrhenian area, Palinuro Seamount, has an E–W-trending long axis. Some of the peri-Tyrrhenian seamounts are nonvolcanic and consist instead of sedimentary and metamorphic rocks (Selli, 1974) (Fig. 2). Baronie, the largest seamount of this province, is a 90-km-long, 1200-m-high, nonvolcanic uplift which has yielded Permian (Verrucano) pebbles from a submarine terrace at 550–740 m (Selli and Fabbri, 1971). The same authors believe that this seamount was emergent during the Middle Pliocene and subsided below sea level subsequently. Vercelli Seamount appears to be granitic (Gallignani, 1973), and low-grade metamorphic to anatectic stage rocks crop out on the Secchi Seamount (Selli, 1974).

III. STRUCTURAL FRAMEWORK

Bathymetric and gravimetric maps show four main tectonic trends in this region (Selli, 1970c; 1974) (Fig. 3): 1) A N–S trend occurs in the Bathyal Plain along the Corso-Sardinian Slope. 2) An E–W trend is seen mainly on the northern slopes of Sicily and on the Campanian Slope south of Naples. (Similarly oriented structures of transcurrent type are found on the mainland in northern Calabria, and from Gargano across the southern Adriatic Sea; they can be traced as far as Dalmatia.) 3) A NE–SW trend is found along the western slopes of Sicily, Campania, and Calabria, as well as in the Bathyal Plain, northward and eastward of the Magnaghi Seamount. Here, several faults,

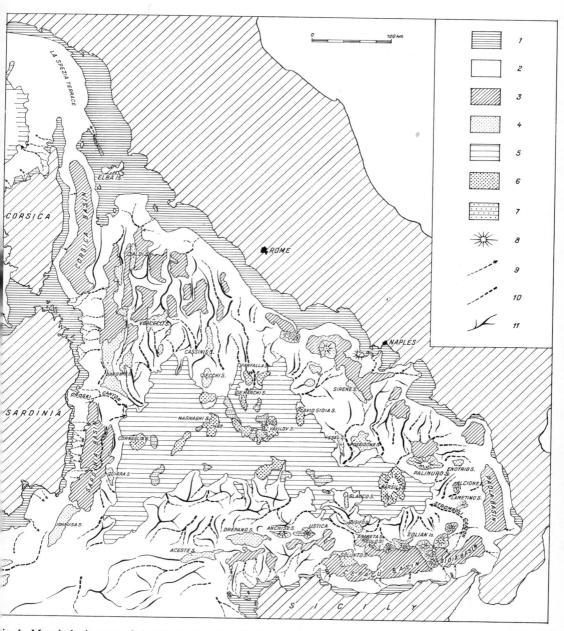

ig. 1. Morphologic map of the Tyrrhenian Sea. 1) Continental shelf; 2) upper and lower continental slopes; 3) eri-Tyrrhenian basins; 4) peri-Tyrrhenian seamounts; 5) Bathyal Plain; 6) central Tyrrhenian seamounts; 7) athyal Plain hills; 8) volcanoes; 9) canyons; 10) valleys and depressions; 11) relief and ridges. (After Selli, 070a, 1974; and Carter *et al.*, 1972.)

up to 150 km long, divide a western area ranging in depth from 2200 to 3000 m
from an eastern area where depths below 3300 m occur (Selli, 1974). 4) North-
west-trending structures parallel to the Apennines are found in the northern
Tyrrhenian Sea and along the eastern edge of the southern Tyrrhenian Sea as
far as the Eolian Islands (see Moullade, this volume).

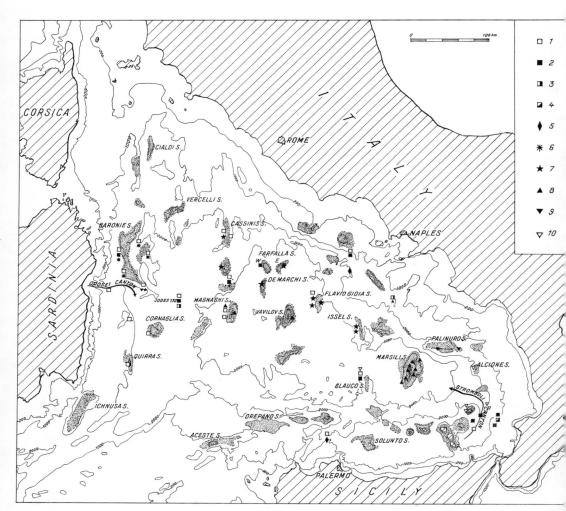

Fig. 2. Map of the pre-Quaternary rocks dredged in the southern Tyrrhenian up to 1973. 1) Ooze with scattere
sand layers, sometime "Trubi" facies (Upper–Middle p.p. Pliocene); 2) "Trubi" facies (Middle p.p.–Lower Pliocene)
3) bituminous marls, evaporitic series (Messinian) at Site 132 of JOIDES; 4) gray marls (Tortonian); 5) chert
limestones, red marlstones, dolomitic limestones (Mesozoic: Jurassic–Cretaceous); 6) conglomerates, Verrucan
quarzarenites (Paleozoic); 7) phyllites, micaschists, gneiss, and granites of metamorphic basement (Paleozoic); 8
tholeitic basalts and differentiated rocks; 9) alkali-olivine basalts and differentiated rocks; 10) basalts s.s. Th
smaller symbols indicate rocks in secondary attitude. (After Selli, 1974.)

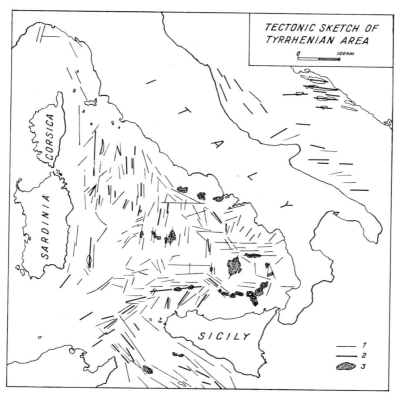

Fig. 3. Tectonic sketch of the Tyrrhenian area. 1) Normal or transcurrent vertical faults; 2) tectonic relief (horst, anticlinorium, anticline); 3) volcanoes. (After Selli, 1970c; 1974.)

IV. NATURAL SEISMICITY

Shallow, intermediate, and deep earthquakes have been detected in the southeastern part of the southern Tyrrhenian Sea (Peterschmitt, 1956; Ritsema, 1969, 1970; Caputo et al., 1970, 1972; Caputo and Postpischl, 1973) (Fig. 4). The focus of the deepest known earthquake in this region lies at a depth of 487 km in the center of the Bathyal Plain; the foci of intermediate and deep earthquakes are scattered along a plane dipping WNW at 50° (Ritsema, 1970) or 60° (Caputo et al., 1970, 1972). The earthquakes can be related to under-thrusting of a lithospheric slab beneath the Tyrrhenian Sea (Ritsema, 1970; Caputo et al., 1970, 1972). Two aseismic zones between 100–200 km and 350–450 km, respectively, break the continuity of the distribution of foci. The shallower aseismic zone, according to Caputo et al. (1972) is due to the high heat flow, which reduces the rigidity of the lithospheric slab in this sector. Keller (1974) thinks that the lithospheric slab may be broken into two pieces, one

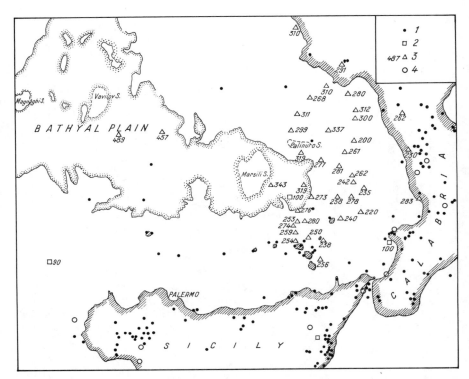

Fig. 4. Seismicity map of the southern Tyrrhenian area based on the epicenters from 1900 to 1970. 1) Normal ipocenters; 2) ipocenters between 60 and 100 km; 3) ipocenters > 100 km; 4) macroseismic data. (After Caputo and Postpischl, 1973.)

lying above 100 km and the other below 200 km (Fig. 5). Such an interpretation, based on models proposed by Oliver *et al.* (1973), is consistent with the evolution of magmatism in the Eolian Islands.

Earthquake foci in the northern Tyrrhenian Sea are both scarce and superficial, and are concentrated along the coasts. Widespread shallow focus seismicity affects the contiguous northern Apennines (Caputo and Postpischl, 1973).

V. HEAT FLOW

The only available heat-flow measurements in the Tyrrhenian Sea are those reported by Erickson (1970). Taking into account the sedimentation factor, the average heat flow for 12 stations is 2.83 HFU (Loddo and Mongelli, 1974a). Loddo and Mongelli (1974b) prepared a preliminary heat-flow map of Italy and adjoining areas (Fig. 6), showing a remarkable increase in the

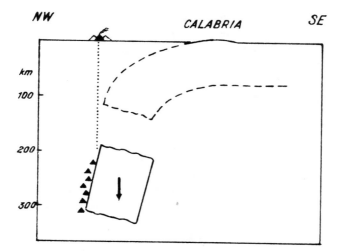

Fig. 5. Model of the detached slab in the Eolian Arc. (After Keller, 1974.)

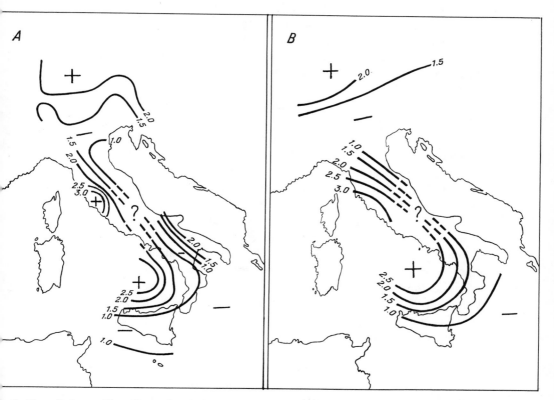

. 6. Map of observed heat-flow values in Italy (A); map of corrected heat-flow values in Italy (B). (After Loddo and Mongelli, 1974.)

Ionian Trough toward the southern Tyrrhenian Basin, where heat-flow values are similar to those measured in some trenches and active marginal basins of the western Pacific (Herzen and Lee, 1969; Horai and Uyeda, 1969; Sclater, 1972).

VI. MAGMATISM

Post-Mesozoic igneous rocks belonging to two different comagmatic series outcrop extensively in the Tyrrhenian area and around its margins (Fig. 7). The different types of magmatism present can be associated with Neogene and Recent geodynamic evolution of the Tyrrhenian area. Volcanic activity still affects the southern Tyrrhenian area. The distribution of volcanicity is summarized below.

A. Eolian Islands

The Eolian volcanic arc is made up of seven main islands, two of which are active volcanoes (Stromboli and Vulcano). Four seamounts are also connected with this volcanic arc (Selli, 1974). Recently, the volcanic rocks of these islands have attracted the attention of several investigators who have described their geographic, geochemical, and geophysical aspects (Jakob, 1958; Pichler, 1967; Keller, 1967, 1974; Caputo et al., 1970, 1972; Latter, 1971; Villari, 1972; Barberi et al., 1973, 1974a, 1974b; Bigazzi and Bonadonna, 1973; Bini et al., 1973; Romano, 1973).

Barberi et al. (1974c) summarized the present state of knowledge of Eolian magmatism. It is younger than 1 m.y. and occurred in two main stages. During the Early–Middle Pleistocene, Alicudi, Filicudi, Panarea, and part of Salina and Lipari built up, and after a low in activity during the Upper Pleistocene, Salina and Lipari were completed while Vulcano and Stromboli were formed.

The volcanic products of this activity can be grouped into two main comagmatic series. A typical calcalkaline series during the first stage, followed at the beginning of the second stage by a high-K calcalkaline series which passes into the shoshonitic series found as recent products of Vulcano, Stromboli, and Lipari. Leucitic–tephritic lavas have been observed on Vulcano and Vulcanello (Bini et al., 1973; Barberi et al., 1974a). The K_2O content of the recent lavas is in agreement with focal depths of the earthquakes according to Dickinson's relationship (Dickinson, 1972), which would suggest that the Eolian magmatism is at present in a senile stage (Barberi et al., 1974a). Ninkovich and Hays (1971, 1972) were the first authors to interpret the Eolian volcanism as being due to fusion of the underthrust lithospheric slab. More recently, Barberi et al. (1974b), on the basis of trace elements and isotope ratios, suggested that the magma of both the calcalkaline series and the sho-

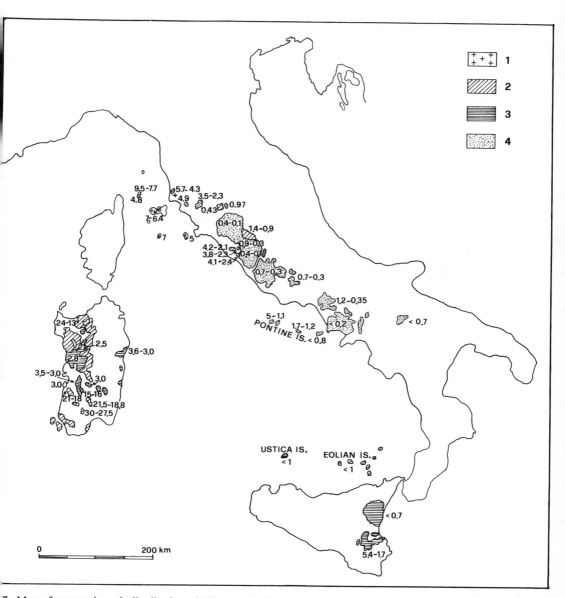

7. Map of magmatic rock distribution of Oligocene to Recent outcropping in areas surrounding the Tyrrhenian
1. 1) Intrusive rocks mainly granitic; 2) volcanic rocks of the calcalkaline series; 3) alkali-olivine basalts and
rentiated rocks; 4) volcanics of the K-alkaline series (Mediterranean suite). Radiometric ages of the rocks are
ated. (Data from Barberi *et al.*, 1971, 1974*b*; Borsi *et al.*, 1967; Civetta *et al.*, 1970; Coulon, 1974; Di Paola
, 1975; Lauro, 1973; Lombardi *et al.*, 1974; Savelli and Pasini, 1974; Savelli, 1975; and Selli, 1974.)

shonitic series were derived from partial mantle fusion over the inclined seismic zone without significant crustal contamination. Klerkx *et al.* (1974) from a detailed investigation of $^{87}Sr/^{86}Sr$, rare earth elements, and some other trace elements concur with the subcrustal origin of the calcalkaline magmatism associated with a subduction zone. They doubt, however, that the shoshonitic series can be derived from the mantle and suggest a lower crustal anatexis or at least crustal contamination.

B. Submarine Volcanoes and Ustica

Several submarine volcanoes rise from the Tyrrhenian Bathyal Plain (Fig. 8). The volcanoes are generally located along faults trending N–S and NNE–SSW. The dates of volcanic activity may reflect multiple periods of volcanism as well as change from place to place. For Magnaghi and Marsili Seamounts, Selli and Fabbri (1971) suggest an age of 4.5–5.0 m.y., while Selli (1974) suggests an age of 2.4 m.y. The age of the volcanics from the Vavilov Seamount were determined by Selli (1974) to fall in the range of 1.2 m.y. Barberi *et al.* (1973), on the basis of magnetic anomalies, suggest that Vavilov Seamount may have been formed during the Matuyama reversed epoch (0.7–2.4 m.y.) and that the latest activity occurred after 0.7 m.y. Absolute age data on the tholeitic basaltic samples collected in the Bathyal Plain during the DSDP cruise (Leg 42A, site 373A) range from 7.5 to 3.5 m.y. (Barberi *et al.*, 1978).

Ustica Island, north of Palermo, has a typical alkali–basalt trachytic series with an age of less than 1 m.y. (Barberi *et al.*, 1969; Romano and Sturiale, 1971). Quirri and Anchise Seamounts (Selli, 1974) and Palinuro Seamount (Del Monte, 1972) might be expected to be like Ustica, with alkali-basaltic rocks. Barberi *et al.* (1974*c*), however, on the basis of sparse data for Palinuro petrology, suggest a calcalkaline nature. Marsili Seamount is made of tholeitic basalts at its base, while the later products are more alkaline (Maccarone, 1970; Del Monte, 1972). The few $^{87}Sr/^{86}Sr$ data from Ustica and Marsili point to an uncontaminated mantle provenance of the lavas (Barberi *et al.*, 1969). The Marsili volcano shows a magnetic anomaly pattern similar to that of the spreading oceanic ridges (Selli, 1974*b*).

C. Pontine Islands

On the eastern side of the southern Tyrrhenian Sea lies a NW–SE-trending archipelago of five principal islands (Ponza, Palmarola, Zannone, Ventotene, and St. Stefano). All of these islands are volcanic with the exception of a small part of Zannone, where metamorphic and sedimentary rocks occur. The westernmost islands (Ponza, Palmarola, and Zannone) consist mainly of sodic rhyolites (5 m.y.) with a "Middle Pacific" type of serial character and with a trachytic dome (1.1 m.y.) of the latest products (Barberi *et al.*, 1967).

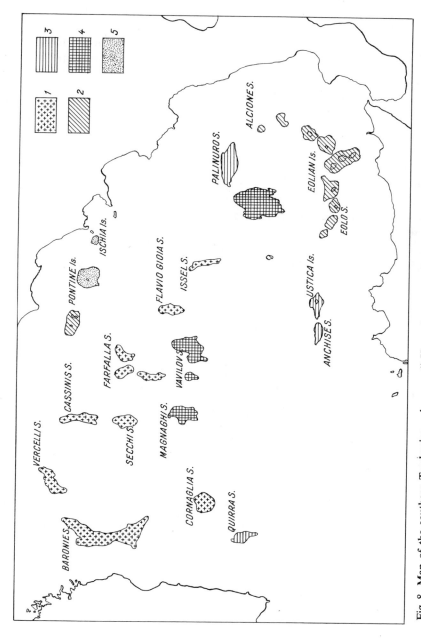

Fig. 8. Map of the southern Tyrrhenian volcanoes. 1) Hercynian metamorphic rocks; 2) calcalkaline and shoshonitic series in the Eolian Islands; 3) alkali-olivine basalts and differentiated rocks; 4) central Tyrrhenian seamounts, probably tholeitic; 5) K-alkaline series (Mediterranean suite). (After Selli, 1974, modified.)

The easternmost islands (Ventotene and St. Stefano) have undersaturated trachybasaltic and phonolitic lavas, as well as phonolitic pyroclasts ranging in age from 1.2 to 1.7 m.y. The $^{87}Sr/^{86}Sr$ ratio of these rocks (average 0.7073) remains unchanged through the entire series, suggesting a single magmatic origin. The rocks probably formed either by contamination of a subcrustal magma with crustal material of high $^{87}Sr/^{86}Sr$ ratio or from a magma from the lower crust where Rb/Sr and $^{87}Sr/^{86}Sr$ are intermediate between those of the mantle and the upper crust (Barberi et al., 1969).

D. Tuscan Archipelago

The Tuscan Archipelago, which divides the southern from the northern Tyrrhenian Sea, consists of seven main islands which from north to south are: Gorgona, Capraia, Elba, Pianosa, Montecristo, Giglio, and Giannutri (Fig. 9). Magmatic rocks of late and post-Miocene age crop out on Capraia, Elba, and Montecristo. Effusive activity, which is well seen on Capraia, can be divided into two main stages (Franzini, 1964; Barberi et al., 1971). The first stage lasted from 9.50 to 7.74 m.y. ago (Borsi, 1967) and is characterized by latitic lavas and latitic and quartz-latite pyroclasts. The second phase, which occurred between 4.9 and 4.8 m.y. ago, consists of trachybasalts, the most basic products of

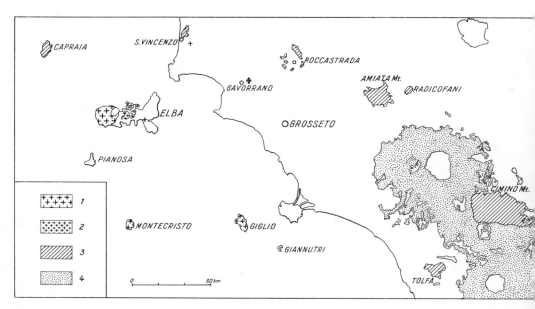

Fig. 9. Map of magmatic rocks in the Tuscan Archipelago, Tuscany, and Latium. 1) Granitic rocks; subvolcanic acid rocks; 3) volcanic rocks of the calcalkaline series (Tusco-Latium province); 4) volcanic rock of the K-alkaline series (Roman province). (After Barberi et al., 1971.)

post-Miocene magmatism known in the area. From a genetic standpoint the Capraia magmatism appears to be related to the Oligo-Miocene magmatism of Sardinia (Marinelli, 1961; Franzini, 1964) rather than to the Tuscany–Latium petrographic province.

On the other three islands (Elba, Montecristo, and Giglio) only small intrusive stocks or subeffusive bodies crop out. Their composition is granitic, granodioritic, and, more rarely, quartz-monzonitic. The age becomes younger eastward (Fig. 7): 7 m.y. at Montecristo and Elba, 5 m.y. at Giglio. Genetically connected with these intrusions are small granodioritic stocks located on the mainland together with the quartz-latitic and trachytic lavas of the Tuscany–Latium petrographic province. The age of the latest, easternmost products is 0.43 m.y. at Monte Amiata (Evernden and Curtis, 1965). Chemical, petrographic, and isotope data suggest that all the magmatism is of crustal origin (Marinelli, 1967). The southernmost products (Tolfa and Cimini Mts.) are areally associated with the northern edge of the Roman potassic province (Mediterranean series). The latter is developed uninterruptedly along the Tyrrhenian coast as far as southern Latium, where it gives way to the Campanian potassic province with the Roccamonfina, Volcano, the Phlegrean Fields, and Ischia Island.

The age of magmatites in the Tuscany–Latium province is generally older than that of the Roman potassic province, though they partially interfinger (Gasparini and Adams, 1969).

VII. GRAVITY AT SEA

Gravity measurements carried out by the Istituto Geofisico Sperimentale of Trieste allowed Finetti and Morelli (1973) to compile a detailed free-air gravity map and a Bouguer gravity map of the whole Tyrrhenian Sea (Fig. 10). The main feature of the latter map is the presence, in the central part of the Bathyal Plain, of positive anomalies reaching a maximum value of +260 mgal. Minor negative anomalies occur along a small sector of the eastern coast of Corsica. In the northern part of the Tyrrhenian Sea, positive Bouguer anomalies are lower than in the south and reach only +100 mgal in the slope and +40–+10 mgal along the coast.

The free-air anomalies are generally low and positive in the Bathyal Plain. High positive anomalies on the seamounts and values of about +50 mgal have been detected along the shelf and the slope. In the latter areas the negative values are located within the peri-Tyrrhenian basins filled with low density sediments. The low positive values of the Bathyal Plain have been interpreted by Colombi et al. (1973) as being due to the sinking of a basin not yet in equilibrium.

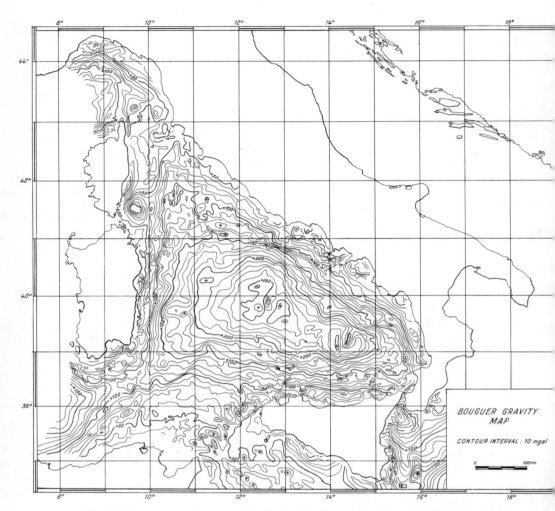

Fig. 10. Bouguer gravity map of the Tyrrhenian Sea. (After Finetti and Morelli, 1973.)

VIII. MAGNETISM

Magnetic data have been published by Morelli (1970), Selli (1970b), Allan and Morelli (1971), Vogt *et al.* (1971), Storetvedt (1973) and Finetti and Morelli (1973). The last have recently published a map of total field intensity (Fig. 11). The Mediterranean, in general, lacks magnetic features. However, in the Tyrrhenian Sea, strong magnetic anomalies are connected with the volcanic seamounts (Vavilov and Marsili) or islands (Ustica). On the Marsili Seamount magnetic polarity bands have been observed (Selli, 1970b). Most magnetic

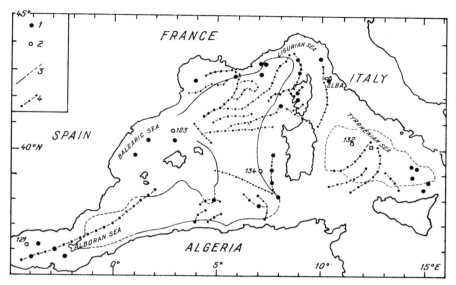

Fig. 12. Schematic map of the Western Mediterranean, showing the distribution of the magnetic lineations. 1) Magnetic centers; 2) location of JOIDES sites; 3) boundary between oceanic and continental crust; 4) axis of the magnetic anomalies. (After Storetvedt, 1973.)

lineations trend NNE and ENE, corresponding to an elongate ridge and what are interpreted as transform faults (Fig. 12; Storetvedt, 1973).

IX. CRUSTAL THICKNESS

Only a few measurements of crustal thickness have been carried out in the Tyrrhenian Sea. The earliest refraction profiles are those of Ewing and Ewing (1959) located 60 miles north of Sicily. Subsequently, Fahlquist and Hersey (1969) reported a profile (193) 80–90 km long, located in the Bathyal Plain. The section shows a thin layer of unconsolidated sediments (0.15–0.30 km) with a seismic velocity 1.8 km/sec, followed by 2.66–3.05 km of volcanics with 3.9–4.3 km/sec velocity, and finally by a layer with a 7.0–7.3 km/sec seismic velocity. The last values would suggest the presence of rocks having a density of 2.95–3.00 g/cm³, which could be derived, at least in part, from the mantle (Giese and Morelli, 1973). According to Menard (1967), however, it represents a layer overlying the mantle and different from it.

Moskalenko (1967) reported three seismic profiles located around the Vavilov Seamount, but these do not reach the Moho Discontinuity. Colombi et al. (1973) describe a refraction profile located between Pantelleria and Gargano, across the southernmost part of the Tyrrhenian Sea (Fig. 13). Between Pantelleria and the southern coast of Sicily the crust is of continental type,

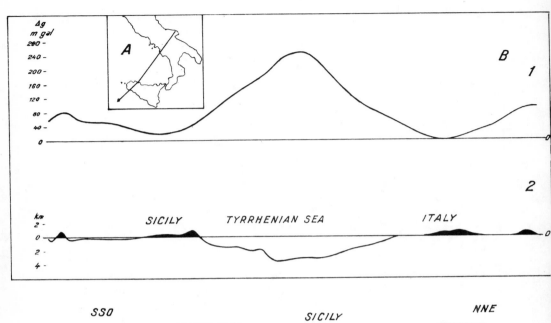

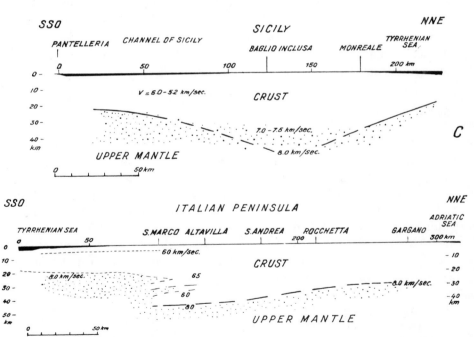

Fig. 13. Simplified cross section along the profile Gargano–Salerno–Pantelleria, reported in A. B-1 shows the Bouguer anomalies and B-2 the topographic section. C shows an interpretative seismic refraction cross section (After Colombi *et al.*, 1973, simplified.)

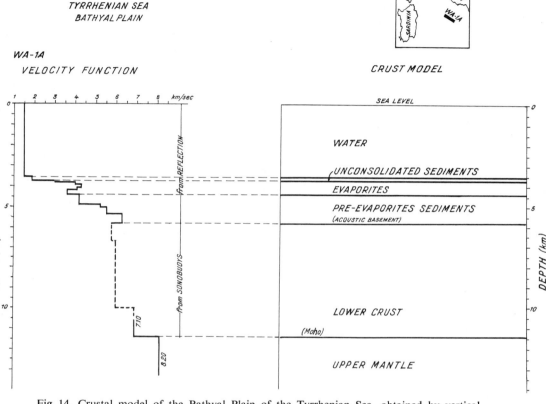

Fig. 14. Crustal model of the Bathyal Plain of the Tyrrhenian Sea, obtained by vertical seismic reflection and sonobuoy profiling interpretation. (After Finetti and Morelli, 1973.)

generally about 20–21 km thick, but with a maximum of 35 km in Sicily. In the southern part of the Tyrrhenian Sea the Moho lies at a depth of 20–25 km, and along the coast close to Salerno it is at 20 km. In Campania the crustal thickness is locally doubled by tectonic processes.

Finetti and Morelli (1973), on the basis of a reflection–sonobuoy profile, proposed a crustal model for the Bathyal Plain area (Fig. 14) in which the lower crust has a seismic velocity of 7.1 km/sec, the Moho lying at a depth of 12 km. Such a crust, one of the thinnest in the Mediterranean, is similar to typical oceanic crust. However, the 7.0 km/sec layer, between sediments and the Moho, appears to be limited in the Tyrrhenian Bathyal Plain to a thin layer in the lowest part of the crust (Finetti and Morelli, 1973).

The results of the crustal studies in the Italian region have been summarized by Giese and Morelli (1973) (Fig. 15). According to these authors, the

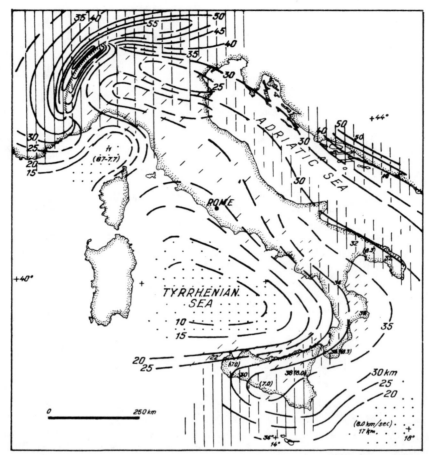

Fig. 15. Moho depth and types of the crust in the Italian regions. 1)▥, Zone with normal continental crust and mantle with $V_{pn} > 8$ km/sec; 2)▨, zone with anomalous continental crust and mantle with $V_{pn} < 8$ km/sec; 3)▦, crust of intermediate type; 4)⋮, oceanic crust. (After Giese and Morelli, 1973.)

northern Tyrrhenian area has a crustal thickness ranging between 15 and 25 km.

X. SEISMIC STRATIGRAPHY

Several reflection profiles have been carried out recently in the Tyrrhenian Sea (data in Ryan *et al.*, 1965; Selli, 1970*c*, 1974; Selli and Fabbri, 1971; Finetti and Morelli, 1972, 1973; Ryan *et al.*, 1973; and Morelli and Finetti, 1974). The profiles reported by Selli and Fabbri (1971) refer to four areas of the southern Tyrrhenian Sea, the Gioia and Sardinia Basins, and the eastern and western parts of the Bathyal Plain. Four main acoustic units, A_1, B_1, B_2, and C, bounded by reflectors X, Y, and Z, have been identified and successively dated with dredged and cored samples (Selli, 1974). Unit A, characterized by nearly continuous horizontal reflections, can be traced over long distances and is well developed in the Bathyal Plain, the peri-Tyrrhenian basins, and even on the shelf. It has a maximum thickness of 1100 m and overlies conformably, or nearly conformably, Unit B_1 and in some places Unit C. The age of Unit A is Middle–Upper Pliocene to Recent. At the base, Horizon X corresponds to transgressive conglomerates with littoral faunas and represents the "Middle Pliocene" transgression, which extends onto the Italian Peninsula. The B_2 unit is marked by less continuous and less regular reflections and contains folded and faulted layers. The age assigned is Lower Pliocene. Horizon Y, which is very well defined, marks the top of the evaporitic sequence. The M reflector of Ryan *et al.* (1971) corresponds, in general, to Horizon Y, but sometimes also to Horizon Z (Selli, 1974). Unit C is the substratum. This unit is Tortonian in age, and taking into account the relationship with the surrounding landmass probably includes all the Miocene from Tortonian to Aquitanian. Horizon Z marks an unconformity possibly corresponding to the Lower Miocene transgression, which represents an important event in the southern Apennines and Sicily. The thickness of the sedimentary sequence overlying the substratum ranges from 0 to 800–1200 m in the Bathyal Plain, to 3500 m in the Paola Basin (Selli and Fabbri, 1971), and 8000 m in the Corsica Basin (Finetti and Morelli, 1973).

Finetti and Morelli (1973) reinterpreted several seismic profiles on the basis of the JOIDES data and identified three main reflecting horizons: Horizon A (Horizon Y of Selli, 1974) is the base of Plio-Quaternary unconsolidated sediments or top of the evaporite sequence; Horizon B represents the base of the Miocene evaporites; Horizon S (Horizon Z of Selli, 1974) corresponds to the acoustic basement.

In the northern Tyrrhenian Sea Profile MS-38 of Finetti and Morelli (1973) (Fig. 16), taken normal to the continental slope in the Tuscan Archipelago zone, showed about 500 m of Plio-Quaternary sediments and 1200 m of mainly Miocene sediments lying on the acoustic basement. The Upper Miocene evaporites are absent. In the direction of the Ligurian Sea a positive structure, the northward extension of the Corsica Block, is encountered. Between Elba and Corsica, Profile MS-9 (Finetti and Morelli, 1973; Fig. 17) showed the Corsica Basin filled with 8 km of mostly Miocene sediments and bounded to the east by a plutonic body and a ridge 150 km long.

In Orosei Canyon, which separates the Corsica and Sardinia Basins (Fig. 1), Selli and Fabbri (1971) dated transgressive conglomerates as Middle Pliocene. These conglomerates contain basaltic pebbles with a radiometric age of 3 to

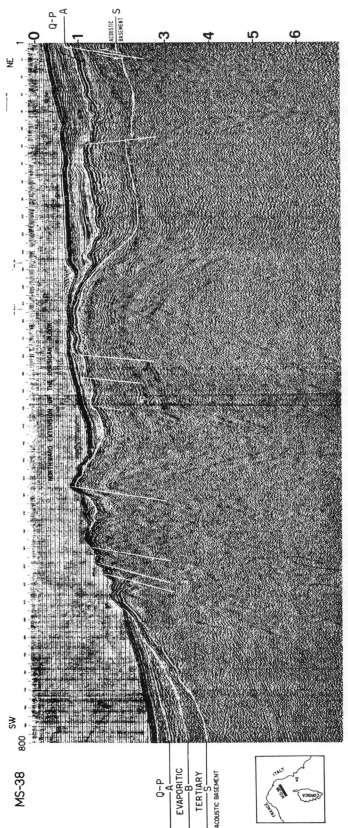

Fig. 16. Example of seismic reflection line on the northern Tyrrhenian Sea. In the eastern part a Quaternary–Pliocene layer (about 500 m thick) and about 1200 m of reflecting sediments overlay the acoustic basement. The marked positive structure in the central part represents the northward extension of the Sardinia–Corsica block. In the western part there is the Ligurian–Balearic Basin. (After Finetti and Morelli, 1973.)

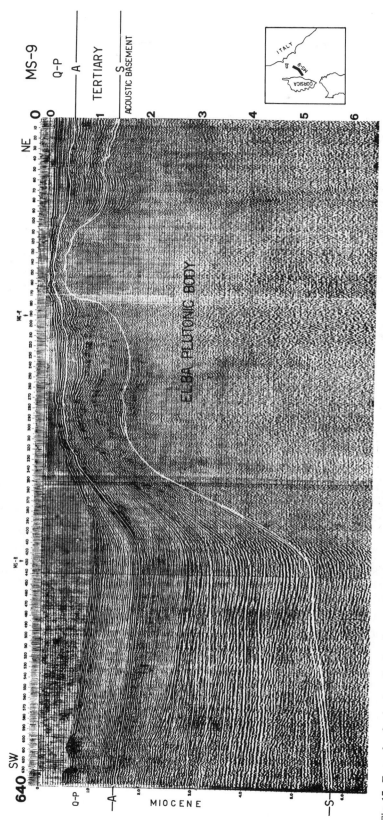

Fig. 17. Example of seismic reflection line between Corsica and Elba Island. In the western part below a thin Quaternary–Pliocene layer there appears to be a very thick pre-Pliocene sedimentary basin (Corsica Basin) with at least 8000 m of sediments, mostly Miocene in age. (After Finetti and Morelli, 1973.)

3.4 m.y., and are similar in age and composition to those cropping out in eastern Sardinia (Savelli and Pasini, 1974). In the Sardinia Basin evaporites have been identified. Evaporites are absent in the Corsica Basin and this is the most striking difference between the two basins.

The Middle Pliocene transgression, according to Selli and Fabbri (1971), occurred everywhere in the southern Tyrrhenian Sea. The Corsica and Sardinia Basins with their fill of Miocene sediments were displaced by distensive tectonic movements (Fig. 18; Finetti and Morelli, 1973), while the peri-Tyrrhenian basins of Calabria and Sicily show marked horizontal displacements and a thick fill of mainly Plio-Quaternary age (Fig. 19; Finetti and Morelli, 1973).

In most places on the lower continental slope, a thin Plio-Quaternary layer overlies the evaporitic sequence but may transgress onto older sediments or even onto the basement. Several horizons can be traced for long distances in the Bathyal Plain (Fig. 20). Unit A of Selli and Fabbri (1971) reaches a maximum thickness of 600 m, but is usually 200–300 m thick or less. Unit B_1 is thinner and never exceeds 200 m.

Finetti and Morelli (1973) presented a map of the Upper Miocene evaporitic interval in which the seismic velocity is 4 to 4.9 km/sec (Fig. 21). In the southern Tyrrhenian Sea the interval is present mainly in the northern and western peri-Tyrrhenian slopes and in the deeper southwestern areas, where it reaches a maximum thickness of about 800 m. The seismic evidence, tested with some wells of DSDP, Leg XIII (Sites 122, 124, 132, and 134) suggests that the evaporitic rocks are present in the Tyrrhenian Sea in the sulfate phase.

The genesis of the Mediterranean evaporites has been strongly debated since the JOIDES Expedition. The sedimentology of these deposits points to shallow environment similar to the present day sabkhas (Friedman, 1973; Nesteroff, 1973; Hsü et al., 1973). The isotopic data confirm that the evaporites were formed in a desiccated inland basin (Lloyd and Hsü, 1972). Hsü et al. (1973) and Cita (1974) state that the Tyrrhenian Sea was formed before the salinity crisis in the Middle–Upper Miocene and, therefore, propose a model of a desiccated deep basin separated from the Atlantic. According to this model, the evaporites would have been deposited, in playas, sabkhas, or alkaline lakes several thousand meters below the Atlantic sea level. This process occurred repeatedly in the Upper Miocene, thus allowing formation of a thick evaporite section. According to Finetti and Morelli (1973), a first subsidence took place during evaporite deposition, mainly in the peri-Tyrrhenian area. The sinking rate was compensated by evaporite deposition (1 km/m.y.) which took place, therefore, during the Upper Miocene in shallow water basins.

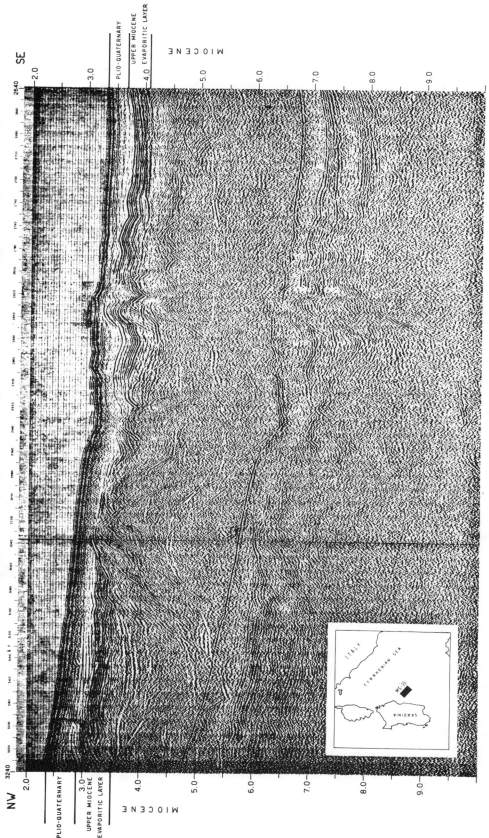

Fig. 18. Example of seismic reflection line on the eastern margin of Sardinia. Some small diapiric structures attributable to Miocene evaporitic series are present. The positive feature which affected the Miocene is associated with a long N-S ridge located east of Sardinia. (After Finetti and Morelli, 1973.)

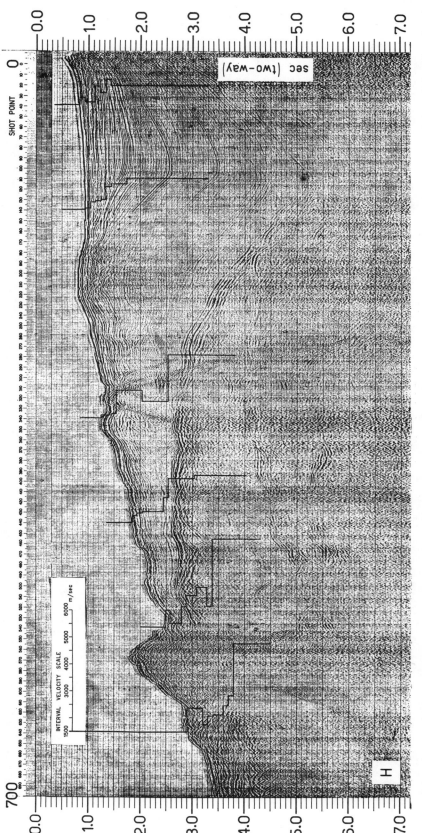

Fig. 19. Example of seismic reflection line on Calabrian margin. It shows a very thick basin filled by sediments, for the most part Quaternary–Pliocene in age. (After Finetti and Morelli, 1973.)

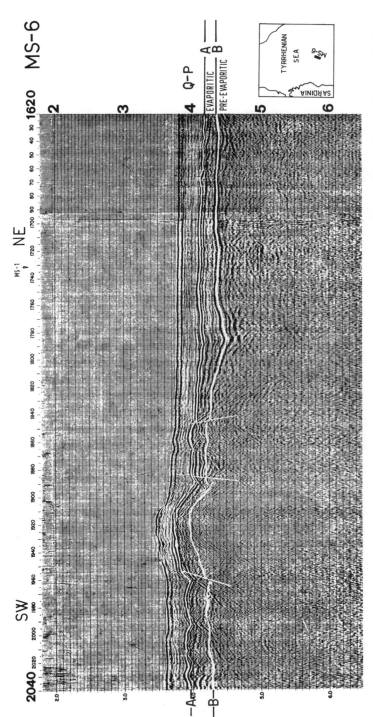

Fig. 20. Example of seismic reflection line on the Bathyal Plain. It shows a thin Quaternary–Pliocene layer, thin evaporites with no evidence of salt-phase, and relatively thin preevaporitic sediments. (After Finetti and Morelli, 1973.)

XI. STRATIGRAPHY

Hole 132 was drilled in the southern Tyrrhenian Sea during JOIDES Leg XIII (Ryan *et al.*, 1973), and is located between the slope and the Bathyal Plain at 2835 m. The drilled section is 223 m thick. It may be divided into three lithostratigraphic units (Fig. 22). Unit 1, of Pleistocene age, is 50 m thick and consists of foraminiferal marl ooze with intervening thin ash beds (tephra) mainly in its upper part, as well as structureless foraminiferal and terrigenous sand layers, probably reworked by gentle bottom current activity. Unit 2 is composed of foraminiferal ooze with rare, volcanic ash layers. The upper 20 m are of Pleistocene age, while the underlying 113 m are Pliocene. The Plio-Pleistocene boundary was located 70 m below the sea floor, where *Discoaster broweri* becomes extinct.

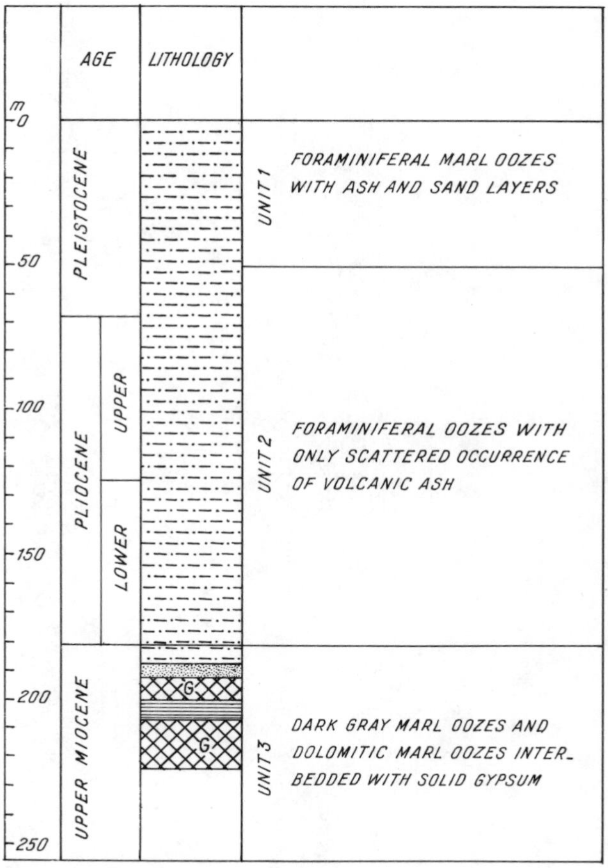

Fig. 22. Schematic lithostratigraphic column of Hole 132 (Leg 13, DSDP). (After Ryan *et al.*, 1973, modified.)

In the Pliocene section a facies similar to the "Trubi" of Sicily and Calabria has been encountered. The volcanic ash intercalations of Units 1 and 2 contain sanidine and biotite, and appear therefore to be of trachytic composition like some ashes of the Campania Province volcanoes (Nesteroff *et al.*, 1973). The sedimentation rate in the Plio-Pleistocene section ranges from 3.2 cm/1000 y. at the base to 3.8 cm/1000 y. at the top.

The contact between Units 2 and 3 is rather sharp in lithology and grain size as well as in calcium carbonate percentage, and points to a switch in the sedimentary environment.

Unit 3, 40 m of which have been drilled, consists of dark marl ooze, dolomitic marls with abundant gypsum crystals and pyrite, together with thick gypsum layers. The upper section of the evaporites displays marls, arkosic sands, and sandy clay with cross laminations. Gypsum with characteristic algal mat laminations is widespread in the lower drilled section (Nesteroff *et al.*, 1973). A single deep drilling, however, cannot give a picture of the sedimentology and lithology of the whole basin.

XII. SURFACE MINERALOGY AND SEDIMENTOLOGY

Piston and gravity cores and dredgings give a good knowledge of the surface sediments. Three thousand such stations over the Mediterranean served as a base for the compilation of a surface sediment distribution map by the Scripps Institution of Oceanography (Frazer *et al.*, 1970), which furnishes much of the information used here. In the southern Tyrrhenian Sea, muds and clay predominate in the north and along the Sicily coast. Calcareous oozes are widespread in the Bathyal Plain and along the eastern border of the sea. Calcareous mud or clay occurs between the Calabria–Campania coast and the Bathyal Plain. Along the circum-Tyrrhenian margins some sand and silt is also present.

Emelyanov (1972) published a comprehensive study of Mediterranean surface sediment mineralogy (topmost 20 cm) based on 340 samples collected during several cruises between 1959 and 1969. On the basis of quantitative data (size fraction 0.1–0.5 mm) he identified three mineralogic provinces— eastern Tyrrhenian, southern Sardinia, and Corsica. Tyrrhenian sediments generally have a high $CaCO_3$ content (up to 45–55%). Illite (50–70%) and montmorillonite (10–30%) predominate among the clay minerals, while kaolinite is more common in the marginal areas. Tomadin (1970) found the following average composition for 24 cores located in the central Tyrrhenian seamounts: illite 57%, montmorillonite and expandable minerals 14.5%, kaolinite 19.8%, and chlorite 8.7%. The illite–montmorillonite ratio increases northward away from the centers of volcanic activity.

According to Emelyanov (1972) the eastern Tyrrhenian province is marked by large amounts of glass and clinopyroxenes. In the southern Sardinia province epidote, zircon, and pyroclasts predominate, but in the Corsica province they are subordinate to metamorphic minerals (alkaline and fibrose amphiboles, basaltic hornblende, kyanite, micas, and epidote).

Few detailed mineralogical studies were carried out on cores retrieved from the Tyrrhenian Sea. The most detailed investigations are those stemming from the pioneering work of the Swedish Deep Sea Expedition 1946–1949 (Norin, 1958). More recently Ryan et al. (1965), Lenardon (1969), and Bartolini et al. (1974) published some mineralogical data on Bathyal Plain and Bathyal Slope sediments in the southern Tyrrhenian Sea. These sediments appear to be mostly volcanic with subordinate amounts of biogenic material.

Ryan et al. (1965), from the study of three long cores in the Bathyal Plain collected in 1961 by the R/V Chain, found two types of coarse-grained layers— eolian transported tephra and turbidity current deposits. Fierro and Passega (1965), analyzing the upper part of 32 cores collected on the slope of the northern section of the southern Tyrrhenian Sea, infer a pelagic origin for the finer samples, while concluding that a few coarser samples could have been laid down by bottom currents and, perhaps, turbidity currents.

Charnok et al. (1972) examined several cores and grab samples collected on the eastern margin near Naples and on the Bathyal Plain around Vavilov Seamount. They found only a few coarse-grained layers composed mainly of volcanic and biogenic material. Sediment dispersion appears to be due principally to bottom currents controlled by bottom morphology.

Bartolini et al. (1974) made a detailed study of an 8-m core retrieved from the Bathyal Plain by Saclant of La Spezia in 1965. It was correlated with 20 other cores located in the same area. A turbidity origin was assigned to the coarser layers by means of component analyses (Sarnthein and Bartolini, 1973), while the deposition process of the fine-grained layers appeared questionable. However, the comparison of the Quaternary volcanic products outcropping on land with the cored sediments of the Bathyal Plain apparently excluded a direct volcanic origin for some of the latter.

XIII. OUTLINE OF THE GEOLOGICAL SETTING OF LAND MASSES SURROUNDING THE TYRRHENIAN BASIN

The development of the Tyrrhenian Basin is closely connected to the geological evolution of the adjoining regions. Therefore, we shall attempt a brief geological outline of the structural units which make up the land masses surrounding the basin (Fig. 23).

A. Sardinia–Corsica Massif

Sardinia is a crystalline massif affected by two orogenic cycles, Caledonian and Hercynian, with compressive phases especially important during the first cycle. The Alpine Cycle had a very limited effect. Between the Triassic and the Eocene, local sedimentation took place, related to unimportant subsidences. From the Oligocene to the Quaternary, molasse troughs contemporaneous with the outer Alpine molasse were formed. These events were accompanied by calcalkaline magmatic activity during the Upper Oligocene–Miocene (from 24–22 m.y. to 13 m.y. ago) (Coulon, 1974), followed by alkali-basaltic magmatism during the Pliocene–Quaternary (3.6–2 m.y. ago) (Savelli and Pasini, 1974). Local displacements toward the NW of the Mesozoic carbonate and Tertiary conglomeratic cover, related by Chabrier (1970) to compressive tectonics, have been interpreted by other authors (Dieni and Massari, 1970; Cocozza et al., 1974) as being due to gravitational sliding during the Oligocene, associated with important vertical movements which led to the formation· of horst–graben structures. Three N–S-trending belts with different tectonic behavior can be distinguished in Sardinia—a western, a central, and an eastern belt (Cocozza et al., 1974). The western belt reacted in a somewhat elastic fashion to the vertical movements of the crust, which began in the Late Tertiary and which were related to those affecting the Ligurian–Balearic Basin. These movements resulted in marine transgression and the deposition of thick continental sequences in the grabens. The central belt seems to have had a more stable history, and was only locally submerged for short periods of time. Finally, the eastern belt reacted more rigidly and seems to be the one which was most affected by the Alpine orogenic cycle, with distensive movements related to the Tyrrhenian movements lasting until the Pliocene–Quaternary. The Alpine belt, present in the northeastern part of Corsica, must originally have been at least 50–100 km from the east coast of Sardinia (Alvarez and Cocozza, 1974).

B. The Italian Peninsula and Sicily

The Italian Peninsula can be subdivided geologically into three principal regions. From north to south these are—the northern Apennines Arc, the carbonate central–southern Apennines, and the Calabrian Arc, which continues westward into northern Sicily (Calabro-Sicilian Arc).

1. The Northern Apennines Arc*

The northern Apennines are geologically bounded to the NW by the Sestri–Voltaggio line and the Piemontais Basin, to the NE by the Padan

---

* See also Moullade, this volume.

Plain and the Tertiary Adriatic Basin, and to the SE by the Ancona–Anzio line. The arc consists of a series of great nappes accompanied by fold structures (Baldacci *et al.*, 1967; Abbate *et al.*, 1970; Dallan Nardi and Nardi, 1974). The overthrusting is prevalently Miocene in age and was followed in Pliocene–Quaternary times by late distensive phases with a constant NW–SE trend (Apennine trend). These trends cut across the earlier structures, including the Ancona–Anzio line, which separates the northern Apennines from the central–southern Apennines. In palinspastic reconstructions, the single tectonic units have been grouped into western units (Internides or eugeosynclinal sequences), which for the most part belong to a paleo-oceanic realm (Boccaletti *et al.*, 1971; Laubscher, 1971) and have undergone great displacement, and eastern units (Externides or miogeosynclinal sequences), which make up the cover of continental areas. These can be allochthonous (internal Tuscan sequence), parautochthonous, or autochthonous. The westernmost areas of the Internides realm (area of deposition of the westernmost Ligurian units) underwent an early tectogenesis, with folding during Cretaceous and Eocene times. The easternmost Ligurian units do not seem to have been affected by these early phases, and were involved only in later tectogenetic phases together with the units of the external domain (Baldacci *et al.*, 1967). The polarity of the early tectonic phases is still uncertain. Some authors maintain an eastward Apennine polarity (early Apennine phases), as does the early literature; others relate these events to Alpine phases, and suppose a prevalently westward Alpine polarity for them (Boccaletti and Guazzone, 1970, 1972, 1974*a*, *b*; Dallan Nardi *et al.* 1971; Haccard *et al.*, 1972; Elter and Pertusati, 1973; Dal Piaz, 1974).

Originally interposed between the Ligurian and the miogeosynclinal complexes (Tuscan–Umbrian sequences) was the area of deposition of the sub-Ligurian unit (Canetolo Complex). This unit is at present tectonically superposed on the Tuscan units (Macigno, Cervarola Sandstones) and is itself covered by the Ligurides. It can be related to the Sicilide units of the southern Apennines and Calabro-Sicilian Arc (Ogniben, 1969).

Two cross-sections representing the two principal Alpine and Apennine phases, Eocene and Tortonian in age, are shown in Fig. 24 (Boccaletti *et al.*, 1971). Important distensive phases (Ponto-Plio-Quaternary phases) affected the Tuscan area after the Tortonian. This led to the formation of horst and graben structures, accompanied by magmatic activity which migrated from west to east. At the same time the external Padan margin was affected by compressive phases. These distensive phases can be related to crustal extension taking place in the north Tyrrhenian area from Messinian time onward (Boccaletti and Guazzone, 1972, 1974*a*, *b*; Giglia, 1974).

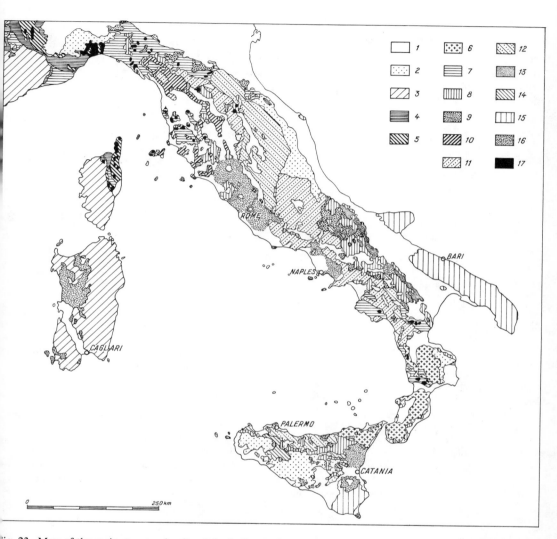

Fig. 23. Map of the main structural units of the Italian region. 1) Continental and marine sediments (post-Middle
Pliocene); Neogene of Tuscany; 2) "Gessoso solfifera" Formation; molasse (Tortonian); 3) external and foreland
"alpine" and "corso-sarde" sequences; 4) Briançonnais Zone (*s.l.*); 5) Piemontais Zone (*s.l.*); 6) Calabride complex; 7) Liguride (*s.l.*) complex; 8) Sicilide (*s.l.*) complex; 9) "Apennine" late orogenic flysch; 10) Tuscan nappe;
11) Tuscan autochthonous; Panormide (*s.l.*) complex, etc.; 12) Umbrian sequences; Lagonegrese and Imerese
units; 13) Daunia and Serra Palazzo flysch; Numidian flysch of Lucania; 14) "Trapanese–Sicani" sequence;
15) external and foreland "Apennine" sequences; 16) Cenozoic and Quaternary magmatism; 17) Mesozoic
ophiolites. (After Ogniben *et al.*, 1973, slightly modified.)

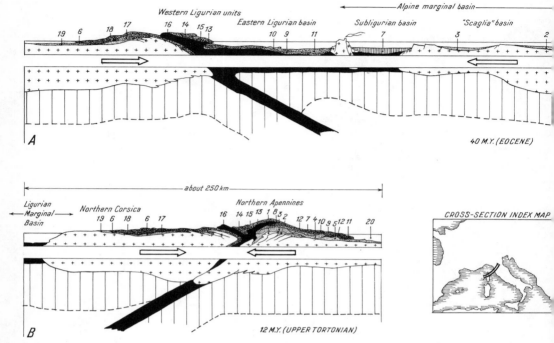

Fig. 24. Schematic sequences of sections across northern Corsica illustrating the Tertiary development of the Alps–Apennines structures. (After Boccaletti *et al.*, 1971.) 1) Massa Zone; 2) Apuane autochthonous cover; 3) Tuscany nappe; 4) Monte Cervarola unit; 5) Umbro-Marchigiana Zone; 6) Briançonnais and sub-Briançonnais Zone; 7) Canetolo unit; 8) Monte Caio unit; 9) basal complex of Ligurian flysch; 10) Monte Cassio unit; 11) Monte Sporno unit; 12) Oligo-Miocene Ranzano sequence; 13) Monte Gottero unit; 14) Monte Antola unit; 15) Bracco ophiolitic unit; 16) Massiccio di Voltri ophiolitic unit; 17) schistes lustrés unit; 18) Balagne Ligurian unit; 19) Corsica autochthonous cover; 20) Po Valley post-orogenic formations. Black zone—oceanic crust; crosses—continental crust; large vertical hatched zone—upper mantle; white—asthenosphere; light stipple—"miogeosynclinal" covers; heavy stipple—"eugeosynclinal" covers; dense vertical hatched zone— sub-Ligurian unit.

2. The Central–Southern Apennines

The central–southern Apennines also consists of a series of nappes which can be grouped into structural–stratigraphic units (Selli, 1962; Accordi, 1966; Pieri, 1966; D'Argenio *et al.*, 1973). These originated both from oceanic realms (internal units) and from continental realms (external units). The internal units are common also to the Calabro-Sicilian Arc and will, therefore, be described later.

In the external units the remains of three carbonate platforms have been recognized paleogeographically. These are now found piled upon one another.

They were originally bounded by basins in which pelagic carbonates and clastic (flysch) sedimentation took place. Only limited transition zones between the platforms now remain as evidence of the existence of these basins (D'Argenio, 1970; D'Argenio et al., 1973). Compressive phases began affecting the westernmost carbonate platform (Campano-Lucanian platform) during the Burdigalian, and activity migrated eastward until by Middle Pliocene time the other platforms were also involved.

From the Upper Pliocene onward, movements were prevalently vertical, causing limited gravity sliding in the internal zones, which continued until the Calabrian in the Bradanic fore-trough.

In the central Apennines also, distinct Bahamas-type carbonate platforms have been recognized, piled up on one another, and separated by basins characterized by pelagic carbonate sediments followed by flysch (Devoto and Praturlon, 1973). Units belonging to a more western region are found tectonically superposed directly upon these sediments. These units can be correlated with the Sicilides (Ogniben, 1969; Ogniben et al., 1973). Here too the tectogenetic phases are Miocene–Pliocene in age (Accordi, 1966; D'Argenio et al., 1973).

3. The Calabro-Sicilian Arc

The Sicilian Arc consists of a nappe structure built up in several orogenic phases, the last of which was very recent (Ogniben, 1960; 1969; 1973). The tectonic units which make up this arc can be separated into "internal" and "external" units. Geologically, the Calabro-Lucanian Apennines are very similar to the central–southern Apennines, especially with regard to the Mesozoic limestones and dolomites of the carbonate shelves (D'Argenio et al., 1973), which also continue into Sicily.

In recent years, numerous general studies have been carried out by Italian and by French workers (Caire et al., 1961; Dubois, 1970; Haccard et al., 1972). Ogniben proposed a synthesis of the geological evolution of the Lucania–Calabria boundary area (1969, 1973) and Sicily (1960, 1970), distinguishing various complexes belonging to different paleogeographic realms. From east to west, these are: a) Apulian or Iblean foreland; b) miogeosyncline (Basal and ex-Basal Complex), respectively consisting of siliceous–calcareous sequences and of flysch, including Numidian flysch; c) intermediate shelf (Panormide Complex, represented by the Mesozoic calcareous dolomitic sequences and by Miocene transgressive sediments); d) eugeosyncline, divided in two parts—"internal" (Liguride Complex, which does not outcrop in Sicily and which in the Calabro-Lucania area includes flysch-type sequences containing ophiolites, sometimes metamorphosed, and Calpionella limestones) and "external" (Sicilide Complex, consisting of variegated clays and flysch); and e) internal massif (Calabride Complex, consisting of Hercynian and pre-Hercynian basement, e.g., the dioritico-kinzigite formation).

According to Ogniben (1969), these units were piled upon one another in five principal tectonic phases, which started during the Eocene and propagated from the internal to the external zones until after the Tortonian. More recently, Scandone (1972) and D'Argenio *et al.* (1973) presented an outline of the tectonic evolution limited to the external zones, in which three principal phases are defined (Burdigalian, Tortonian, and Middle Pliocene), with a polarity from the interior to the exterior (that is, toward the SE). The evolution of the internal zones is more complex and controversial.

According to Dietrich and Scandone (1972) two principal types of ophio- litic units can be distinguished in the Ligurian Complex of the Lucania–Ca- labria area: a) Unit I, affected by high pressure metamorphism dated as Early Cretaceous (Haccard *et al.*, 1972) and correlated with the Piemontais units present in the western Alps and Alpine Corsica, and b) Unit II, with only slightly metamorphosed ophiolites correlated with those of the Ligurian units present in the northern Apennines and in a few areas of the Alps (e.g., Ubaye– Embrunais and Simme Nappes) (Elter *et al.* 1966; Dal Piaz, 1974). The Li- gurian-type units tend to disappear southward. The polarity of the first tectonic phases of the internal zones is still uncertain. Some authors (Boccaletti and Guazzone, 1972; 1974*a, b*; Dietrich and Scandone, 1972; Haccard *et al.*, 1972), on the basis of structural data presented by earlier authors, suggest that the first tectonic movements which affected the internal zones (during Cre- taceous–Eocene times) were from east to west (Alpine polarity), so that the Calabride Complex is considered to be of African origin and not of European origin.

Recent tectonic activity affecting the Lucania–Calabria area is charac- terized by vertical movements (Bousquet, 1973) and correlated with the recent structural evolution of the whole southern Tyrrhenian area. This area is still tectonically active, with compressive movements taking place in the Ionian zone, external to the Calabrian Arc, as shown by seismic reflection studies (Finetti and Morelli, 1973; Fig. 25).

XIV. PALEOMAGNETIC DATA ON THE LAND MASSES SURROUNDING THE TYRRHENIAN BASIN

Anticlockwise rotations of Corsica and Sardinia, as well as of other sialic massifs of the western Mediterranean (e.g., Iberian Peninsula), were proposed by Argand (1922) and later by Carey (1955) on the basis of geodynamic con- siderations. Recent paleomagnetic studies carried out in the area have brought new evidence in favor of this hypothesis.

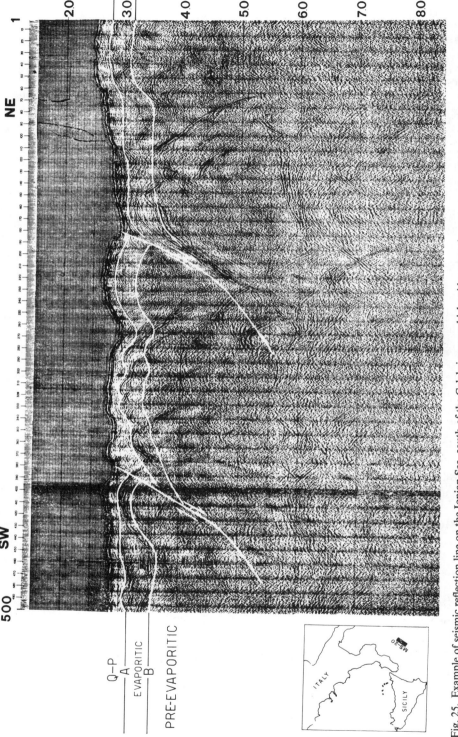

Fig. 25. Example of seismic reflection line on the Ionian Sea, south of the Calabrian arc which evidences active compressional movements with overthrusts. (After Finetti and Morelli, 1973.)

A. Corsica and Sardinia*

Paleomagnetic data in favor of an anticlockwise rotation of Corsica with respect to stable (extra-Alpine) Europe have been collected by Ashworth and Nairn (1964) and Nairn and Westphal (1968) from Permo-Carboniferous volcanic rocks and dikes. Values found indicate an anticlockwise rotation of about 25° since the Upper Permian. More data are available for Sardinia. De Jong *et al.* (1969) obtained paleomagnetic results from pre-Helvetian trachy-andesitic rocks near Alghero, which indicate a late Oligocene–early Miocene rotation of Sardinia with respect to stable Europe. Zijderveld *et al.* (1970a) reported paleomagnetic data on Permian rocks of Sardinia, which show a counterclockwise rotation of the island of about 50°. Subsequently, Bobier and Coulon (1970) confirmed these results by sampling other volcanic rocks from the same area, and showed that the paleomagnetic direction of the post-Tortonian (generally Pliocene) basalts was reasonably close to the present axial-dipole direction, which would mean that Sardinia rotated before the Pliocene, in contrast with the hypothesis advanced by Alvarez (1972) of a post-Tortonian rotation. Stanley and Mutti (1968) have shown, on sedimentological grounds, that Corsica as well as Sardinia must still have been attached to Europe during the Paleogene.

More recently De Jong *et al.* (1973) extended paleomagnetic research to Tertiary volcanic rocks (andesites, trachytes, and liparites) from near Castel Sardo, on the north coast of Sardinia. The results confirmed a counterclockwise rotation of the island of an angle of approximately 50° with respect to stable Europe during the early Miocene. New data have been gathered by Orsini *et al.* (1974) on Permian rocks sampled along the Campidano Graben. These confirm the previous results (see Manzoni, 1974b). A reasonably similar value for the rotation was obtained by Westphal *et al.* (1973) by fitting Corsica and Sardinia against the Franco-Spanish coast.

The different values of rotation for Corsica and Sardinia suggested to Alvarez (1974) the idea of a differential rotation of the two islands.

B. The Italian Peninsula

There are few paleomagnetic data relating to the Italian Peninsula. In the northern Apennines, Lowrie and Alvarez (1974) sampled the Umbrian carbonate sequence of Upper Cretaceous–Paleogene age and obtained paleomagnetic results which indicated an anticlockwise rotation of approximately 40° with respect to stable Europe during the Cretaceous. Total rotation by mid-Eocene time would then have totaled about 70° (Lowrie and Alvarez, 1974). Because these values concern the sedimentary cover, it has yet to be

---

* See also Moullade, this volume.

ascertained whether the proposed rotation also affected the crustal substratum or not. A similar but more complex global rotation for the same area has been suggested by Premoli Silva *et al.* (1974) on the basis of a detailed study of magnetic stratigraphy. Further research in the same Umbrian area (Napoleone *et al.*, 1974) has revealed details in the stratigraphic sequence, but has not confirmed the differential rotation proposed by Lowrie and Alvarez (1974). All these paleomagnetic values are in general agreement with the hypothesis of an anticlockwise rotation of the northern Apennines flysch basins suggested by Boccaletti and Guazzone (1970), based on the rotation of directions of paleocurrents.

Paleomagnetic data indicating an anticlockwise rotation have also been collected for Permian and Triassic rocks of the southern Alps (Zijderveld *et al.*, 1970*b*; Manzoni, 1970) and for Tertiary rocks of the southern Alpine foothills (Soffel, 1972, 1974). The last author, sampling Eocene and Oligocene volcanic rocks of the Colli Euganei, has shown an anticlockwise rotation of about 50° for the Upper Eocene rocks. Lowrie and Alvarez (1974) point out that paleomagnetic studies in the southern Alpine region cannot be used as evidence for the rotation of the Italian Peninsula because the two areas are separated by the Po Basin, whose sediments have been subjected to important compressional tectonic activity during the Pliocene–Quaternary. For the southern Apennines, no detailed paleomagnetic data have yet been published. Research on carbonate rocks and associated bauxites in Gargano, Matese, and Cilento has been started by Channel, D'Argenio, Napoleone, and Tarling and is still in progress. The first results indicate a counterclockwise rotation of approximately 13° with respect to stable Europe since the Middle Cretaceous for the Apulia platforms, and of about +32° for the Campania platform (Channel and Tarling, 1975).

Paleomagnetic studies have been carried out by Schult (1973) and Barberi *et al.* (1974*b*) on volcanic rocks of Upper Cretaceous age in the Sicilian foreland (eastern Sicily). Data obtained indicate that there have been no significant rotational movements of this part of Sicily with respect to Africa. According to Barberi *et al.* (1974*b*), Sicily rotated anticlockwise together with Africa, showing a northward drift relative to Europe during the Tertiary. Since the end of the Miocene, Sicily has not moved with respect to stable Europe. Other paleomagnetic studies have been carried out on carbonate rocks of tectonic units of Mesozoic age in northwestern Sicily (near Palermo) by Catalano, Channel, D'Argenio, and Napoleone. From unpublished results, this area appears to have undergone clockwise rotation relative to stable Europe (Napoleone, 1975, personal communication).

It is possible, from presently available data, to support the rotation, in varying degrees, of landmasses surrounding the Tyrrhenian Basin. Data are still scarce and further research in this field is necessary to determine the sense and the degree of rotation for each area, as well as the exact age of these movements in order to distinguish Cretaceous–Eocene (Alpine age) rotations from Neogene (Apennine age) rotations. Furthermore, it must be ascertained whether the rotational movements affected only the sedimentary cover which makes up the single tectonic units, and must consequently be considered only as effects of the single detachments, or whether they also affected the crustal substratum.

XV. HYPOTHESIS ON THE GENESIS OF THE TYRRHENIAN BASIN

Various and contrasting hypotheses have been advanced for the genesis of the Tyrrhenian Basin. Argand (1922) and Carey (1955) have suggested the presence of Recent oceanic crust in the Tyrrhenian Basin. Glangeaud (1962) considered the Tyrrhenian as the residue of a paleo-ocean which formed after Permian time. Dercourt (1970) and Smith (1971) suggested that the Mediterranean basins, including the Tyrrhenian, are remnants of an oceanic zone which formed by extension during the Jurassic. Other authors, such as Van Bemmelen (1969) and Muratov (1972), believe the Tyrrhenian Basin to have developed through the collapse of a craton, the oceanization process occurring through subcrustal erosion processes. According to Selli (1974) the oceanization of the Tyrrhenian area and the consequent thinning of the sialic layer is due to a mechanical process related to deep magmatic currents, a process causing the displacement of the basal sial from the center of the Tyrrhenian Sea toward the Apennines, where it formed the "orogenic" roots.

From comparison of geophysical, petrological, and structural data of the Tyrrhenian Basin with those of some western Pacific marginal basins, Boccaletti and Guazzone (1972) and Bousquet (1973) suggested that the Tyrrhenian Basin was formed through an expansion mechanism similar to that proposed for the western Pacific marginal basins (Karig, 1971). The very position of the Tyrrhenian Basin within an orogenic arc is in agreement with this hypothesis, which was later confirmed also by Barberi et al. (1973, 1974b), Dewey et al. (1973), and Loddo and Mongelli (1974a). The southern Tyrrhenian Basin, according to this hypothesis, belongs to a Neogene arc–trench marginal basin system. In this system the outer trench is represented by the Ionian Trough, the folded arc by the southeastward migrating Calabrian Arc, and the magmatic arc by the Eolian Islands with their calcalkaline and shoshonitic magmatism. The southern Tyrrhenian Sea, with its basaltic magmatism in the Bathyal Plain and with its geophysical characteristics (heat flow, gravimetric anomalies, crustal thickness, etc.), represents the back-arc marginal basin. In the same context, the Tyrrhenian peripheral basins (Cefalù, Gioia, and Paola Basins), which are present along the Siculo-Calabrian margin, could represent basins developed within the arc–trench gap. Similarly, the northern Tyrrhenian Sea (= SE Ligurian Sea) has been considered as a back-arc marginal basin which is still in a pre-spreading stage behind the folded arc of the northern Apennines (Boccaletti and Guazzone, 1972).

The superposition of the tensional tectonic phases which affected Tuscany since Messinian time on the pre-Tortonian and Tortonian compressive phases (Giglia, 1974), the eastward migration of the post-Messinian intramontane basins, and the eastward migration of the Tuscany–Latium magmatism are all events related to the spreading of the Tyrrhenian sea floor behind the Apennine folded arc, as this latter migrated to the east.

The different degrees of opening of the southern and northern parts of the basin would then have caused the fragmentation and displacement of the older structural units, in particular those of the Cretaceous–Eocene (Boccaletti and Guazzone, 1974; Alvarez et al., 1974). This is suggested mainly by the present-day position of the outcrops of the Ligurids and Sicilids (Fig. 23).

The ophiolites (present in Ligurian units) are, therefore, remnants of a portion of Mesozoic oceanic crust, which, during the Cretaceous–Eocene, represented a geosuture and which was subsequently broken up as a result of Neogene evolution of the arcs and the opening of the back-arc marginal basins.

Naturally, the evolution of the Tyrrhenian Basin must be seen in the context of the evolution of the whole of the western Mediterranean area. An outline of the development of the Tyrrhenian area in this light has recently been proposed by Boccaletti and Guazzone (1974a) and Boccaletti, Guazzone, and Manetti (1974). They consider the western Mediterranean as consisting of a set of back-arc marginal basins in different stages of development, spreading behind the Rif–Tellian and Apennine Arcs, which migrated during Miocene and post-Miocene times (Apennine polarity phase). These arcs developed during the Cretaceous and Eocene.

In Fig. 26a–e, a reconstruction of the structural evolution of the western Mediterranean area since the Cretaceous, with the paleogeographic positions of the principal basins, is attempted by Boccaletti, Guazzone, and Manetti (1974). Fig. 26a represents the late Cretaceous situation, with an oceanic zone in an advanced phase of consumption. The detachment of sialic micro-continents from the African continental margin, the migration of the arcs, and the spreading of marginal basins behind the arcs are related to "Alpine" crustal subduction, with S–SE-dipping Benioff zones (Boccaletti et al., 1971), and ocean trenches where deposition of Alpine flysch (Helminthoid flysch and equivalents) was most active. The ocean floor external to the trenches is considered as the realm of the Schistes Lustrés units, which crop out in the Alps, in Alpine Corsica, and in Calabria and which are characterized by a high pressure metamorphism related to Alpine subduction processes (Ernst, 1971; Dal Piaz, 1974; Dietrich et al., 1974). The oceanic zone between the inner wall of the

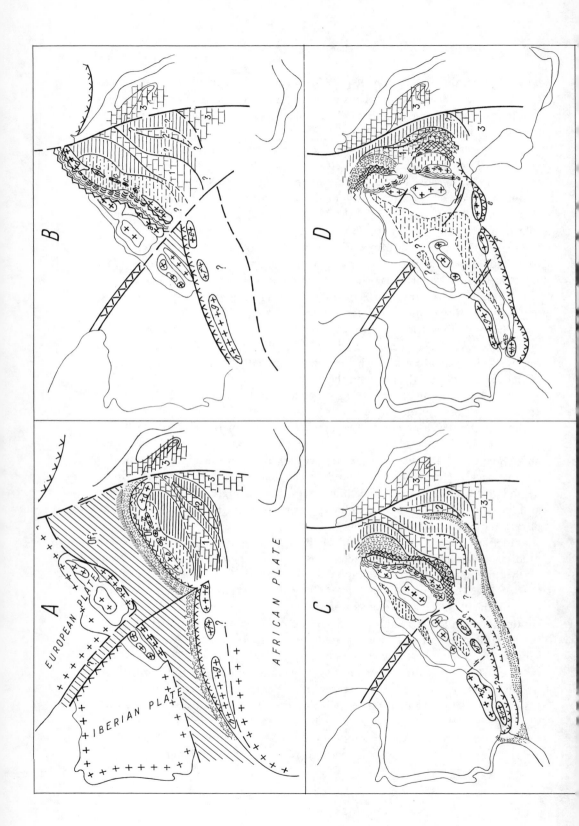

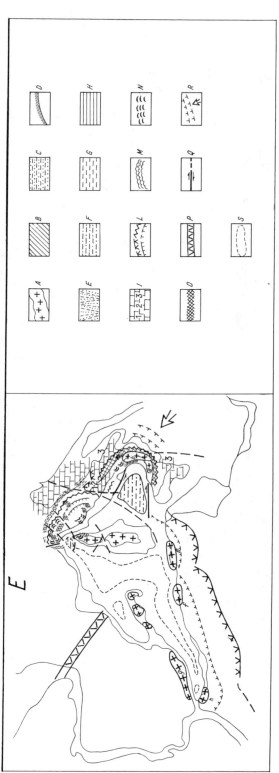

Fig. 26. Sequence of sketch maps showing the geodynamic evolution of the main tectonic units in the central-western Mediterranean area since the Upper Cretaceous. (After Boccaletti *et al*., 1974.) Many sedimentary covers are not drafted. No distinction between emerged and submerged zones has been made. No scale. A) Continental margins and sialic elements; B) ocean floor; C) flysch deposition in oceanic trenches during the Alpine phases (Helminthoides flysch p.p.) and/or in pseudo-trenches during the Apennine phases (Macigno and Marnoso-Arenacea Formations; Petrignacola, Aveto, and Tusa Formations and equivalents); D) Piemontais units in the scraping area along the Betic–Alpine subduction zones; E) flysch deposition on the continental rise (Numidian flysch and equivalents) and eventual extension onto borders of carbonate platforms; F) sediment deposition in back-arc marginal basins, both in Alpine and Apennine phases (Canetolo Complex and equivalents; Sicilide Complex; sediments of Balearic–Ligurian Sea and of Tyrrhenian Sea); G) sediment deposition in the arc–trench gaps, both in the Alpine and Apennine phases (Mt. Piano marls; Farini d'Olmo Formation; Ranzano–Bismantova sequences; Reitano Formation; Capo d'Orlando and Gorgoglione Formations, etc.); H) pelagic sediments on thinned and fragmented substratum and particularly into basins between carbonate platforms (Scaglia Formation; Scisti Policromi Formation; Scaglia cinerea Formation; pelagic sediments in the Lagonegro Basin, Molisano Basin, Imerese Basin); I) carbonate platforms belonging to the African continental margin, more or less detached and fragmented during the Mesozoic oceanic spreading and the later marginal basin expansion. They partially represent first the Alpine backland and successively the Apennine foreland [tentatively: 1) Panormide–Campania–Lucania platform; 2) Abruzzese–Campania platform; 3) Iblean–Apulian platforms and external Dinaric platforms]; L) folded arcs and main overthrusts of the Piemontais units and related flysch with Apennine polarity; O) main overthrusts of Sicilide complexes with Apennine polarity; P) Pyrenean suture; Q) oceanic and intracontinental transform faults and transcurrent lines; R) Ionian active subduction zone; S) inactive marginal basins borders. a) paleo-Calabrian basement; b) Calabride sialic cover; c) sialic fragments of the hypothetical Insubric "Ridge" alignment; d) Austroalpine p.p. sialic cover; e–f) Kabylides p.p. sialic cover; g) Alpujarrides p.p. sialic cover; h) Sebtides sialic covers.

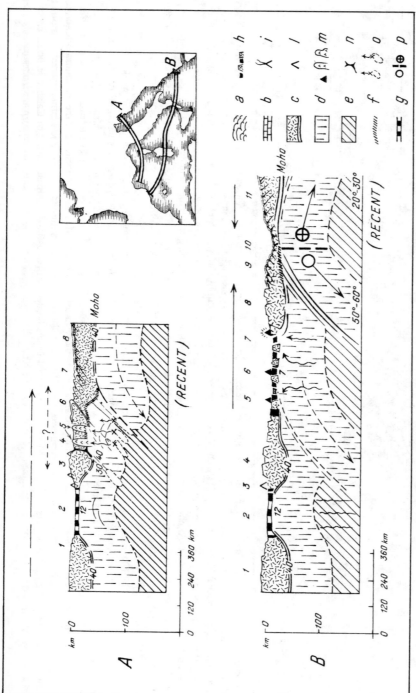

Fig. 27. Schematic sections across northern (A) and southern (B) Apennine arcs. (After Boccaletti and Guazzone, 1974b.) a) Plastic cover; b) carbonate covers; c) continental crust; d) upper mantle; e) low viscosity asthenosphere; f) detachment zone; g) oceanized floor of the Ligurian–Balearic marginal basin; h) spread floor of the southern Tyrrhenian marginal basin; i) "Alpine" magmatic paleo-arc; l) Oligo-Miocene "Apennine" magmatic arc; m) post-Miocene "Apennine" magmatic arc; n) geosuture zone; o) magmatic arc migration; p) triple point; A) section across Corsica, from Provence to Yugoslavia—1) European continental margin; 2) present Ligurian Sea; 3) Corsica; 4) north Tyrrhenian–Tuscany Sea; 5) Elba's and Monte Amiata-Roccastrada's magmatism; 6) present ridge of the Apennines; 7) Adriatic Sea; 8) Yugoslavian coast. B) section from Provence across Sardinia and Calabria to external Hellenic Arc—1) European continental margin; 2) present Ligurian Sea; 3–4) Sardinia remnant arc with magmatism; 5–6) basin magmatism by the spreading of the Tyrrhenian Basin; 7) calcalkaline magmatism of the Eolian Islands; 8) Calabria; 9) folded Calabrian Arc;

trenches and the African continental margin was the area of deposition of the Ligurian units, with their ophiolitic members characterized by an oceanic-type low-pressure metamorphism, which crop out in some parts of the Alps (e.g., Simme Nappe) in the northern Apennines, Calabria, and northeastern Corsica.

During the Eocene (Fig. 26b) an arc–continent collision with northward polarity took place, causing the final up-building of the Alpine Arc by collision and overthrusting onto the paleo-European margin of "oceanic" tectonic units (the Piemontais and Ligurian-type units) as well as sialic parts of African origin (the Austroalpine p.p., Calabrides, Kabylides, etc.). During this phase, the "Alpine" marginal basins reached their maximum expansion and the carbonate shelves underwent their maximum fragmentation, with an increase in the dimension of the pelagic basins between the shelves. Sicilide-type sedimentation took place in the wider basins ("Kalk-ton Series").

Fig. 26c, represents the situation in Lower Miocene time, when new arc–trench systems having a reversed Apennine–Rif–Tellian polarity were developing at the expense of the former Alpine marginal paleo-basins. The deposition of flysch, such as the Macigno of the northern Apennines, began in the "pseudo-trenches," and the Numidian flysch began to be deposited on the continental rise off the African neo-margin (Wezel, 1970). The first Apennine–Rif–Tellian overthrusting began at this time. New marginal basins related to the new arc–trench systems began to develop in the western Mediterranean. One of these was the Balearic–Ligurian Basin, which reached its maximum expansion in Burdigalian–Tortonian times (Fig. 26d), when maximum overthrusting was affecting the Apennine Arcs which were migrating south, east, and northeast. The Ligurides and Sicilides also were involved in dislocations during this stage.

The last stage (Fig. 26e) represents the situation in Recent and present times. It is during this stage that the maximum expansion was reached by the Tyrrhenian Basin behind the generally eastward migrating Apennine Arcs (Fig. 27). According to Alvarez *et al.* (1974), an initial phase of expansion affected the Tyrrhenian Basin during the Upper Miocene up to the Anzio–Le Galite scarp line, where evaporites were deposited during the Messinian.

In the Tyrrhenian area, two major basins can be distinguished—the North Tyrrhenian Basin (= E Ligurian area) behind the northern Apennines Arc, and the south Tyrrhenian Basin, in a more advanced stage of spreading, behind the still active Calabrian Arc.

ACKNOWLEDGMENT

We wish especially to acknowledge the critical discussion of Dr. C. Bartolini (Geological Institute of Florence) about the sedimentology of the Tyrrhenian Basin.

REFERENCES

Abbate, E., Bortolotti, V., Passerini, P., Sagri, M., and Sestini, G., 1970, Development of the Northern Apennines Geosyncline, *Sediment. Geol.*, v. 4(314), p. 207–642.

Accordi, B., 1966, La componente traslativa nella tettonica dell'Appennino Laziale–Abruzzese, *Geol. Romana*, v. 5, p. 355–406.

Allan, T. D., and Morelli, C., 1971, A geophysical study of the Mediterranean Sea, *Boll. Geol. Teor. Appl.*, v. 13(50), p. 99–141.

Alvarez, W., 1972, Rotation of the Corsica–Sardinia Microplate, *Nature Phys. Sci.*, v. 235, p. 103–105.

Alvarez, W., 1974, Sardinia and Corsica: one microplate or two? in: *Paleogeografia del Terziario Sardo nell'Ambito del Mediterraneo Occidentale*, Cagliari, 23–27 July, 1973, *Rend. Semin. Fac. Sci. Univ. Cagliari*, Suppl., v. 43, p. 1–4.

Alvarez, W., and Cocozza, T., 1974, The tectonics of central-eastern Sardinia and the possible continuation of the Alpine Chain to the South of Corsica, in: *Paleogeografia del Terziario Sardo nell'Ambito del Mediterraneo Occidentale*, Cagliari, 23–27 Juli, 1973, *Rend. Semin. Fac. Sci. Univ. Cagliari*, Suppl., v. 43, p. 5–34.

Alvarez, W., Cocozza, T., and Wezel, F. C., 1974, Fragmentation of the Alpine orogenetic belt by microplate dispersal, *Nature*, v. 248, p. 309–314.

Argand, E., 1922, La téctonique de l'Asie, *3rd Int. Geol. Congr.*, Liège, v. 1, p. 171–372.

Ashworth, T. P., and Nairn, A. E. M., 1964, An anomalous Permian pole from Corsica, *Paleogeogr. Palaeoclimatol. Palaeoecol.*, v. 1, p. 119–125.

Baldacci, F., Elter, P., Giannini, E., Giglia, G., Lazzarotto, R., Nardi, R., and Tongiorgi, M., 1967, Nuove interpretazioni sul problema della Falda Toscana e sulla interpretazione dei flysch arenacei tipo "Macigno" dell'Appennino Settentrionale, *Mem. Soc. Geol. Ital.*, v. 11, p. 367–390.

Barberi, F., Borsi, S., Ferrara, G., and Innocenti, F., 1967, Contributo alla conoscenza vulcanologica e magmatologica delle Isole dell'Arcipelago Pontino, *Mem. Soc. Geol. Ital.*, v. 4, p. 581–606.

Barberi, F., Borsi, S., Ferrara, G., and Innocenti, F., 1969, Strontium isotopic composition of some recent basic volcanites of the Southern Tyrrhenian Sea and Sicily Channel, *Contrib. Mineral. Petrol.*, v. 23, p. 157–172.

Barberi, F., Innocenti, F., and Ricci, C. A., 1971, Il magmatismo nell'Appennino Centro Settentrionale, *Rend. Soc. Ital. Mineral. Petrol.*, v. 37 (Special Paper), p. 169–210.

Barberi, F., Gasparini, P., Innocenti, F., and Villari, L., 1973, Volcanism of the Southern Tyrrhenian Sea and its geodynamic implications, *J. Geophys. Res.*, v. 78, p. 5221–5232.

Barberi, F., Innocenti, F., Ferrara, G., Keller, J., and Villari, L., 1974a, Evolution of Eolian Arc volcanism (Southern Tyrrhenian Sea), *Earth Planet. Sci. Lett.*, v. 21, p. 269–276.

Barberi, F., Civetta, L., Gasparini, P., Innocenti, F., Scandone, R., and Villari, L., 1974b, Evolution of a section of the Africa–Europe plate boundary: paleomagmatic and volcanological evidences from Sicily, *Earth Planet. Sci. Lett.*, v. 22, p. 123–132.

Barberi, F., Innocenti, F., Marinelli, G., and Mazzuoli, R., 1974c, Vulcanismo e tettonica a placete: esempi nell'area mediterranea, 67 Congr. Soc. Geol. Ital., Parma, 27–31 October 1974, in: *L'Italia nell'Ambito dell'Evoluzione del Mediterraneo*, *Mem. Soc. Geol. Ital.*, Suppl., 89 p.

Barberi, F., Bizouard, H., Capaldi, G., Ferrara, G., Gasparini, P., Innocenti, F., Joron, J. L., Lambret, B., Treuil, M., and Allègre, 1978, Age and nature of basalts from the Tyrrhenian Abyssal Plain, in: *Initial Reports of the Deep Sea Drilling Project*, Ryan, W. B. F., Hsü, K. J., *et al.*, eds., Washington, D.C.: U.S. Government Printing Office.

Bartolini, C., De Giuli, C., and Gianelli, G., 1974, Turbidites of the Southern Tyrrhenian Sea, *Boll. Soc. Geol. Ital.*, v. 93, p. 3–22.

Basilone, P., and Civetta, L., 1975, Datazione K–Ar dell'attività vulcanica dei Mt. Ernici (Latina), *Rend. Soc. Ital. Mineral. Petrol.*, v. 31, p. 175–179.

Bemmelen, R. W. Van, 1969, Origin of Western Mediterranean Sea, *Verh. Kon. Ned. Geol. Mijnb.*, v. 51, p. 548–573.

Bigazzi, G., and Bonadonna, F. P., 1973, Fission track dating of the obsidian of Lipari Island (Italy), *Nature*, v. 242, p. 322–323.

Bini, C., Faraone, D., and Giaquinto, S., 1973, Isola di Vulcano: le latiti di Vulcanello, *Per. Mineral.*, v. 42, p. 535–581.

Bobier, C., and Coulon, C., 1970, Résultats préliminaires d'une étude paléomagnétique des formations volcaniques tertiaires et quaternaires du Logudoro (Sardaigne Septentrionale), *C. R. Acad. Sci. Paris*, v. 270, p. 1434–1437.

Boccaletti, M., and Guazzone, G., 1970, La migrazione terziaria dei bacini toscani e la rotazione dell'Appennino Settentrionale in una "zona di torsione" per deriva continentale, *Mem. Soc. Geol. Ital.*, v. 9, p. 177–195.

Boccaletti, M., and Guazzone, G., 1972, Gli archi appenninici, il Mar Ligure ed il Tirreno nel quadro della tettonica dei bacini marginali retro-arco, *Mem. Soc. Geol. Ital.*, v. 11, p. 201–216.

Boccaletti, M., and Guazzone, G., 1974a, Remnant areas and marginal basins in the Cainozoic development of the Mediterranean, *Nature*, v. 252, p. 18–21.

Boccaletti, M., and Guazzone, G., 1974b, Plate tectonics in the Mediterranean Region, in: *Geology of Italy*, Squyres, C., ed., Tripoli: *Petrol. Soc. Expl. Lybia*, 23 p.

Boccaletti, M., Elter, P., and Guazzone, G., 1971, Plate tectonic models for the development of the Western Alps and Northern Apennines, *Nature Phys. Sci.*, v. 234, p. 108–111.

Boccaletti, M., Guazzone, G., and Manetti, P., 1974, Evoluzione paleografica e geodinamica del Mediterraneo: I Bacini marginali, in: *L'Italia nell'Ambito dell'Evoluzione del Mediterraneo*, 67th Congr. Soc. Geol. Ital., Parma, 27–31 October, 1974, *Mem. Soc. Geol. Ital.*, Suppl.

Borsi, S., 1967, Contributo alla conoscenza dell'età e della origine magmatica del vulcanismo dell'Isola di Capraia (Arcipelago toscano), *Atti Soc. Toscana Sci. Nat.*, v. 74, p. 232–243.

Borsi, S., Ferrara, G., and Tongiorgi, E., 1967, Determinazione con il metodo K/Ar della età delle rocce magmatiche della Toscana, *Boll. Soc. Geol. Ital.*, v. 86, p. 403–410.

Bousquet, S. C., 1973, La téctonique récente de l'Apennin Calabro-Lucarien dans son cadre géologique et géophysique, *Geol. Romanica*, v. 12, p. 1–103.

Caire, A., Glangeaud, L., and Grandjacquet, C., 1961, Les grands traits structuraux et l'évolution du territoire calabro-sicilien (Italie méridionale), *Bull. Soc. Géol. Fr.*, Ser. 7, v. 2, p. 915–938.

Caputo, M., and Postpischl, D., 1973, Carta della sismicità (epicentri dal 1900 al 1970), in: *Modello Strutturale d'Italia, Scala 1/1,000,000*, Rome: Consiglio Nazionale delle Ricerche.

Caputo, M., Panza, G. F., and Postpischl, D., 1970, Deep structure of the Mediterranean Basin, *J. Geophys. Res.*, 75: 4919–4923.

Caputo, M., Panza, G. F., and Postpischl, D., 1972, New evidence about the deep structure of the Lipari arc, *Tectonophysics*, v. 15, p. 219–231.

Carey, S. W., 1955, The orocline concept in geotectonics, *Proc. Roy. Soc. Tasmania*, v. 89, p. 255–288.

Carter, T. G., Flanagan, J. P., Jones, C. R., Marchant, F. L., Murchison, R. R., Rebman, J. H., Sylvester, J. C., and Whitney, J. C., 1972, A new bathymetric chart and physiography of the Mediterranean Sea, in: *The Mediterranean Sea: A Natural Sedimentation Laboratory*, Stanley, D. J., ed., Pennsylvania: Dowden, Hutchinson and Ross, p. 1–23.

Chabrier, G., 1970, Téctonique de socle d'âge alpin en Sardaigne centro-orientale, *C. R. Acad. Sci. Paris*, v. 271, p. 1251–1255.

Channel, J. E. T., and Tarling, D. H., 1975, Palaeomagnetism and the rotation of Italy, *Earth Planet. Sci. Lett.*, v. 25, p. 177–188.

Charnock, H., Rees, A. I., and Hamilton, N., 1972, Sedimentation in the Tyrrhenian, in: *The Mediterranean Sea: A Natural Sedimentation Laboratory*, Stanley, D. J., ed., Pennsylvania: Dowden, Hutchinson and Ross, p. 615–629.

Cita, M. B., 1974, I pozzi profondi perforati nel 1970 nel quadro paleogeografico e geodinamico del Mediterraneo occidentale, in: *Paleogeografia del Terziario Sardo nell'Ambito del Mediterraneo Occidentale*, Cagliari, 23–27 July, 1973, *Rend. Semin. Fac. Sci. Univ. Cagliari*, Suppl., v. 43, p. 91–143.

Civetta, L., Gasparini, P., and Adams, A. S., 1970, Geochronology and geochemical trends of volcanic rocks from Campania, S. Italy, *Eclogae Geol. Helv.*, v. 63/1, p. 57–68.

Cocozza, T., Jacobacci, A., Nardi, R., and Salvadori, I., 1974, Schema stratigrafico–strutturale del massiccio sardo-corso e minerogenesi della Sardegna, *Mem. Soc. Geol. Ital.*, v. 13, p. 85–186.

Colombi, B., Giese, P., Luongo, G., Morelli, C., Riuscetti, M., Scarascia, S., Schutte, K. G., Strowald, J., and De Visintini, G., 1973, Preliminary report on the seismic refraction profile Gargano–Salerno–Palermo–Pantelleria (1971), *Boll. Geofis. Teor. Appl.*, v. 15(59), p. 225–254.

Coulon, C., 1974, Données géochronologiques, géochimiques et paléomagnétisme sur le volcanisme cenozoique calco-alcalin de la Sardaigne nord-occidentale. Le problème de la derive de la Sardaigne, in: *Paleogeografia del Terziario Sardo nell'Ambito del Mediterraneo Occidentale*, Cagliari, 23–27 Juli, 1973, *Rend. Semin. Fac. Sci. Univ. Cagliari*, Suppl., v. 43, p. 163–169.

Dallan Nardi, L., and Nardi, R., 1974, Schema stratigrafico e strutturale dell'Appennino Settentrionale. *Mem. Accad. Lunigianese Sci. "G. Capellini,"* v. 42, p. 1–212.

Dallan Nardi, L., Elter, P., and Nardi, R., 1971, Considerazioni sull'arco dell'Appennino Settentrionale e sulla "linea" Ancona–Anzio, *Boll. Soc. Geol. Ital.*, v. 90, p. 203–211.

Dal Piaz, G. V., 1974, Le métamorphisme de haute pression et basse température dans l'évolution structurale du bassin ophiolitique alpino-apenninique (1ere partie: considérations paléogéographiques), *Boll. Soc. Geol. Ital.*, v. 93, p. 437–468.

D'Argenio, B., 1970, Evoluzione geotettonica comparata tra alcune piattaforme carbonatiche dei Mediterranei Europeo e Americano. *Atti Acad. Pontaniana Napoli*, v. 20, 34 p.

D'Argenio, B., Pescatore, T., and Scandone, P., 1973, Schema geologico dell'Appennino meridionale (Campania e Lucania), in: *Atti del Convegno sul Tema: Moderne Vedute sulla Geologia dell'Appennino*, Rome, 16–18 February, 1972, *Accad. Naz. Lincei*, Quaderno no. 183, p. 49–81.

De Jong, K. A., Manzoni, M., and Zijderveld, J. D., 1969, Palaeomagnetism of the Alghero Trachyandesites. *Nature*, v. 224, p. 67–69.

De Jong, K. A., Manzoni, M., Stavenga, T., Van Dijk, F., Van der Voo, R., and Zijderveld, J. D. A., 1973, Palaeomagnetic evidence for rotation of Sardinia during the Early Miocene, *Nature*, v. 243, p. 281–283.

Del Monte, M., 1972, Il vulcanismo del Mar Tirreno; nota preliminare sui vulcani Marsili e Palinuro, *Giorn. Geol.*, v. 38, p. 231–252.

Dercourt, J., 1970, L'expansion océanique actuelle et fossile; ses implications géotectoniques. *Bull. Soc. Géol. Fr.*, v. 12, p. 261–317.

Devoto, G., and Praturlon, A., 1973, L'Appennino Centrale, in: *Atti del Convegno sul tema: Moderne vedute sulla geologia dell'Appennino*, Rome, 16–18 February, 1972, *Accad. Naz. Lincei*, Quaderno no. 183, p. 83–95.

Dewey, J. F., Pitman III, W. C., Ryan, W. B. F., and Bonnin, J., 1973, Plate tectonics and the evolution of the Alpine system, *Bull. Geol. Soc. Amer.*, v. 84, p. 3137–3180.

Dickinson, W. R., 1972, Evidence for plate-tectonic regimes in the rock record, *Amer. J. Sci.*, v. 272, p. 577–602.

Dieni, I., and Massari, F., 1970, Tettogenesi gravitativa di età oligocenica della Sardegna centro-orientale. Nota preliminare, *Boll. Soc. Geol. Ital.*, v. 89, p. 57–64.

Dietrich, D., and Scandone, P., 1972, The position of the Basic and Ultrabasic Rocks in the Tectonic Units of the Southern Apennines, *Atti Accad. Pontaniana Napoli*, v. 21, 15 p.

Dietrich, V., Vuagnat, M., and Bertrand, J., 1974, Alpine metamorphism of maphic rocks, *Mitt. Schweiz. Mineral. Petrogr.*, v. 54, p. 291–332.

Di Paola, G. M., Puxeddu, M., and Santacroce, R., 1975, K–Ar Ages of Monte Arci volcanic complex (Central-Western Sardinia), *Rend. Soc. Ital., Mineral. Petrol.*, v. 31, p. 181–190.

Dubois, R., 1970, Phases de serrage, nappes de socle et métamorphisme alpin à la jonction Calabre-Apennin: la suture calabro-appenninique. *Rev. Géogr. Phys. Géol. Dyn.*, v. 12, p. 221–254.

Elter, G., Elter, P., Sturani, C., and Weidmann, M., 1966, Sur la prolongation du domaine ligure de l'Apennin dans le Monferrat et les Alpes et sur l'origine de la Nappe de la Simme s.l. des Préalpes romandes et chablaisiennes, *Arch. Sci. Genève*, v. 19, p. 279–377.

Elter, P., and Pertusati, P., 1973, Considerazioni sul limite Alpi-Appennino e sulle relazioni con l'arco delle Alpi Occidentali, *Mem. Soc. Geol. Ital.*, v. 12, p. 339–375.

Emelyanov, E. M., 1972, Principal types of recent bottom sediments in the Mediterranean Sea: their mineralogy and geochemistry, in: *The Mediterranean Sea: A Natural Sedimentation Laboratory*, Stanley, D. J., ed., Pennsylvania: Dowden, Hutchinson and Ross, p. 355–399.

Erickson, A. J., 1970, Heat-flow measurements in the Mediterranean, Black and Red Seas, Unpublished Ph.D. thesis, M.I.T., Cambridge, Massachusetts.

Ernst, W. G., 1971, Metamorphic zonations on presumably subduced lithospheric plates from Japan, California and the Alpes, *Contrib. Mineral. Petrol.*, v. 34, p. 43–59.

Evernden, J. F., and Curtis, G. H., 1965, The potassium–argon dating of Late Cenozoic Rocks in East Africa and Italy, *Curr. Anthropol.*, v. 6, p. 343–364.

Ewing, J., and Ewing, M., 1959, Seismic refraction measurements in the Atlantic Ocean basins, in: *The Mediterranean Sea, Bull. Geol. Soc. Amer.*, v. 70, p. 291–305.

Fahlquist, D. A., and Hersey, J. B., 1969, Seismic refraction measurements in the western Mediterranean Sea, *Bull. Inst. Océanogr. Monaco*, v. 67, 52 p.

Fierro, G., and Passega, R., 1965, Studio sedimentologico di 32 carote prelevate nel Nord Tirreno, *Atti 24° Convegno della Ass. Geofis. Ital.*, Rome, 8–20 February, 1965, p. 1–12.

Finetti, I., and Morelli, C., 1972, Wide scale digital seismic exploration of the Mediterranean Sea, *Boll. Geofis. Teor. Appl.*, v. 14, p. 291–342.

Finetti, I., and Morelli, C., 1973, Geophysical Exploration of the Mediterranean Sea, *Boll. Geofis. Teor. Appl.*, v. 15, p. 263–342.

Franzini, M., 1964, Studio mineralogico e litologico dell'Isola di Capraia, *Atti Soc. Toscana Sci. Nat.*, v. 71, p. 326–386.

Frazer, J. Z., Arrhenius, G., Hanor, J. S., and Hawkins, D. L., 1970, Surface sediment distribution; Mediterranean Sea, Scripps Inst. Oceanogr., Defense Language Inst., Syst. Develop. Agency, Presidio of Monterey, California, 8 p.

Friedman, G. M., 1973, Petrographic data and comments on the depositional environment of the Miocene sulfates and dolomites at Sites 124, 132, and 134, Western Mediterranean Sea, in: *Initial Reports of the Deep Sea Drilling Project*, Ryan, W. B. F., Hsü, K. J., et al., eds., Washington, D.C.: U.S. Government Printing Office, v. 13, p. 695–708.

Gallignani, P., 1973, I sedimenti della cima del Monte Vercelli (Mar Tirreno), *Giorn. Geol.*, v. 39, p. 1–9.

Gasparini, P., and Adams, J. A. S., 1969, K/Ar dating of Italian plio-pleistocene volcanic rocks, *Earth Planet. Sci. Lett.*, v. 6, p. 225–230.

Giese, P., and Morelli, C., 1973, La struttura della crosta terrestre in Italia, in: *Atti Convegno sul Tema: Moderne Vedute sulla Geologia dell'Appennino*, Rome, 16–18 February, 1972, *Accad. Naz. Lincei*, Quaderno No. 183, p. 317–366.

Giglia, G., 1974, L'insieme Corsica–Sardegna e i suoi rapporti con l'Appennino settentrionale: rassegna di dati cronologici e strutturali, in: *Paleogeografia del Terziario Sardo nell'ambito del Mediterraneo Occidentale*, Cagliari, 23–27 July, *Rend. Semin. Fac. Sci. Univ. Cagliari*, Suppl., v. 43, p. 245–275.

Glangeaud, L., 1962, Paleogéographie dynamique de la Méditerranée et de ses bordures. Le rôle des phases Ponto-Plio-Quaternaires, in: *Oceanographie Géologique et Géophysique de la Méditerranée Occidentale, Colloq. C.N.R.S. Villefranche*, April, 1961, p. 125–161.

Haccard, D., Lorenz, C., and Grandjacquet, C., 1972, Essai sur l'évolution téctonogénetique de la liason Alps–Apennines (de la Ligurie à la Calabre), Mem. Soc. Geol. Ital., v. 11, p. 309–341.

Heezen, B. C., Gray, C., Segre, A. G., and Zarudski, E. F. K., 1971, Evidence of foundered continental crust beneath the Central Tyrrhenian Sea, *Nature*, v. 229, p. 327–329.

Herzen, R. P., von, and Lee, W. H. K., 1969, Heat flow in oceanic regions, in: *The Earth's Crust and Upper Mantle*, Hart, P. J., ed., *Amer. Geophys. Union Geophys. Monogr.*, v. 13, p. 88–95.

Horai, K., and Uyeda, S., 1969, Terrestrial heat flow in volcanic areas, in: *The Earth's Crust and Upper Mantle*, Hart, P. J., ed., *Amer. Geophys. Union Geophys. Monogr.*, v. 13, p. 95–109.

Hsü, K. J., Ryan, W. B. F., and Cita, M. B., 1973, Late Miocene desiccation of the Mediterranean, *Nature*, v. 242, p. 240–244.

Jakob, R., 1958, Zur petrographie von Vulcano, Vulcanello und Stromboli (Aeolische Inseln, Italien), *Stiftung 'Vulkaninstitut I. Friedlaender'* Zurich, v. 7, p. 1–117.

Karig, D. E., 1971, Origin and development of Marginal Basins in the Western Pacific, *J. Geophys. Res.*, v. 76, p. 2542–2561.

Keller, J., 1967, Alter und Abfolge der vulkanischen Ereignisse auf den Aeolischen Inseln, Sizilien, *Ber. Naturf. Ges. Freiburg*, v. 57, p. 33–67.

Keller, J., 1974, Petrology of some volcanic rock series of the Aeolian Arc, Southern Tyrrhenian Sea: calc-alkaline and shoshonitic associations, *Contrib. Mineral. Petrol.*, v. 46, p. 29–47.

Klerkx, J., Deutsch, S., Hertogen, J., De Winter, J., Gijbels, R., and Pichler, H., 1974, Comments on "Evolution of Eolian Arc volcanism (Southern Tyrrhenian Sea)" by F. Barberi, G. Ferrara, F. Innocenti, J. Keller, and L. Villari, *Earth Planet. Sci. Lett.*, v. 23, p. 297–303.

Latter, J. H., 1971, Near-surface seismicity of Vulcano, Aeolian Islands, *Bull. Volcanol.*, v. 35, p. 117–142.

Laubscher, H. P., 1971, The large-scale kinematics of the Western Alps and the Northern Apennines and its palinspastic implications, *Amer. J. Sci.*, v. 271, p. 193–226.

Lauro, C., 1973, Magmatismo terziario e recente, plutonico e vulcanico, della Italia peninsulare e moderne vedute sulla geologia dell'Appennino, in: *Atti del Convegno sul Tema: Moderne Vedute sulla Geologia dell'Appennino*, Rome, 16–18 February, 1972, *Accad. Naz. Lincei*, Quaderno No. 183, p. 251–271.

Lenardon, G., 1969, Ricerche sedimentologiche e petrografiche su due carote prelevate dal fondo marino al largo di Napoli, *Arch. Oceanogr. Limnol.*, v. 16, p. 129–161.

Lloyd, R. M., and Hsü, K. J., 1972, Preliminary isotopic investigations of samples from Deep-Sea Drilling Cruise to the Mediterranean, in: *The Mediterranean Sea: A Natural Sedimentation Laboratory*, Stanley, D. J., ed., Pennsylvania: Dowden, Hutchinson and Ross, p. 681–686.

Loddo, M., and Mongelli, F., 1974a, On the relation of heat flow to elevation in Western Mediterranean, *Riv. Ital. Geofis.*, v. 23, p. 145–148.

Loddo, M., and Mongelli, F., 1974b, Tentativo di una mappa del flusso geotermico in Italia in: *L'Italia nell'Ambito dell'Evoluzione del Mediterraneo*, 67th Congr. Soc. Geol. Ital., Parma, 27–31 October, 1974, *Mem. Soc. Geol. Ital.*, Suppl., p. 12.

Lombardi, G., Nicoletti, M., Petrucciani, C., 1974, Età delle vulcaniti acide dei complessi Tolfetano, Cerite e Manziatte (Lazio nord-occidentale), *Per. Mineral.*, v. 43, p. 351–375.

Lowrie, W., and Alvarez, W., 1974, Rotation of the Italian Peninsula, *Nature*, v. 251, p. 285–288.

Maccarone, E., 1970, Notizie petrografiche e petrochimiche sulle lave sottomarine del Seamount 4 (Tirreno Sud), *Boll. Soc. Geol. Ital.*, v. 89, p. 159–180.

Manzoni, M., 1970, Paleomagnetic data of middle and upper Triassic age from the Dolomites (Eastern Alps, Italy), *Tectonophysics*, v. 10, p. 411–424.

Manzoni, M., 1974a, Paleomagnetism of Tertiary volcanics from Campidano and associated grabens (Sardinia), Eur. Geophys. Soc. 2nd Int. Meeting, Trieste, 20–26 September, 1974 (Abstract).

Manzoni, M., 1974b, Un'interpretazione dei dati paleomagnetici del Terziario della Sardegna ed alcuni nuovi risultati, in: *Paleogeografia del Terziario Sardo nell'Ambito del Mediterraneo Occidentale*, Cagliari, 23–27 July, 1973, *Rend. Semin. Fac. Sci. Univ. Cagliari*, Suppl., v. 43, p. 283–295.

Marinelli, G., 1961, Genesi e classificazione delle vulcaniti recenti toscane, *Atti Soc. Toscana Sci. Nat.*, v. 68, p. 74–116.

Marinelli, G., 1967, Genèse des magmas du volcanisme plio-quaternaire des Apennines, *Geol. Rundschau*, v. 57, p. 127–141.

Menard, H. W., 1967, Transitional types of crust under small ocean basins, *J. Geophys. Res.*, v. 72, p. 3061–3073.

Morelli, C., 1970, Physiography, gravity and magnetism of Tyrrhenian Sea, *Boll. Geofis. Teor. Appl.*, v. 12(48), p. 275–308.

Morelli, C., and Finetti, I., 1974, Stato attuale delle conoscenze geofisiche nel Mediterraneo, in: *L'Italia nell'Ambito dell'Evoluzione del Mediterraneo*, 67th Congr. Soc. Geol. Ital., Parma, 27–31 October, 1974, *Mem. Soc. Geol. Ital.*, Suppl., 28 p.

Moskalenko, V. N., 1967, Structure of the sedimentary series in the Tyrrhenian Sea by seismic methods, *Proc. Acad. Sci. U.S.S.R. Geol. Ser.*, v. 6, p. 49–56.

Muratov, M. V., 1972, Formation history of the Black Sea abyssal basin and its comparison with some basins of the Mediterranean, *Geotectonics*, v. 5, p. 269–278.

Nairn, A. E. M., and Westphal, M., 1968, Possible implications of the Late Palaeozoic igneous rocks of Northwestern Corsica, *Palaeogeogr. Palaeoclimatol. Palaeoecol.*, v. 5, p. 179–204.

Napoleone, G., Premoli Silva, I., Roggenthen, W., and Fischer, A. G., 1974, An example of paleomagnetic stratigraphy from the Tethys, Eur. Geophys. Soc., 2nd Int. Meeting, Trieste, 20–26 September (Abstract).

Nardi, R., 1968, Le unità alloctone della Corsica e loro correlazione con le unità delle Alpi e dell'Appennino, *Mem. Soc. Geol. Ital.*, v. 7, p. 323–344.

Nesteroff, W. D., 1973, Mineralogy, petrography, distribution, and origin of the Messinian Mediterranean evaporites, in: *Initial Reports of the Deep Sea Drilling Project*, Ryan, W. B. F., Hsü, K. J., et al., eds., Washington D.C.: U.S. Government Printing Office, v. 13, p. 673–694.

Nesteroff, W. D., Wezel, F. C., and Pautot, G., 1973, Summary of lithostratigraphic findings and problems, in: *Initial Reports of the Deep Sea Drilling Project*, Ryan, W. B. F., Hsü, K. J., *et al.*, eds., Washington, D.C.: U.S. Government Printing Office, v. 13, p. 1021–1040.

Ninkovich, D., and Hays, J. D., 1971, Tectonic setting of the Mediterranean volcanoes, in: *Acta 1st Int. Sci. Congr. Volcano of Thera*, Athens, 15–23 September 1969, p. 111–135.

Ninkovich, D., and Hays, J. D., 1972, Mediterranean island arcs and origin of high-potash volcanoes, *Earth Planet. Sci. Lett.*, v. 16, p. 331–345.

Norin, E., 1958, The sediments of the Central Tyrrhenian sea, in: *Reports of the Swedish Deep-Sea Expedition*, v. 8, *Sediment Cores from the Mediterranean Sea and the Red Sea*, Göteborg, Sweden: Elanders Bronryckerei Aktiebolag, p. 1–136.

Ogniben, L., 1960, Nota illustrativa dello schema geologico della Sicilia nord-orientale, *Riv. Min. Siciliana*, v. 11, p. 183–212.

Ogniben, L., 1969, Schema introduttivo alla geologia del confine calabro lucano, *Mem. Soc. Geol. Ital.*, v. 8, p. 453–763.

Ogniben, L., 1970, Paleotectonistic history of Sicily, in: *Geology and History of Sicily*, Alvarez, W., and Gohrbandt, H. A., eds., Tripoli: Petrol. Soc. Expl. Lybia, p. 133–143.

Ogniben, L., 1973, Schema geologico della Calabria in base ai dati odierni, *Geol. Romana*, v. 12, p. 243–585.

Ogniben, L., Martinis, B., Rossi, P. M., Fuganti, A., Pasquaré, G., Sturani, C., Nardi, R., Cocozza, T., Praturlon, A., Parotto, M., D'Argenio, B., Pescatore, T., Scandone, P., Vezzani, L., Agip-Mineraria, Finetti, I., Morelli, C., Caputo, M., Postpischl, D., and Giese, P., 1973, *Modello strutturale d'Italia, Scala 1:1,000,000*, Rome: Consiglio Nazionale delle Ricerche.

Oliver, J., Isacks, B., Barazangi, M., and Mitronovas, W., 1973, Dynamics of the Down-Going Lithosphere, *Tectonophysics*, Vol. 19, p. 133–147.

Orsini, J., Vellutini, P., and Westphal, M., 1974, The initial fit of Corsica and Sardinia: Paleomagnetic, petrologic and structural matching, Eur. Geophys. Soc., 2nd Int. Meeting, Trieste, 20-26 September, 1974 (Abstract).

Peterschmitt, E., 1956, Quelques données nouvelles sur les seismes profonds de la Mer Tyrrhénienne, *Ann. Geofis.*, v. 9, p. 305–334.

Pichler, H., 1967, Neue Erkenntnisse über Art und Genese des Vulkanismus der Aeolischen Inseln (Sizilien), *Geol. Rundschau*, v. 57, p. 102–126.

Pieri, M., 1966, Tentativo di ricostruzione paleogeografico–strutturale dell'Italia centro-meridionale. *Geol. Romana*, v. 5, p. 407–424.

Premoli Silva, I., Napoleone, G., and Fischer, A. G., 1974, Risultati preliminari sulla stratigrafia paleomagnetica della Scaglia cretaceo–paleocenica della sezione di Gubbio (Appennino centrale), *Boll. Soc. Geol. Ital.*, v. 93, p. 647–659.

Ritsema, A. R., 1969, Seismic data of the West Mediterranean and the problem of oceanization, *Verh. Kon. Ned. Geol. Mijnb. Gen.*, v. 26, p. 105–120.

Ritsema, A. R., 1970, On plate tectonics in the Mediterranean Region, *1st Draft Paper E.S.C. Luxembourg*.

Romano, R., 1973, Le isole di Panarea e Basiluzzo. Contributo alla conoscenza geo-vulcanologica e magmatologica delle isole Eolie, *Riv. Minerar. Siciliana*, v. 24 (no. 139–141), p. 1–40.

Romano, R., and Sturiale, C., 1971, L'isola di Ustica, *Riv. Minerar. Siciliana*, v. 12, p. 3–61.

Ryan, W. B. F., Workum, F. Jr., and Hersey, J. B., 1965, Sediments on the Tyrrhenian Abyssal Plain, *Bull. Geol. Soc. Amer.*, v. 76, p. 1261–1282.

Ryan, W. B. F., Stanley, D. J., Hersey, J. B., Fahlquist, D. A., and Allan, T. D., 1971, The tectonics and geology of the Mediterranean Sea, in: *The Sea*, Maxwell, A. E., ed., New York: Wiley–Interscience, Vol. 4, p. 387–492.

Ryan, W. B. F., Hsü, K. J., *et al.*, 1973, *Initial Reports of the Deep Sea Drilling Project*, Washington, D.C.: U.S. Government Printing Office, v. 13, 1447 p.

Sarnthein, M., and Bartolini, C., 1973, Grain size studies on turbidite components from Tyrrhenian deep-sea cores, *Sedimentology*, v. 20, p. 425–436.

Savelli, C., 1975, Datazioni preliminari col metodo K–Ar di vulcaniti della Sardegna sud-occidentale, *Rend. Soc. Ital. Mineral. Petrol.*, v. 31, p. 191–198.

Savelli, C., and Pasini, G., 1974, Nota preliminare sull'età K/Ar di Basalti della Sardegna orientale e del Canyon sottomarino di Orosei (Mar Tirreno), in: *Paleogeografia del Terziario Sardo nell'Ambito del Mediterraneo Occidentale*, Cagliari, 23–27 July, 1973, *Rend. Semin. Fac. Sci. Univ. Cagliari*, Suppl., v. 43, p. 321–325.

Scandone, P., 1972, Studi di geologia lucana: carta dei terreni della serie calcareo–siliceo-marnosa e note illustrative, *Boll. Soc. Nat. Napoli*, v. 81, p. 225–299.

Schult, A., 1973, Palaeomagnetism of Upper Cretaceous volcanic rocks in Sicily, *Earth Planet. Sci. Lett.*, v. 19, p. 97–100.

Sclater, J. G., 1972, Heat flow and elevation of the marginal basins of the Western Pacific, *J. Geophys. Res.*, v. 77, p. 5705–5719.

Segre, A. G., 1955, Morphologie Sousmarine de la Mer Tyrrhénienne, *Bull. Inform. Comité Central Océanogr. Étude Côtes.*, v. 8, p. 346–350.

Selli, R., 1962, Il Paleogene nel quadro della geologia dell'Italia centro-meridionale, *Mem. Soc. Geol. Ital.*, v. 3, p. 737–789.

Selli, R., 1970*a*, Cenni morfologici generali sul Mar Tirreno, in: *Ricerche Geologiche Prelimi-nari nel Mar Tirreno*, Selli, R., ed., *Giorn. Geol.*, v. 37(1), p. 5–24.

Selli, R., 1970*b*, III. Profili magnetometrici in: *Ricerche Geologiche Preliminari nel Mar Tir-reno*, Selli, R., ed., *Giorn. Geol.*, v. 37(1), p. 43–54.

Selli, R., 1970*c*, XIII. Discussione dei risultati e conclusioni, in: *Ricerche Geologiche Preli-minari nel Mar Tirreno*, Selli, R., ed., *Giorn. Geol.*, v. 37(1), p. 43–54.

Selli, R., 1974, Appunti sulla geologia del Mar Tirreno, in: *Paleogeografia del Terziario Sardo nell'Ambito del Mediterraneo Occidentale*, Cagliari, 23–27 July, 1973, *Rend. Semin. Fac. Sci. Univ. Cagliari*, Suppl., v. 43, p. 327–351.

Selli, R., and Fabbri, A., 1971, Tyrrhenian: a Pliocene deep sea, *Accad. Naz. Lincei Rend. Classe Sci. Fis. Mat. Nat.*, Ser 8, v. 50, p. 104–116.

Smith, A. G., 1971, Alpine deformation and the oceanic areas of the Tethys, Mediterranean and Atlantic, *Bull. Geol. Soc. Amer.*, v. 82, p. 2039–2070.

Soffel, H., 1972, Anticlockwise rotation of Italy between Eocene and Miocene, Palaeomagnetic evidence, *Earth Planet. Sci. Lett.*, v. 17, p. 207–210.

Soffel, H., 1974, Palaeomagnetism of the Tertiary Monti Lessini and Monti Berici Volcanites and its implication for the rotation of Northern Italy, *Eur. Geophys. Soc.*, 2nd Int. Meeting, Trieste, 20–26 September (Abstract).

Stanley, D. J., and Mutti, E., 1968, Sedimentological evidence for an emerged land mass in the Ligurian Sea during Palaeogene, *Nature*, v. 218, p. 32–36.

Storetvedt, K. M., 1973, Genesis of West Mediterranean Basins, *Earth Planet. Sci. Lett.*, v. 21, p. 22–28.

Tomadin, L., 1970, VI. Mineralogia dei sedimenti politici, in: *Ricerche Geologiche Preliminari nel Mar Tirreno*, Selli, R., ed., *Giorn. Geol.*, v. 37(1), p. 89–108.

Villari, L., 1972, L'isola di Filicudi ed il suo significato magmatologico, *Rend. Soc. Ital. Mineral. Petrol.*, v. 38, p. 475–506.

Vogt, P. R., Higgs, R. H., and Johnson, G. L., 1971, Hypotheses on the origin of the Medi-terranean basin: magnetic data, *J. Geophys. Res.*, v. 76, p. 3207–3228.

Westphal, M., Bardon, C., Bossert, A., and Hamzeh, R., 1973, A computer fit of Corsica and Sardinia against Southern France, *Earth Planet. Sci. Lett.*, v. 18, p. 137–140.

Wezel, F. C., 1970, Geologia del Flysch Numidico della Sicilia Nord-orientale, *Mem. Soc. Geol. Ital.*, v. 9, p. 225–280.

Zijderveld, J. D. A., De Jong, K. A., and Van der Voo, R., 1970a, Rotation of Sardinia: Paleomagnetic evidence from Permian rocks, *Nature*, v. 226, p. 933–934.

Zijderveld, J. D. A., Hazeu, G. J. A., Nardin, M., and Van der Voo, R., 1970b, Shear in the Tethys and the Permian Palaeomagnetism in the Southern Alps, including new results, *Tectonophysics*, v. 10, p. 639–661.

Chapter 4A

THE CENTRAL MEDITERRANEAN MOUNTAIN CHAINS IN THE ALPINE OROGENIC ENVIRONMENT

A. Caire

Département de Géologie Structurale
Université de Paris
Paris, France

I. INTRODUCTION

The present article provides general information on Alpine structure and evolution, as a framework within which to consider various regional studies. The relatively short text is conceived as forming the link between the figures, which, more clearly than words, illustrate various aspects of Alpine geology. Thus, a rapid review of the present structure may be obtained by successively studying Figs. 1, 3, 8, 13, 19, 14, 16, 10, 11, 4, 6, 7, and 12; of the retrotectonics and paleogeography from Figs. 2, 3, 4, 5, and 6; while Figs. 2, 15, 17, 18, and 19 illustrate various recent hypotheses.

The Mediterranean is a recent, postorogenic sea (Fig. 1) whose present margins are essentially a reflection of Plio-Quaternary vertical movements. It covers parts of the European and African epi-Paleozoic platforms and conceals part of the Alpine chains (Alpine orogen) formed during the Mesozoic and Cenozoic. The formation of the Mediterranean will be discussed after retracing Mesozoic and Cenozoic Alpine evolution.

Fig. 1. European framework and situation of the studied regions: 1) orogenic axis of the Alpine system; 2) median massifs (Zwischengebirge)—cores of pre-Alpine rocks or massifs of early Alpine tectonization, e.g., pre-Paleogene tectonization beneath the Pannonian Plain; 3) internal zone of the Alpidic and Dinaric branches; in the western Mediterranean this same cross-hatching represents the known or presumed ophiolitic zones; 4) marginal amygdule of the Continental Rise; 5) Alpine foredeep; 6) As part of the hypothesis for a monogenic Alpine deformation, the large arrows indicate the direction of basic pressures caused by the movement of Africa to the northwest or to a sinistral rotation of Europe; 7) folds and movement resulting from regional adaptation to the basic pressures of 6; 8) contour lines of anteclises and synclises; the teeth point to the low areas; 8a) graben, rifts; 9) areas of the Variscan (Hercynian) Chain overlain by Mesozoic and Cenozoic sediments; 10) outcrops of the folded Variscan complexes; 11) pre-Caledonian blocks that were locally reactivated by Caledonian and Variscan folding; 12) outcrops of Caledonian folded

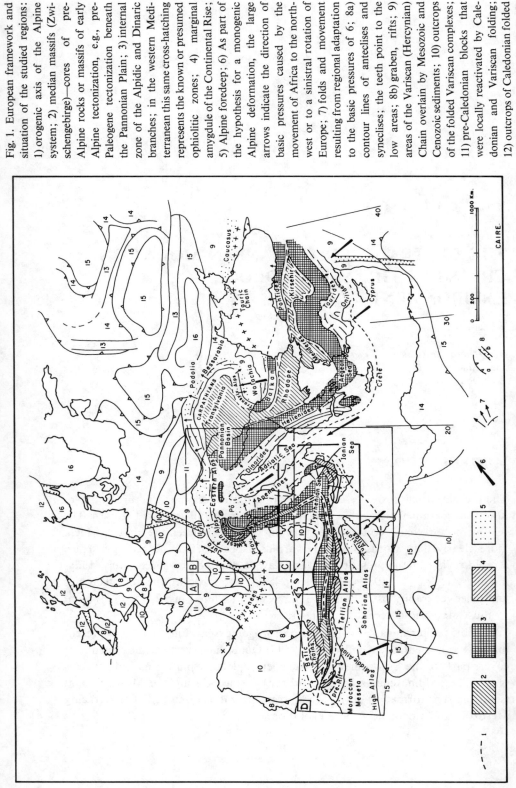

complexes; 13) aulacogenes; 14) syneclises (deep parts); 15) transition slopes (deep parts); 15) transition slopes from shields to syneclises; 16) shields (Caledonian basement outcrops); A, B, C) areas covered by the following maps—A = Fig. 13, B = Fig. 10, C = Fig. 9, D = Fig. 8. For the structures outside the Alps consult Bogdanoff (1962, p. 900).

II. THE EUROPEAN AND AFRICAN PLATFORMS

The platforms are continental areas formed of Hercynian or older basement unconformably overlain by a cover of Triassic and younger rocks. The cover is an epicontinental facies (terrigenous, neritic, evaporitic, etc.), and is sometimes folded along the margins of the Alpine chain (see below). However, over most of its extent significant tangential overthrusting is absent, the principal movements being dominantly vertical, epeirogenic deformation with normal (extensional) faulting or wrench faulting.

Epeirogenic deformation resulted in the formation of cuvettes and basins (syneclises) subsequently filled with sediments, whose thickness may sometimes be greater than that found in orogenic trenches (10–20 km). The basins are separated by anteclises over which sediments are thin, contain many breaks, and reveal repeated transgression. In the more elevated regions of the platform the older rocks, Paleozoic or Precambrian, appear in massifs, e.g., Armorican Massif, Iberian Meseta, Moroccan Meseta, Anti-Atlas, and the Corso-Sardinian Massif. Faulting resulted in the formation of trenches, which may have remained empty (grabens) or have been filled with sediments (aulacogens). Certain of the fault-troughs may be regarded as embryonic or aborted rifts, e.g., the Oslo–Tibesti rift system, which traverses part of the Mediterranean (Provençal Basin, Gallican Sea, and the Campidano of Sardinia), and parts of the African platform, the Pelagian Sea between Tunisia and Sicily and Libya. This rift system is the repetition of the system formed by Levantine graben, Red Sea, and East African rift valleys. It results from a NW–SE dextral offset. Its limit in the west is the present seismic and volcanic arc of Sicily–Calabria–southern Apennines. This arc results from continental shearing and subduction under the Tyrrhenian Sea. The subduction increases the Alpine symmetry (see Fig. 14) of the Calabro-Sicilian Arc.

III. THE VARIOUS CHAINS AND THEIR CHARACTER

A. Intracratonic Chains

In the absence of thrusting, platforms are considered as indurated blocks, which resist folding and are called cratons. The narrower, less evolved chains between two platform elements are, therefore, referred to as intracratonic chains, e.g., the Caucasus, the Pyrenees, and their pyreneo-provençal extension. In these chains initial extension-created facies zones later compressed to form a fan-shaped structure.

B. Intermediate Chains

As noted earlier, the marginal regions of the platform which abut the Alpine orogen show effects of tectonic activity, as, for example, the Jura Mountains and part of the sub-Alpine and Provençal chains. Along the border of the African platform the same is true of the southern Atlas (or pre-Saharan Atlas), the High and Middle Atlas of Morocco, the Saharan Atlas of Algeria, and the Tunisian Atlas (see Fig. 8).

These chains warrant the term "marginal" since they form the platform margin; they could also be called "intermediate" chains for they are intermediate in character between a platform and an alpine geosynclinal chain with its typical facies development and tectonic features (nappes and thrusting). They are the "chaînes de couverture," for the sedimentary cover has suffered decollement with folding and fracturing. In this context two types of basement sedimentary cover should be distinguished in marginal chains:

(1) A stratigraphic cover resting unconformably on the basement, beginning with Triassic of Germanic facies (Andalusian in Spain, Tellian in North Africa), followed by the rest of the Mesozoic and Cenozoic beds.

(2) A tectonic cover formed by beds which have undergone decollement and folding independent of the substratum. Decollement occurred within the Germanic Triassic facies in the evaporites (gypsum, rock salt) of the Muschelkalk or Keuper. The tectonic cover thus consists of the Upper Triassic and younger beds. The tectonic basement comprises: a) the "tegument," of sandy Lower Triassic (Buntsandstein), and sometimes the Muschelkalk; and b) the underlying beds.

C. Alpine, Geosynclinal Chains (Alpine Orogen *s.s.*)

In reconstructing those parts of the Alpine orogen now hidden under the Mediterranean or eroded or engulfed by tectonic movement, it can be seen (Fig. 1) that the Alpine chains, as a whole, form an essentially east–west band folded upon itself. This undulating band can be broken up into loops where arcs linked by more or less rectilinear segments (wings) can be differentiated. There are, from west to east (Fig. 1): 1) the Betico-Rifian loop with the Gibraltar Arc, the Betico-Balearic and the Rifo-Tellian forming the wings of the loop; 2) the Tyrrhenian loop with the Calabro-Sicilian Arc and the Apennines and Rifo-Tellian wings; and 3) the Alpine loop (*s.s.*) with the Alpine Arc (western Alps) and the Apennine and eastern Alp wings, etc.

The concave, internal part of the loops are in accord with the location of recent basins (end-Miocene or younger) occupied by seas or filled with sediments—the Alboran Sea in the Betico-Rifian loop, the Tyrrhenian Sea in the

Tyrrhenian loop, and the Po Basin with more than 6 km of Pliocene sediments in the Alpine loop. All are the result of postorogenic subsidence.

All the rectilinear segments or wings follow the same basic model in the sense that they are formed of nappes which tend to cover the adjacent platform; these nappes arose from paleogeographic zones parallel or subparallel to the platform margin. Each wing, however, has its own specific character, either paleographically or structurally (see Figs. 2, 9, 10, and 11). Thus, the Rifo-Tellian wing, which will be described below, contains a paleogeographic zone, the Mesozoic–Eocene Tellian trough, which is not found in the Betics, nor in the Calabro-Sicilian Arc, nor in the Apennines (see Fig. 2).

There are structural and paleogeographic links within the arcs. Thus, within the Calabro-Sicilian Arc there are facies zones identical with those of the Rifo-Tellian wing (flysch and Kabyle facies) and zones of facies similar to those in the southern Apennines (Panormide = calcareous Apennines; Sclafani zone = Lagonegro–San Fele zone) (see Figs. 2, 9, and 11). However, on the structural level, there are many analogies along a loop such as the Tyrrhenian loop, readily seen in the sections given in Fig. 11. These similarities extend to the structure of the internal zones, the position of the flysch, foredeep, and various features of the external or intermediate zones. The fundamental unity of the paleogeographic pattern and tectonic evolution of the diverse chains leads to remarkable symmetries. The symmetry on both sides of a median line through the Tyrrhenian Sea is illustrated in Fig. 14, and the symmetry of the Alpine–Tyrrhenian loop is shown in Fig. 16.

The arcs are the principal regions of major overthrusting. They are also the regions of superposed nappes, and of secondary modification by peripheral uplift with compensatory downward movements in the internal zones.

D. Alpine Axis, Branches, and Fanlike Structure (Fig. 1)

Within the orogenic Alpine belt an axis of divergence can be traced (Alpine Axis). In the central and western Mediterranean, the axis divides into two branches (see below). Each branch is characterized by major overthrusting over the margin of the adjoining platform, the platforms acting as foreland with respect to the thrusting.

Schematically then, the Alpine orogen has a fanlike structure with divergent overthrusting (Fig. 1). A Neogene foredeep runs between each branch of the Alpine orogen and the neighboring platform or marginal chain (see Fig. 13). The branch bordering the European platform (Alpidic branch) (Figs. 1 and 2) comprises the Betic Cordillera (Southern Spain), the Balearic Islands, northeastern Corsica, western Alps (or Alpides), the northern margins of the eastern Alps (or Austrides), the Carpathians, the Balkans, and Pontides (northern Turkey). The principal thrusting is directed toward the European platform.

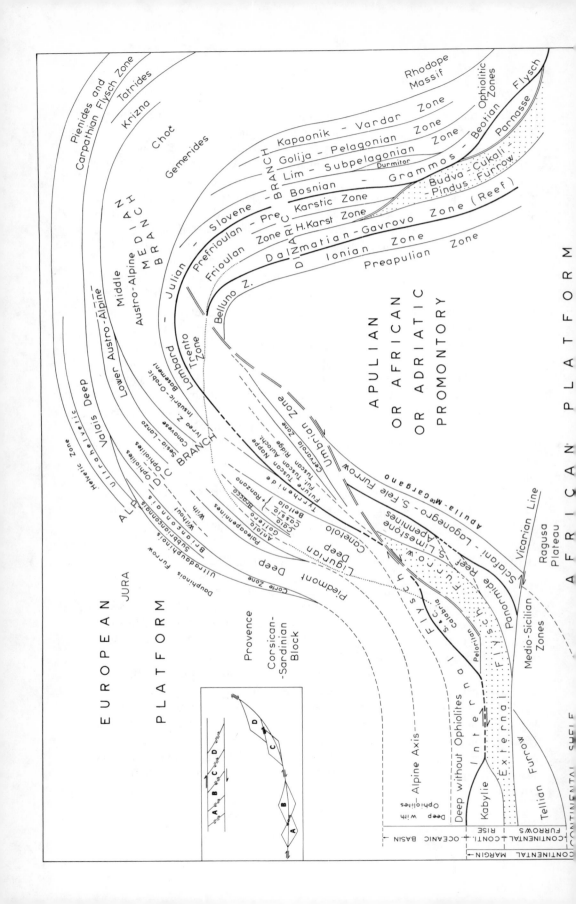

Fig. 2. Attempt to reconstruct the pretectogenic paleogeographic Alpine zones. This sketch, of necessity without a scale, limits itself to showing the contiguous relationships of the various Alpine zones before their tectonization (crushing, digestion, or becoming allochthonous). The internal zones were ejected by the tectonic processes before the external zones; as a consequence units of different age are shown together. Each one is marked with its tendency during its separate existence—distinction between uplifted zones (ridges, plateaus, and saddles) and downwarped zones (furrow and basin). At the bottom, the left-hand side of the figure shows a paleo-oceanographic interpretation based on the definitions of the present-day knowledge of the eastern American continental borders. Thus we recognize an "oceanic basin" (with the sea bottom of oceanic crust) comparable to the the Vicarian Line, bordered by continental "margins" with "rise," "furrow," and "shelf." The eastern prolongation of these elements is interrupted by a line of discontinuity, form. This line was probably part of a system of dextral wrench faults that allowed east–west mobility between the African and European plates. To the north of the orogenic African border and whole Alpine area, which was probably already united with the European plat-Adriatic Promontory, a base surely much wider than we see it today, as its borders are, to a great extent, covered by the Sicilian, Apennine, and Dinaride nappes. It is possible that another line, parallel or oblique to the Vicarian Line, separates the Kabylias and the Peloritan (passing south of the Peloritan). Another could exist to the north of the ancient Calabrian Massif ("Catanzaro wrench" now overlapped by the movement of this massif to the north), separating the ophiolitic nappes to the north from the nappes to the south where ophiolites are absent. To the north of the Vicarian Line we see the alternation of furrows (Sclafani–Lagonegro–San Fele, Monte Soro external flysch) and "ridges or reefs" (Panormide, southern Limestone Apennines, Peloritano–Calabrian amygdule), which were the continental margin of the Adriatic Promontory. The double line is an imaginary boundary which acts as a link between the Sicily–southern Apennine zones and those of the northern Apennines. Its prolongation could perhaps separate the Alps from the Car-pathians. West of this line, only zones of condensed sedimentation, of the Briançonnais type (Kabylias, Peloritan, central southern Calabria, Tuscan Ridge, and Trento Zone) are found. To the east the equivalent zones show fairly thick neritic and reef sedimentation of the Gavrovo type, such as the Frioulan, Karstic, Parnasse, and Gavrovo Zones (see Aubouin). A certain symmetry from one side to the other of the Adriatic Promontory is suggested with the paleogeographic extremities of the external flysch of Barbary–Sicily and the Pindus furrow. In the oceanic domain a difference is made between the relatively external flysch Zones without ophiolites (external flysch of Algeria–Sicily, Lombard–Julian flyschs, etc., of Dinarides, and on the Alpine side the external Piedmont Zone); and the relatively internal zones where we find ophiolitic complexes (Ligurides, internal Piedmont Zone, Lim Zone, and sub-Pelagonian Zone). The link between the Alps and Apennines is delicate in areas where the Helminthoid flysch seems to have spread without meeting a barrier and where as a consequence, in Upper Cretaceous time, the paleo-Apennines and Tyrrhenide could no longer have existed, any more than the Insubrian and Sesia–Lanzo zones. With this figure we can see the important difference in the distribution of *the preorogenic Alpine flysch*, undoubtedly linked with Mesozoic deformation that did not have the same pattern as the Alpine tectonics; this is the case for the internal and external Cretaceous flysch of North Africa and Sicily; the "internal" Bosnian Cretaceous flysch, external flysch of Pindus, etc.; and *the synorogenic Alpine flysch*, which migrated from the interior of the Alpine system toward the platforms. The dotted line suggests an isochronous curve for the first Eocene flysch. Lines parallel to this imply that the first synorogenic flysch of the external Dinarides was Eocene, while they were Oligocene then Miocene from Tuscany to Umbria, and in the southern Apennines. The synorogenic flysch furrow has thus swept over the Apennines later than in the Dinarides. On the left, the sketch attempts to explain the separation between the Kabyle and Calabro-Peloritan Massifs, by right-slip fracturing and sliding.

The Dinaric branch (Figs. 1 and 2) forms the branch bounding the African platform. It consists of the Rif and Tellian chains of North Africa, Sicily, and Calabria (excluding the Ragusa platform, Apulia and Monte Gargano) (see Figs. 9, 13, and 14), the Apennines, part of the eastern Alps, the Dinarides of Yugoslavia, the Hellenides, and the chains of southern Turkey. These chains cover the margins of the African platform (Fig. 1).

In the eastern Mediterranean, intermediate massifs (Zwischengebirge) lie between the two branches. On paleogeographic grounds, a median branch is recognized (Fig. 2). The Alpidic and Dinaric branches connect in the region of Gibraltar (Figs. 1 and 8).

E. Internal and External Zones, Allochthon and Autochthon

Within each branch an internal part (the Alpine axis part) and an external part on the side nearest the platform can be distinguished. Thus, there is a progression to increasingly external zones with decreasing distance from the platform. This concept is widely used by Alpine geologists and is so used in this chapter.

Complementary geometric and kinematic relationships are expressed by the terms *overthrusting* and *underthrusting*, *obduction* and *subduction*, *allochthon* and *autochthon*. Classically, nappes are thrust over the autochthon; that is to say, they are obducted. The terms will be used in that sense in this article. However, it could equally be claimed that the autochthon passes under the allochthon by subduction or underthrusting for the motion is only relative, and from a dynamic standpoint the latter is a better description. In effect, the major Alpine overthrusts seem to be due to the approach of the European and African platforms, crushing the geosyncline and passing under the Alpidic and Dinaric branches (excluding the Austro-Alpine area, see below). Mechanically, the platforms act as the jaws of a giant vise which compressed the plastic Alpine zones, creating a major uplift. The margins of the uplift progressively covered the neighboring parts of the platforms with material directed toward the external part of the orogen. At certain periods thrusting in the opposite sense (retro-overthrusting), toward the internal zones, occurred. This was particularly the case with precursory and late phase movements in the Alps or Apennines. It is readily understandable that during the compression of the Alps, thrusting occurred in the depressed zones. Vertical movements depressed or uplifted parts of the chain between phases of orogenic compression.

F. The African or Adriatic Promontory and the Eastern Alps

The Dinaric branch overrides part of the African platform, which Argand called the African promontory. Following the rule, the margins of the promontory are overthrust by the Apennines and Dinarides (Fig. 1). Parts of the promontory crop out, a case in point being the Ragusa platform in Sicily and Apulia and Monte Gargano (see Figs. 9, 13, and 14). Since at present the greater part of the promontory is covered by the Adriatic, it could equally be called the Adriatic promontory, or Apulian promontory, from the principal southern outcrop (Fig. 2). That part of the Dinaric branch to the north, which links the Apennines and Dinarides, has been pushed by the promontory until it now forms the Austrian Alps, that is the greater part of the eastern Alps, thrusting northward over the Alpidic branch (Figs. 1 and 10). This thrusting is thus the exception in the Mediterranean Alpine chains, and is explained by the marked northerly projection of the African promontory. The history of this promontory will be discussed later (see Figs. 18 and 19).

If allowance is made for the region covered by nappes in Sicily, Calabria, the Apennines, and Dinarides, the promontory at the beginning of the Alpine orogeny was much larger than it is today (compare Figs. 1 and 2), although it is still convex to the north. Taking the Mediterranean as a whole, the sinuous trend of the Alpine orogenic band reflects, qualitatively at least, the form of the bounding platforms, and the sinuous character of the orogen results from its being formed between the jaws of a crushing vise.

IV. RETROTECTONIC ANALYSIS

As a result of the Alpine tectonic convulsions, and in particular major overthrusting, the present structure of the Alpine chains is a poor reflection of the original state of the Mesozoic paleogeographic zones. To reconstruct the succession of zones, a method called retrotectonic analysis is used. Starting with a knowledge of the present state, the effects of the successive tectonic phases must be removed by going back through time and reconstructing the different stages of evolution. Such tentative attempts form the basis of Figs. 2, 3b, 4c, 5, and 6.

Obviously, uncertainties increase as progressively older periods are considered, and in the present case it is impossible to discuss all the problems which arise in the analysis. Instead, reconstructions will be synthesized on the basis of paleogeographic arguments which indicate the main evolutionary features in chronological order. Examples will be chosen primarily from the Atlas segment and from the Calabro-Sicilian Arc.

A. The Beginning of the Alpine Cycle

At the end of the Hercynian episode, that is, during the Permian, Europe and Africa are presumed to have formed a single continental block. During

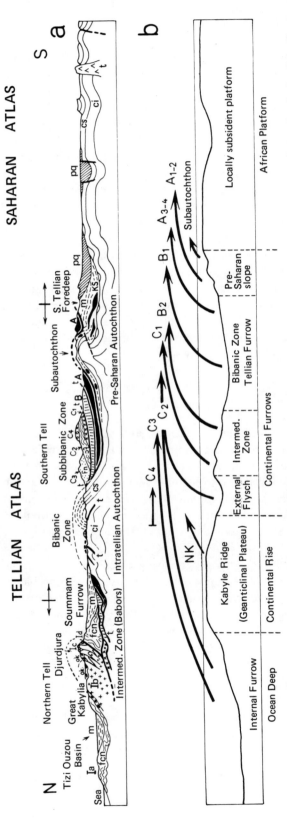

Fig. 3. Structure and evolution of the Algerian Tell: *a.* Schematic, composite north–south section through the Tellian and Saharan Atlas at the longitude of the Great Kabylia. From south to north: t) dome of Triassic salt; ci) Lower Cretaceous; cs) Upper Cretaceous; pq) Plio-Quaternary basin (characteristic of the Tunisian Atlas) and Plio-Quaternary subsident zone on the border of the Saharan Atlas; m) Lower Miocene of the south Tellian foredeep with "Sedimentary Klippes" (KS); A) Unit A (= nappe A); t) Triassic cushions supporting Units B and C; B) Unit B (= nappe B); C₁) Plate C₁ (conglomeratic Senonian derived from the Intermediate Zone, Babors type); C₂) Cretaceous flysch (fc); C₃) Oligo-Miocene, Numidian flysch (fn) or lateral equivalents; C₄) synorogenic Lower Miocene; cs, ci) Cretaceous of the parautochthonous Bibanic Zone; m) Lower and Upper Miocene (postnappes sequence of the Sheliff–Soummam furrow); fcn) Cretaceous to Paleocene allochthonous flysch; t) Oligo-Triassic; Id) Southern Kabyle flysch; Ic) Limestone Chain; Ib) basement of the Kabylia; Ia) Northern Kabyle flysch; ec) splinters of the Limestone Chain; ok) Oligo-Miocene of the Kabylia (Kabyle Oligo-Miocene); m) postnappe Miocene of the Tizi Ouzou Basin. *b.* Senonian paleogeography: The arrows show the origin, the sense of displacement, and the order of superposition of the nappes (overthrust units). *Note:* One must not confuse the expressions "Southern Kabyle" and "Northern Kabyle," which concern the present geographic position of the flysch relative to the Kabyle Massif, with the expressions "Sub-Kabyle" and "North or Ultra-Kabyle," which refer to the paleogeographic position of the flysch, external or internal (see Fig. 2) relative to the Limestone Chain, prior to the overthrusting.

the Triassic this was disrupted, giving rise to two cratonic plates which moved apart, leaving between them less extensive fragments. The plates were probably separated to the point of exposing oceanic crust. In certain internal Alpine zones an ophiolitic complex of ultrabasic (peridotites) and basic rocks was formed. These rocks represent scales or slices of oceanic crust and magmatic rocks akin to those found along present oceanic ridges. On either side of the belt of oceanic crust, the principal part of the plates formed the continental areas of Europe and Africa. Close to the newly formed ocean the continental margins were subjected to deformation more or less parallel to the lines of initial rupture.

These deformations determined the location of the facies belts which formed the earliest Alpine, paleogeographic zones. Under transgressive seas a platform facies of Germano-type Triassic, a detrital (Verrucano) and flysch facies, and finally an Alpine carbonate facies formed.

B. The Evolution of the Tellian Atlas Segment during the Mesozoic

As a type section the region crossing the Grande Kabylia* may be considered. In the tectonic analysis of this complex region, reference will be made to Figs. 3 through 7. Figure 3a is a schematic section, and Fig. 3b a reconstruction, of the paleogeographic zones of the Senonian. In particular, an underthrusting or subduction of the external autochthon (pre-Saharan and intra-Tellian) under the Kabyle Massif and the Limestone Chain can be recognized. There is also overthrusting of nappes of relatively plastic material from the Tellian trough and flysch trenches. Figure 4 illustrates some of the details; a scale section of the region (Fig. 4a) and an expanded section (Fig. 4b) give more stratigraphic detail of the diverse units, and another section (Fig. 4c) shows the situation prior to the major Miocene overthrusting. Figure 4c, in effect, illustrates the tectonic zones in pre-Miocene time. In particular, it shows the Intermediate Zone between the Tellian furrow and the external sub-Kabyle flysch, broken up during the Cretaceous with the rise of the Triassic, and the Kabyle domain tectonized during the Eocene. The scale of the section in Fig. 4c is not the same as in Figs. 4a and 4b, for the tectonic shortening of the Tellian Atlas during the Miocene was of the order of a hundred kilometers.

In the Triassic only two facies are distinguishable, the Tellian and Kabyle (Fig. 4c). The former is of Germanic type, with evaporites which later control decollement and rise as salt domes, diapirs, and cushions lubricating the base of nappes. This Triassic facies developed in that part of the region which later became the Cretaceous Tellian trough, but also extended southward into the

* In English usage, Kabylia refers to the location; the adjectival form is Kabyle.

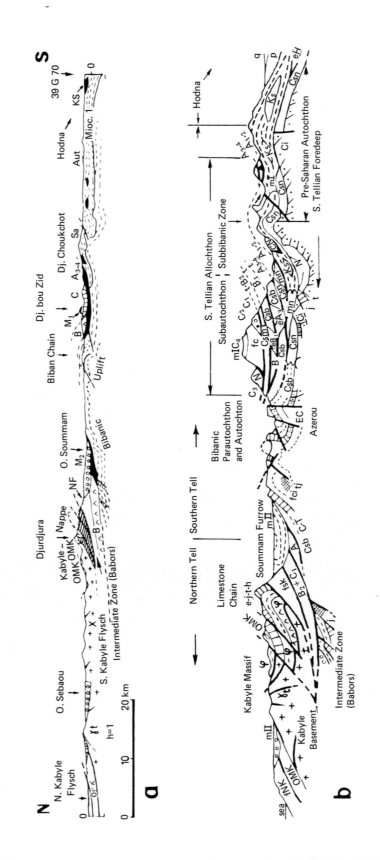

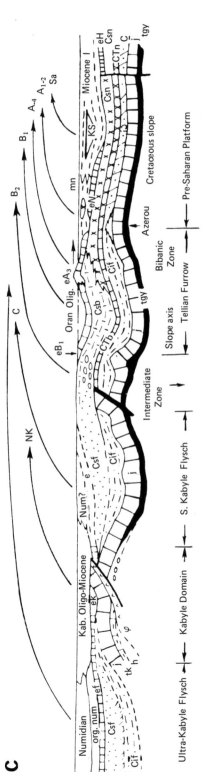

Origin, side and width of displacement, and mode of superposition of the allochthonous units

Fig. 4. Present structure of the Tellian Atlas and reconstruction prior to Miocene overthrusting: *a.* Present condition (no vertical exaggeration). From south to north: KS) sedimentary klippe; Mioc. 1) Miocene I = Lower Miocene of the foredeep, (prenappe Miocene); Aut) autochthonous Miocene of the foredeep; Sa) subautochthon; A_{3-4}) plates of Unit (nappe) A, B, C) Units (nappes) B and C; M_1) synorogenic Lower Miocene, deposited on nappe C during the displacement of the nappes; M_2) postorogenic Miocene (postnappe Miocene); NF) flysch nappe; OMK) Kabyle Oligo-Miocene; X) pre-Alpine basement of the Kabylia; γt) granitic intrusions. *b.* Present condition, with expansion of the tectonic units and indication of their contents: From south to north: 3, p) Quaternary, Pliocene; eH) Eocene of the Hodna; Csn) neritic Upper Cretaceous (Senonian); KS) sedimentary klippe; Ci) Lower Cretaceous; CN) neritic Cretaceous; t, j) Triassic, Jurassic; 2, mn) northern Miocene of the foredeep; mI) Miocene prenappes of the South Tellian foredeep; A_{1-2}, A_{3-4}) diverticulate plates of nappe A; CN) neritic Upper Cretaceous (Senonian); Ci) Lower Cretaceous; t, j) Triassic, Jurassic; 2, mn) northern Miocene of the foredeep; Csb) bathyal Upper Cretaceous of the Tellian foredeep; eA$_3$) Eocene of Plate A$_3$ (with *Nummulites*); eB$_1$) Eocene of Plate B$_1$ (with *Globigerina*); CSm) conglomeratic Upper Cretaceous; C$_1$) Plate C1 (macro-conglomerates of the Intermediate Zone); fc) Cretaceous flysch of the Bibanic Zone; C-T, Csb) Cenomanian, Turonian, and Senonian horst—splinter of the Azerou; tj) diapiric Triassic with Jurassic; fci) Lower Cretaceous flysch of Plate C2; mIC$_4$) synorogenic Lower Miocene forming Plate C4; N) Numidian of Plate C3; EC) of the Bibanic Zone; A) nappe A (internal part); mII) Miocene II—postorogenic Miocene = postnappe Miocene of the Shelif-Soummam furrow; B$_2$ + C$_1$) Plates B2 and C1; fsk) south Kabyle flysch; φ) plastic phyllites of the Kabyle basement; e-j-t-h) Eocene, Jurassic, Triassic, and Carboniferous of the Limestone Chain; OMK) Kabyle Oligo-Miocene; γt) intrusive granites; fNK) north Kabyle flysch; mII) postnappe Miocene of the Tizi-Ouzou Basin. *c.* Reconstitution before the Miocene overthrusting: From south to north: tgy) evaporitic Triassic; j, C) Jurassic and Lower Cretaceous; CTn) neritic Cenomanian and Turonian of the pre-Saharan platform; Csn) neritic Senonian; eH) Eocene of the Hodna; KS) sedimentary klippes coming from Csn; eN) Eocene of Nador; Cif) Lower Cretaceous flysch; CTb) bathyal Cenomanian and Turonian; Csb) bathyal Senonian; eA$_3$, eB$_1$) Eocene of Plates A3 and B1; Num?) presumed Numidian; ek, j, tk) Eocene, Jurassic, and Triassic of the future Limestone Chain (Alpine Kabyle domain); h, φ) Carboniferous and phyllites of the pre-Alpine Kabyle basement; ef) Eocene flysch.

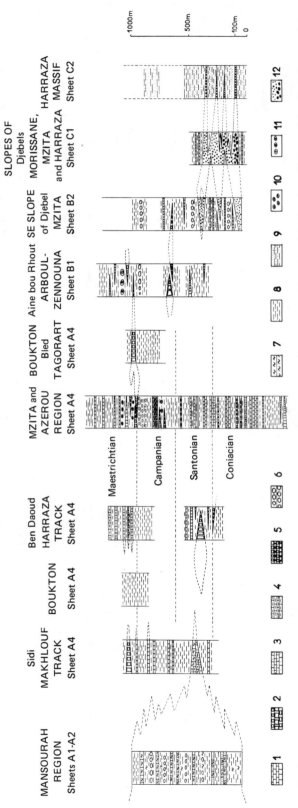

Fig. 5. Succession and correlation of the Senonian facies in the Tellian plates: In the sub-Bibanic Zone (see Figs. 3 and 4) there is a piling up of the allochthonous plates with many different Senonian series. From this stacking, one can reconstitute a sequence of facies zones, which reveals a certain paleogeographic logic. From left to right: The neritic Senonian of Plates A_1–A_2 is identical with the external pre-Saharan platform. It passes into the bathyal Senonian of Plates A_4 and B_1, which characterizes the Tellian furrow (see Fig. 4c). This bathyal Senonian contains more and more breccias in Plate B_2. Plate C_1 is characterized by megabreccias from the Intermediate Zone (northern part of the Tellian furrow). Plate C_2, with its flysch facies (external flysch), contains the same elements as B_2.

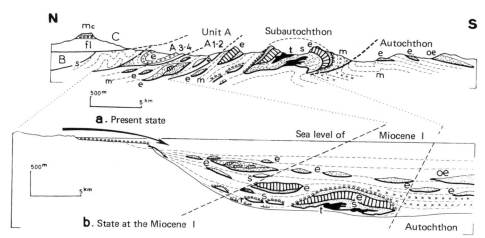

Fig. 6. Synthetic cross sections of the southern Tell and the Pre-Saharan autochthon in the Hodna Region, and reconstruction of the South Tellian foredeep, prenappe Lower Miocene: *a. Present state*, from south to north: oe) Oligocene, Eocene; m) molassic Lower Miocene of the South Tellian foredeep; e) Eocene; s) neritic Senonian; t) evaporitic Triassic; fl) Cretaceous external flysch; mc) synorogenic Lower Miocene; A_{1-2}, A_{3-4}) plates of Nappe A; B, C) Nappe B and C. *b. State at the beginning of the prenappe Lower Miocene*, the Miocene foredeep after submarine sliding of sedimentary klippes (resedimentation). The Miocene (dashes) contains exotic fragments of t) Triassic; T) reefal Turonian; s) neritic Senonian; e) Eocene; o) Oligocene. Noteworthy is the presence in the Miocene of Eocene fragments in a normal or overturned position. The Turonian, Senonian, Eocene, and Oligocene fragments come from the pre-Saharan Slope and Bibanic Zone (see Fig. 3).

southern Atlas. The Kabyle Triassic, with sandstone and limestone, is practically without evaporites, which explains why the decollement level of the Mesozoic to Eocene beds of the Kabyle Massifs occurs at a lower horizon, for example, in the phyllites of the pre-Alpine basement.

Limestone deposition occurred in all zones during the Jurassic, with facies reflecting slight local differences. Everywhere, the Jurassic forms a band of competent limestones several hundred meters thick, one which in some zones gives rise to significant relief such as the Limestone Chain, and the Babors Chain south of the Gulf of Bougie (Figs. 8 and 9).

As a result of tectonic movements there was a more marked paleogeographic differentiation during the Cretaceous. On both sides of the oceanic rift the continental margins were broken into ridges and troughs, which extend the length of the Alpidic and Dinaric branches (Fig. 2). In the troughs thick terrigenous sediments (which acted as a plastic material during orogenic compression) were deposited.

During the Lower Cretaceous there developed within the Riffo-Tellian Chain a major trough, the Tellian furrow, which received terrigenous sediments.

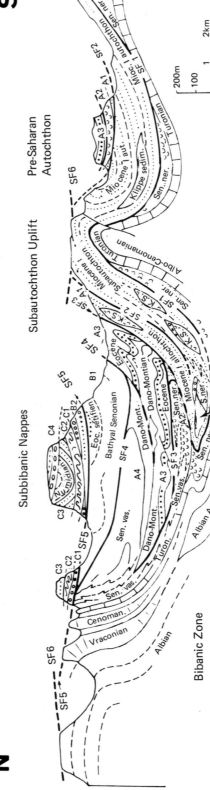

Fig. 7. This cross section shows different surfaces covered by transgressive series (SF1 and SF6), or by allochthonous units (SF2 to SF5). Surfaces SF1 and SF6 are evidently erosion surfaces. Surfaces SF2 to SF5 have to be interpreted as pediments developed in front of the moving, allochthonous beds. Each of these pediments only existed at a particular geological moment as a narrow level zone, controlled by the nappe front, and rapidly overridden by the nappe. This instantaneous pediment progressively moves downstream at the foot of the moving talus constituted by the nappe front. Thus, the exposed part of this pediment moves at the same time and with the same rhythm as the allochthon. The order of succession of the erosion pediments indicates that Plates A_{1-2} were formed first, without overloading, on the ancient pre-Saharan Slope. Then, there arrived successively the group numbered A_3-A_4, Plate B_1, then group B_2-C. A decollement first affects the unstable ridges and slope (pre-Saharan Slope, Plates A_1-A_2, Kabyle ridge), then the furrows (Tellian furrow, flysch furrows). Abbreviations: Turon = Turonian; S. ner. or Sen. ner. = neritic Senonian; Sen. vas. = bathyal Senonian; Dano-Mont. = Dano-Montian; Eoc. Nad. = Eocene of Nador; Eoc. sétifien = Setifian Eocene; Mioc. I = molassic Lower Miocene of the South Tellian foredeep.

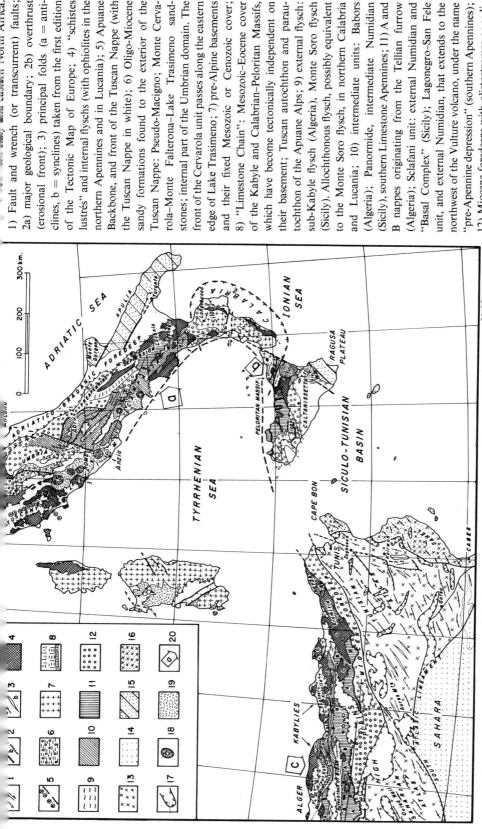

CAIRE

1) Fault and wrench (or transcurrent) faults; 2a) major geological boundary; 2b) overthrust (erosional front); 3) principal folds (a = anticlines, b = synclines) taken from the first edition of the Tectonic Map of Europe; 4) "schistes lustrés" and internal flyschs (with ophiolites in the northern Apennines and in Lucania; 5) Apuane Backbone, and front of the Tuscan Nappe (with the Tuscan Nappe in white); 6) Oligo-Miocene sandy formations found to the exterior of the Tuscan Nappe: Pseudo-Macigno; Monte Cervarola–Monte Falterona–Lake Trasimeno sandstones; internal part of the Cervarola unit passes along the eastern edge of Lake Trasimeno; 7) pre-Alpine basements and their fixed Mesozoic or Cenozoic cover; 8) "Limestone Chain": Mesozoic–Eocene cover of the Kabyle and Calabrian–Peloritan Massifs, which have become tectonically independent on their basement; Tuscan autochthon and parautochthon of the Apuane Alps; 9) external flysch: sub-Kabyle flysch (Algeria), Monte Soro flysch (Sicily). Allochthonous flysch, possibly equivalent to the Monte Soro flysch, in northern Calabria and Lucania; 10) intermediate units: Babors (Algeria); Panormide, intermediate Numidian (Sicily), southern Limestone Apennines; 11) A and B nappes originating from the Tellian furrow (Algeria); Sclafani unit: external Numidian and "Basal Complex" (Sicily); Lagonegro–San Fele unit, and external Numidian, that extends to the northwest of the Vulture volcano, under the name "pre-Apennine depression" (southern Apennines); 12) Miocene foredeeps with olistostromes or sedimentary klippes of various dimensions. In the northern Apennines the same includes the Umbrian series, and the Middle to Upper Miocene of the "Marnoso–Arenacea" Formation; 13) pre-Saharan Atlas: Saharan Atlas and Tunisian Atlas; 14) Sahara: platform limited to the north by the north Saharan "Fault" (south Atlas lineament); 15) Ragusa plateau, Apulia and Monte Gargano, forming the foreland of Sicily and the Apennines; 16) Tunisian Sahel, Kerkennah High, and Cape Bon; 17) intramontane basins and depressions that were posterior to the overthrusting tectonics (postnappe Mio-Plio-Quaternary); 18) volcanoes (the volcanic cones of the Eolian or Lipari Islands, and Linosa, have not been given this symbol); 19) Tertiary and Quaternary volcanic material; 20) indication of the transversal sections of Fig. 6 of Caire (1973, p. 18).

This trough maintained its identity up to Oligocene–Miocene time despite frequent deformation of its margins. It separated a platform region to the south, the region of the future southern Atlas, where deposits were continental sands and neritic beds, from the Kabyle region to the north. The latter region was characterized by reduced thickness, gaps, and condensed sequences of pelagic carbonates of geanticlinal type. The Kabyle domain corresponds to the sedimentary cover over the ancient Kabyle Massifs, which now form the Limestone Chain (Chaîne Calcaire). The chain is discontinuous and its various segments (Kabylias, Peloritan–Calabria) are surrounded by flysch (Figs. 2, 8, and 9).

Thus the northern continental margin of Africa, during the Lower Cretaceous, may be divided into three major zones (Fig. 2):

(1) Continental shelf with epicontinental facies. A differentiated platform sank irregularly and from it rose the pre-Saharan southern Atlas.

(2) Continental troughs or furrows, in particular the Tellian furrow and the external flysch.

(3) Continental rise, the continental margin, irregularly uplifted, next to the oceanic basin.

The junction between the Tellian furrow and the external flysch furrows was a mobile zone (the Intermediate Zone in Figs. 3 and 4). It was subjected to important deformation at the end of the Jurassic, with folding, metamorphism, and the injection of ultrabasic rocks continuing into the Lower Cretaceous. It was reactivated by Upper Cretaceous tectonic events. As will be seen later, it can interpreted as a wrench fault zone and is continued in Sicily as the Vicarian line (Fig. 2).

In the Upper Cretaceous there was no change in the major paleogeographic zones. The Senonian is of particular interest for beds of this age are the best represented in the various structural units of the present Tellian chains (Fig. 4b). The restoration of the Senonian series of the Tellian nappes (Fig. 5) leads to a slightly different Senonian paleogeography (Fig. 4c).

Figure 5 was constructed on the assumption that the allochthonous units (A_1, A_2, to C_2) were more internal in origin the higher they occur in the structure. This leads to a succession and a correlation between the various stratigraphic columns, the harmony of which serves to justify the original assumption. This paleogeographic harmony serves as a key to the reconstruction of the tectonic evolution.

During the Senonian, tectonic movements affecting the Intermediate Zone resulted in folding, discordance, resedimentation of Triassic olistostromes and various olistolites, and the formation of conglomerates. The movements also affected the transition zone between the Tellian trough and the pre-Saharan platform, that is the pre-Saharan slope (Fig. 3b), resulting in folding followed by minor discordances. The flysch bordering the Kabyle Massif was also deformed. These movements are manifestations of Cretaceous tectonic activity also known in many parts of the Mediterranean Alpine chains.

C. Cretaceous or Paleo-Alpine Tectonics

Tectonic activity during the Cretaceous, which may have begun in the Upper Jurassic, is explained by the oblique approach of the African and European plates. The movement was related to the opening of the South Atlantic, with the African plate moving eastward with respect to the European plate. The continental borders have a complicated form (the African promontory is an example) and compression, crushing, folding, slip, etc., vary in importance from region to region. The strongest compression occurred along the northern margin of the African promontory, which explains the Jurassic and Cretaceous folding and overthrusting in the eastern Alps.

Along the flanks of the promontory the Apennines and Dinarides were compressed, and thrusts directed toward the interior of the chain can be seen. The oceanic deep (Fig. 2) was crushed and the ophiolites tectonized, producing nappes and scales by Upper Jurassic time. In the Tellian orogen the Intermediate Zone was folded during the Albian and overturned to the north. After this time the oceanic rift closed, with the sinuous continental margins in direct contact. Longitudinal sinistral displacement of Africa with respect to Europe then became difficult, and this explains the development of slip or shear zones in the continental margins themselves. The Vicarian line (Fig. 2) is one such shear zone. The Catanzaro Zone (Fig. 14) forms the southern outcrop limit of the Apennine ophiolites. In the Cenozoic, Alpine tectogenesis, in the strict sense, can be explained by dextral shear and slip between Europe and Africa (see Figs. 18 and 19).

D. Evolution of the Tellian Wing during the Cenozoic

The Eocene, up to the Lutetian, generally follows the Cretaceous concordantly except in the Kabyle region, which continued to be affected by pulsatory movements.

In the Tellian trough paleogeographic differentiation was more marked than in the Senonian. In the Ypresian, limestones with flints form a fairly constant facies, but marls, phosphate beds, shell beds, and sandstones are more variable in distribution.

The first Cenozoic phase had its main activity in Lutetian time, with pulsations which extend from Ypresian to Upper Eocene. The Kabyle Zone was broken up and even overthrust to the south. The flysch border of Tellian trough, and southern Atlas, was folded, but without strong overturning. Fol-

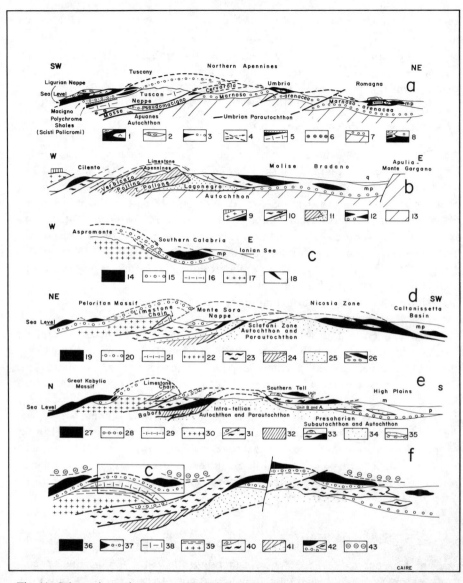

Fig. 11. Schematic sections across the Barbary and Apennine orogens. These sections are shown by the arrows "a" to "f" of Fig. 13. They are to show the relationships between the major structural units which are not always represented everywhere in their real development. Post-overthrust sediments, deformation, and faulting are omitted, except for the external zones to the right of the sections. *Section a: northern Apennines:* 1) Ligurian nappe; 2) neoautochthonous Miocene of the Ligurian domain, transported with the debris of the Ligurian and Tuscan nappes as far as the external Apennine zones; 3) Macigno (Oligo-Miocene flysch) with Ligurian series sedimentary klippes and olistostromes; 4) detachment of the Tuscan nappe fragments carrying with them Ligurian nappe debris. These elements were transported,

torn up, and resedimented as far away as Romagna; 5) Tuscan nappe: Macigno, polychrome schists, dominantly calcareous Jurassic and Cretaceous, gliding level of Norian age; 6) phyllites of Carnian age with Verrucano quartz conglomerates, of the Tuscan nappe or parautochthon of the Massa Zone; 7) the whole thing corresponds to the Tuscan autochthon (with a posthumous arching (Apuane Backbone) which is not shown here), to the parautochthon of Marches and Umbria, and to the autochthon of Romagna. During the Miocene, the Pseudo-Macigno of the Massa Zone, the Monte Cervarola–Monte Falterona–Lake Trasimeno flyschs, then the "Marnoso–Arenacea" Formation, indicate migration of clastic facies toward the exterior. The upper part of the section shows schematically the flysch brought forward by the orogenic wave; 8) the Ligurian "Argille Scagliose," which enclose exotic rags of the Alberese and Macigno, are indicated in the external resedimentation basin, together with the Tuscan Miocene neoautochthon which was displaced to the exterior of the chain. *Section b: Lucania:* 9) "Argille Scagliose" and, seen further south, borders overthrusting the Calabrian Massif; 10) Lucania and Cilento flysch, and metamorphic flysch; 11) southern Limestone Apennines nappes, and Lagonegro unit; 12) Miocene resedimentation basin with rags of the 9 and 10 series; 13) external parautochthon and autochthon (Apulia–Monte Gargano domain). *Section c: southern Calabria:* 14) internal flysch: "Argille Scagliose" complex enclosing rags of Cretaceous and Numidian flysch, and Eocene limestones; 15) transgressive Oligo-Miocene upon 16 and 17; 16) splinters of the Limestone Chain that have slid onto the phyllites; 17) Peloritan–Calabrian metamorphic basement; 18) Mio-Pliocene resedimentation basin of southern Calabria. *Section d: northeast Sicily:* 19) internal flysch (identical to 14) with the Reitano series, where the Numidian is replaced by a molassic Oligo-Miocene; 20) Oligo-Miocene molasse of the Peloritan, transgressive on 21 and 23; 21) splinters and slices ("écailles et lambeaux") of the Limestone Chain enclosed in, or slipped over, the phyllites of the basement; 22) Peloritan metamorphic basement; 23) Monte Soro external flysch; 24) external and intermediate Numidian with Panormide sedimentary klippes (cf. 11); 25) Sclafani Zone; 26) Caltanissetta resedimentation basin with olistostromes and sedimentary klippes of internal flysch, external Numidian, etc.; *Section e: Algeria:* 27) internal flysch (here ultra-Kabyle), essentially Numidian; 28) Oligo-Miocene transgressive on 29 and 30, with sedimentary klippes of basement, Limestone Chain, and flysch; 29) Limestone Chain, ancient Mesozoic–Eocene cover of 30; 30) Kabyle basement; 31) external flysch and associated detrital Tertiary formations; 32) Algerian Intermediate Zone (Babors Chain and its deep prolongation); 33) external (or internal) flysch that has been dismembered, then transgressively overlain by syncronappe Lower Miocene; 34) A and B nappes, consisting of sedimentary cover, detached from the intra-Tellian autochthon and parautochthon; 35) south Tellian foredeep with sedimentary klippes coming from the A and B nappes. *Section f: composite section grouping the basic elements of sections "a" to "e":* 36) internal series, with the Ligurian Nappe, Lucania and Cilento Nappes, Calabrian and Sicilian internal flysch, and the ultra-Kabyle flysch of North Africa; 37) transgressive Oligo-Miocene over the Tuscan nappe and autochthon, over the Peloritan–Calabrian Massif, over the Kabyle Massifs of North Africa. This Oligo-Miocene includes debris of more internal units; 38) detached covers ahead of the no. 36 internal series: Tuscan Nappe; Calabrian, Sicilian, and North African Limestone Chains; 39) pre-Alpine basement; 40) Monte Soro and sub-Kabyle external flysch, and related Tertiary; 41) southern Limestone Apennines, Panormide and Algerian Intermediate Zone, transgressively overlain by the intermediate or external Numidian, Lagonegro, Sclafani, and intra-Tellian units; 42) pre-Apenninic, Adriatico-Bradanic, external Calabrian, central Sicilian, and south Tellian resedimentation basins (Neogene foredeep); 43) postnappe formations, omitted in the internal areas of sections "a" to "e."

lowing this first phase of tectonism, the external zones (southern Atlas, Tellian trough) were leveled by an erosion surface (SFI in Fig. 7).

The Oligo-Miocene marks a period during which tectonic activity diminished and detrital sediments were laid down. The most important formations are (Fig. 4e):

(1) Oligo-Miocene of Oran (or Boghari) with the Boghari Sandstone characteristic of the median part of the Tellian trough. It is concordant with the underlying series.

(2) Oligo-Miocene of the Kabylias, consisting of sandy, conglomeratic beds discordant upon the basement of the Kabyle Massif and the detached cover scales (of the Limestone Chain).

(3) Oligo-Miocene or Numidian discordantly covers Cretaceous–Eocene flysch and consists of sandstones and shales, and in particular "grès à dragées," that is, sandstones with small quartz pebbles. In the internal flysch zone the Numidian shales, variously colored (green, lie de vin, red) and rich in *Tubotomaculum*, form a plastic matrix which is often scaly (Argille Scagliose) and in which, during the Miocene, there are fragments of both older and younger series. The Numidian sequence continues into Sicily (see Fig. 12) and ends in the Apennines.

The South Tellian foredeep formed in the Lower Miocene (prenappe) between the Tellian Atlas and the southern Atlas. It filled with fossiliferous molassic marls and sandstones, and soon received sedimentary klippe, then nappes sliding in from the north (Fig. 6).

The major tectonic phase occurred during the Lower Miocene. It was responsible for the major nappes (Figs. 3, 4, and 8) and profoundly disturbed the Tellian Chain. Morphotectonic analysis of this chain permits reconstruction of the successive events (Fig. 3). The old slope, sill, and ridge zones were the first to be tectonized. From the pre-Saharan slope and Tellian furrow were derived the sedimentary klippe and scales now incorporated in the Lower Miocene of the South Tellian foredeep (Fig. 6). Major obduction resulted in the Kabyle domain largely covering the flysch to the south (Figs. 3a and 11). The Tellian trough then furnished massive plastic nappes which slid southward as far as the South Tellian foredeep. Thus, in the southern Tell, there is a nappe structure comparable with the Chablais Prealps in the western Alps (Figs. 3a, 4b, 8, and 11). In the southern Atlas, the movements result in structures of a tectonic style similar to the Jura (Figs. 8 and 9) with folds, fold-faults, decollement, and wrench faults.

The slide nappes in Algeria have been divided into three groups, identified by A, B, and C (Figs. 3, 4, 6, 7, and 11). The A nappes (locally further subdivided into slices A_1, A_2, A_3, A_4) are the deepest and the least displaced. In the Southern Tell, they originated on the old pre-Saharan slope (Fig. 3b) and from Mesozoic–Eocene beds in the southern part of the Tellian trough. They are often differentiated (diverticules, according to Lugeon); e.g., A_3 in the Southern Tell is formed essentially of Eocene, and is overlain A_4 mainly of Senonian age. The Senonian and Eocene are part of the same series, thus A_3 slid over the Paleocene to form its tectonic cover. A_1 and A_2 are themselves formed from a series in which Senonian was covered by transgressive Lower Miocene beds; A_1 is essentially Miocene and is covered by the Senonian A_2.

The B group nappes are intermediate between A and C. In the Tell they were derived from the central and northern parts of the Tellian trough, reflecting in their lithology the paleogeographic conditions of that trough. They are essentially made up of terrigenous Cretaceous and/or Eocene, and Oligo-Miocene of Boghari (Figs. 3, 4, and 8). The C nappes, the most internal in origin, are composed of both internal and external Cretaceous flysch and the Numidian or its lateral equivalents. There is also, on occasion, synorogenic Lower Miocene deposited during the displacement of the nappe. The C nappes form the allochthonous superstructure of the Tell, with the relief due to the Numidian Sandstone or the synorogenic Lower Miocene (Figs. 3 and 4).

This subdivision of nappes may be found all over North Africa, from northern Tunisia to the Riff of Morocco, and it could even be extended to the Betic Cordillera in southern Spain (Fig. 8). It provides evidence that all the chains have the same basic structural form, including the same prior paleo-geographic organization.

In the Calabro-Sicilian Arc, the C nappes are still clearly distinguishable, for the external and internal flysch are the same as in North Africa. However, the A and B nappes are no longer recognizable, for the Tellian trough has been replaced by other facies zones (in Sicily by the Sclafani, Vicari, Campofiorito-Cammarata, and Sciacca Zones) (Figs. 2, 11, and 12).

E. Posttectonic (Tarditectonic or Neotectonic) Evolution from the Upper Miocene to Recent

In the Middle Miocene vertical movements began, and have continued as the dominant activity up to the present time. It is because of this that the tectonic activity is referred to as posttectonic, neotectonic, or tarditectonic. The earliest movements have the same predominantly E–W trend, as in earlier periods. Thus, the Cheliff–Soummam trough in Algeria formed and received postnappe detrital deposits. This was followed by regional elevation or subsidence about axes of variable but dominantly WSW–ENE and SW–NE trends. The intra-Tellian autochthon or para-autochthon reappeared at the surface as a result of uplifts piercing the slide nappes. Paleotectonic reconstructions must rely both on the allochthonous slide nappes and upon deeper parts of the tectonic structure exposed as a result of subsequent uplift (Figs. 3, 4, and 8, for example).

During the period which followed the major overthrusting, compressional movements are relatively unimportant and affect principally the external zones and the orogenic shield. This confirms the notion of closure and accentuation

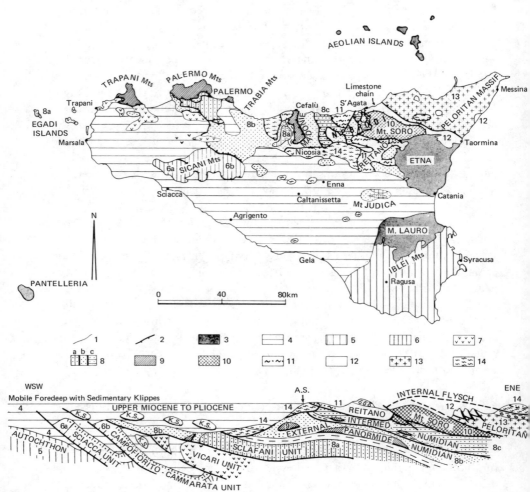

Fig. 12. Schematic map and cross section of Sicily. The section is composite, and shows the various relations between major structural units, not all of which would appear on a single line of section. 1) Geologic con and faults; 2) structural contacts; 3) Recent volcanic rocks; 4) Mio-Pliocene and Quaternary basins (K.S.: imentary klippe); 5) Ragusa–Monti Iblei Plateau; 6) Sciacca and Campofiorito–Cammarata Zones; 7) V Zone; 8) Sclafani Zone (a: Mesozoic–Eocene; b: external Numidian), intermediate Numidian Zone (c); 9) Pa mide Zone; 10) Monte Soro Nappe (external flysch and discordant molassic Oligo-Miocene); 11) Reitano Na (intermediate between the Peloritan domain and the internal flysch); 12) Limestone Chain; 13) Peloritan pre-A basement; 14) Argille Scagliose Nappe (internal flysch), with internal Numidian.

of the orogenic arcs. The situation leads progressively through the vicissitudes of the Plio-Quaternary, where vertical movements predominated and were largely responsible for the present-day shorelines and basins, to the present condition of the Tellian chains.

F. Special Features of the Calabro-Sicilian Arc

The Calabro-Sicilian Arc is more complex than the Rifo-Tellian wing or segment. Sicily provides the paleogeographic link between the Tellian zones described above and the Apennine zones (Fig. 2). The most internal zones in Sicily equate with those in Algeria, with the flysch and the Peloritan domain equivalent to the Kabyle domain. The external zones, however, have different characteristics (Figs. 2, 11, and 12). Among the latter, the Panormide has the same significance as the southern Calcareous Apennines. There is no equivalent of the Tellian trough, and only the external border of the external flysch presents Senonian with the character of that found in the Intermediate Zone of Algeria. There is no longer Triassic of Germanic facies.

The flysch and the Peloritan domain (with the continuation in Calabria) were expelled from their internal location (Fig. 14) over the more external zones to form an entirely allochthonous assemblage. This expulsion of the Peloritan assemblage is analogous to the westerly displacement of the Rif–Betic Massif in the Gibraltar Arc. In both cases, this phenomenon is a part of the closing of the arcs. All the external zones of western and central Sicily, including the Peloritan, form part of a pellicular nappe structure, covering each other like the slates on a roof (imbricate style). The internal flysch, which includes the Argille Scagliose and Numidian Sandstone, is spread over all the external units, and it is often secondarily incorporated at the base of the external nappes (Fig. 12) due to a renewal of tectonic activity.

G. Outline of the Evolution of the Calabro-Sicilian Arc and the Tyrrhenian Loop

In Sicily, Eocene and Lower Miocene tectonic phases are found. Here, more clearly than in Algeria, a wave of folding originating in the internal zones at the beginning of the Miocene moved across Sicily until Pliocene time, progressively affecting more external zones. The foredeep was itself displaced externally preceding the wave of folding. The same evolution is characteristic of the Apennine chains.

The pre-Alpine terrains of Peloritan and Calabria formed a single geotectonic unit, the Calabro-Peloritan Massif, in Alpine time. The massif was fractured, metamorphosed, and thrust at various stages during the Hercynian orogeny, the Eocene, and the Miocene. In Alpine paleogeography it plays the

Fig. 13. The rhegmatic field of Italy and its neighboring areas. Thanks to the many publications in the bibliograpl it has been possible to assemble a complete rhegmatic field showing the major fractures (longest and greatest thro many of which were active wrench faults. They are associated in networks, with dominant directions ranging fr northwest–southeast, to southwest–northeast, to north–south, and east–west, or in fans (with rectilinear fracture or star patterns. Depending on the example, they may have had the pattern from the beginning, or may have veloped it through a combination of fortuitous and successive fractures resulting in distinct tectonic patter The visible throw is a result of several movements or of a final, often late, movement, commonly affecting fold and overthrust (late Hercynian wrenches in the Hercynian massifs, late Alpine elsewhere). The oldest movemer seen in the paleogeography or early tectonic phases, are not shown here, at least not in their pure state. Aside fr the fractures, *sensu stricto*, there are various guidelines such as flexures, supple wrenches, and alignments showi deep-seated ruptures or discontinuities. As a guide, we have shown the residual foredeeps, postorogenic depos and the Hercynian massifs of the Alpine foreland. The arrows—a, b, c, d, and e—show the position of the sectic of Fig. 11.

Fig. 14. Structure of the circum-Tyrrhenian ranges. The Atlas–Apennine system is symmetrical about an axial zone, which corresponds approximately to the major axis of the Tyrrhenian Basin as delimited by the 3000-m isobath. This axis appears to be marked by the Catanzaro Zone, which can be considered the limit between the Atlas and Apennine orogens and is affected by a sinistral wrench fault. The Tyrrhenian Basin is of recent origin. Before the overthrusting it was occupied, at least partially, by the allochthonous units which now surround it, and it collapsed after the Alpine tectogenesis. 1) Internal limit of the Saharan domain, the Ragusa Plateau, and the Apulo-Garganic domain; 2) structural trends in Campania and in the Sicily–Tunisia Basin; 3) anticlines and synclines; 4) displacement of plastic nappes and their resedimented portions; 5) Neogene foredeeps; 6) Zaghouan Fault and the Umbria–Abruzzi limit (Anzio–Ancona line); 7) erosional front of the North African nappes, present front of the Tuscan Nappe, and in more external position, the thrust front of the sandstones of Monte Falterona and Lake Trasimeno (after Giannini and Tongiorgi, 1962; Giannini *et al.*, 1962); approximate limit of the "Littoral Flysch" nappes of Lucania, Calabria, and Sicily, excluding their resedimented portions in the external basins; 8) Kabyle and Panormide Backbone, along which are found the principal outcrops of the metamorphic autochthon; 9) axis of the Tyrrhenian Basin, and its prolongation in the Catanzaro Zone.

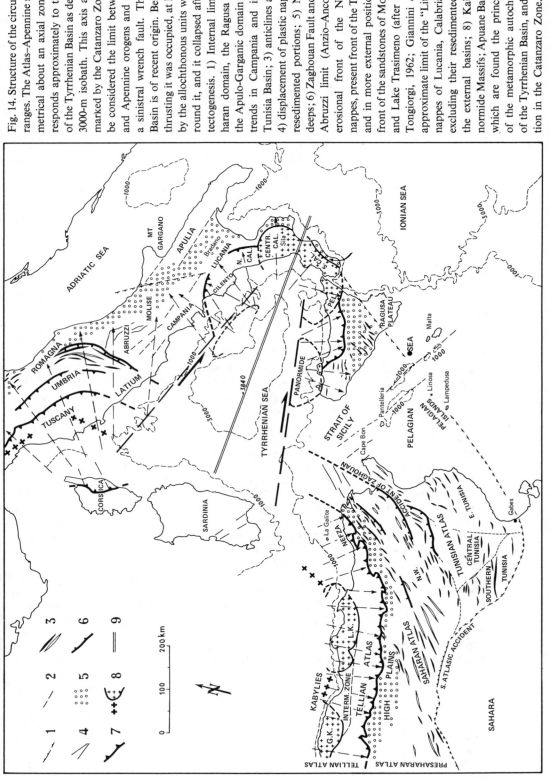

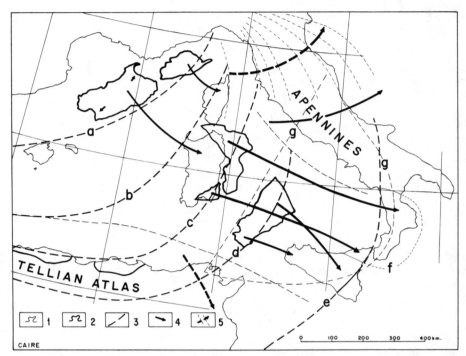

Fig. 15. Paleotectonics of the central Mediterranean in the style of Argand. Starting with the present-day structural framework found in the Tyrrhenian loop, we propose to reconstitute the various parts of Sicily and central–southern Calabria to their respective positions prior to Tertiary tectogenesis. This reconstruction takes into consideration several factors: (1) The most internal parts of the Sicilian orogen (flysch and Peloritan domain) are similar if not identical to the corresponding Atlas units (flysch and Kabyle domain). Today, they are geographically separated by more external zones that differ from each other (northern Tunisian Atlas, Panormide, Sclafani series). Consequently, to join the Kabylias to the Peloritan in their present position requires us to accent a very sinuous, and thus artificial, axis. (2) The Peloritan–Calabrian Massif and the neighboring flysch are today isolated at the base of the Tyrrhenian loop, and almost touch the most external zone (Apulia, and Ragusa Plateau) of the orogen. However, they would find themselves conveniently placed, paleotectonically speaking, between eastern Algeria and the northern Apennines in the prolongation of the Kabyle domain toward the Tuscan ridge. (3) The Peloritan–Calabrian basement, with its Limestone Chain (of relatively internal character) overlies the southern part of the southern Limestone Apennines (of relatively external character). Thus, it does not have a logical place in its present-day environment, which can be explained by a "Catanzaro wrench" prior to overthrusting, or by drifting. (4) Several authors since Argand (1922, Fig. 26) admit that, before the "great Alpine disjunctions," Corsica and Sardinia were placed in an area near Provence and the Pyrenees. Recently the movement of the Corsican–Sardinian block was linked to the sinistral rotation of the northern Apennine sedimentary basin (Boccaletti and Guazzone, 1968, Fig. 3, p. 255). In addition, this angular movement is confirmed by paleomagnetic studies. According to van Bemmelen, the Pyrenean Chain has suffered, compared to its borders, a wrench to the

east–southeast which is in agreement with the presumed movement of Sardinia. (5) In putting middle Sicily to the northeast of Bizerte, we bring together the very special Numidian series of northern Tunisia (Rouvier) and western Sicily: Roccapalumba–Margana Castle regions (Broquet). We even bring near to each other the Lower Miocene (or Oligo-Miocene) glauconitic series recognized exterior to the Numidian, as noted above: Bejaoua (Burollet, Jauzein) in northern Tunisia; Vicari and Campofiorito–Cammarata Zones (Broquet, Mascle) in western Sicily. In this light the Cherichira Sandstone (*sensu lato*) and perhaps the Numidian Sandstone appear to be of the pre-Saharan series, with an African basement source paleogeographically probable. (6) That part of the Tyrrhenian Sea near the Calabrian–Sicilian coast shows signs of major volcanic activity and some important magnetic anomalies that are easily explained if we accept that the Calabrian–Sicilian Arc continues to move to the east–southeast. If, as this figure suggests, there is a flux to the east–southeast which induces a progressive closing of the Tyrrhenian loop, one can imagine that during the Mesozoic the Calabrian–Sicilian paleogeographic arc was in a more acceptable and open form than that seen today. Besides, the hypothesis presented here is not as advanced as that of Argand, for whom Sicily was between the Baleares and the Oran area before the "great Alpine disjunctions," and was carried by the extremity of the Apennines pushed to the west. From this Mesozoic situation of the Peloritan–Calabrian Massif, it is only necessary to admit that the massif, at the time of its displacement to the east–southeast, was squeezed several times between the Atlasides and the Apennines. The orientation, the direction and level at which the successive overthrusts operated, is then explained by the geographical situation and the relative height of the two squeezing jaws. We may consider that the interpretation proposed here essentially reconciles the points of view of various allochthonist geologists of the first half of this century: "Sicilian" Nappe and overthrust Calabrian Arc (Lugeon and Argand, 1906*a, b, c*), big displacement of the Peloritan–Calabrian Massif from the Tyrrhenian Sea (Limanovski, 1913), divergent tectonic overlaps (Quitzow, 1935). Of course, one must allow, in these articles, for the ideas "à la mode" of each period: interpretation of tectonic overlap by major recumbent folds; essential role of compression and minimum importance of gravity slides and gliding....
Legend: 1) Present-day coastlines. 2) Presumed Mesozoic positions of the Corsican–Sardinian block, the median parts of the Peloritan–Calabrian group, and middle Sicily, marked by their actual coastlines. The present-day outline of the coast is given only as a geographical reference and clearly has no paleotectonic significance. Evidently it would have been impossible to spread and split the several middle Sicily zones, to make them touch at one and the same time the Peloritan (Line c) and the Ragusa Plateau (Line e). It can be done similarly, in varying degrees, to the regions shortened by Alpine tectonics. We have indicated in Sardinia the Alpine tectonic displacements (work of G. Chabrier) corresponding to those of the Pyrenees. 3) Mesozoic paleogeographic directions deduced from the fixed positions of 2. 4) Movements of the median branches of the "Tyrrhenian fan." We have shown here neither the transverse or oblique displacements, certainly present, such as the Eocene overthrust of the Peloritan–Calabrian block to the north and to the south nor the late displacements, such as that in Tunisia to the northeast. 5) Displacement of the outer branches of the fan: migration of the Northern Apennine sedimentary basin and nappes, from the Cretaceous to present (Boccaletti and Guazzone, 1968); movement of the northern Tunisian nappes and overthrusting units. a) External limit of the European epi-Paleozoic platform; b) Mesogean orogenic axis separating the "Alpidic" and "Dinaric" branches; c) a rising zone marked by the Kabyle–Peloritan–Calabrian Massifs and possibly the Tuscan ridge (see Fig. 2); d) arcs marked by the external Tell and middle Sicily; e) external edge of the African platform and its Adriatic Promontory (see Fig. 2); f) presumed external limit of the Peloritan–Calabrian block.

same role as the Kabyle Massifs. It was, in effect, the site of Permo-Triassic detrital (Verrucano) sedimentation, later covered by carbonates of Mesozoic to Eocene age. These sediments, separated from the basement, form a Limestone Chain as in the Kabylias. After the Eocene phase of tectonism, the basement and cover were discordantly covered by an Oligo-Miocene molasse. Finally, the basement and cover were thrust over more external zones, and in turn overthrust by the internal flysch nappe.

By the Middle Miocene, the allochthonous, internal structures of the Calabro-Sicilian Arc had been formed. These subsequently were subjected to flexure and normal faulting. The external part, however, continued to be affected by tangential movements during the Upper Miocene, Pliocene, and even the Quaternary. Overthrusting affected particularly the southwest of Sicily (Sicani region) and the pre-Apennine region. The foredeep migrated toward the Pelagian Sea and the Adriatic. At the present time, the Adriatic Sea, with molasse sedimentation and active olistostromes, may represent the residual foredeep common to both the Apennines and Dinarides. Based on seismic evidence, it is probable that the Calabro-Sicilian unit continues its obduction toward the southeast. This present active obduction evidently corresponds to subduction along a plane plunging under the Calabro-Sicilian Arc and the Tyrrhenian Sea.

Apart from this Quaternary evolution, the Cenozoic presents the progressive closure of the Tyrrhenian loop (Figs. 14, 15), which can be superposed along the margin of the African promontory (Ragusa platform, Apulia). If it is unfolded and the nappes restored to their original positions, a pattern similar to that on Fig. 15 is generated. Sardinia and the Corsican basement, which had the same characteristics as the European platform, must have lain to the northwest of their present locations prior to the Cenozoic. Here retrotectonics is in agreement with paleogeographic and paleomagnetic results, which would attach it initially to the Provençal region.

H. The Two Tyrrhenian Arcs

Various facts noted earlier suggest that the Tyrrhenian loop is made up of two distinct arcs (Fig. 14): 1) an external arc comprising the southern Atlas (Saharan Atlas, Tunisian Atlas) and the southern Apennines of Lucania, Campania, and the Abruzzi; 2) an internal arc comprising the Tellian Atlas and the Kabylias, Umbria, and Tuscany.

The central part of the internal arc, made up of the Calabro-Peloritan Massif and associated flysch, has been displaced toward the southeast, crossing what is now occupied by the central part of the Tyrrhenian Sea and becoming compressed against and overthrusting the external arc in Calabria and Sicily (see arrows in Fig. 14).

Prior to Cenozoic thrusting, both arcs were more open, and the facies zones developed as shown by the curves a to e in Fig. 15.

The present form of the Calabro-Sicilian Arc is the result of recent (Plio-Quaternary) movements, which have reshaped the Cenozoic nappe structure. The recent movements consist of a narrow zone of peripheral uplift compensated by a sinking of the floor of the Tyrrhenian Sea.

One further feature of the Tyrrhenian loop is the existence of an external metamorphosed zone. This zone originates in Morocco, includes the massifs of Oran, Bou Maad, Cheliff, Blida, and Babors in Algeria (see Fig. 8), passes through northern Tunisia, and is found again in the Tuscan dorsal. It is probably hidden in the Calabro-Sicilian Arc under the displaced Peloritan Massif and flysch, according to the model illustrated in Fig. 14. The eastern continuation of the Tellian trough may have remained "behind," not having been tectonically transported to Sicily. If this were so, the Tellian trough could have changed into the Umbrian Zone (which would necessitate a correction to Fig. 2).

I. Role of Lateral Movements in Alpine Tectonic Evolution

The great tectonic overturn of the Cenozoic has been explained by the close approach of Europe and Africa, with the resultant crushing of the Alpine orogen. The crushing of some Mediterranean chains may, to a great extent, also result from subhorizontal thrusts and shear (strike-slip) movements, which have played an important part in the evolution of the Alpine orogen at various times and on various scales. Three stages in this process will be discussed here, indicating the dates and the hypotheses. The essential facts regarding wrench faults are illustrated in Fig. 13.

Regionally, the displacements can be grouped into arcs, bundles, or whirls. Thus, recent right lateral strike-slip faults, subsequent to thrusting, trend W–E across Algeria and bend northward, passing through Tunisia, Sicily, and into the Apennines. They form a wide arc concave to the northwest. The left lateral strike-slip faults of the western Alps form a whirl (Fig. 17) around the western Po plain. They are conjugate with a bundle of dextral strike-slip faults, which develop much further to the south (Fig. 17). These movements accent the curvature of the orogenic arc and form part of the orogenic dynamics. They also intervene in the symmetric relations between loops, as is indicated in Fig. 16.

J. Dispersion of the Kabyle–Peloritan–Calabrian Massifs

The Kabyle and Peloritan–Calabrian Massifs, which had the same evolution during the Alpine events and which are allochthonous, are clearly

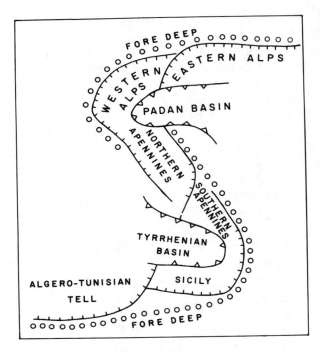

Fig. 16. Alpine symmetry in the central Mediterranean. In this schema we show a symmetry between the Tyrrhenian and Padan (Po) loops. Each of these loops encloses a young basin (as is the case of other loops such as Gibraltar and in the Carpathians). Each loop consists of two arcs fitted together and surrounded by a Miocene foredeep. The supposed axis of this symmetry passes between the northern and southern Apennines along the Anzio–Ancona line. The barbed lines schematically represent the major overthrusts or tectonic overlaps. They clearly do not all have the same importance. The symmetry could be partially due to the opposing movements of the Alpidic and Dinaric internal zones: displacement of the Calabrian–Peloritan Massif to the southeast (Fig. 14), and Alpine displacements to the northwest.

separated in present-day geographical terms. There are strongly divergent opinions as to whether this discontinuity was original or the result of the vicissitudes of the evolution of Alpine tectonics. Some geologists consider that the Kabyle–Peloritan–Calabrian Massifs formed a continuous zone at the beginning of the Mesozoic. Subsequently, certain segments were either swallowed up tectonically, remained behind, or were eroded. According to others, the massifs were originally separated and represent nuclei, microcratons, or microplates resulting from the Triassic separation of Europe and Africa. An intermediate hypothesis can also be envisioned in terms of which the massifs, originally continuous, were broken up and progressively separated during various phases of Cretaceous tectonic activity. The initial separation was due to movement along wrench faults (Fig. 2) similar to those of the Vicarian line. During the Cretaceous, the massifs were strung out like beads, and during the Cenozoic they were thrust over the most external zones. Such a process is probably also applicable in other tectonic arc situations and in various mountain chains. The hypothesis has one advantage in that it explains the communication between the internal and external flysch belts during the Cretaceous (Fig. 2) and the Oligo-Miocene.

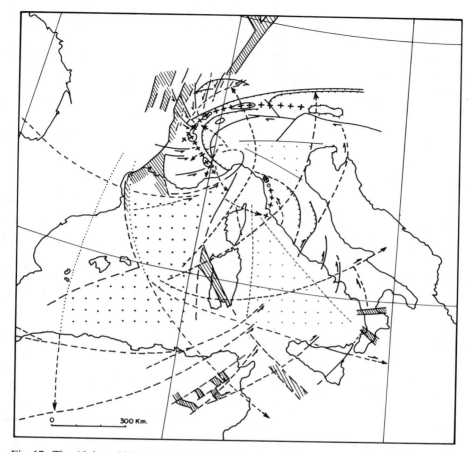

Fig. 17. The Alpine whirl of the central Mediterranean. This sketch tries to link the tectonic knot of the western Alps to the distensions, wrenches, folds, and overthrusts of the adjacent zones. The helicoidal sinistral flux of the Alpino-Apenninic regions seems to have drained material from the peripheral areas, and resulted in a torsion and migration of orogenic basins. The drainage would be affected by distensions, tearing, and slip-strike fault zones, which sometimes penetrate far into the platform. In continental areas it is seen as foundering of mainly Oligocene grabens such as those of Limagnes, Saône, Alsace; the Miocene graben of Campidano in Sardinia; recent foundering between Tunisia and Sicily (shaded on the map), and the late small basins in Italy and Tunisia. In deep marine areas it was the tearing and breaking up which opened the basins around the Corsican–Sardinian block, which no doubt had their paroxysm in the Upper Miocene and Pliocene. We can add to them the downwarping of the Padan Basin. The correlative accumulation of material in the folded zones corresponds to the compressive tectonic phenomena in the Oligocene to Mio-Pliocene Alpine chains: folding and overthrusting of Ponto-Pliocene age in Tunisia; Miocene and Pliocene over-thrusting in Sicily, and wrenching of the platform massif; southern Apennine Miocene and Pliocene overthrusting, migrating from the Oligocene to Pliocene tectogenesis in the northern Apennines, external Alps and Dinarides, and Pontian tectonics of the Jura and sub-Alpine chains. The wrench paths suggest one of the forms that could explain the general evolution. It is hard to believe that such correlations are due to chance and one can imagine that, among the many Alpine processes, some well-organized forces were at the heart of this spiral tectonics. This tendency toward a rolled-up helicoidal shape is not restricted to the Mediterranean, and a similar situation is found in the Carpathians, Turkey, Iran, and in Indonesia around the Celebes, though in a less advanced age.

Along the line of the Kabyle–Peloritan–Calabrian Massifs, the direction of strike-slip movement could be either right or left lateral. Left lateral movement, however, appears the more logical in the Cretaceous, but right lateral movement occurred during Nummulitic time (Fig. 18).

V. EVOLUTION OF THE AFRICAN PROMONTORY

A question arises as to whether the African (or Adriatic) promontory occupied the same location during the Mesozoic as at the present time. It has already been shown that it was formerly much more extensive than the presently visible segment. It is possible that it projected less toward the north (Fig. 2). If allowance is made for all the lateral (wrench) movements in the Alpine orogen, an attempt can be made to restore the promontory along the African margin. Various solutions are possible and not all can be considered here. The uncertainties can be illustrated by considering a limiting case, one which although tentative shows clearly the manner of investigation.

In this limiting case, one Alpine structural feature, the sinuosity of which, in plan, reflects the outline of the promontory, will be considered. This feature is the mid-Alpine lineament, a major zone of dislocation (Fig. 18a), which, bounded by the limit between the internal and external zones of the Betic Cordillera, continues to the Alps and Apennines in the vicinity of Genoa, then follows the peri-Adriatic lineament (Fig. 10), and continues via the Vardar and the North Anatolian fault into the Zagros, etc. Excluding the effects of faults which cut it locally, its global character is that of a right lateral fault. In the western Mediterranean, its trace must be sought in the Tyrrhenian Sea, unless it was covered by the Corso-Sardinian block during Neogene displacement of the latter.

The sinuous form of this feature (Fig. 18a) is scarcely consistent with significant strike-slip movements, although originally it may have been less sinuous. The present sinuosity would then be attributed to compression between Europe and Africa. It is here suggested that the original trend of the lineament, at an early Alpine stage (Fig. 18b), was distorted during the Nummulitic and the Miocene, adopting its present form as the result of two associated mechanisms:

(1) A dextral slip produced by a couple changing in trend from E–W (Fig. 18b) to WNW–ESE (Fig. 18c), which resulted in regional movements oriented NW–SE to NNW–SSE (directions with respect to the present north). During the course of this change, some wrench faults or fault sections may have become inactive, their motion being taken up by other faults or by reactivation of the older faults (e.g., North Anatolian strike-slip faults).

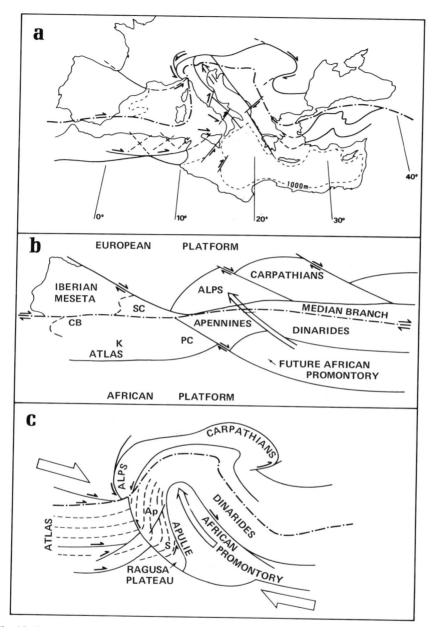

Fig. 18. Retrotectonic analysis of the Alps on the scale of the Mediterranean. On the three maps, the dashed line represents the mid-Alpine lineament. *a.* System of displacement, linked to Cenozoic tectonism on both sides of the African Promontory. The large arrows indicate the sense of final displacement of the promontory. *b.* Hypothetical conditions of the Alpine domain at the end of the Jurassic. The map does not attempt to reproduce the relative dimensions of the Alpine regions lying between the African and European plates. The small arrows and the dislocations suggest the slip trends which developed during the Cretaceous. SC = Sardinia–Corsica; CB = Betic Cordillera; K, P, C = Kabylias, Peloritan, Calabria. *c.* Intermediate state between *a* and *b* in the area bordering the African Promontory, with an indication of the movement trends. The sketch represents a stage of the Cretaceous or Nummulitic. The dotted lines suggest the disposition of the Atlas and Apennine paleogeographic zones. Ap = Apennines; S = Sicily.

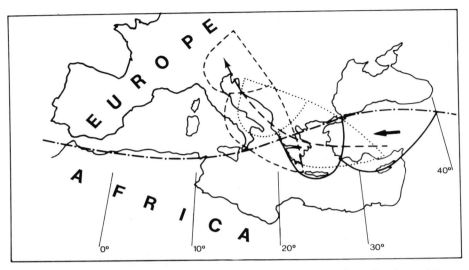

Fig. 19. Role of microplates applied to the hypothesis of Fig. 18b, c. Continuous lines—
present shorelines. Heavy dotted lines—trans-Mediterranean line = approximate present
limit between the African and European plates (McKenzie, 1970). Fine lines—present boundaries of the Aegean and Turkish plates. Fine dotted lines— intermediate positions of the
preceding. Dotted arrows—movements of the paleomicroplates. Heavy arrow: initial displacement of the paleomicroplates.

(2) "Roll-up" phenomena along both edges of the African promontory
(cf. Fig. 17), giving rise to displacements in the form of sheaves and whirls
(see above).

The principal torsion which results is symmetrically disposed in the Alps and
Carpathians (Fig. 18a). It is linked with WNW–ESE displacement affecting
the Iberian Meseta, the Calabro-Peloritan and the Corso-Sardinian blocks,
and the folding of the middle Atlas, Saharan Atlas, and Tunisian Atlas. Shaped
by the movement of these blocks, the more plastic zones of the Alpine geosyncline were crushed.

The question can also be raised as to whether the mid-Alpine lineament is
the reflection of a lithospheric fracture or of a plane of weakness deeper than
the beds affected by Alpine folding. If this is the case, the deep source of the
lineament may act along the length of the trans-Mediterranean line, that is,
along the present margin between the African and Eurasian plates. The mid-
Alpine lineament is then generally displaced with respect to the deeper part in
the whole of the sinuous course it describes from the Atlas to Turkey. Its
prolongation to the west is presumably the Azores fault. The sinuosity of the
trans-Mediterranean line from Greece to Turkey could then reflect the form of
the embryonic African promontory from which the Alpine displacement of the

promontory could be measured (Fig. 19). At the same time, there would be a quantitative answer to the initial question concerning the disposition of the Alpine zone in the Tyrrhenian loop.

REFERENCES

Abbate, E., and Sagri, M., 1967, Suddivisioni lithostratigrafiche nei calcari ad Elmintoidi Auctt. della placca dell'Ebro–Antola e correlazioni con terreni simili affioranti tra Voghera e Castelnovo Ne'Monti (Appennino Settentrionale), *Mem. Soc. Geol. Ital.*, v. 6, (1), p. 23–65.

Abbate, E., and Sagri, M., 1970, The eugeosynclinal sequences, *Sediment. Geol.*, v. 4 (3–4), p. 251–340.

Abbate, E., Bortolotti, V., Passerini, P., and Sagri, M. 1970a, Introduction to the geology of the Northern Apennines, *Sediment. Geol.* v. 4 (3–4), p. 207–250.

Abbate, E., Bortolotti, V., Passerini, P., and Sagri, M., 1970b, The geosyncline concept and the Northern Apennines, *Sediment. Geol.*, v. 4 (3–4), p. 625–736.

Abbate, E., Bortolotti, V., Passerini, P., and Sagri, M., 1970c, The Northern Apennines geosyncline and continental drift, *Sediment. Geol.*, v. 4 (3–4), p. 637–642.

Abbate, E., Bortolotti, V., and Passerini, P., 1970d, Olistostromes and olistoliths, *Sediment. Geol.*, v. 4 (3–4), p. 521–558.

Abbate E., Bortolotti V., and Passerini P., 1972, Studies on mafic and ultramafic rocks. 2—Paleogeographic and tectonic considerations on the ultramafic belts in the Mediterranean area, *Boll. Soc. Geol. Ital.*, v. 91 (2), p. 239–282.

Accordi, B., 1963, Lineamenti strutturali del Lazio e dell'Abruzzo meridionali, *Mem. Soc. Geol. Ital.*, v. 4 (1), p. 595–633.

Accordi, B., 1966, La componente traslativa nella tettonica dell'Appennino laziale–abruzzese, *Geol. Romana*, v. 5, p. 355–406.

Afchain C., 1962, Observations sur la région de Longbucco (Italie méridionale), *Bull. Soc. Géol. France*, (7), v. 8, p. 719–720.

AGIP Direzione Mineraria, 1969, Dati e notizie sulle tecniche e sui risultati della ricerca di idrocarburi in mare, *Rend. Soc. Geol. Ital.*, S.D. Milanese (September 17–19, 1969), 49 p.

Allan, T. D., and Morelli, C., 1971, A geophysical study of the Mediterranean Sea, *Boll. Geofis. Teor. Appl.*, v. 13 (50), p. 99–141.

Alvarez, W., 1972, Rotation of the Corsica–Sardinia microplate, *Nature Phys. Sci.*, v. 235 (58), p. 103–105.

Alvarez W., and Lowrie, W., 1974, Paleomagnetismo della Scaglia rossa Umbra e rotazione della penisola italiana, *Boll. Soc. Geol. Ital.*, v. 93 (4), p. 883–892.

Andreieff P., Broquet P., Duée G., and Mascle, G., 1974, Les rapports du flysch du Monte Soro et des séries panormides en Sicile, *Bull. Soc. Géol. France*, (7), v. 16 (3), p. 295–302.

Andrieux J., 1973, Sur le métamorphisme des zones externes du Rif, *Bull. Soc. Géol. France*, (7), v. 15 (2), p. 105–108.

Andrieux, J., Bles, J. L., and Lepvrier, C., 1973, Tectoniques superposées et métamorphisme épizonal dans la zone externe de l'orogène alpin d'Afrique du Nord, *Réun. Ann. Sci. Terre, Paris*, p. 44.

Andrieux J., and Mattauer, M., 1973, Précisions sur un modèle explicatif de l'arc de Gibraltar. *Bull. Soc. Géol. France*, (7), v. 15 (2), p. 115–118.

Andrusov, D., 1963, Les principaux plissements alpins dans le domaine des Carpathes occidentales, *Livre Mém. P. Fallot*, v. 2, p. 519–528.

Argand, E., 1922, La tectonique de l'Asie, *Congr. Géol. Int.*, 13th Sess., Belgium, Fasc. 1, p. 171–372.

Argyriadis, I., 1970, La position des Alpes carniques dans l'orogène alpin et le problème de la limite alpino-dinarique, *Bull. Soc. Géol. France*, (7), v. 12, p. 473–480.

Argyriadis, I., 1973, La Mésogée et le grand bouleversement tectonique du Crétacé, *Réun. Ann. Sci. Terre*, Paris, p. 48.

Argyriadis, I., 1974, Sur l'orogenèse mésogéenne des temps crétacés, *Rev. Géogr. Phys. Géol. Dynam.*, v. 16 (1), p. 23–60.

Argyriadis, I., 1975, Mésogée permienne, chaîne hercynienne et cassure téthysienne, *Bull. Soc. Géol. France*, (7), v. 17 (1), p. 56–70.

Aubouin, J., 1959, A propos d'un centenaire: les aventures de la notion de géosynclinal, *Rev. Géogr. Phys. Géol. Dynam.*, v. 2 (3), p. 135–188.

Aubouin, J., 1960, Essai sur l'ensemble italo-dinarique et ses rapports avec l'arc alpin, *Bull. Soc. Géol. France*, (7), v. 9, p. 487–526.

Aubouin, J., 1963, Esquisse paléogéographique et structurale des chaînes alpines de la Méditerranée moyenne, *Geol. Rundschau*, v. 53, p. 480–534.

Aubouin, J., 1965, *Geosynclines*, New York Elsevier Publ. Co.

Aubouin, J., *et al.*, 1970, Essai sur la géologie des Dinarides, *Bull. Soc. Géol. France*, (7), v. 12, p. 1060–1095.

Auzende, J. M., Olivet, J. L., and Bonnin, J., 1973, Le détroit sardano-tunisien et la zone de fracture nord-tunisienne, Contrib. no. 141, Dept. Sci. Centre Océanol. Bretagne.

Auzende, J. M., and Pautot, G., 1970, La marge continentale algérienne et le phénomène de subsidence: exemple du golfe de Bougie, *C.R. Acad. Sci. Paris*, v. 271, p. 1945–1948.

Auzende, J. M., Bonnin, J., Olivet, J. L., Pautot, G., and Mauffret, A., 1971, Upper Miocene salt layer in the Western Mediterranean Basin, *Nature Phys. Sci.*, v. 230 (12), p. 82–84.

Auzende, J. M., Bonnin, J., and Olivet, J. L., 1973a, The origin of the Western Mediterranean Basin, *J. Geol. Soc. London*, v. 129, p. 607–620.

Auzende, J. M., Bonnin, J., and Olivet, J. L., 1973b, Hypotheses on the origin of the Western Mediterranean Basin, Dept. Sci., Centre Océanol. Bretagne, 110 p.

Auzende, J. M., Bonnin, J., and Olivet, J. L., 1975a, La marge nord-africaine considérée comme une marge active, *Bull. Soc. Géol. France*, (7), v. 17 (4), p. 486–495.

Auzende, J. M., Rehault, J. P., Pastouret L., Szep, B., and Olivet J. L., 1975b, Les bassins sédimentaires de la mer d'Alboran, *Bull. Soc. Géol. France*, (7), v. 17 (1), p. 98–107.

Azéma, J., Bourrouilh, R., Champetier, Y., Fourcade, E., and Rangheard Y., 1974, Rapports stratigraphiques, paléogéographiques et structuraux entre la Chaîne ibérique, les Cordillères bétiques et les Baléares, *Bull. Soc. Géol. France*, (7), v. 16 (2), p. 140–160.

Badoux, H., 1969, Réflexions et hypothèses à propos de la limite alpino-dinarique, *Eclogae Geol. Helv.*, v. 62 (2), p. 543–545.

Baldacci, F., Elter, P., Giannini, E., Giglia, G., Lazzarotto, A., Nardi, R., and Tongiorgi, M., 1967, Nuove osservazioni sul problema della Falda Toscana e sulla interpretazione dei flysch arenacei tipo "Macigno" dell'Appennino settentrionale, *Mem. Soc. Geol. Ital.*, v. 6 (2), p. 213–244.

Baldacci, F., Cerrina Feroni, A., Elter, P., Giglia, G., and Patacca, E., 1972, Il margine del paleocontinente nord-appenninico dal Cretaceo all'Oligocene: nuovi dati sulla ruga insubrica, *Mem. Soc. Geol. Ital.*, v. 11 (4), p. 367–390.

Baltenberger P., and Recq, M., 1965, Etudes sismiques du socle dans le val de Loire, *C.R. Acad. Sci. Paris*, v. 261, p. 1053–1056.

Bartolini, C., De Giuli, C., and Gianelli, G., 1974, Turbidites of the Southern Tyrrhenian Sea, *Boll. Soc. Geol. Ital.*, v. 93 (1), p. 3–22.

Bayer, R., Le Mouel, J. L., and Le Pichon, X., 1973, Magnetic anomaly pattern in the western Mediterranean, *Earth Planet. Sci. Lett.*, v. 19 (2), p. 168–176.

Beccaluva L., Macciotta, G., and Venturelli, G., 1974, Nuovi dati e considerazioni petrogenetiche sulle serie vulcaniche plio-quaternarie del Montiferro (Sardegna centro-occidentale), *Mem. Soc. Geol. Ital.*, v. 13 (4), p. 539–548.

Bellaiche, G., and Recq, M., 1973, Off-shore seismological experiments south of Provence in relation to "Glomar Challenger" deep sea drillings (JOIDES-DSDP, Leg 13), *Marine Geol.*, 15, p. 49–52.

Bellaiche G., Recq, M., and Rehault, J. P., 1973, Nouvelles données sur la structure du hautfond du Méjean obtenues par la "sismique-réfraction," *C.R. Acad. Sci. Paris*, v. 276, p. 1529–1532.

Bemmelen, R. W. van, 1953, Gravity field and orogenesis in the West-Mediterranean region, *Geol. Mijnb.*, v. 14, p. 306–313.

Bemmelen, R. W. van, 1955, Tectogenèse par gravité, *Bull. Soc. Belge Géol.* v., 64, p. 95–123.

Bemmelen, R. W., van, 1965, Mega-undations as cause of continental drift, *Geol. Mijnb.*, v. 44, p. 320–333.

Bemmelen, R. W. van, 1969a, Origin of the Western Mediterranean Sea, *Geol. Mijnb.*, v. 26, p. 13–52.

Bemmelen, R. W. van, 1969b, The Alpine loop of the Tethys zone, *Tectonophysics*, v. 8 (2), p. 107–113.

Bemmelen, R. W. van, 1972, Driving forces of Mediterranean orogeny, *Geol. Mijnb.*, v. 51, (5), p. 548–573.

Beneo, E., 1943, Schema tettonico dell'Abruzzo nord-orientale, *Boll. Uff. Geol. Ital.*, v. 68 (1), VI, 1945, 7 p.

Beneo, E., 1949, Tentativo di sintesi tettonica dell'Italia peninsulare ed insulare, *Boll. Soc. Geol. Ital.*, v. 68, p. 66–80.

Beneo, E., 1956a, Accumuli terziari da risedimentazione (Olistostroma) nell'Appennino centrale e frane sottomarine. Estensione tempospaziale del fenomeno, *Boll. Serv. Geol. Ital.*, v. 78, p. 291–321.

Beneo, E., 1956b, Il problema delle "Argille scagliose," "Flysch" in Italia e sua probabile risoluzione. Nuova nomenclatura, *Boll. Soc. Geol. Ital.*, v. 75 (3), p. 53–68.

Bernoulli, D., 1972, North Atlantic and Mediterranean mesozoic facies: a comparison, in: *Initial Rept. Deep Sea Drilling Project*, v. 11, p. 801–871.

Bernoulli, D., and Jenkyns, H. C., 1974, Alpine, Mediterranean, and Central Atlantic mesozoic facies in relation to the early evolution of the Tethys, in: *Modern and Ancient Geosynclinal Sedimentation*, Dott., R. H., Jr., and Shaver, R. H., eds., Soc. Econ. Paleont. Mineral. Spec. Publ., no. 19, p. 129–160.

Biju-Duval B., 1974, Commentaires de la carte géologique et structurale des bassins tertiaires du domaine méditerranéen, *Rev. Inst. Français Pétr.*, v. 29 (5), p. 607–639.

Biju-Duval, B., Letouzey, J., Montadert, L., Courrier, P., Mugniot, J. F., and Sancho, J., 1974, Geology of the Mediterranean Sea basins, in: *The Geology of the Continental Margins*, Burk, C. A., and Drake, C. L., eds., New York: Springer-Verlag, p. 695–721.

Biscaye, P. E., Ryan, W. B., and Wezel, F. C., 1972, Age and nature of the Pan-Mediterranean subbottom reflector M, in: *The Mediterranean Sea—A Natural Sedimentation Laboratory*, Stanley, D. J., ed., Stroudsburg, Pa.: Dowden, Hutchinson and Ross, p. 83–90.

Bishop, W. F., 1975, Geology of Tunisia and Adjacent Parts of Algeria and Libya, *Bull. Amer. Assoc. Petrol. Geol.*, v. 59 (3), p. 413–450.

Boccaletti, M., and Guazzone, G., 1968, A contribution to the regional study of the structural–stratigraphic units by systematic analysis of structural trends. I—Serie toscana, *Mem. Soc. Geol. Ital.*, v. 7 (2), p. 247–259.

Boccaletti, M., and Guazzone, G., 1970, La migrazione terziara dei bacini toscani e la rotazione dell'Appennino settentrionale in una "zona di torsione" per deriva continentale, *Mem. Soc. Geol. Ital.*, v. 9 (2), p. 177–195.

Boccaletti, M., and Guazzone, G., 1972, Gli archi appenninici, il mar Ligure ed il Tirreno nel quadro della tettonica dei bacini marginali retro-arco, *Mem. Soc. Geol. Ital.*, v. 11 (2), p. 201–216.

Boccaletti, M., Elter, P., and Guazzone, G., 1971*a*, Plate tectonic models for the development of the western Alps and northern Apennines, *Nature Phys. Sci.*, v. 234 (49), p. 108–111.

Boccaletti, M., Elter, P., and Guazzone, G., 1971*b*, Polarità strutturali delle Alpi e dell'Appennino settentrionale in rapporto all'inversione di una zona di subduzione nord-tirrenica, *Mem. Soc. Geol. Ital.*, v. 10 (4), p. 371–378.

Bogdanoff, A., 1962, Sur certains problèmes de structure et d'histoire de la plate-forme de l'Europe orientale, *Bull. Soc. Géol. France*, (7), v. 4, p. 898–911.

Bonneau, M., 1969, Contribution à l'étude géologique de la Judicarie, au Nord-Ouest du lac de Garde (Alpes méridionales, province de Trente, Italie), *Bull. Soc. Géol. France*, (7), v. 11, p. 816–829.

Bonnin, J., Auzende, J. M., and Olivet, J. L., 1973, L'extrémité orientale de la zone Açores–Gibraltar. Structure et Evolution, *Réun. Ann. Sci. Terre, Paris*, p. 91.

Bonnin, J., Olivet, J. L., and Auzende, J. M., 1975, Structure en nappe à l'Ouest de Gibraltar, *C.R. Acad. Sci. Paris*, v. 280, p. 559–562.

Booy, T. de, 1967, Neue Daten für die Annahme einer sialischen Kruste unter den frühgeosynklinalen Sedimenten der Tethys, *Geol. Rundschau*, v. 56, p. 94–102.

Booy, T. de, 1969, Repeated disappearance of continental crust during the geological development of the Western Mediterranean area, *Verh. K. Ned. Geol. Mijnb. Genoot*, v. 26, p. 79–103.

Borsi, S., and Dubois, R., 1968, Données géochronologiques sur l'histoire hercynienne et alpine de la Calabre centrale, *C.R. Acad. Sci. Paris*, v. 266, p. 72–75.

Bortolotti, V., 1966, La tettonica trasversale dell'Appennino. I—La linea Livorno–Sillaro, *Boll. Soc. Geol. Ital.*, v. 85 (2), p. 529–540.

Bortolotti, V., Passerini, P., Sagri, M., and Sestini, G., 1970*a*, The miogeosynclinal sequences, *Sediment. Geol.*, v. 4 (3–4), p. 341–444.

Bortolotti, V., Sagri, M., Abbate, E., and Passerini, P., 1970*b*, Geological map of the Northern Apennines and adjoining areas (1/500,000), Cons. naz. Ricerche, Centr. studi geol. Appennino, sej. Firenze, *Sediment. Geol.*, v. 4 (3–4).

Bottari, A., and Girlanda, A., 1974, Some results for the Middle Mediterranean basin from the study of *Pn* waves, *Bull. Seism. Soc. Amer.*, v. 64 (2), p. 427–435.

Bottari, A., and Lo Giudice, E., 1974, On the P-Wave velocity and plate-tectonics implications for the Tyrrhenian deep-earthquake zone, *Tectonophysics*, v. 25, p. 187–200.

Bouillin, J. P., 1975, Un modèle de l'évolution tectonique alpine anté-burdigalienne du Nord du Constantinois sur la transversale du Moul de Demamène (Algérie), *Bull. Soc. Géol. France*, (7), v. 17 (4), p. 582–593.

Bouillin, J. P., and Kornprobst, J., 1974, Associations ultrabasiques de Petite Kabylie: péridotites de type alpin et complexe stratifié; comparaison avec les zones internes bético-rifaines, *Bull. Soc. Géol. France*, (7), v. 16 (2), p. 183–194.

Bouillin, J. P., Durand-Delga, M., Gélard, J. P., Leikine, M., Raoult, J. F., Raymond, D., Tefiani, M., and Vila, J. M., 1973, Les olistostromes d'âge miocène inférieur liés aux flyschs allochtones kabyles de l'orogène alpin d'Algérie, *Bull. Soc. Géol. France*, (7), v. 15 (3–4), p. 340–344.

Bousquet, J. C., 1961, Position des diabases–porphyrites dans la région de Cetraro–Intavolata et de Sangineto (Calabre, Italie méridionale), *Bull. Soc. Géol. France*, (7), v. 3, p. 603–609.

Bousquet, J. C., 1962, Age de la série des diabase–porphyrites (roches vertes du flysch calabro-lucanien; Italie méridionale), *Bull. Soc. Géol. France*, (7), v. 4, p. 712–718.

Bousquet, J. C., 1966, Sur l'allure et la mise en place des formations allochtones de la bordure orientale des massifs calabro-lucaniens, *Bull. Soc. Géol. France*, (7), v. 7, p. 937–945.

Bousquet, J. C., 1971, La tectonique tangentielle des séries calcaréo-dolomitiques du Nord-Est de l'Apennin calabro-lucanien, *Geol. Romana*, v. 10, p. 23–52.

Bousquet, J. C., 1972, *La tectonique Récente de l'Apennin Calabro-Lucanien dans son Cadre Géologique et Géophysique*, Thèse Sci. Montpellier.

Bousquet, J. C., and Dubois, R., 1967, Découverte de niveaux anisiens et caractères du métamorphisme alpin dans la région de Lungro (Calabre), *C.R. Acad. Sci. Paris*, sér. D, v. 264, p. 204–207.

Bousquet, J. C., and Grandjacquet, C., 1969, Structure de l'Apennin calabro-lucanien (Italie méridionale), *C.R. Acad. Sci. Paris*, v. 268, p. 13–16.

Bousquet, J. C., and Gueremy, P., 1968, Quelques phénomènes de néotectonique dans l'Apennin calabro-lucanien et leurs conséquences morphologiques. I—Bassin du Mercure et haute vallée du Sinni, *Rev. Géogr. Phys. Géol. Dynam.*, v. 10 (3), p. 225–238.

Broquet, P., 1964a, Observations stratigraphiques, tectoniques et sédimentaires sur le flysch numidien des Madonies (Sicile), *Ann. Soc. Géol. Nord*, v. 84, p. 141–152.

Broquet, P., 1964b, Remarques sur la géologie de la bordure orientale des monts Sicani (Sicile), *Ann. Soc. Géol. Nord*, v. 84, p. 303–308.

Broquet, P., 1967, L'âge de la mise en place de la nappe numidienne dans la région de Valledolmo (Sicile centro-septentrionale), *C.R. Somm., Soc. Géol. France*, v. 6, p. 239–240.

Broquet, P., 1968, *Etude Géologique de la Région des Madonies (Sicile)*, Thèse Fac. Sci., Lille, 796 p.

Broquet, P., 1969, La notion d'olistostrome et d'olistolite. Historique et étude critique, *Ann. Soc. Géol. Nord*, v. 90, p. 77–86.

Broquet, P., 1970a, The geology of the Madonie mountains of Sicily, in: *Geology and History of Sicily*, Tripoli: Petrol. Expl. Soc. Libya, p. 201–230.

Broquet, P., 1970b, Observations on gravitational sliding: the concept of olistostrome and olistolite, in: *Geology and History of Sicily*, Tripoli: Petrol. Expl. Soc. Libya, p. 255–259.

Broquet, P., Caire, A., and Mascle, G., 1966, Structure et évolution de la Sicile occidentale (Madonies et Sicani), *Bull. Soc. Géol. France*, (7), v. 8, p. 994–1011.

Brunn, J. H., 1960, Les zones helléniques internes et leur extension. Réflexions sur l'orogenèse alpine, *Bull. Soc. Géol. France*, (7), v. 2, p. 470–486.

Brunn, J. H., 1961, Les sutures ophiolitiques, *Rev. Geogr. Phys. Géol. Dynam.*, v. 4, p. 89–96, 181–202.

Brunn, J. H., 1964, Sur la disposition originelle du système alpin en trois rameaux, *C.R. Acad. Sci. Paris*, v. 259, p. 4739–4741.

Brunn, J. H., 1965, A la recherche du rameau médian des chaînes périméditerranéennes, *Ann. Soc. Géol. Nord*, v. 84, p. 227–230.

Brunn, J. H., 1967, Recherche des éléments majeurs du système alpin, *Rev. Géogr. Phys. Géol. Dynam.*, v. 9 (1), p. 17–34.

Burollet, P. F., 1962, Signification géologique du détroit de Sicile, *87e Congr. Soc. Savantes*, p. 849–853.

Burollet, P. F., 1966, Remarques sur la stratigraphie du Jebel Nefusa (Tripolitaine–Libye), *Riv. Ital. Paleont.*, v. 72 (4), p. 1313–1317.

Burollet, P. F., 1967, General geology of Tunisia and tertiary geology of Tunisia, in: *Guidebook to the Geology and History of Tunisia*, Amsterdam: Holland Breum, p. 51–58, 215–225.

Burollet, P. F., and Dufaure, P., 1972, The neogene series drilled by the Mistral, no. I well in the Gulf of Lion, in: *The Mediterranean Sea—A Natural Sedimentation Laboratory*, Stanley, D. J., ed., Stroudsburg, Pa.: Dowden, Hutchinson and Ross, p. 91–98.

Burollet, P. F., Dumestre, A., Keppel, D., and Salvador, A., 1952, Unités stratigraphiques en Tunisie centrale, *XIXᵉ Congr. Géol. Int.*, Alger, A.S.G.A., fasc. 21, p. 243–254.

Burollet, P. F., and Rouvier, H., 1971, La Tunisie, *Unesco 1971, Tectonique de l'Afrique, Sci. Terre*, v. 6, p. 91–99.

Busson, G., 1975, Le trias évaporitique d'Afrique du Nord et d'Europe occidentale: données sur la paléogéographie et les conditions de dépôt, *Bull. Soc. Géol. France*, (7), v. 16 (6), p. 653–665.

Cadet, J. P., 1970, Sur la géologie des confins méridionaux de la Bosnie et de la Serbie, *Bull. Soc. Géol. France*, (7), v. 12, p. 967–972.

Cadet, J. P., and Charvet, J., 1973, Sur la structure de la zone bosniaque en Bosnie herzégovine méridionale, *Réun. Ann. Sci. Terre, Paris*, p. 113.

Caflisch, L., 1966, La geologia dei Monti di Palermo, *Mem. Riv. Ital. Paleont. Strat.*, v. 12, p. 1–108.

Caflisch, L., and Schmidt di Friedberg, P., 1967, L'evoluzione paleogeografica della Sicilia e sue relazioni con la tettonica e la naftogenesi, *Mem. Soc. Geol. Ital.*, v. 6, p. 449–474.

Caire, A., 1957, *Les Glissements sous-marins dans le Sillon Miocène sud-tellien et la Mise en Place de l'Unité A*, Publ. Serv. Carte Géol. Algérie, n.s., Bull. 20, (Trav. collab.), p. 21–34.

Caire, A., 1960, Problèmes de tectonique et de morphologie jurassiennes, *Livre Mém. P. Fallot*, v. 2, p. 105–158.

Caire, A., 1964, Comparaison entre les orogènes berbère et apenninique, *Ann. Soc. Géol. Nord*, v. 84, p. 163–176.

Caire, A., 1970a, Sicily in its Mediterranean setting, in: *Geology and History of Sicily*, Tripoli: Petrol. Expl. Soc. Libya, p. 145–170.

Caire, A., 1970b, Tectonique de la Méditerranée centrale, *Ann. Soc. Géol. Nord*, v. 90 (4), p. 307–346.

Caire, A., 1971, Chaînes alpines de la Méditerranée centrale (Algérie et Tunisie septentrionales, Sicile, Calabre et Apennin méridional). Notice dét. Carte tect. Afrique, *Unesco, Sci. Terre*, v. 6, p. 61–90.

Caire, A., 1972, Discontinuités structurales et paléogéographiques dans l'arc tyrrhénien, in: *Livre Jubilaire Prof. Andrusov, Tectonic Problems of the Alpine System*, Veda, Slovak Acad. Sci., p. 175–187.

Caire, A., 1973, Italy in its Mediterranean setting, in: *Geology of Italy*, Tripoli: Petrol. Expl. Soc. Libya, p. 11–74.

Caire, A., 1974, Eastern Atlas, in: *Mesozoic–Cenozoic Belts*, Spencer, A. M., ed., p. 47–59.

Caire, A., and Mattauer, M., 1960, Comparaison entre la Berbérie et le territoire siculo-calabrais, *C.R. Acad. Sci. Paris*, 251, p. 1084–1086.

Caire, A., Glangeaud, L., and Grandjacquet, C., 1960, Les grands traits structuraux et l'évolution du territoire calabro-sicilien (Italien méridionale), *Bull. Soc. Géol. France*, (7), v. 2, p. 915–938.

Caire, A., Glangeaud, L., and Grandjacquet, C., 1961, L'orogenèse ponto-plio-quaternaire de l'arc calabro-sicilien et ses caractères géodynamiques, *C.R. Acad. Sci. Paris*, v. 252, p. 915–938.

Caire, A., and Truillet, R., 1962, Remarques sur la tectonique des phyllades et de la chaîne calcaire des monts péloritains (Sicile), *C.R. Somm., Soc. Géol. France*, v. 7, p. 188–190.

Caire, A., and Truillet, R., 1963, A propos de la phase tectonique antérieure au dépôt de l'Oligo-Miocène dans les Monts Péloritains (Sicile), *C.R. Acad. Sci. Paris*, v. 256, p. 2446–2447.

Caire, A., and Truillet, R., 1967, Les relations entre le domaine péloritain et le flysch du Monte Soro aux environs de Rocella Valdèmone, et le problème du charriage des Péloritains orientaux (Sicile), *Bull. Soc. Géol. France*, (7), v. 9, p. 255–260.

Carey, S. W., 1955, The orocline concept in geotectonics, *Ann. Soc. Géol. Belg.*, v. 89, p. 255–288.

Carmignani, L., Dessau, G., and Duchi, G., 1972, I giacimenti minerari delle Alpi Apuane e loro correlazioni con l'evoluzione del gruppo montuoso, *Mem. Soc. Geol. Ital.*, v. 11, (4), p. 417–432.

Carraro, F., Dal Piaz, G. V., Franceschetti, B., Malaroda, R., Sturani, C., and Zanella, E., 1967, Carta geologica del Massiccio dell'Argentera. Note illustrative, p. 1–105.

Cassinis, R., Lechi, G. M., and Tonelli, A. M., 1974, Contribution of space platforms to a ground and airborne remote-sensing programme over active Italian volcanoes, *C.N.R. Lab. Geofis. Litosfera*, Publ. no. 95, p. 186–197.

Castany, G., 1948, Les fosses quaternaires d'effondrement de Tunisie, *Ann. Mines Géol.*, Tunis, v. 3.

Castany, G., 1951a, Etude géologique de l'Atlas tunisien oriental. Thèse Sc. Paris, *Ann. Mines Géol.*, Tunis, v. 8, p. 1–632.

Castany, G., 1951b, L'orogenèse de l'Atlas tunisien, *Bull. Soc. Géol. France*, (6), v. 1, p. 701–720.

Castany, G., 1955, Le Haut-Bassin siculo-tunisien, *Bull. Station Océanogr. Salambô* (Tunisie), v. 52, p. 1–17.

Castany, G., 1956, Essai de synthèse géologique du territoire Tunisie–Sicile, *Ann. Mines Géol.*, Tunis, v. 16, p. 1–101.

Cestari, G., 1967, Lineamenti strutturali del Cilento (Campania meridionale), *Boll. Soc. Geol. Ital.*, v. 86 (1), p. 9–20.

Chabrier, G., 1969, Sur la stratigraphie et la structure des Monts d'Oliena (Sardaigne orientale), *C.R. Somm., Soc. Géol. France*, v. 6, p. 218–219.

Chabrier G., and Mascle, G., 1975, Comparaison des évolutions géologiques de la Provence et de la Sardaigne, *Rev. Geogr. Phys. Géol. Dynam.*, (2), v. 17 (2), p. 121–136.

Charnock, H., Rees, A. I., and Hamilton, N., 1972, Sedimentation in the Tyrrhenian Sea, in: *The Mediterranean Sea—A Natural Sedimentation Laboratory*, Stanley, D. J., ed., Stroudsburg, Pa.: Dowden, Hutchinson and Ross, p. 615–629.

Cocco, E., 1973, Correlazione tra alcune successioni sedimentarie del Cretacico Sup.–Paleocene–Eocene Inf. delle zone interne della "Geosinclinale" sudappenninica, *Boll. Soc. Geol. Ital.*, v. 92 (4), p. 841–860.

Cocco, E., Cravero, E., Ortolani F., Pescatore, T., Russo, M., Torre, M., and Coppola, L., 1974, Le Unità Irpine nell'area a nord di Monte Marzano, Appennino Meridionale, *Mem. Soc. Geol. Ital.*, v. 13 (4), p. 607–654.

Cocozza, T., Jacobacci, A., Nardi, R., and Salvadori, I., 1974, Schema stratigrafico–strutturale del massiccio sardo-corso e minerogenesi della Sardegna, *Mem. Soc. Geol. Ital.*, v. 13, (2), p. 85–186.

Colacicchi, R. 1964, La facies di transizione della Marsica nord-orientale. I, *Geol Romana*, v. 3, p. 93–124.

Colacicchi, R., 1966, Le caratteristiche della facies abruzzese alla luce delle moderne indagini geologiche, *Mem. Soc. Geol. Ital.*, v. 5 (1), p. 1–18.

Colacicchi, R., 1967, Geologia della Marsica orientale, *Geol. Romana*, v. 6, p. 189–316.

Colacicchi, R., and Praturlon, A., 1965, Stratigraphical and paleogeographical investigation on the mesozoic shelf-edge facies in Eastern Marsica (Central Apennines, Italy), *Geol. Romana*, v. 4, p. 89–118.

Colonna V., Lorenzoni, S., and Zanettin Lorenzoni, E., 1973, Sull'esistenza di due complessi metamorfici lungo il bordo sud-orientale del massiccio "granitico" delle Serre (Calabria), *Boll. Soc. Geol. Ital.*, v. 92 (4), p. 801–830.

Conedera, C., Dieni, I., Piccoli, G., and Saccardi, P., 1969, Studio fotogeologico dei Colli Euganei, *Cons. naz. Ric., Centr. Stud. Geol. Petrogr.*, IA Sez. Geol., p. 1–20.

Coque, R., and Jauzein, A., 1967, The geomorphology and quaternary geology of Tunisia, in: *Guide Book to the Geology and History of Tunisia*, Tripoli: Petrol. Expl. Soc. Libya, p. 227–257.

Cotecchia, V., 1959, Le argille scagliose ofiolitifere della valle del Frido a Nord del M. Pollino, *Boll. Soc. Geol. Ital.*, v. 77 (3), p. 205–245.

Cousin, M., 1970, Esquisse géologique des confins italo-yougoslâves: leur place dans les Dinarides et les Alpes méridionales, *Bull. Soc. Géol. France*, (7), v. 12, p. 1034–1047.

Cousin, M., 1973, Sur les relations entre les Alpes méridionales et les Dinarides, *Réun. Ann. Sci. Terre, Paris*, p. 145.

Crostella, A., and Vezzani, L., 1964, La geologia dell'Appennino foggiano, *Boll. Soc. Geol. Ital.*, v. 83 (1), p. 121–142.

Dallan Nardi, L., Elter, P., and Nardi, R., 1971, Considerazioni sull'arco dell'Appennino settentrionale e sulla "linea" Ancona–Anzio, *Boll. Soc. Geol. Ital.*, v. 90 (2), p. 203–211.

D'Amico C., Messina A., Puglisi G., Rottura A., and Russo S., 1973, Confronti petrografici nel cristallino delle due sponde dello Stretto di Messina. Implicazioni geodinamiche, *Boll. Soc. Geol. Ital.*, v. 92 (4), p. 939–953.

D'Argenio, B., 1966, Zone isopiche e faglie trascorrenti nell'Appennino centro-meridionale, *Mem. Soc. Geol. Ital.*, v. 5 (4), p. 279–299.

Debelmas, J., 1955, Les zones subbriançonnaise et briançonnaise occidentale entre Vallouise et Guillestre (H.A.), *Mém. Carte Géol. France*, Paris, p. 164.

Debelmas, J., 1957, Quelques remarques sur les conceptions actuelles du terme "cordillère" dans les Alpes françaises, *Bull. Soc. Géol. France*, (6), v. 7, p. 463–474.

Debelmas, J., and Lemoine, M., 1964, La structure tectonique et l'évolution paléogéographique de la chaîne alpine d'après les travaux récents, *L'Information Scientifique*, I, p. 1–33.

Debelmas, J., and Lemoine, M., 1970, The western Alps, paleogeography and structure, *Earth Sci. Rev.*, v. 6, p. 222–256.

Demangeot, J., 1950, Rôle des déformations récentes dans la cordillère abruzzaise, *C.R. 16ᵉ Congr. Int. Géogr.*, Lisbonne, p. 223–227.

Demangeot, J., 1951, L'arc abruzzais externe, *Ric. Scient.*, v. 21 (6), p. 904–939.

Demangeot, J., 1952, Sur la continuité de la surface villafranchienne du Tibre à l'Adriatique, *Rend. Accad. Naz. Lincei*, v. 8, XII, 2, p. 175–178.

Demangeot, J., 1965, Géomorphologie des Abbruzes Adriatiques, *Mém. Doc. C.N.R.S.*, in 4 vols., (hors sér.) p. 404.

Dercourt, J., 1970, L'expansion océanique actuelle et fossile: ses implications géotectoniques, *Bull. Soc. Géol. France*, (7), v. 12, p. 261–317.

Desio A., Ronchetti, C. R., and Pozzi, R., 1966, Osservazioni alla nota di P. F. Burollet, "Remarques sur la stratigraphie du Jebel Nefusa," *Riv. Ital. Paleont.*, v. 72, (4), p. 1319–1322.

De Sitter, L. U., 1960, La structure des Alpes lombardes, *Livre Mém. P. Fallot*, v. 2, p. 245–256.

Devaux, J., 1969, Recherches de l'organisation des contraintes dans le tréfonds de l'Algérie du Nord. Le rôle des failles de décrochement obliques sur l'Ouest, *Bull. Serv. Géol. Algérie*, v. 39, p. 41–69.

Dewey, J. F., Pitman III, W. C., Ryan, W. B. F., and Bonnin, J., 1973, Plate tectonics and the evolution of the Alpine system, *Bull. Geol. Soc. Amer.*, v. 84 (10), p. 3137–3180.

Didon, J., 1973, Accidents transverses et coulissages longitudinaux dextres dans la partie nord de l'arc de Gibraltar (Cordillères bétiques occidentales Espagne), *Bull. Soc. Géol. France*, (7), v. 15 (2), p. 121–128.

Didon, J., Durand-Delga, M., and Kornprobst, J., 1973, Homologies géologiques entre les deux rives du détroit de Gibraltar, *Bull. Soc. Géol. France*, (7), v. 15 (2), p. 77–105.

Dieni, I., and Massari, F., 1971, Scivolamenti gravitativi ed accumuli di frana nel quadro della morfogenesi plio-quaternaria della Sardegna centro orientale, *Mem. Soc. Geol. Ital.*, v. 10, p. 313–345.

Dubois, R., 1967, La suture calabro-apenninique, *C.R. Somm., Soc. Géol. France*, v. 6, p. 236–237.

Dubois, R., 1969, Le passage latéral des prasinites de la Rose-Fuscaldo aux épanchements jurassiques de Malvito et ses conséquences sur l'interprétation de la suture calabro-apenninique, *C.R. Acad. Sci. Paris*, v. 269, p. 1815–1818.

Dubois, R., 1970, Phases de serrage, nappes de socle et métamorphisme alpin à la jonction Calabre-Apennin: la suture calabro-apenninique, *Rev. Géogr. Phys. Géol. Dynam.*, (2), v. 12 (3), p. 221-254.

Dubois, R., 1973, L'origine et l'interprétation des nappes de socle en Calabre, *Réun. Ann. Sci. Terre, Paris*, p. 171.

Dubois, R., and Truillet, R., 1966, L'évolution pétrographique des phyllades à l'approche des masses granitiques et la tectonique du cristallin des monts Péloritains (Sicile), *C.R. Acad. Sci. Paris*, v. 263, p. 101–104.

Dubois, R., and Truillet, R., 1971, Le polymétamorphisme et la structure du domaine péloritain (Sicile). La notion de socle péloritain antéhercynien, *C.R. Acad. Sci. Paris*, v. 272, p. 2134–2136.

Dubourdieu, G., 1956, Etude géologique de la région de l'Ouenza (confins algéro-tunisiens), *Publ. Serv. Carte Géol. Algérie*, n.s., Bull. 10.

Dubourdieu, G., 1961, Dynamique wegenérienne de l'Afrique du Nord, *Livre Mém. P. Fallot*, v. 1, p. 627–644.

Dubourdieu, G., 1962, Déplacement et déformation du pourtour de la Méditerranée occidentale depuis la fin du Miocène. *C.R. Acad. Sci. Paris*, v. 254, p. 2029–2031.

Duée, G., 1964, Relations paléogéographiques entre flysch, chaîne calcaire et complexe de base dans les Monts Nebrodi (Sicile), *Ann. Soc. Géol. Nord*, v. 84, p. 153–162.

Duée, G., 1965, La paléogéographie des Monts Nebrodi (Sicile), *Bull. Soc. Géol. France*, (7), v. 7, p. 889–899.

Duée, G., 1966, La succession des zones de facies dans la partie occidentale de la chaîne bordière des monts Péloritains (Sicile nord-orientale), *Ann. Soc. Géol. Nord*, v. 86, p. 35–46.

Duée, G., 1968a, Distinction d'une série schisto-quartzitique dans les monts Nebrodi (Sicile), *C.R. Somm., Soc. Géol. France*, p. 18–19.

Duée, G., 1968b, Analyse structurale et stratigraphique de la nappe du Monte Soro, *Publ. Serv. Géol. Algérie*, n.s., Bull. 39, 1969, p. 73–87.

Duée, G., 1969, *Etude Géologique des Monts Nebrodi (Sicile)*, Thèse Sci., Paris.

Duée, G., 1970, The geology of the Nebrodi mountains of Sicily, in: *Geology and History of Sicily*, Tripoli: Petrol. Expl. Soc. Libya, p. 187–200.

Dufour, T., 1970, Sur la tectonique de la région de Spoleto (Ombrie, Italie), *Bull. Soc. Géol. France*, (7), v. 12, p. 431–434.

Durand Delga, M., 1960, Le sillon géosynclinal du flysch tithonique–néocomien en Méditerranée occidentale, *Rend. Accad. Naz. Lincei*, (8), v 29, p. 579–585.

Durand Delga, M., 1961a, Au sujet du sillon méso-méditerranéen du flysch au Crétacé et au Nummulitique, *C.R. Somm., Soc. Géol. France*, v. 2, p. 45–47.

Durand Delga, M., 1961b, Le sillon du flysch éocène en Méditerranée occidentale, *C.R. Acad. Sci. Paris*, v. 252, p. 296–298.

Durand Delga, M., 1961c, Le sillon géosynclinal des flyschs oligocènes en Méditerranée occidentale, *C.R. Acad. Sci. Paris*, v. 252, p. 431–433.

Durand Delga, M., 1961d, Le sillon des flyschs du Crétacé supérieur en Méditerranée occidentale, *Rend. Accad. Naz. Lincei*, (8), v. 30 (1), p. 62–66.

Durand Delga, M., 1967, Structure and geology of the Northeast Atlas mountains, in: *Guide Book to the Geology and History of Tunisia*, Tripoli: Petrol. Expl. Soc. Libya, p. 59–84.

Durand Delga, M., 1969, Mise au point sur la structure du Nord-Est de la Berbérie, *Publ. Serv. Géol. Algérie*, v. 39, p. 89–131.

Durand Delga, M., 1972, La courbure de Gibraltar, extrémité occidentale des chaînes alpines, unit l'Europe et l'Afrique, *Eclogae Geol. Helv.*, v. 65 (2), p. 267–278.

Durand Delga, M., 1973, Hypothèses sur la genèse de la courbure de l'arc de Gibraltar, *Bull. Soc. Géol. France*, (7), v. 15 (2), p. 119–120.

Ellenberger, F., 1958, *Etude Geologique du Pays de Vanoise*, Mém. Carte Géol. France, Thèse Sci., Paris.

Elmi, S., 1973a, Décrochements et mouvements atlasiques dans la région frontalière algéro-marocaine (Monts de Rhar Roubane), *C.R. Acad. Sci. Paris*, v. 276, p. 1521–1524.

Elmi, S., 1973b, Structure et paléogéographie tronçonnée des Hautes Plaines oranaises (Ouest algérien), *Réun. Ann. Sci. Terre, Paris*, p. 183.

Elter, G., 1965, Osservazioni sulla tettonica del Monferrato orientale, *Publ. Cons. Naz. Ricerche*, p. 22.

Elter, G., 1971, Schistes lustrés et ophiolites de la zone piémontaise entre Orco et la Doire Baltée (Alpes Graies). Hypothèses sur l'origine des ophiolites, *Géol. Alp.*, v. 47, p. 147–169.

Elter, G., Elter, P., Sturani, C., and Weidmann, M., 1966, Sur la prolongation du domaine ligure de l'Apennin dans le Monferrat et les Alpes et sur l'origine de la Nappe de la Simme s.l. des Préalpes romandes et chablaisiennes, *Arch. Sci. Genève*, v. 19 (3), p. 279–377.

Elter, P., and Raggi, G., 1965, Contributo alla conoscenza dell'Appennino ligure, 3—Tentativo di interpretazione delle brecce ofiolitiche cretacee in relazione con movimenti orogenetici nell'Appennino ligure, *Boll. Soc. Geol. Ital.*, v. 84 (5), p. 1–12.

Elter, P., Giannini, E., Tongiorgi, M., and Trevisan, L., 1961, Le varie unità tettoniche della Toscana e della Liguria orientale, *Rend. Accad. Naz. Lincei, Cl. Sci. Fis. Math. Nat.*, (8), v. 29 (6), p. 497–502.

Emery, K. O., Uchupi, E., Phillips, J. D., Bowin, C. O., Bunce, E. T., and Knott, S. T., 1970, Continental rise of eastern North America, *Bull. Amer. Assoc. Petrol. Geol.*, v. 54, p. 44–108.

Fabiani, R., and Segre, A. G., 1952, Schema strutturale della regione italiana, *Contrib. Sci. Geol., Suppl. Ric. Sci. C.N.R.*, Rome, p. 7–23.

Fabiani, R., and Trevisan, L., 1940, Prove dell'esistenza di uno stile tettonico a falde di ricoprimento nei monti di Palermo, *C.R. Accad. Italia*, v. 11, p. 435–488.

Faculty of Science, University of Libya, 1971, *Symposium on the Geology of Libya*, C., Gray, ed., p. 522.

Fahlquist, D. A., and Hersey, J. B., 1969, Seismic refraction measurements in the western Mediterranean Sea, *Bull. Inst. Océanogr. Monaco*, v. 67 (1386), 52 p.

Fancelli, R., Ghelardoni, R., and Pavan, G., 1966, Considerazioni sull'assetto tettonico dell'Appennino calcareo centro-meridionale, *Mem. Soc. Geol. Ital.*, v. 5 (1), p. 67–90.

Fanucci, F., Fierro, G., Gennesseaux, M., Rehault, J. P., and Tabbò S., 1974, Indagine sismica sulla piattaforma litorale del Savonese (Mar Ligure), *Boll. Doc. Geol. Ital.*, v. 93 (2), p. 421–436.

Farinacci, A., and Radoičić, R., 1964, Correlazione fra serie giuresi e cretacee dell'Appennino centrale e delle Dinaridi esterne, *Ric. Sci.*, v. 34, (II-A; 1965), p. 269–300.

Faure-Muret, A., 1955, Etudes géologiques sur le massif de l'Argentera–Mercantour et ses enveloppes sédimentaires, *Mém. Expl. Carte Géol. France.*

Fenet, B., and Magné, J., 1973, Aperçu sur le bassin miocène synchro-nappes et les conditions de mise en place des unités allochtones dans les Monts du Tessala (dép. d'Oran, Algérie) *Bull. Soc. Géol. France*, (7), v. 15 (3–4), p. 345–351.

Finetti, I., and Morelli, C., 1972, Wide scale digital seismic exploration of the Mediterranean Sea, *Boll. Geofis. Teor. Appl.*, v. 14 (56), p. 291–342.

Finetti, I., and Morelli, C., 1973, Geophysical exploration of the Mediterranean Sea, *Boll. Geofis. Teor. Appl.*, v. 15 (60), p. 263–340.

Finetti, I., Morelli, C., and Zarudzki, E., 1970, Reflection seismic study of the Tyrrhenian Sea, *Boll. Geofis. Teor. Appl.*, v. 12 (48), p. 311–346.

Flores, G., 1959, Evidence of slump phenomena (Olistostromes) in areas of hydrocarbons exploration in Sicily, *Proc. Fifth World Petrol. Congr.*, New York, Sect. 1/13.

Galli, M., Bezzi A., Piccardo, G. B., Cortesogno, L., and Pedemonte, G. M., 1972, Le ofioliti dell'Appennino ligure: un frammento di crosta-mantello "oceanici" dell'antica Tetide, *Mem. Soc. Geol. Ital.*, v. 11, p. 467–502.

Gasperi, G., 1970, Lineamenti tettonici dei rilievi di Castiglione della Pescaia e Scarlino (Grosseto), *Mem. Soc. Geol. Ital.*, v. 9 (1), p. 91–105.

Gélard, J. P., and Sigal, J., 1974, Le flysch massylien de Tagdinnt (Grande Kabylie: étude stratigraphique et structurale), *Bull. Soc. Géol. France*, (7), v. 16 (5), p. 526–536.

Gelmini, R., 1966, Studio fotogeologico dell'Appennino settentrionale tra il Valdarno e la Romagna, *Boll. Soc. Geol. Ital.*, v. 84 (6), p. 167–212.

Gennesseaux, M., Auzende, J. M., Olivet, J. L., and Bayer, R., 1974, Les orientations structurales et magnétiques sous-marines au Sud de la Corse et la dérive corso-sarde, *C.R. Acad. Sci. Paris*, v. 278, p. 2003–2006.

Gennesseaux, M., and Rehault, J. P., 1975, La marge continentale corse, *Bull. Soc. Géol. France*, (7), v. 17 (4), p. 505–518.

Ghelardoni, R., 1965, Osservazioni sulla tettonica trasversale dell'Appennino settentrionale, *Boll. Soc. Geol. Ital.*, v. 84 (3), p. 277–290.

Giannini, E., and Tongiorgi, M., 1958, Osservazioni sulla tettonica neogenica della Toscana marittima, Publ. Centr. Stud. Geol. dell'Appennino, 9, no. 51–59, *Boll. Soc. Geol. Ital.*, v. 77, p. 147–170.

Giannini, E., and Tongiorgi, M., 1962, Les phases tectoniques néogènes de l'orogenèse alpine dans l'Apennin septentrional, *Bull. Soc. Géol. France*, (7), v. 4, p. 682–690.

Giannini, E., Nardi, R., and Tongiorgi, M., 1962, Osservazioni sul problema della falda toscana, *Boll. Soc. Geol. Ital.*, v. 81 (2), p. 17–98.

Giese, P., Günther, K., and Reutter, K. J., 1970, Vergleichende geologische und geophysikalische Betrachtungen der Westalpen und des Nordapennins, *Zeitschr. Deutsch. Geol. Ges.*, v. 120, p. 151–195.

Giglia, G., 1967, Geologia dell'Alta Versilia settentrionale (Tav. M. Altissimo), *Mem. Soc. Geol. Ital.*, v. 6 (1), p. 67–95.

Giglia, G., and Radicati di Brozolo, F., 1970, K/Ar Age of Metamorphism in the Apuane Alps (Northern Tuscany), *Boll. Soc. Geol. Ital.*, v. 89 (4), p. 485–497.

Giunta, G., and Liguori, V., 1970, Geologia della penisola di Capo San Vito (Sicilia nord-occidentale), *Lav. Ist. Geol. Univ. Palermo*, v. 9, 21 p.

Giunta, G., and Liguori, V., 1973, Evoluzione paleotettonica della Sicilia nord-occidentale, *Boll. Soc. Geol. Ital.*, v. 92 (4), p. 903–924.

Glangeaud, L., 1951, Interprétation tectono-physique des caractères structuraux et paléogéographiques de la Méditerranée occidentale, *Bull. Soc. Géol. France*, (6), v. 1, p. 735–762.

Glangeaud, L., 1961, Paléogéographie dynamique de la Méditerranée et de ses bordures; le rôle des phases ponto-plio-quaternaires, in: *Océanographie de la Méditerranée Occidentale*, Villefranche-sur-Mer: Coll. Nat. du C.N.R.S.

Glangeaud, L., 1966, Les grands ensembles structuraux de la Méditerranée occidentale d'après les données de Géomède I, *C.R. Acad. Sci. Paris*, v. 262, p. 2405–2408.

Glangeaud, L., 1968, Les méthodes de la géodynamique et leurs applications aux structures de la Méditerranée occidentale, *Rev. Géogr. Phys. Géol. Dynam.*, (2), v. 10 (2), p. 83–135.

Goguel, J., 1963, L'interprétation de l'arc des Alpes occidentales, *Bull. Soc. Géol. France*, (7) v. 5, p. 20–33.

Grandjacquet, C., 1963a, Données nouvelles sur la tectonique tertiaire des massifs calabrolucaniens, *Bull. Soc. Géol. France*, (7), v. 4, p. 695–706.

Grandjacquet, C., 1963b, Importance de la tectonique tangentielle en Italie méridionale, *Rev. Géogr. Phys. Géol. Dynam.*, (2), v. 5 (2), p. 107–111.

Grandjacquet, C., 1963c, Schéma structural de l'Apennin campano-lucanien, *Rev. Géogr. Phys. Géol. Dynam.*, (2), v. 5 (3), p. 185–202.

Grandjacquet, C., 1969, Les phases tectoniques et le métamorphisme tertiaire de la Calabre du Nord et de la Campanie du Sud (Italie), *C.R. Acad. Sci. Paris*, v. 269, p. 1819–1822.

Grandjacquet, C., 1971, Les séries transgressives d'âge oligo-miocène inférieur de l'Apennin méridional, *Bull. Soc. Géol. France*, (7), v. 13, p. 315–320.

Grandjacquet, C., and Glangeaud, L., 1962, Structures mégamétriques et évolution de la mer Tyrrhénienne et des zones périthyrréniennes, *Bull. Soc. Géol. France*, (7), v. 4, p. 760–773.

Grandjacquet, C., and Grandjacquet, M.-J., 1962, Géologie de la zone de Diamante–Verbicaro (Calabre), *Geol. Romana*, v. 1, p. 297–312.

Grandjacquet C., and Haccard, D., 1975, Analyse des sédiments polygéniques néogènes à faciès de cargneules associés à des gypses dans les Alpes du Sud. Extension de ces faciès au pourtour de la Méditerranée occidentale, *Bull. Soc. Géol. France*, (7), v. 17 (3), p. 242–259.

Grandjacquet, C., Haccard, D., and Lorenz, C., 1972, Essai de tableau synthétique des principaux événements affectant les domaines alpin et apennin à partir du Trias, *C.R. Somm., Soc. Géol. France*, v. 4, p. 158–160.

Guazzone, G., and Malesani, P., 1970, Osservazioni sulla provenienza dei clasti e sulle modalità di sedimentazione della formazione marnoso–arenacea tosco-romagnola, *Mem. Soc. Geol. Ital.*, v. 9 (1), p. 107–120.

Guillaume, A., 1967, *Contribution à l'Etude Géologique des Alpes Liguro-Piémontaises*, Thèse Sci., Paris.

Guiraud, R., 1973, *Evolution Post-Triasique de l'Avant-Pays de la Chaîne Alpine en Algérie*, Thèse Sci., Nice.

Haccard, D., Lorenz, C., and Grandjacquet, C., 1972, Essai sur l'évolution tectogénétique de la liaison Alpes–Apennins (de la Ligurie à la Calabre), *Mem. Soc. Geol. Ital.*, v. 11 (4), p. 309–342.

Heezen, B. C., Gray, C., Segre, A. G., and Zarudski, E. F. K., 1971, Evidence of foundered continental crust beneath the Central Tyrrhenian Sea, *Nature*, v. 229, p. 327–329.

Hinz, K., 1972, A low velocity layer in the upper crust of the Ionian Sea, *22ᵉ Congr. CIESM, Rome* (Abstract).

Hinz, K., 1972, Results of seismic refraction investigations (Project *Anna*) in the western

Mediterranean Sea, South and North of the Island of Mallorca, in: *Results of the Anna Cruise*, Leenhardt, O., *et al.*, eds., *Bull. Centre Rech. Pau, SNPA*, v. 6 (2), p. 405–426.

Iaccarino, E., 1968, Attività sismica in Italia dal 1893 al 1965, *C.N.E.N.*, Rome, p. 1–16.

Ietto, A., Pescatore, T., and Cocco, E., 1965, Il flysch mesozoico–terziaro del Cilento occidentale, *Boll. Soc. Nat. Napoli*, v. 74, p. 396–402.

Ippolito, F., 1950, Contributo alla geologia del Monte Pisano e delle Alpi Apuane, *Mem. Note Ist. Geol. Appl. Univ. Napoli*, v. 3.

Ippolito, F., Ortolani, F., and Russo, M., 1973, Struttura marginale tirrenica dell'Appennino Campano: reinterpretazione di dati di antiche ricerche di idrocarburi, *Mem. Soc. Geol. Ital.*, v. 12 (2), p. 227–250.

Ippolito, F., Ortolani, F., and Di Nocera, S., 1974, Alcune considerazioni sulla struttura profonda dell'Apennino Irpino: reinterpretazione di ricerche di idrocarburi, *Boll. Soc. Geol. Ital.*, v. 93 (4), p. 861–882.

Jacobacci, A., 1962, Evolution de la fosse mio-pliocène de l'Apennin apulo-campanien (Italie méridionale), *Bull. Soc. Géol. France*, (7), v. 4, p. 691–694.

Jacobacci, A., 1965, Frane sottomarine nelle formazioni geologiche, *Boll. Serv. Geol. Ital.*, v. 86, p. 65–85.

Jacobacci, A., and Martelli, G., 1969, Età e giacitura del flysch nella Toscana meridionale, *Boll. Soc. Geol. Ital.*, v. 88, p. 621–636.

Jauzein, A., 1962, *Contribution à l'Etude Géologique des Confins de la Dorsale Tunisienne*, Thèse Sci., Paris, and *Ann. Mines Géol.*, Tunis, no. 22, 1967.

Jong, K. A., de, 1967, Tettonica gravitativa e raccorciamento crostale nelle Alpi meridionali, *Boll. Soc. Geol. Ital.*, v. 86 (4), p. 749–776.

Juteau, T., Lapierre, H., Nicolas, A., Parrot, J. F., Ricou, L. E., Rocci, G., and Rollet, M., (H. Mésorian), (1973), Idées actuelles sur la constitution, l'origine et l'évolution des assemblages ophiolitiques mésogéens, *Bull. Soc. Géol. France*, (7), v. 15 (5–6), p. 478–493.

Ksiazkiewicz, M., 1960, Evolution structurale des Carpathes polonaises, *Livre Mém. P. Fallot*, v. 2, p. 529–562.

Kuenen, P. H., 1960, Les formations de turbidites des Apennins du Nord, *Livre Mém. P. Fallot*, v. 2, p. 227–235.

Lanteaume, M., 1958, Schéma structural des Alpes maritimes franco-italiennes, *Bull. Soc. Géol. France*, (6), v. 8, p. 651–674.

Lajat, D., Gonnard, R., Letouzey, J., Biju-Duval, B., and Winnock, E., 1975, Prolongement dans l'Atlantique de la partie externe de l'arc bético-rifain, *Bull. Soc. Géol. France*, (7), v. 17 (4), p. 481–485.

Laubscher, H. P., 1971*a*, The large-scale kinematics of the Western Alps and the Northern Apennines and its palinspastic implications, *Amer. J. Sci.*, v. 271, p. 193–226.

Laubscher, H. P., 1971*b*, Das Alpen-Dinariden Problem und die Palinspastik der südlichen Tethys, *Geol. Rundschau*, v. 60 (3), p. 813–833.

Laval, F., 1974, Schéma structural de l'Est de la Grande Kabylie, *Bull. Soc. Géol. France*, (7), v. 16 (3), p. 303–310.

Laval, F., 1974, Précisions sur la tectonique des flyschs dans l'Est de la Grande Kabylie (Algérie), *C.R. Acad. Sci. Paris*, v. 279, p. 1609–1612.

Leikine, M., and Velde, B., 1974, Les transformations post-sédimentaires des minéraux argileux du Sénonien dans le NE algérien. Existence probable d'un épi-métamorphisme anté-éocène, *Bull. Soc. Géol. France*, (7), v. 16 (2), p. 177–182.

Lemoine, M., 1972, Rythme et modalités des plissements superposés dans les chaînes subalpines méridionales des Alpes occidentales françaises, *Geol. Rundschau*, v. 61, p. 975–1010.

Le Pichon, X., Pautot, G., Auzende, J. M., and Olivet, J. L., 1971, La Méditerranée occidentale depuis l'Oligocène. Schéma d'évolution, *Earth Planet. Sci. Lett.*, v. 13, p. 145–152.

Le Pichon, X., Pautot, G., and Weill, J. P., 1972, Opening of the Alboran Sea, *Nature Phys. Sci.*, v. 236 (67), p. 83–85.

Limanovski, M., 1913, Die grosse kalabrische Decke, *Bull. Acad. Sci. Cracovie, Cl. Sci. Math. Nat.*, A, p. 370–385.

Lorenz, C., 1969, *Contribution à l'Etude Stratigraphique de l'Oligocène et du Miocène Inférieur des Confins Liguro-Piémontais (Italie)*, Thèse Sci., Paris, 1967, and *Att. Ist. Geol. Univ. Genova*, v. 6 (2).

Lort, J. M., 1972, *The Crustal Structure of the Eastern Mediterranean*, Thesis, Univ. Cambridge, p. 117.

Lort, J. M., Limond, W. Q., and Gray, F., 1974, Preliminary seismic studies in the eastern Mediterranean, *Earth Planet. Sci. Lett.*, v. 21 (4), p. 355–366.

Losacco, U., 1963, Osservazioni geologiche sulla parte centrale e settentrionale della catena di Pratomagno, *Boll. Soc. Geol. Ital.*, v. 82 (2), p. 291–404.

Lowrie, W., and Alvarez, W., 1974, Rotation of the Italian Peninsula, *Nature*, v. 251, p. 285–288.

Lucchetti, L., Albertelli, L., Mazzei, R., Thieme, R., Bongiorni, D., and Dondi, L., 1962, Contributo alla conoscenze geologiche del Pedeappennino padano, *Boll. Soc. Geol. Ital.*, v. 81 (4), p. 5–245.

Lucini, P., 1959a, Considerazioni sulle ricerche geologiche nella parte occidentale della regione calabro-lucana, *Boll. Soc. Geol. Ital.*, v. 77 (3), p. 153–160.

Lugeon, M., and Argand, E., 1906a, Sur de grands phénomènes de charriage en Sicile, *C.R. Acad. Sci. Paris*, v. 142, p. 966–968.

Lugeon, M., and Argand, E., 1906b, Sur la grande nappe de recouvrement de la Sicile, *C.R. Acad. Sci. Paris*, v. 142, p. 1001–1003.

Lugeon, M., and Argand, E., 1906c, La racine de la nappe sicilienne et l'arc de charriage de la Calabre, *C.R. Acad. Sci. Paris*, v. 142, p. 1107–1109.

Magné, J., and Raymond, D., 1974, Le Néogène "post-nappes" de la région de Dellys–Tizi Ouzou (Algérie); un enregistreur de l'évolution dynamique du NW de la Grande Kabylie après le Burdigalien, *Bull. Soc. Géol. France*, (7), v. 16 (5), p. 537–542.

Malaroda, R., and Raimondi, C., 1957, Linee di dislocazione e sismicità in Italia. Parte 1ª— Le linee di faglia e di scorrimento in Italia, *Boll. Geod. Sci. Affini*, v. 16 (3), p. 273–289.

Maluski, H., Mattauer, M., and Matte, P., 1973, Sur la présence de décrochements alpins en Corse, *C.R. Acad. Sci. Paris*, v. 276, p. 709–712.

Marchetti, M. P., 1957, The occurrence of slide and flowage materials (olistostromes) in the Tertiary Series of Sicily, *Congr. Geol. Int. XX Sess.*, Mexico (1956), Sect. V (1), p. 209–225.

Martinis, B., 1962, Lineamenti strutturali della parte meridionale della Penisola Salentina, *Geol. Romana*, v. 1, p. 11–23.

Martinis, B., 1964, Osservazioni sulla tettonica del Gargano orientale, *Boll. Soc. Geol. Ital.*, v. 85, 1965, p. 45–93.

Mascle, G., 1967, Remarques stratigraphiques et structurales sur la région de Palazzo–Adriano (Monts Sicani, Sicile), *Bull. Soc. Géol. France*, (7), v. 9, p. 104–110.

Mascle, G., 1968, Structure et morphologie de la région de Montevago (Sicile occidentale), *Bull. Soc. Géol. France*, (7), v. 10, p. 653–657.

Mascle, G., 1970, Geological sketch of western Sicily, in: *Geology and History of Sicily*, Tripoli: Petrol. Expl. Soc. Libya, p. 231–243.

Mascle, G., 1973, *Etude Géologique des Monts Sicani (Sicile)*, Thèse Sci., Paris.

Mascle, G., 1974, Les grands traits de l'évolution géologique des Monts Sicani (Sicile), *Bull. Soc. Géol. France*, (7), v. 16 (2), p. 161–170.

Mattauer, M., 1963, Le style tectonique de la chaîne tellienne et rifaine, *Geol. Rundschau,* v. 53, p. 269–313.

Mattauer, M., 1968, Les traits structuraux essentiels de la chaîne des Pyrénées, *Rev. Géogr. Phys. Géol. Dynam.,* v. 10, p. 2–12.

Mattauer, M., and Henry, J., 1974, Pyrénées, in: *Mesozoic–Cenozoic Orogenic Belts, Data for Orogenic Studies,* Spencer, A. M., ed., Geol. Soc. London Spec. Publ. 4, p. 3–21.

Mauffret, A., Fail, J. P., Montadert, L., Sancho, J., and Winnock, E., 1973, North-Western Mediterranean sedimentary basin from seismic reflection profile, *Bull. Amer. Assoc. Petrol. Geol.,* v. 57 (11), p. 2245–2262.

McKenzie, D. P., 1970, Plate tectonics of the Mediterranean region, *Nature,* v. 226, p. 239–243.

McKenzie, D. P., 1972, Active tectonics of the Mediterranean region, *Geophys. J.R. Astr. Soc.,* v. 30 (2), p. 109–185.

Mercier, J., Bosquet, B., Delifassis, N., Drakopoulos, I., Keraudren, B., Korroneou, V., Lemeille, F., Pechoux, P. Y., Pegoraro, O., Philip, J., Sauvage, J., and Sorel, D., 1973, Déformations superficielles en compression et en extension dans la "plaque égéenne" (Grèce) au cours du Pliocène et du Quaternaire, *Réun. Ann. Sci. Terre.,* Paris, p. 297.

Merla, G., 1951, Geologia dell'Appennino settentrionale, *Boll. Soc. Geol. Ital.,* v. 70 (1), p. 92–382.

Migliorini, C. I., 1948, I cunei compositi nell'orogenesi, *Boll. Soc. Geol. Ital.,* v. 67, p. 29–142.

Montadert, L., and Letouzey, J., 1975, Structure de la marge méditerranéenne d'après les études de détail effectuées pour le leg 42 du *Glomar Challenger. Bull. Soc. Géol. France,* (7), v. 17 (4), p. 519–520.

Montenat, C., Bizon, G., and Bizon, J. J., 1975, Remarques sur le Néogène du forage Joides 121 en Mer d'Alboran (Méditerranée occidentale), *Bull. Soc. Géol. France,* (7), v. 17 (1), p. 45–51.

Morelli, C., 1948, La rete geofisica e geodetica in Italia nello stato attuale nei suoi rapporti con la struttura geologica superficiale e profonda. Pt. I: Stato attuale della rete geofisica e geodetica in Italia, *Ist. Geofis. Trieste,* Publ. no. 238, p. 101.

Morelli, C., 1951, Rilievo gravimetrico e riduzione isostatica nell'Italia nord-orientale, *Osserv. Geofis. Trieste,* Publ. no. 20, p. 47.

Morelli, C., 1970, Physiography, gravity and magnetism of the Tyrrhenian Sea, *Boll. Geofis. Teor. Appl.,* v. 12 (48), p. 275–308.

Mutti, E., and Ricci Lucchi, F., 1972, Le torbiditi dell'Appennino settentrionale: introduzione all'analisi di facies, *Mem. Soc. Geol. Ital.,* v. 11 (2), p. 161–200.

Nairn, A., and Westphal, M., 1968, Possible implications of the paleomagnetic study of late Paleozoic igneous rocks of Northwestern Corsica, *Palaeogeogr. Palaeoclimatol. Palaeoecol.,* v. 5, p. 179–204.

Nardi, R., 1963, La "zona degli scisti sopra i marmi" nelle Alpi Apuane e i terreni che la costituiscono, *Boll. Soc. Geol. Ital.,* v. 82, (2), p. 505–522.

Nardi, R., 1965, Schema geologico dell'Appennino tosco-emiliano tra il Monte Cusna e il Monte Cimone e considerazioni sulle unità tettoniche dell'Appennino, *Boll. Soc. Geol. Ital.,* v. 84 (5), p. 35–92.

Nardi, R., 1968a, Le unità alloctone della Corsica e loro correlazione con le unità delle Alpi e dell'Appennino, *Mem. Soc. Geol. Ital.,* v. 7 (2), p. 323–344.

Nardi, R., 1968b, Contributo alla geologia della Balagne (Corsica nord-occidentale), *Mem. Soc. Geol. Ital.,* v. 7 (4), p. 471–489.

Nardi, R., and Tongiorgi, M., 1962, Contributo alla geologia dell'Appennino tosco-emiliano: 1—Stratigrafia e tettonica dei dintorni di Pievepelago (Appennino Modenese), *Boll. Soc. Geol. Ital.,* v. 81 (3), p. 1–76.

Nesteroff, W. D., and Ryan, W. B. F., 1973, Séries stratigraphiques et implications tectoniques du forage Joides 121 en mer d'Alboran, *Bull. Soc. Géol. France*, (7), v. 15 (2), p. 113–114.

Nesteroff, W. D., Ryan, W. B. F., Hsü, K. J., Pautot, G., Wezel, F. C., Lort, J. M., Cita, M. B., Maync, W., Stradner, H., and Dumitrica, P., 1972, Evolution de la sédimentation pendant le Néogène en Méditerranée d'après les forages JOIDES-DSDP, in: *The Mediterranean Sea—A Natural Sedimentation Laboratory*, Stanley, D. J., ed., Stroudsburg, Pa.: Dowden, Hutchinson and Ross, p. 47–62.

Obert, D., 1972, Déplacements tectoniques est–ouest dans les Babors, *Bull. Soc. Hist. Nat. Afrique Nord*, v. 63 (3–4), p. 135–140.

Obert, D., 1974, Phases tectoniques mésozoïques d'âge anté-cénomanien dans les Babors (Tell nord-sétifien, Algérie), *Bull. Soc. Géol. France*, (7), v. 16 (2), p. 171–176.

Ogniben, L., 1960, Nota illustrativa dello schema geologico della Sicilia nord-orientale, *Riv. Min. Sic.*, no. 64–65, p. 183–212.

Ogniben, L., 1965, Le Argille Scagliose del Crotonese, *Mem. Ist. Geol. Appl. Napoli*, v. 6, p. 72.

Ogniben, L., 1969, Schema introduttivo alla geologia del confine calabro-lucano, *Mem. Soc. Geol. Ital.*, v. 8 (4), p. 453–763.

Ogniben, L., 1970, Paleotectonic history of Sicily, in: *Geology and History of Sicily*, Tripoli: Petrol. Expl. Soc. Libya, p. 133–143.

Ogniben, L., 1972, Gli elementi strutturali della regione appenninica, *Mem. Ist. Geol. Min. Univ. Padova*, v. 29, p. 1–29.

Olivet, J. L., Pautot, G., and Auzende, J. M., 1972, Alboran Sea, in: *Initial Rep. Deep Sea Drilling Project*, v. 13, 48-1–48-3, p. 1417–1447.

Olivet, J. L., Auzende, J. M., and Bonnin, J., 1973, Structure et évolution tectonique du bassin d'Alboran, *Bull. Soc. Géol. France*, (7), v. 15 (2), p. 108–112.

Ortolani, F., 1974, Faglia trascorrente pliocenica nell'Appennino campano, *Boll. Soc. Geol. Ital.*, v. 93 (3), p. 609–622.

Pannekoek, A. J., 1969, Uplift and subsidence in and around the western Mediterranean since the Oligocene: a review, *Verh. K. Ned. Geol. Mijnb. Genoot.*, v. 26, p. 53–77.

Paquet, J., 1974, Tectonique éocène dans les Cordillères bétiques; vers une nouvelle conception de la paléogéographie en Méditerranée occidentale, *Bull. Soc. Géol. France*, (7), v. 16 (1), 58–73.

Pervinquière, L., 1903, *Etude Géologique de la Tunisie Centrale*, Thèse Sci., Paris.

Pescatore, T., 1967, Miocenic turbidites of Sorrento Peninsula; Western Cilento Flysch, in: *Sedimentological Characteristics of Some Italian Turbidites*, Angelluci et al., eds., *Geol. Romana*, v. 6, p. 391–395.

Pescatore, T., and Ortolani, F., 1973, Schema tettonico dell'Appennino campano–lucano, *Boll. Soc. Geol. Ital.*, v. 92 (3), p. 453–472.

Pescatore, T., and Sgrosso, I., 1973, I rapporti tra la piattaforma campano–lucana e la piattaforma abruzzese–campana nel Casertano, *Boll. Soc. Geol. Ital.*, v. 92 (4), p. 925–938.

Pescatore, T., Sgrosso, I., and Torre, M., 1970, Lineamenti di tettonica e sedimentazione nel Miocene dell'Appennino campano–lucano, *Mem. Soc. Nat. Napoli*, Suppl. v. 78, p. 337–406.

Peterschmitt, E., 1956, Quelques données nouvelles sur les seïsmes profonds de la mer Tyrrhenienne, *Ann. Geofis.*, v. 9 (3), p. 305–334.

Piccareta, G., Amodio Morelli, L., and Paglionico, A., 1973, Evoluzione metamorfica delle rocce in facies granulitica nelle Serre nord-occidentali (Calabria), *Boll. Soc. Geol. Ital.*, v. 92 (4), p. 861–890.

Pichler, H., 1967, Neue Erkenntnisse über Art und Genese des Vulkanismus der Aeolischen Inseln (Sizilien), *Geol. Rundschau*, v. 57 (1), p. 102–126.

Pieri, M., 1966, Tentativo di ricostruzione paleogeografico–strutturale dell'Italia centro-meridionale, *Geol. Romana*, v. 5, p. 407–424.

Quitzow, H. W., 1935, Der Deckenbau des Kalabrischen Massivs und seiner Randgebiete, *Beitr. Geol. Westl. Mediterrangebiete* (*Abh. Ges. Wis. Göttingen, Math. Phys. Kl*), III F., H. 13, p. 63–179.

Rampnoux, J. P., 1970, Regards sur les Dinarides internes yougoslaves (Serbie, Montenegro oriental), *Bull. Soc. Géol. France* (7), v. 12, p. 948–966.

Rampnoux, J. P., 1973, Essai de reconstitution géotectonique des Dinarides internes yougo-slaves (Serbie) au Jurassique et au Crétacé, *Réun. Ann. Sci. Terre*, p. 353.

Rangin, C., 1975, Contribution à l'étude paléogéographique et structurale des Monts Termini Imerese (Sicile), *Bull. Soc. Géol. France*, (7), v. 17 (1), p. 38–44.

Raoult, J. F., 1975, Evolution paléogéographique et structurale de la chaîne alpine entre le golfe de Skikda et Constantine (Algérie orientale), *Bull. Soc. Géol. France*, (7), v. 17 (3), p. 394–409.

Recq, M., 1967, Structure de la croute terrestre en Provence d'après les expériences du Revest et du Lac Nègre, *C.R. Acad. Sci. Paris*, v. 264, p. 1588–1591.

Recq, M., 1970, Courbes d'égale profondeur de la discontinuité de Mohorovicic en Provence, *C.R. Acad. Sci. Paris*, v. 270, p. 11–13.

Recq, M., 1972*a*, Sur la stabilité récente du massif des Maures et son indépendance par rapport aux régions environnantes, *C.R. Acad. Sci. Paris*, v. 275, p. 333–336.

Recq, M., 1972*b*, Profils de réfraction en Ligurie, *Pure Appl. Geophys.*, v. 101 (9), p. 155–161.

Recq, M., 1972*c*, La structure profonde de la croute terrestre sous le massif de l'Estérel, *Boll. Geofis. Teor. Appl.*, v. 14 (55), p. 253–268.

Recq, M., 1973*a*, The Pn velocity under the gulf of Genoa, *Earth Planet. Sci. Lett.*, v. 20, p. 447–450.

Recq, M., 1973*b*, Contribution à l'étude de l'évolution des marges continentales du Golfe de Gênes, *Tectonophysics*, v. 22, p. 363–375.

Rehault, J. P., Olivet, J. L., and Auzende, J. M., 1974, Le bassin nord-occidental méditer-ranéen: structure et évolution, *Bull. Soc. Géol. France*, (7), v. 16 (3), p. 281–294.

Richert, J. P., 1971, Mise en évidence de quatre phases tectoniques successives en Tunisie, *Notes Serv. Géol. Tunisie*, no. 34, *Trav. Géol. Tunisie*, no. 4, p. 115–125.

Righi, F. Rigo de, 1956, Olistostromi neogenici in Sicilia, *Boll. Soc. Geol. Ital.*, v. 75, p. 1–33.

Ritsema, A. R., 1963, Seismic data of the West Mediterranean and the problem of oceani-zation, *Verh. K. Ned. Geol. Mijnb. Genoot.*, v. 26, p. 105–120.

Ruggieri, G., 1966, Primi risultati di ricerche sulla tettonica della Sicilia occidentale, *Geol. Romana.*, v. 5, p. 453–456.

Ryan, W. B. F., Hsü, K. J., *et al.*, eds., 1973, *Initial Reports of the Deep Sea Drilling Project*, Washington, D.C.: U.S. Government Printing Office, v. 13, p. 1447.

Ryan, W. B. F., Stanley, D. J., Hersey, J. B., Fahlquist, D. A., and Allan, T. D., 1971, The tectonics and geology of the Mediterranean Sea, in: *The Sea*, II, Maxwell, A. E., ed., New York: Wiley–Interscience, p. 387–492.

Scandone, P., and Bonardi, G., 1968, Synsedimentary tectonics controlling deposition of Mesozoic and Tertiary carbonatic sequences of areas surrounding Vallo di Diano (Southern Apennines), *Mem. Soc. Geol. Ital.*, v. 7, (1), p. 1–10.

Schmidt Di Friedberg, P., 1959, La geologia nel gruppo montuoso delle Madonie nel quadro delle possibilità petrolifere della Sicilia centro-settentrionale, *Atti 2 Conv. Int. Studi* "*Petr. di Sicilia*," p. 130–136.

Schmidt Di Friedberg, P., 1962, Introduction à la géologie pétrolière de la Sicile, *Rev. Inst. Franc. Petrole*, and *Ann. Comb. Liq.*, v. 17, p. 635–668.

Schmidt Di Friedberg, P., 1964–65, Litostratigrafia petrolifera della Sicilia, *Riv. Min. Sic.*, v. 15 and v. 16 (88–90) and (91–93), p. 198–217 and 50–71.

Schmidt Di Friedberg, P., and Trovo, A., 1962, Contribution à l'étude du groupe de Monte Judica (Sicile orientale), *Bull. Soc. Géol. France*, (7), v. 4, p. 754–759.

Schmidt Di Friedberg, P., Barbieri, F., and Giannini, G., 1960, La geologia del gruppo montuoso delle Madonie (Sicilia centro-settentrionale), *Boll. Serv. Geol. Ital.*, v. 81 (1), p. 73–140.

Selli, R., 1962, Il Paleogene nel quadro della geologia dell'Italia centro-meridionale, *Mem. Soc. Geol. Ital.*, v. 3, p. 737–789.

Selli, R., and Fabbri, A., 1971, Tyrrhenian: a Pliocene deep sea. *Rend. Accad. Naz. Lincei, Cl. Sci. Fis.*, v. 8, (50), 5, p. 580 (104)–592 (116).

Semenza, E., 1974, La fase giudicariense, nel quadro di una nuova ipotesi sull'orogenesi alpina nell'area italo-dinarica, *Mem. Soc. Geol. Ital.*, v. 13 (2), p. 187–225.

Sestini, G., 1970, Development of the Northern Apennines Geosyncline: Sedimentation of the late geosynclinal stage, postgeosynclinal deposition, flysch facies and turbidite sedimentology, *Sediment. Geol.*, v. 4 (3–4), p. 445–480, 481–520, 559–598.

Sgrosso, I., 1974, I rapporti tra la piattaforma carbonatica campano-lucana e la piattaforma abruzzese–campana al Monte Massico (Caserta), *Boll. Soc. Geol. Ital.*, v. 93 (4), p. 1197–1210.

Solignac, M., 1927, Etude géologique de la Tunisie septentrionale, *Publ. Serv. Géol. Tunisie*.

Spadea, P., 1968, Pillow-lavas nei terreni alloctoni dell'Appennino lucano, *Atti Accad. Gioenia Sci. Nat. Catania*, (6), v. 20, p. 105–142.

Stanley, D. J., ed., 1972, *The Mediterranean Sea—A Natural Sedimentation Laboratory*, Stroudsburg, Pa.: Dowden, Hutchinson and Ross, p. 765.

Storetvedt, K. M., 1973, The rotation of Iberia: Caenozoic paleomagnetism from Portugal, *Tectonophysics*, v. 17, p. 23–39.

Suess, E., 1904–1924, *The Face of the Earth* (English translation of *Das Antlitz der Erde*), Oxford: Clarendon Press, 5 v.

Teichmuller, R., and Quitzow, H. W., 1935, Die Beziehungen zwischen den Nordapenninen und dem kalabrischen Deckenbau, *Beitr. Geol. Westl. Mediterrangebiete* (*Abh. Ges. Wiss. Göttingen, Math. Phys. Kl.*), III F., H. 13, p. 181–186.

Termier, P., 1903, Les nappes des Alpes orientales et la synthèse des Alpes, *Bull. Soc. Géol. France*, (4), v. 3, p. 711–765.

Thomas, G., 1974, La phase de compression pleistocène en Algérie nord-occidentale: âge, premiers éléments cinématiques, relations avec les mouvements en distension, *C.R. Acad. Sci. Paris*, v. 279, p. 311–314.

Tollmann, A., 1963a, Résultats nouveaux sur la position, la subdivision et le style structural des zones helvétiques, penniques et austro-alpines des Alpes orientales, *Livre Mém. P. Fallot*, v. 2, p. 477–491.

Tollmann, A., 1963b, *Ostalpensynthese*, Vienna: Deuticke, p. 256.

Tollmann, A., 1966–1968, Tektonische Karte der nördlichen Kalkalpen, *Mitt. Geol. Gesellsch.*, Vienna (1967), v. 59, p. 231–253, 1969, v. 61, p. 124–181.

Tollmann, A., 1969, Die Bruchtektonik in den Ostalpen. *Geol. Rundschau*, v. 59 (1), p. 278–288.

Trevisan, L., 1955, Les mouvements tectoniques récents en Sicile, hypothèses et problèmes, *Geol. Rundschau*, v. 43, p. 207–221.

Trevisan, L., 1956, Aspetti e problemi del complesso delle "argille scagliose ofiolitifere" nei suoi affioramenti occidentali (Toscana marittima e Liguria), *Boll. Soc. Geol. Ital.*, v. 75, p. 3–20.

Trevisan, L., 1958, *Geologia delle Alpi Apuane. Guida ai Monti d'Italia*, Milano: Ed. C.A.I.

Trevisan, L., 1961, La paléogéographie du Trias de l'Apennin septentrional et central et ses rapports avec la tectogenèse, *Livre Mém. P. Fallot*, v. 2, p. 217–225.

Trevisan, L., 1962, Considérations sur deux coupes à travers l'Apennin septentrional, *Bull. Soc. Géol. France*, (7), v. 4, p. 675–681.

Trevisan, L., and Tongiorgi, E., 1957, La Tyrrhénide, *Scientia*, v. 6, p. 1–15.

Truillet, R., 1961, Remarques stratigraphiques et tectoniques sur la région de Novara di Sicilia (Monts Péloritains, Sicile), *Bull. Soc. Géol. France*, (7), v. 3, p. 559–567.

Truillet, R., 1962, Détails structuraux de la chaîne calcaire dans les environs de Rocella Valdèmone (Monts Péloritains, Sicile), *C.R. Acad. Sci. Paris*, v. 254, p. 2032–2034.

Truillet, R., 1968, *Etude Géologique des Péloritains Orientaux (Sicile)*, Thèse Sci, Paris.

Truillet, R., 1970, The geology of the eastern Peloritani Mountains of Sicily, in: *Geology and History of Sicily*, Tripoli: Petrol. Expl. Soc. Libya, p. 171–186.

Trümpy, R., 1958, Remarks on the pre-orogenic history of the Alps, *Geol. Mijnb.*, v. 20, p. 340–352.

Trümpy, R., 1960a, Paleotectonic evolution of the central and western Alps, *Bull. Geol. Soc. Amer.*, v. 71, p. 843–908.

Trümpy, R., 1960b, Sur les racines des nappes helvétiques, *Livre Mém. P. Fallot*, v. 2, p. 419–428.

Trümpy, R., 1969, Die helvetischen Decken der Ostschweiz: Versuch einer palinspastischen Korrelation und Ansätze zu einer Kinematischen Analyse, *Eclogae Geol. Helv.*, v. 62 (1), p. 105–142.

Valle, P. E., 1951, Sulla struttura della crosta terrestre nel Mediterraneo centro-occidentale e nell'Adriatico, *Ann. Geofis.*, v. 4, p. 399–409.

Van der Voo, R., 1969, Paleomagnetic evidence for the rotation of the Iberian peninsula, *Tectonophysics*, v. 7, p. 5–56.

Vardabasso, S., 1952, Analogie geologiche tra la Sardegna e l'Africa Minore, *Rend. Accad. Naz. Lincei*, (8), v. 14, 1953, p. 178–183.

Vecchia, O., 1954, Lineamenti geofisici e geologia profonda nella Sicilia ed aree circostanti, *Riv. Geofis. Appl.*, v. 15, p. 15–46.

Vecchia, O., 1956, La Sicilia e le aree circostanti: lineamenti geofisici e geologia profonda, *Boll. Soc. Geol. Ital.*, v. 75, p. 61–87.

Vecchia, O., 1963, The gravity field of Italy: a geotectonic interprétation, *Boll. Geod. Sci. Affini*, v. 22 (3), p. 19.

Vercellino, J., 1970, Here's what's known about the geology of the Italian Adriatic, *Oil Gas Int.*, v. 10 (11), p. 70–78.

Viola, C., 1891, Appunti geologici sulla regione miocenica di Stigliano (Basilicata), *Boll. R. Com. Geol. Ital.*, v. 22, p. 85–98.

Viola, C., 1892, Nota preliminare sulla regione dei gabbri e delle serpentine nell'alta valle del Sinni in Basilicata, *Boll. R. Com. Geol. Ital.*, v. 23, p. 105–125.

Vogt, P. R., and Higgs, R. H., 1969, An aeromagnetic survey of the eastern Mediterranean Sea and its interpretation, *Earth Planet. Sci. Lett.*, v. 5 (7), p. 439–448.

Vogt, P. R., Higgs, R. H., and Johnson, G. L., 1971, Hypotheses of the origin of the Mediterranean basin: magnetic data, *J. Geophys. Res.*, v. 76 (14), p. 3207–3228.

Wezel, F. C., 1966, La sezione tipo del flysch numidico: stratigrafia preliminare della parte sottostante al complesso Panormide (Membro di Portella Colla), *Atti Accad. Gioenia Sci. Nat. Catania*, (6), v. 18, p. 71–92.

Wezel, F. C., 1967, Lineamenti sedimentologici del flysch numidico della Sicilia Nord-orientale, *Mem. Ist. Geol. Min. Univ. Padova*, v. 26, p. 29.

Wezel, F. C., 1968, Osservazioni sui sedimenti dell'Oligocene–Miocene inferiore della Tunisia settentrionale, *Mem. Soc. Geol. Ital.*, v. 7 (4), p. 417–439.

Wezel, F. C., 1970a, Geologia del flysch numidico della Sicilia nord-orientale, *Mem. Soc. Geol. Ital.*, v. 9, p. 225–280.

Wezel, F. C., 1970b, Numidian flysch: an Oligocene–early Miocene continental rise deposit off the African Platform, *Nature*, v. 228, p. 275–276.

Wezel, F. C., 1973, Diacronismo degli eventi geologici oligo-miocenici nelle Maghrebidi, *Riv. Min. Sic.*, no. 142–144, p. 219–232.

Wezel, F. C., 1974, Flysch successions and the tectonic evolution of Sicily during the Oligocene and early Miocene, in: *Geology of Italy*, Squyres, C., ed., Tripoli: Petrol. Expl. Soc. Libya, p. 1–23.

Zarudski, E. F. K., and Phillips, J. D., 1969, Geophysical study of the Ionian Sea, *Symp. Geol. Libya* (1971), p. 301–305.

Chapter 4B

THE STRUCTURE OF THE IONIAN SEA, SICILY, AND CALABRIA–LUCANIA

C. Grandjacquet and G. Mascle

Département de Géologie Structurale
Université Pierre et Marie Curie
Paris, France

PART I

THE STRUCTURE OF THE IONIAN SEA AND SICILY

I. INTRODUCTION

A. Limits

The Ionian Sea forms the westernmost part of the eastern Mediterranean basin. The eastern limit of the sea lies along the western margin of the East Mediterranean Ridge, which ends at about the longitude of Syrte. To the northeast it washes the shores of Greece, but geologically the Hellenic Trough (or Matapan Trench) forms part of the Hellenic Arc, and thus does not belong to the Ionian Sea. To the north the Ionian Sea is linked to the Adriatic by the relatively narrow and shallow (65 km wide, less than 100 m deep) straits of Otranto. The narrow Messina Straits (3.5 km) provide a link with the Tyrrhenian Sea and form the northwestern shore of the Ionian Sea. The Ionian Sea is

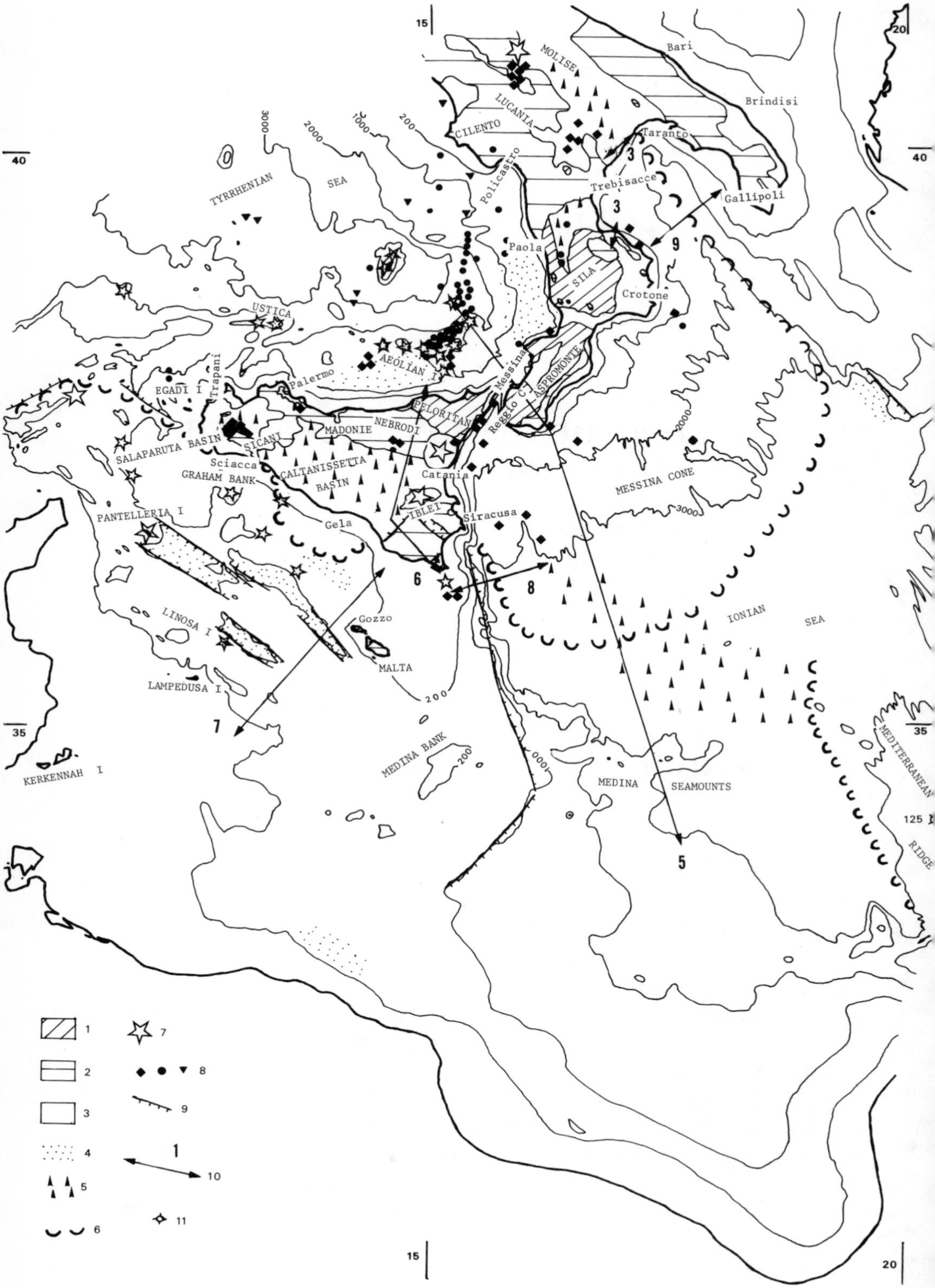

TYRRHENIAN SEA
MOLISE
LUCANIA
CILENTO
Policastro
Bari
Brindisi
Taranto
Trebisacce
Gallipoli
Paola
SILA
Crotone
USTICA
AEOLIAN I
Palermo
Messina
EGADI I
Trapani
PELORITANI
Reggio C.
ASPROMONTE
MADONIE
NEBRODI
SALAPARUTA BASIN
SICANI
Sciacca
GRAHAM BANK
CALTANISSETTA BASIN
Catania
MESSINA CONE
IBLEI
Siracusa
PANTELLERIA I
Gela
IONIAN SEA
LINOSA I
Gozzo
MALTA
LAMPEDUSA I
MEDINA BANK
MEDINA SEAMOUNTS
KERKENNAH I
MEDITERRANEAN RIDGE
200
1000
2000
3000
15
20
40
35
125
1
2
3
4
5
6
7
8
9
10
11

in open communication to the west with a region sometimes regarded as its extension and sometimes distinguished as the Pelagian Basin. The latter connects with the western Mediterranean through the Sicilian Channel, which, though 140 km wide, seldom attains a depth in excess of 300 m. To the south the Ionian Sea ends against the Libyan shore.

B. Bathymetry

The Ionian Sea (*s.s.*) is dominated by a roughly triangular abyssal plain, the center of which contains the deepest part of the basin (-4103 m), except for the Matapan Trench, which should be considered as part of the Hellenic Arc. To the south the depth decreases slowly and regularly along the African continental slope, interrupted only by a few shoals such as the Medine Mountains (Fig. 1). The western limit with the Pelagian Basin is marked by a series of scarps extending up to the coast of Sicily (Fig. 1). The Calabrian coast to the north is separated from the abyssal plain by the Messina Cone. This is a region of complex topography (Fig. 1), and a zone of sediment accumulation. The Pelagian Basin is shallow and characterized by narrow grabens trending northwest–southeast (Linosa, Pantelleria, and Malta grabens). The grabens appear at the center of the channel separating Sicily and the islands to the south and west, from Tunisia. Several banks are well developed on the continental shelf off southern Sicily (Iblean–Maltese Zone), western Sicily (Graham and Aventure banks), and east of Tunis (Kerkennah and Pelagian Islands).

C. Recent Sedimentation

Recent sedimentation is fairly well known as a result of numerous cores (Emelyanov, 1972) and seismic profiles, which indicate significant thicknesses in only a few spots (Fig. 1; Biju-Duval *et al.*, 1974). The most important of these are the grabens in the Sicilian Channel, a zone in the center of the Syrte Basin and the extension of the Matapan Trench, which trap sediments con-

---

Fig. 1. Map of the Ionian Sea. For the marine region, use has been made of the results of Finetti and Morelli (1971); Defense Mapping Agency, Oceanographic Center (1972); Mulder (1973); and Biju Duval *et al.* (1974). For the land areas, reference was made to the maps of the Servizio Geologico d'Italia, 1:1,000,000, and Sicily, 1:500,000. (1) Paleozoic; (2) Mesozoic–Tertiary (up to Burdigalian); (3) Neogene (post-Burdigalian) to Recent; (4) zone of heavy sedimentation; (5) zones where Messinian evaporites are present; (6) marine limit of allochthon; (7) recent volcanism; (8) earthquake epicenters—shallow, intermediate, and deep; (9) faults and fault zones (downthrow side marked); (10) line of sections in following figures; (11) deep-sea drilling site.

sisting of detrital materials, fine sands, and clays with which volcanic ash is mixed in the Sicilian Channel. Other zones of appreciable sedimentary thicknesses include the Messinian Cone and the marine continuation of the Salaparuta and Caltanissetta Basins. In seismic profiles they appear as highly disturbed zones where reflectors are hard to trace. Mulder (1973), and Biju-Duval et al. (1974) consider flows of allochthonous material to be responsible for these reflecting horizons.

Over the rest of the region where the recent sediment cover is thin, a twofold division is possible. One is the Ionian abyssal plain, characterized by fine detrital and pelagic deposits (Chamley, 1971). The other is the Pelagian platform and the shoal areas of the Sicilian Channel, covered essentially by carbonate muds (Blanc, 1968). Large, possibly reworked glauconite nodules are found off Marsala (Peronne, 1967).

II. SICILY–SOUTHERN CALABRIA AND THE IONIAN SEA FROM MIDDLE PLIOCENE TO QUATERNARY TIMES

There is only a small amount of information on marine horizons of Middle Pliocene and Quaternary age, although two DSDP holes have been drilled. Neither Hole 126, on the northern foot of the East Mediterranean Ridge, and thus within the domain of the Hellenic Arc, nor Hole 125 (see Fig. 1) at the end of that ridge, is characteristic of the Ionian Sea. This is readily demonstrable in the seismic section from the Malta escarpment to Site 125, (DSDP Site 125, Fig. 4, in Ryan et al., 1973). At Site 125 the Upper Pliocene–Quaternary consists of about 60 m of nannoplanktonic clays with several sapropelic horizons and a level of tuff near the top. The first 17 m are relatively poor in foraminifera, and thus less calcareous than the remainder of the section. In contrast, the published seismic profiles (Biju-Duval et al., 1973; Mulder, 1973; Finetti and Morelli, 1971; Renard et al., 1973; Ryan et al., 1973) show a much greater thickness of Plio-Quaternary on the Ionian abyssal plain and in the south Sicilian Basin. Since the Ionian Basin is only the marine continuation of the Molise, and the South Sicilian Basin that of the Caltanissetta Basin, the study of the terrestrial basins can give some indication of the probable sediment distribution in the deep basins.

Younger Quaternary (Tyrrhenian) deposits are found along the Sicilian coast, where they consist of yellow, calcareous coastal sands (Panchina facies), which are sometimes fossiliferous (Strombus). Older Quaternary (Sicilian and Calabrian) and Middle–Upper Pliocene beds consist essentially of argillaceous–arenaceous beds whose distribution shows the progressive filling of the basin. In consequence, in the Caltanissetta Basin the following sequence of events can be seen. In the north, in the region of Enna–Caltanissetta–Calascibetta, the sequence begins with bluish or gray Middle Pliocene marls which give way upward to more sandy marls with sand intercalations. The sequence ends with a band of argillaceous–calcareous sandstone rich in Upper Pliocene shell frag-ments. The Calabrian is marked by thin discordant calcareous sand. Further to the south in the region of Agrigento, the Middle–Upper Pliocene bluish marls pass without break into sandy marls followed by calcareous sandstones of Calabrian age. The latter contain two to four sand horizons separated by sandy marls. In the Salaparuta Basin the same phenomenon continues into the Sicilian, which rests discordantly on the Calabrian and begins with a thick

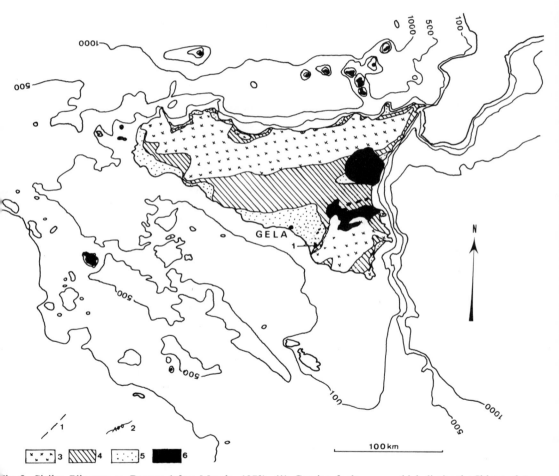

Fig. 2. Sicily: Pliocene to Recent (after Mascle, 1973). (1) Comiso fault zone, which limits the Iblean plateau to the NW; (2) isobaths (fathoms) after U.S. Oceanographic Office (1967); (3) emergent land in Pliocene; (4) regions emerging during the Pliocene; (5) regions emerging during the Quaternary (3, 4, 5 after Trevisan, 1955); (6) recent volcanism.

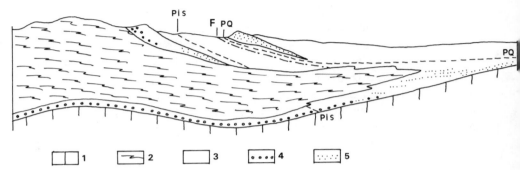

Fig. 3. Section of the region Montalbano Ionico (after Mostardini *et al.*, 1963). For location, see Fig. 1. (1) Autochthonous platform series; (2) allochthonous masses; (3) Pliocene and Quaternary; (4) conglomerate, transgressive over the platform; (5) zones rich in detrital material; F = limit of allochthons; Pis = limit of Lower Pliocene–Upper Pliocene; PQ = Pliocene–Quaternary boundary.

marly sequence overlain by highly fossiliferous argillaceous–calcareous sand-stones (Trevisan and Di Napoli, 1938; Aruta *et al.*, 1973), succeeded in turn by renewed marl deposition.

Within the Plio-Quaternary sequence, locally allochthonous masses are found. These olistostromes consist of older material such as the Lower Pliocene chalks, Messinian gypsum and marl, Middle Miocene marl sands and sand-stones, and even older flysch horizons. Olistostromes have also been found in wells (Beneo, 1950) and in some places are visible in the field (Regione Siciliana, 1962–64) in rocks as young as Quaternary (Bianchini and Mascle, 1969). The evolution of the Caltanissetta Basin can thus be summarized as characterized by sedimentary filling simultaneous with uplift of one margin (in this case, the northern margin) (Fig. 2). As a consequence, the shallow-water coastal facies becomes older as the rising margin is approached. This situation is a variant of progradation, and it is reasonable to assume that the part still submerged was subjected to similar sedimentation.

In the Calabrian Basin the same type of evolution has been described by Mostardini *et al.* (1966), who were able to follow progressive emplacement of allochthonous material in interstratified beds which become progressively younger eastward. At the same time, younger and younger beds are incorporated in the allochthonous element (Fig. 3). There is thus not just a sedimentary progradation but also a progradation of the allochthon. This prevents a precise determination of the age of emplacement of the allochthonous masses found in the Gulf of Taranto (see Fig. 9), (Finetti and Morelli, 1972), in the Ionian Sea (Figs, 1, 5, and 8) (Mulder, 1973; Biju-Duval *et al.*, 1974), and south of Sicily. It is probable that they are younger there when in more "external" zones.

III. THE STRUCTURE OF SICILY AND SOUTHERN CALABRIA

A. The Sicilian Foredeep from Helvetian–Tortonian to Middle Pliocene

The only information on the nature of sediments of this age comes from DSDP Hole 125, which terminated 121 m below the sea bed in Messinian evaporites (Ryan *et al.*, 1973), and from several lucky dredge hauls (Bobier, personal communication). Seismic profiles suggest the existence of evaporites or allochthonous elements at several points.

In both the Caltanissetta and Salaparuta Basins of Sicily, the tectonic phase responsible for the emplacement of the principal structural units occurred a little below the Helvetian–Tortonian boundary. More recent phases may have severely deformed the structures, but their effects are felt in more restricted areas. The corresponding phase in Calabria occurs within the Burdigalian–Helvetian. Thus, the sedimentary slice described below does not everywhere cover the same time interval, a fact which must be borne in mind in interpreting marine data.

The stratigraphic sequence in the Sicilian basins begins with marls, often sandy but always rich in detrital mica and containing rare coral reefs. The thickness of the succession varies from tens to thousands of meters. To the north conglomerates, about 100 m thick with boulders of several cubic meters, may be intercalated. The conglomerates contain not only boulders of Cretaceous and Eocene flysch, and carbonates, but also of basement gneiss and granite. Finally, at various horizons within the marls (see Ogniben, 1955; Marchetti, 1957; Flores, 1959; Regione Siciliana, 1963–64), there are important allochthonous masses of flysch (olistostromes) and limestone (sedimentary klippe) (Broquet *et al.*, 1966; Mascle, 1973).

Resting on the marls with slight unconformity are Messinian beds. They form sedimentary cycles with a succession passing from sand, marl, tripoli (diatomite), evaporitic limestones, finely banded gypsum, crystalline gypsum, to anhydrite, and finally to halite. In general the cycles are incomplete and the number present varies from place to place, but usually gypsum is present (Heimann and Mascle, 1974). This sequence, in turn, is overlain with slight unconformity by Lower Pliocene, white, calcareous marls (chalk) (locally known as "trubi"). In the trubi is a rich planktonic foraminiferal assemblage, locally interstratified with coastal or reefal deposits (Trapani, Sciacca, Lascari, Pachino, Peloritan). In both the Messinian and Lower Pliocene (Ogniben, 1954; Flores, 1959; Mascle, 1973), evidence of the persistence of tectonic activity is seen in the existence of allochthonous materials at various horizons.

The Molise Basin had a similar evolution (Ogniben, 1969), so it seems likely that the submarine parts of the Sicilian and Molise Basins underwent the same type of evolution. In particular, one part of the allochthonous (Figs.

Fig. 4. The structural units of Sicily. (1) Peloritan domain: (a) basement and Oligo-Miocene cover, (b) Limestone Chain; (2) Flysch units: (a) Argille Scagliose, (b) Reitano, (c) Monte Soro, (d) Intermediate Numidian; (3) Panormid units: (a) Monte Acci, (b) Panormid s.s., (c) Erice; (4) Sclafani nappe: (a) Sagana unit, (b) Mesozoic–Eocene of Sclafani unit s.s., (c) external Numidian; (5) External units: (a) Vicari, (b) Campofiorito, (c) Sciacca; (6) Iblean platform: (a) western facies, (b) eastern facies; (7) post-Burdigalian Neogene and older Quaternary; (8) recent volcanism; (9) recent Quaternary; (10) line of sections (Fig. 13).

1, 3, 5, 6, and 7) material of the Messina Cone must have been emplaced between the Tortonian and Middle Pliocene.

Adjoining the basin sequences in the Iblean domain during the Upper Miocene (Helvetian–Tortonian), there was a marno-calcareous sequence with shell fragments (Palazzolo Formation) (Rigo and Barbieri, 1959), passing laterally by interstratification into marls (Tellaro Formation). The first signs of Iblean volcanism appear in the Middle Tortonian. The Messinian with its several evaporitic cycles is known only along the margin of the domain, and the same is true of Lower Pliocene, although at this time the Monte Lauro

NW
SE | NNW
SSE

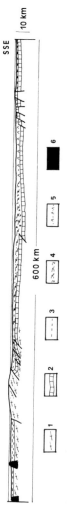

Fig. 5. Section of the Ionian Sea, redrawn with no vertical exaggeration (after Mulder, 1973). For location, see Fig. 1. (1) Paleozoic; (2) Mesozoic; (3) Tertiary; (4) allochthonous masses; (5) zone of Messinian evaporites; (6) recent volcanism.

10 km

600 km

| 1 | 2 | 3 | 4 | 5 | 6 |

NNE
SSW | NW
SE

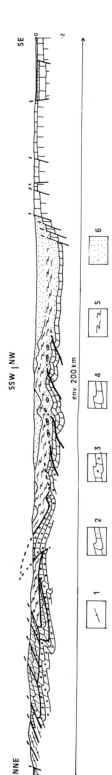

env. 200 km

| 1 | 2 | 3 | 4 | 5 | 6 |

Fig. 6. Simplified cross section of western Sicily (cf. Fig. 5). For location, see Fig. 1. (1) Paleozoic basement with unconformable Oligo-Miocene cover; (2) Mesozoic–Burdigalian (Monte Soro and Intermediate Numidian nappes); (3) Mesozoic–Burdigalian (Panormide and Sclafani nappes); (4) Permian–Burdigalian (external units and Iblean platform) or Permian–Ypresian (Peloritain domain); (5) Mesozoic–Burdigalian (internal flysch: Reitano and Argille Scagliose nappes, and resedimented bodies); (6) Neogene, post-Burdigalian to Quaternary.

SW
FOSSE DE MALTE   HORST DE MALTE
NE

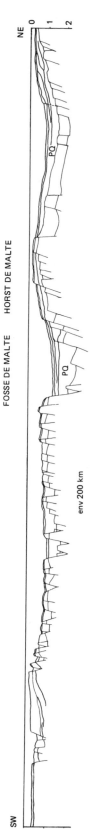

env 200 km

Fig. 7. Seismic section across the Sicilian Channel (modified after Finetti and Morelli, 1971). For location, see Fig. 1. PQ = Plio-Quaternary.

volcanic massif developed to the north. This indicates that from the Tortonian onward the Iblean domain showed the characteristics of a platform.

B. The External Sicilian Domains (Figs. 4 and 10)

In Sicily a series of domains (Figs. 4, 5, and 6), characterized by structural style, position, the age of tectonic activity affecting them, and the stratigraphic succession, can be recognized. The domains form three broad groups—the external ones underlying the flysch nappes, the flysch nappes themselves, and the Peloritan domain.

The Iblean domain is the most external. The structure, well known from exploratory holes, consists of broad, open folds and normal faults. The principal structural trends are NE–SW, NNE–SSW, and WNW–ESE. The structures have been active from Tortonian to Recent times (Mascle, 1974), as is shown in Malta where normal faults trending WNW–ESE and NNE–SSW affect a recent erosion surface.

Seismic sections in the Sicilian Channel or in the Ionian Sea (Figs. 7 and 8) show clearly that these domains are cut by numerous normal faults, some of which are shown in Fig. 1. Some have a major throw, and delimit either the grabens of the Sicilian Channel or the edge of the Malta platform. Some clearly affect Recent series. Excluding the NNW–SSE-oriented faults which limit the Ibleo-Malta plateau to the east, the faults are oriented WNW–ESE to NE–SW, the same trends as those affecting Malta and the Iblean domain.

The rest of the external Sicilian zones present tangential structures from the exterior to the interior (Broquet *et al.*, 1966) in the Sciacca, Campofiorito, Vicari, Sclafani, and Panormide (Fabiani and Trevisan, 1940) Zones. In the Sciacca Zone, thrusts and reverse faults affect the Lower Pliocene. The Campofiorito Zone, in the form of some thrust sheets, is thrust over the preceding and has several generations of structures between the Helvetian and the Middle

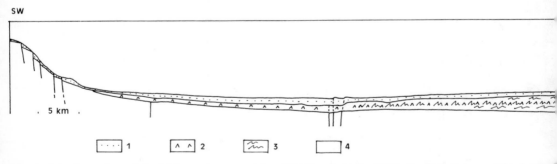

Fig. 8. Seismic section of the Malta escarpment (after Biju-Duval *et al.*, 1974). For location, see Fig. 1. (1) Quaternary; (2) evaporites; (3) allochthonous masses; (4) basement.

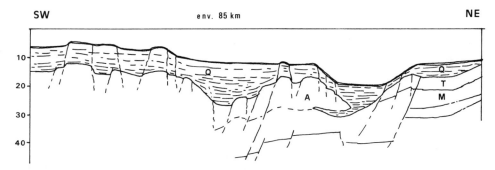

Fig. 9. Seismic section across the Gulf of Taranto (modified after Finetti and Morelli, 1971). For location, see Fig. 1. Q = Quaternary; T = Tertiary; M = Mesozoic; A = allochthonous units.

Pliocene. The Vicari Zone is made of elements of the preceding, reworked during the Helvetian–Tortonian. The Sclafani Nappe overthrusts the Helvetian–Tortonian, and some of the elements are resedimented. The Panormide units, thrust over the Sclafani Zone in the Oligocene–Aquitanian, were also involved in the Helvetian–Tortonian structures affecting the Sclafani Zone. All these deformations occurred above the schistosity front. Fracture cleavage occurs only very locally and is always related to important tectonic fractures.

1. *The Iblean Domain (Figs. 4, 6, and 10)*

Rocks older than Oligo-Miocene are found at only two places in the eastern Iblean domain (Allisson, 1955; Rigo and Barbieri, 1959; Regione Siciliana, 1961)—at Pachino in the extreme south of Sicily and at Priolo near Syracuse. Older beds are known only in wells where, in the extreme south, they comprise a thick sequence of marls, shales, and limestones of Upper Triassic age. Elsewhere, wells bottomed in dolomites or algal limestones assigned to the Middle Triassic and Jurassic. The Lower Cretaceous is represented by shales and cherty limestone and is overlain by the cherty limestones and basic lavas (500 m) of the Upper Cretaceous, which crop out at Pachino and Priolo. These beds are followed by rudist limestones and at Pachino by Lower and Middle Eocene foraminiferal bioclastic limestones.

In the western Iblean domain (Rigo and Barbieri, 1959; Regione Siciliana, 1961), beds older than Oligo-Miocene are found only at Licodia Eubea and Monterosso Almo, so that most of our present knowledge is derived from well information, and in particular from the Ragusa oilfield. The oldest rocks, the reservoir rocks, are Triassic dolomites (probably Middle Triassic), overlain by an alternation of shales, marls, limestones, and microbreccias (calcareous flysch), with intercalations of basic rocks of Upper Triassic (Carnian) age. The

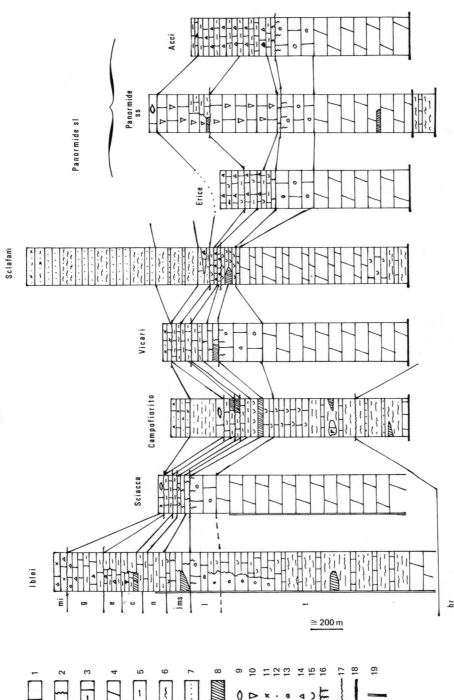

Fig. 10. Lithostratigraphic profiles of the external domains, Sicily. (1) Limestones; (2) nodular limestones (ammonitico rosso); (3) argillaceous limestones (Scaglia); (4) dolomite; (5) marl; (6) shales, pelites; (7) sandstone; (8) hyaloclastites and pillow lavas; (9) macroforaminifera; (10) rudists; (11) glauconite; (12) detrital quartz; (13) oolites, oncholites, pellets; (14) calcarenites, limestone turbidites; (15) chert, radiolarite; (16) neptunian dikes, Mn crusts and nodules; (17) discordance; (18) tectonic contact; (19) sequence known only in wells. Abbreviations: c = Upper Cretaceous; e = Eocene; g = Oligo-Miocene; hr = Carboniferous–Permian; jms = Middle Upper Jurassic; l = Lias; mi = Lower Miocene (pre-

sequence continues with finely banded limestones, sometimes cherty, with shale partings, and with an Upper Triassic and Liassic pelagic fauna. In the uppermost Liassic and the Middle–Upper Jurassic, there is a thin sequence of red, nodular limestones sometimes with manganese coatings. Tuffs and occasional basaltic flows may occur at this level. The ends of the Jurassic and the Lower Cretaceous are marked by the deposition of fine-grained, occasionally cherty limestones first bearing calpionellids and radiolaria, then radiolaria only, ending with marls. The rocks are exposed at Licodia Eubea. The Middle–Upper Cretaceous is partially exposed at Monterosso Almo and is made up of cherty, finely banded limestones and marls (Scaglia facies) comparable with modern foraminiferal nanno-oozes, overlain by rocks of the Ragusa Formation which crop out over a wide area. The formation begins with a calcareous conglomerate followed by alternations of fine grained, occasionally siliceous limestones and yellowish marly calcarenites of Oligocene (and perhaps Middle Eocene) age. The Lower Miocene comprises a sequence of thick-bedded, marly calcarenites and limestones, sometimes containing phosphatic debris (Modica) and bituminous limestones (Ragusa).

The structures (Fig. 6) found in the Iblean domain are young. The Pliocene is downthrown by a normal fault east of Ispica. The Comiso fault system also strongly affects the Pliocene, and part of the displacement was probably of Quaternary age. Wells in the Ragusa and Gela fields, lying respectively east and west of this fault system show that not only are the beds displaced but from the Tortonian onward there are marked thickness variations. It was previously indicated that Lower Miocene sedimentation occurred in a region showing deformations with a large radius of curvature and only rare reverted faults.

2. The Sciacca Domain (Broquet et al., 1966; Mascle, 1970, 1973; Figs. 4, 10, and 13).

The base of the sequence in this domain consists of thick-bedded dolomite with algal remains, assigned to the Upper Triassic. More than 3000 m of such beds have been traversed in wells. They are overlain by thick-bedded, Liassic, agal limestones with Megalodont molds. Locally, crinoidal Middle Liassic limestones have been recorded. The Upper Liassic and Middle–Upper Jurassic sometimes form a condensed succession a few meters thick, lying discordantly over an eroded surface. Red, nodular limestones (ammonitico rosso), ferruginous beds, manganese crusts and nodules, and occasional trachyandesitic flows are found. The overlying cherty, argillaceous limestones are of Upper Jurassic and Lower Cretaceous age. They are followed by Upper Cretaceous sediments sometimes preserved in joints in earlier beds, and by Eocene material represented by fine-grained argillaceous limestones (Scaglia facies). The Oligo-

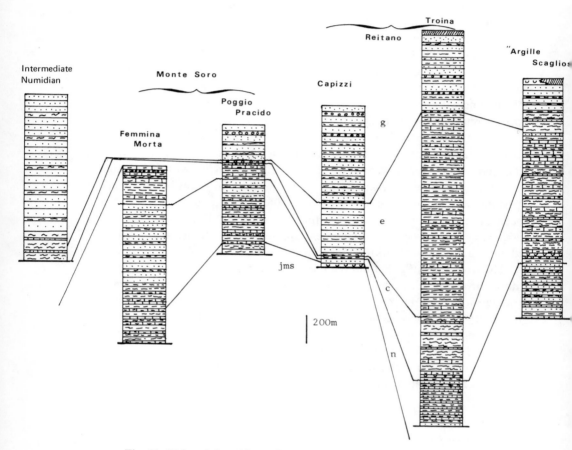

Fig. 11. Lithostratigraphic profiles of the flysch units. Legend as in Fig. 10.

cene consists of Nummulitic limestones also resting discordantly upon the Jurassic. In the Lower Miocene calcarenites with glauconite and phosphatic debris are found.

Deformation in the Sciacca Zone occurred in several stages, of which the most recent was normal faulting, which affects the older Quaternary (Calabrian) (Mascle, 1973, 1974). The principal structures were more formed during the Pliocene, with thrust and reverse faults affecting Lower Pliocene beds upon which the Middle–Upper Pliocene rests discordantly. The Lower Pliocene, in turn, rests discordantly upon the Messinian, evidence of further movements

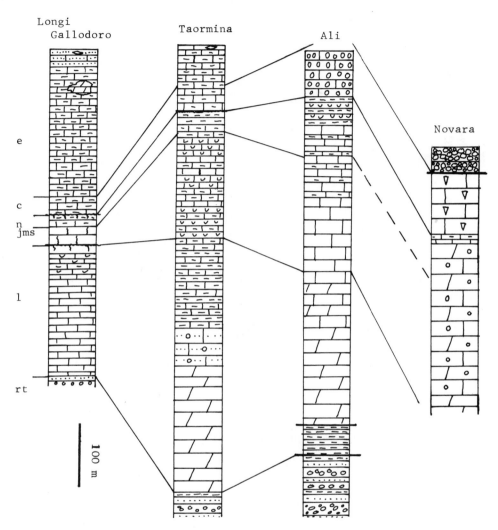

Fig. 12. Lithostratigraphic profiles of the Limestone Chain units. Legend as in Fig. 10.

during the Messinian and in particular near the Helvetian–Tortonian boundary. Deformation at this time resulted in resedimentation in the Miocene basin of some parts of the Sciacca Zone, while at the same time the internal part of the zone was being overthrust by units from the Campofiorito Zone. Prior to that there are traces of early deformation, the style of which is still not clear but which is evidenced by the Lower Miocene, Oligocene, and Upper Cretaceous unconformities. The Upper Liassic–Dogger discordance fixed the date of structures due to a tensional (distension) phase.

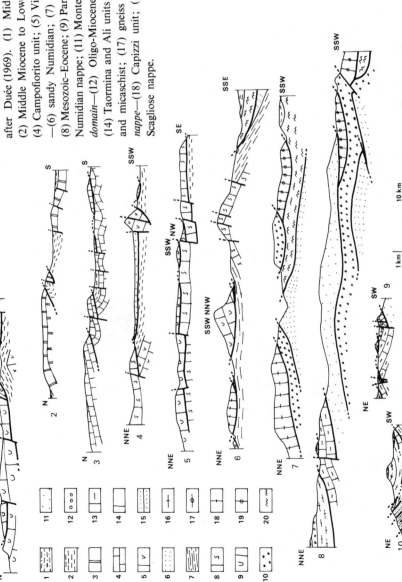

Fig. 13. Structure of Sicily from sections. (Locations shown in Fig. 4.) Sections 2, 3, 4 after Mascle (1970); 5 after Caflisch (1968); 6 after Broquet *et al.* (1966); 7, 8, 9, 10 after Duée (1969). (1) Middle Pliocene to Quaternary; (2) Middle Miocene to Lower Pliocene; (3) Sciacca unit; (4) Campofiorito unit; (5) Vicari unit; (6–8) *Sclafani nappe*—(6) sandy Numidian; (7) argillaceous-pelitic Oligocene; (8) Mesozoic–Eocene; (9) Panormid units; (10) Intermediate Numidian nappe; (11) Monte Soro nappe; (12–17) *Peloritan domain*—(12) Oligo-Miocene molasse; (13) Longi unit; (14) Taormina and Ali units; (15) Verrucano; (16) phyllite and micaschist; (17) gneiss and granite; (18–19) *Reitano nappe*—(18) Capizzi unit; (19) Troina unit; (20) Argille Scagliose nappe.

3. *Campofiorito Domain* (*Broquet et al., 1966; Mascle, 1970, 1973; Figs. 4, 10, and 13*)

The oldest unmetamorphosed rocks known in Sicily belong to this domain. In outcrop they appear as a flysch of varicolored shales with beds of green quartzite, which become sandy toward the top (Fabiani and Trevisan, 1937). Occasionally, microbreccias with Lower Permian fusulinids, blocks of Productid limestone, and diabase are found. As is clearly seen in the field and in drill cores, this sequence is broken up into a series of superposed tectonic slices, the lowest containing at their base limestones assigned to the Carboniferous (Caflisch and Schmidt di Friedberg, 1967). Carnian alternations of variegated shales and finely banded limestone turbidites with rare horizons of sandstone and limestone microbreccias (Carnian flysch) lie discordantly upon the Permian (Fabiani and Trevisan, 1937). They contain blocks of highly fossiliferous Permian reef limestone (the well-known Sosio Permian) and a few fragments of Triassic rocks. The succession continues with further finely banded calcareous turbidites containing *Halobia*. Sometimes rich in chert, sometimes dolomitized, these rocks are assigned to the Upper Carnian–Norian. The Liassic consists of finely banded limestones with brachiopods and radiolaria, followed by marls. In both lithologies, turbidites, and conglomerates of Liassic reef limestones, probably derived from the Sciacca Zone, may occur. The series of tuffs and basaltic pillow-lavas are assigned to the Dogger, and the silicified limestones (false radiolarites) to the Upper Jurassic. The uppermost Jurassic and Lower Cretaceous are represented by more or less siliceous, light-colored, ammonitic, nodular limestones. These are overlain by Neocomian blue marls. Part of the sequence is commonly absent, reflecting Cretaceous movements. The Upper Cretaceous and Eocene successions are incomplete and variable, although homogeneous from a facies standpoint. The section consists of well bedded, fine-grained, cherty argillaceous limestones (Scaglia facies) with intercalated basalts. The first discordance, not often seen, occurs below Upper Albian–Cenomanian, and a second in Upper Campanian–Maestrichtian (or intra-Maestrichtian), where it is sometimes indicated by a conglomerate. The Oligocene begins above with basal uncomformity, with a limestone conglomerate, succeeded by argillaceous limestones and sandy marls, the latter passing up into Lower Miocene (Aquitanian–Burdigalian) glauconitic, calcareous sandstones. The Helvetian consists of marls.

Once again, this domain was subjected to several phases of deformation. The Pliocene and Calabrian found along the southern edge of the domain are folded and faulted (normal faults), and rest with unconformity upon structures in which the Trias is thrust over Lower Pliocene. The Messinian phase of movement is marked by the resedimentation of slices of the Jurassic–Oligocene cover. Then, during the tectonic movements of the Helvetian–Tortonian, slices of the Campofiorito domain were thrust over each other and some were transported as far as the Sciacca Zone. Prior to these movements the Oligocene, Upper Cretaceous, and Upper Triassic unconformities represent evidence of tectonism whose style is still uncertain.

4. *Vicari Domain (Broquet et al. 1966; Mascle 1970, 1973; Figs. 4, 10, and 13).*

A thick series of reef dolomites with megalodonts and solenopores of Middle Triassic age are overlain by Lower–Middle Liassic, white limestones (pelsparites, oosparites, biosparites with algae and crinoids). The Upper Liassic to Upper Jurassic is present in the ammonitico rosso facies of red nodular limestones with manganese crusts and nodules and locally, pillow lavas and basic tuffs (Alcamo trachy-andesites, Roccapalumba, and Vicari basalts) resting unconformably upon the preceding. These rocks seal a paleo-karst topography cut by normal faults and form a remarkable series of sedimentary dikes. The topmost Jurassic–Lower Cretaceous consists of fine-grained, sometimes siliceous limestones with calpionellids (Maiolica facies), then radiolarites and *Nannoconus* (a nanno-ooze) capped by marls. The Upper Cretaceous is unconformable and covers nappes produced toward the end of the Albian–Cenomanian. Together, the Upper Cretaceous and Eocene are represented by a thin Scaglia facies. Commonly, the Oligocene is absent with Lower Miocene glauconite and sandy calcarenites discordant on beds as old as Liassic.

Structurally, elements of the Vicari domain are resedimented in the Miocene basin, often as elements of appreciable size (12 km × 2 km × 2 km). However, some northwest Sicilian massifs, present as southerly directed nappes, may be rooted. Whatever their condition, they were emplaced by tectonic movements occurring toward the Helvetian–Tortonian boundary. Earlier phases of movement occurred in the Lower Miocene, Upper Cretaceous, and Middle Jurassic. The Middle Cretaceous phase of activity raises the question of whether the Vicari domain existed prior to this time, for there are very great similarities between the Triassic to Lower Cretaceous successions in the Vicari and Sciacca domains during the Upper Cretaceous.

5. *Sclafani Domain (Schmidt et al., 1960; Ogniben 1960; Broquet et al., 1966; Broquet, 1968; Figs. 4, 10, and 13)*

At the base a series of variegated- shales and calcareous turbidites, rare bands of carbonate-cemented sandstone and masses of diabase can be observed. They can be dated as Carnian from the presence of *Halobia* and *Trachyceras aon*, and are followed by a succession of gray, finely bedded limestones with halobias and radiolaria in which calcareous turbidites with crinoid debris occur. Above the limestones lies a thick sequence of rhythmically alternating banded, cherty dolomite and detrital dolomites (dolarenites) of Upper Triassic–Lower Liassic age. The Middle–Upper Lias consists of finely banded siliceous limestones passing upward into radiolarites, radiolarian breccias, and "siliceous shales" rich in manganese and locally intercalated diabases, which are regarded as uppermost Lias and Upper–Middle Jurassic. The uppermost Jurassic and basal Cretaceous, which rest discordantly on older beds, comprise limestone turbidites formed of reef material (*Ellipsactinia*, clypeines). Marls in the upper part of the succession represent the Lower–Middle Cretaceous. The latter lithology includes limestone turbidites formed of reef materials (rudists), although in the vicinity of Termini Imerese (Rangin, 1974) rudists are found in place. The Upper Cretaceous and Paleocene, where preserved, also contains limestone turbidites within a dominantly marly limestone (Scaglia). Most of the Eocene is absent. The Upper Lutetian is normally discordant and consists of red, argillaceous, limestone turbidites with *Nummulites*. The Middle Eocene and Oligocene are represented by a relatively thick (700–1000 m) sequence of argillites and pelites with thin quartzite bands. There then follows a thick (more than 2 km) succession of Numidian flysch (external Numidian) consisting of alternations of pelites, quartzites, and masses of coarse sandstone (quartz arenites and quartz rudites), ending with marly, glauconitic Burdigalian sandstones. Between the Oligocene and the Numidian, there are sometimes intercalated large calcareous masses consisting of blocks of Panormid facies resedimented in the Numidian Basin. The Numidian is thus discordant, although no angular discordance is seen where the Panormid intercalations are absent.

The Quaternary and Pliocene movements are represented by broad, open folds and faults. The characteristic feature of the Sclafani domain is nappe emplacement during Helveto-Tortonian movements. This nappe is often differentiated, one part being Numidian–Oligocene with slivers of Eocene, the other essentially Mesozoic. Further, elements are resedimented in the Miocene foredeep, these consist mainly of Numidian, only rarely of Mesozoic, and sometimes Panormid blocks may be resedimented a second time. During the Lower Miocene sedimentary klippe of the Panormid units were emplaced. The form of pre-Eocene deformation is not known, but it did result in the erosion of all or part of the succession between Jurassic and Upper Cretaceous.

6. *Panormid Domain (Fabiani and Trevisan, 1940; Schmidt et al., 1960, Ogniben, 1960; Montanari, 1964–65; Broquet et al., 1966; Broquet, 1968; Broquet and Mascle, 1972; Figs. 4, 10, and 13)*

The succession here begins with varicolored marls and bioclastic limestone (Carnian flysch), followed by massive reef-dolomites with rare basic flows. These Upper Triassic beds are followed by Lower–Middle Liassic oolites, oncholites and bioclastic (algal) limestones. The Upper Liassic and Middle–

Upper Jurassic are present as a condensed "ammonitico rosso" sequence, with manganese nodules, crusts, and sedimentary dikes, except in the Erice unit. The remainder of the succession is diversified according to location and the unit involved, although consisting typically of reef-limestones. The uppermost Jurassic and Lower Cretaceous is represented by one thick band of limestone with rudists, *Ellipsactinia*, trocholinids, and orbitolinids, while the Upper Cretaceous is represented by a second band. In the Eocene there are limestones with alveolinids, flosculines, and *Nummulites*. The exterior Erice and interior Monte Acci units have fine-grained, argillaceous limestones (Maiolica Scaglia), with calcareous turbidites in which the reworked material is reefal and ranges from Upper Jurassic to Upper Cretaceous (or Eocene at Monte Acci). At the base of the Upper Cretaceous sometimes discordant, basaltic tuffs occur in the region of Trapani. In the Madonie Mountains there is a thick argillaceous and argillo-pelitic series overlying these horizons and assigned to the Oligocene, though it contains blocks of Eocene. In the Trapanese, the succession closes with a Lower Miocene, glauconitic conglomerate.

Panormid structures have been deformed by Quaternary and Pliocene faulting, but the essential structures were acquired during two earlier periods of thrusting, Helvetian–Tortonian and about the Oligocene–Miocene boundary. In the Trapani and Palermo Mountains, the piling of the Monte Acci, Monaco, and Erice units over the Vicari Zone or even the Sclafani nappe occurred during the Helvetian–Tortonian as a reactivation of the Madonie. In these latter, the essential structures were acquired during the Oligo-Miocene phase. The Monte Acci unit of Triassic to Eocene rocks was also folded and overturned before its emplacement during the Lower Tortonian. In the Madonie the discordance of the Numidian sandstones on the Panormid blocks suggests that these were emplaced as sedimentary klippe in the Numidian basin. Along the margin of the Panormid elements, shales with blocks (Wildflysch of Ogniben, 1964; Gratteri Formation of Broquet, 1968) occur. Drag folds at the base of the Triassic suggest that a south-to-southwestward motion took place. Earlier deformation of the Panormid domain at the base of the Upper Cretaceous is shown by the presence of Upper Cretaceous sedimentary dikes (in a paleokarst environment) in the Madonie and by an angular discordance of up to 30° at Erice. Sedimentary dikes in the Middle Jurassic are evidence of tensional activity.

C. The Sicilian Flysch (Figs. 4, 6, 11, and 13)

The flysch forms an important zone of outcrops from Etna to Termini Imerese (Fig. 3) in northeastern Sicily. It also occurs more sporadically in the Caltanissetta Basin in the south where there are resedimented elements. The structural work of Duée (1965–69) permits the recognition of four superposed units; the Intermediate Numidian, Monte Soro, Argille Scagliose, and Reitano nappes. There were at least two periods of emplacement. In these marly, argillaceous, pelite beds stratigraphic contacts are rarely preserved; there is thus a somewhat arbitrary element in the importance attributed to some tectonic contacts, and consequently in the reconstruction of the stratigraphic sequence. Contacts considered as secondary by one group may appear fundamental to another. This explains, to a large extent, the divergence in the paleogeographic interpretation of the flysch around the western Mediterranean.

1. *The Intermediate Numidian Nappe (Broquet and Duée, 1967; Broquet, 1968; Duée, 1969; Figs. 6, 11, and 13)*

   This nappe is a structural unit thrust on to the external Numidian (Sclafani) nappe, which underlies the other unit. It has a complex sole of shales and varicolored marls of Upper Cretaceous and Eocene age, and a thick mass of Numidian Sandstone. As a working hypothesis it is considered to be part of the ancient Oligo-Miocene cover of the southern units of the Monte Soro nappe.

2. *Monte Soro Nappe (Ogniben, 1960; Broquet et al., 1963; Duée, 1965, 1969; Andreieff et al., 1974; Figs. 4, 6, 11, and 13).*

   The Monte Soro nappe is divided into a number of tectonic scales. The lowermost consists of slightly sandy Kimmeridgian limestones, which are the oldest rocks known here. Portlandian and Lower Cretaceous beds consist of marly pelites with rare limestone (calpionellid and *Nannoconus*) bands or breccias (with *Aptychus*), followed by a thick argillaceous–arenaceous flysch (Monte Soro Flysch). The flysch consists of regular alternations of gray or brown pelites and fine pale green sandstones comparable with the sequence of the same age in Guerrouch (Algeria) or Tisirhen (Morocco). The probable age is Barremian–Aptian–Albian. The thickness of the series varies greatly in different structural units and is probably related to an Upper Cretaceous unconformity whose existence, given all the tectonic contacts, is hard to demonstrate. The Upper Cretaceous consists of shales and marls with limestone partings and zones of breccia, the material of which was derived from the Peloritan and Panormid domains. The thickness is never great because of the Eocene discordance.

   In the most northerly structural unit, the Middle–Upper Eocene, which consists of marls and sandstones with debris from phyllites, and conglomerates of vein quartz and phyllite, rests discordantly upon structures oriented N–S. The middle units do not contain Eocene. The external units are made up of red and green marly limestones and sandy and conglomeratic nummulitic limestones. The pebbles are of Triassic dolomite, Jurassic limestones, and radiolarites, of which part is certainly derived from the Panormids. Oligo-Miocene beds, which occur only in the most northerly unit, are micaceous sandstones and sandy and conglomerate marls (Reitano Formation). It should be recalled here that the Intermediate Numidian is regarded as representing the old Oligo-Miocene cover of the southern units.

The structure of the Monte Soro nappe is the result of several phases of deformation. In addition to the Eocene and Cretaceous phases at least two Miocene phases are discernible, the first responsible for the emplacement of the four units making up the Monte Soro nappe which overlies the Intermediate Numidian, the second due to the thrusting of the front of the Monte Soro nappe over the Reitano and Argille Scagliose nappes.

3. *The "Argille Scagliose" Nappe (Broquet et al., 1963; Duée, 1965, 1969; Figs. 4, 6, 11, and 13)*

The Argille Scagliose unit is widely found in Sicily in the sense that it was this unit which was mostly resedimented during the Mio-Plio-Quaternary. The term "Argille Scagliose" has been used in a variety of contexts in the literature on Sicily and Italy to designate horizons discussed here. In this article it designates a structural unit characterized by its structural situation and secondarily by the stratigraphic sequence. The oldest horizons consist of a pelitic–quartzitic flysch with bands of microbreccia made up of Barremian–Aptian–Albian orbitolines and identical to the Apto-Albian flysch of Algeria (also known in the Algerian Ziane units, in Morocco in the Mellousa and Djebel Chouamat units, and in Spain in the Facinas unit). The Upper Cretaceous–Paleocene is discordant and comprises varicolored radiolarites (Turonian), varicolored shales with platy, green limestone microbreccias, and rare sandstones. The Lower Eocene is represented by limestones and marly limestones with chert and some bands of microbreccia consisting of foraminiferal limestone which are sometimes silicified (Eocene calcareous flysch of Polizzi). The Middle–Upper Eocene, formed of varicolored shales, limestones, and microbreccias is followed by varicolored *Tubotomaculum* shales whose age is in dispute. The sequence is capped by the Internal Numidian nappe, followed by silicified pelites (silexites, pseudophthanites, diatomites, of various authors) which extend upward to the Burdigalian or tuffaceous sandstones.

Structurally two superposed units can sometimes be distinguished in the region of Nicosia. This nappe rests upon all the other structural units with the exception of the Reitano nappe. It has been resedimented far to the south, but can also be seen in the Peloritan domain where its emplacement took place in Burdigalian time. Prior to its emplacement in the Nicosia region it showed the effects, in terms of northward directed folds, of an important period of deformation. Subsequent to its emplacement it was overthrust by more external units (Monte Soro, Intermediate Numidian) during renewed activity.

4. *Reitano Nappe (Ogniben, 1960; Broquet et al., 1963; Duée, 1965, 1969; Figs. 4, 6, 10, and 13)*

The nappe can be divided into two major structural units, Capizzi and Troina. Both have the same Oligo-Miocene succession of pelites and micaceous sandstones with conglomerate horizons. The same molasse sequence is to be found in the Peloritan domain and it is also known in Algeria (Oligo-Miocene grèso-micacé), Morocco (Beni Ider Formation), and Spain (Algeciras Formation). In Sicily and elsewhere in peninsular Italy, there are andesitic tuff-sandstone horizons (Tusa Sandstone) near the top of the succession. In addition to the Oligo-Miocene, the Capizzi unit contains thin Middle–Upper Jurassic limestones and radiolarites, Lower Cretaceous shales with microbreccias and sandstones ("Albo-Aptian" flysch), Upper Cretaceous to Middle Eocene limestones with conglomeratic horizons, and a relatively thick sequence (400 m) of nodular, sandy, micaceous marls with conglomerate bands (pebbles of phyllite), extending from Middle Eocene (in part) to Upper Eocene. In the Troina unit the Cenomanian–Turonian radiolarites, sometimes with bituminous horizons containing fish and oysters, rest unconformably upon Albo-Aptian pelito-quartzitic flysch. These beds are followed by varicolored shales with black pelite and fine-grained limestone horizons assigned to the Upper Cretaceous–Paleocene. The Middle–Upper Eocene is a succession of thick (more than 1000 m) light-colored, sometimes sandy marls, alternating with fine, argillaceous limestones containing a pelagic microfauna. The sequence, prior to Middle Eocene, resembles the "Argille Scagliose". The equivalents of the Troina Series are found in the Chouamat and Mellousa in Morocco and the Massylian in Algeria.

Two phases of deformation, in addition to those of the Cretaceous and Eocene, can be recognized. One led to overthrusting of the other unit (Sclafani, Panormid, Intermediate Numidian, "Argille Scagliose" nappes), the second to overthrusting of the Troina in its turn by the Monte Soro nappe.

D. The Peloritan Domain and Southern Calabria (Figs. 1, 4, 5, 6, 12 and 13)

It is in the Peloritan and southern Calabrian domains that metamorphosed Paleozoic basement is exposed, allowing the recognition of basement and cover. The domains were affected by important tectonic movements between Middle and Upper Eocene. Tangential movements were first demonstrated by Caire and Truillet (1963) in Sicily and were subsequently noted in Spain (Paquet 1965) and North Africa (Lepvrier, 1967; Raoult, 1967). They were responsible for the principal structures in the Peloritan domain.

1. *The Basement* (*Truillet, 1968; Dubois and Truillet, 1969; Dubois and Truillet, 1969; Duée, 1969; Fig. 13*)

Two major, but still poorly known, groups may be distinguished in the basement, a group of epimetamorphic phyllites and a more altered sequence of micaschist, gneiss, and batholitic granites. The phyllites themselves may be grouped into a lower sericitic chlorite group rich in tourmaline, and an upper complex which has at its base andesitic and rhyolitic flows and tuffs, variegated shales, metamorphosed Lower–Middle Devonian limestones (with *Tentaculites*), and is capped by shales. This upper complex has been interpreted as an (alpine) nappe overlying a phyllite nappe. Detailed studies (Dubois and Truillet, 1969) show that there is a progressive passage of one to the other and that the whole sequence was overturned and granitized after the Devonian and prior to the Permian. The essential structures of the basement, where traces of at least three tectonic phases are discernible (two isoclinal), are thus Hercynian.

2. *The Pre-Lutetian Mesozoic–Eocene Cover* (*Fig. 12*)

Four structural units may be discerned, each with a different stratigraphic sequence—the Longi Gallodoro, Taormina, Ali, and Novara units. Together they form the Limestone Chain.

Longi–Gallodoro Unit (*Truillet, 1968; Duée, 1969; Fig. 12*). At the base of this unit the classic Permo-Triassic Verrucano facies of red sandy-pelites with occasional conglomeratic horizons is found. The Verrucano is often sheared, for the Liassic series is frequently tectonically separated from its basement. The Lias consists of thick, gray limestones often rich in detrital quartz with breccia horizons, crinoidal limestone turbidites, and chert. The Middle–Upper Jurassic is an "ammonitico rosso" facies of red, nodular limestones with the development of sedimentary dikes and often a manganese crust at the base. This is followed by Portlandian–Neocomian, white, argillaceous limestones and gray marls with *Aptychus* and *Calpionella* and, on occasion, with some Albian present (Scaglia facies). The latter forms a sequence of argillaceous, varicolored limestones in which there are numerous lacunae. In places Cenomanian or Turonian may occur, but more commonly the Upper Cretaceous begins with the Senonian. Often a good part of the Jurassic is also missing with Senonian resting directly on the Liassic. The Scaglia facies with some detrital quartz continues into the Paleocene and Ypresian, and at these horizons large blocks of sediment derived from the Taormina unit occur. The whole of this Upper Cretaceous–Lower Eocene sequence gives rise to extraordinary Neptunian dikes which may penetrate the underlying beds down to the phyllites. There may be a second manganese crust at the base of the Upper Cretaceous. The equivalents of this series occur in North Africa (external ridge, Abiod and Tengout units) and perhaps even in northern Calabria (Longobucco).

Taormina Unit (*Truillet, 1968; Duée, 1969; Fig. 12*). The base of this unit is also formed by the Verrucano over which lie massive oolitic limestones and dolomites with algae and molluscs. Their base is Sinemurian. The Middle Liassic which follows consists of somewhat argillaceous, finely banded limestones, with oolitic or detrital quartz turbidites. *Posidonomya* limestones and some beds of red and brown radiolarites represent the Middle–Upper Jurassic, and fine argillaceous limestones with radiolaria, calpionellids, and *Aptychus* form the Portlandian and Neocomian. The Lower Cretaceous ends with radiolarian marls. The varicolored scaglia facies of the Upper Cretaceous is incomplete and variable as in the preceding unit. The Paleocene is absent, consistent with the resedimentation of blocks of the Taormina in the Paleocene–Ypresian of the Longi unit. The Ypresian and Lutetian (in part) exist as foraminiferal calcarenites. The North African equivalents are the Djebel Bou Aded Sebargoud units (internal ridge).

Ali Unit (*Truillet, 1968; Fig. 12*). Only two small outcrops of this unit are known, both in Sicily, one at Ali Terme and the other near Gioiosa Vecchia. At both, the sequence is overturned, strongly cleaved, and epimetamorphosed. Thus, at the top (topographically), a Verrucano facies rests on black pelites and carbonaceous sandstones with plant fragments assigned to the Liassic. These in turn rest upon the principal limestone and dolomite horizons, although sandy beds and radiolarites occur. In this crushed mass Liassic, Middle–Upper Jurassic, Portlandian–Neocomian, and Upper Cretaceous are all represented.

Novara Unit (*Truillet, 1960, 1968; Bonardi et al., 1971; Fig. 12*). The red conglomerates below the Novara unit, for a long time interpreted as Verrucano, contain pebbles of Portlandian–Neocomian age and are now assigned to the Eocene. Resting upon them, in tectonic contact, is the Novara unit. The unit consists of dolomites of unknown age which pass without break into a thick sequence of poorly bedded limestones containing algae, trocholines and *Clypeina jurassica*. The beds are thus a reef facies of the Portlandian–Neocomian. In Calabria, at Stilo (Bonardi *et al.*, 1971), the sequence is completed by limestones with rudist debris. No equivalent is known in Africa. It recalls the platform sequence (of Panormid) as well as the calcareous breccia horizons in Calabria, which contain calpionellids, trocholines, and clypeines and which are associated with ophiolites.

3. *Eocene Tectonics and the Discordant Middle–Upper Eocene–Oligo-Miocene Succession* (*Fig. 13*).

The Taormina Series is a structural unit resting upon the Longi–Gallodoro Series, the two together forming a marginal chain. Fairly commonly an exclusively phyllitic unit covers the Taormina. The Novara unit comprises red conglomerates. Over all these units and contacts, the sandy-micaceous beds of the Oligo-Miocene rest discordantly. At several fortunate locations argillaceous calcareous sandstone horizons, with a phyllitic debris and preserved forami- nifera, may be found below the Oligo-Miocene. The latest horizons tectonically affected being Ypresian, the deformation must date between Ypresian and the Lutetian, i.e., it is Ypreso-Lutetian. The red continental conglomerates of Novara are also assigned to the Middle–Upper Eocene. Above the red conglom- erate, marine or continental, but most frequently transgressing directly onto the basement or onto Mesozoic limestones, is the Peloritan Oligo-Miocene. This is a thick series with massive basal conglomerates containing granite, gneiss, phyllite, and limestone pebbles in variable quantity according to the substrate, followed by a thick, sandy, argillaceous molasse, the top of which is dated as Burdigalian. Resedimentation of facies belonging to the "Argille Scagliose" nappe can also be seen. This Peloritan Oligo-Miocene, identical to the Oligo-Miocene of the Reitano nappe, is equivalent to the Rif or Kabylia Oligo-Miocene.

4. *Miocene Tectonics* (*Fig. 13*)

The whole Peloritan domain was sheared and overrides onto the Monte Soro nappe. The overthrust itself was subsequently flexed and faulted. The Peloritan Oligo-Miocene is in turn overthrust by slices of Cretaceous flysch (Cenomanian–Santonian) of the "Argille Scagliose" nappe, well seen at Flo- resta. On top of these klippe and discordant on the Peloritan domain is a Burdigalian, glauconitic, calcareous molasse. The "Argille Scagliose" slices at Floresta are in the same position as the flysch of southern Calabria, the north Kabylia flysch of Algeria, and the slices of Djebel Zem Zem in Morocco where Numidian occurs.

E. Volcanism and Plutonism

1. *Recent Volcanism* (*Figs. 1 and 4*)

The imposing volcano, Mount Etna, has been active since early Quaternary (Sicilian) time. The oldest rocks (pre-Etna) are alkali basalts, the more recent and present rocks are tephritic trachybasalts (Tanguy, 1966), more or less porphyritic according to the conditions of emplacement and rapidity of cooling. Activity has been paroxysmal with the appearance of important flows linked to the occurrence of fractures (Tazieff, 1973). Outside of such periods, explosive activity is limited to central craters below which there sometimes exists a small flow.

Iblean volcanism (Monte Lauro), also basic, occurred between Upper Miocene and older Quaternary times, and produced an important sequence of hyaloclastites (palagonitic tuffs). Two volcanos rise above sea level in the islands of Linosa and Pantelleria, and a third appears periodically (Graham Bank). In Linosa there are olivine basalts, but a more varied sequence is found in Pantelleria, and in addition to basalts there are rhyodacites and trachytes. According to Rittman (1964), while fundamentally basic, the emplacement gave rise to contact anatexis with a thin crust, the products of which formed the acidic series. On Graham Bank olivine basalts with titanaugite are found (Colantoni *et al.*, 1974).

Other volcanos have been identified in the Sicilan Channel, either by their morphology or by the magnetic anomalies induced. One in particular, lying southwest of Cape Passero, may be of Cretaceous age as are the volcanics found on Cape Passero. Large magnetic anomalies found near the Medine Mountains and over shoal areas near Cyrenaica may also be due to volcanic sequences.

2. Ancient Volcanism (*Figs. 10 and 11*)

Basic volcanism has occurred at many horizons since the Permian. The Permian flysch at Lercara Friddi contains blocks of diabase. Basic volcanics of Upper Triassic age are known, traces of which have been penetrated by wells in the Iblean region in the Sicani Mountains and at Leonforte. Submarine basaltic pillow lavas and tuffs, emplaced during the Middle Jurassic, are particularly well developed in the external domains (Vicari Ridge, Campofiorito Basin) and extend to the Sclafani Basin where they are found intercalated in a radiolarite series. They also occur in the Iblean domain. Cretaceous volcanics occur in Upper Albian–Cenomanian beds in the Panormide domain, and in the Iblean Senonian. These are again basaltic pillow lavas and tuffs, here intercalated in parareef beds. In some wells more than 500 m have been proven. Between the Cretaceous and the Eocene, and during the Eocene itself, some basaltic effusives formed in the Sicani Mountains.

Trachy-andesitic volcanism is much less common. It occurs in the Devonian in the Peloritan region, and then only in the Middle Liassic of Alcamo and Segesta. The Tusa Formation, the age of which is under discussion and which caps the flysch (Reitano and internal Numidian) succession, contains trachy-andesitic tuffs (Ogniben, 1965). It occurs in Calabria and extends into the northern Apennines (Petrignacola Sandstone).

The Peloritan granites, identical to the Calabrian granites, are of Carboniferous age.

F. Interrelationship of the Sicilian Domains

1. *External Domains*

The Iblean Series show no sensible differences until the Upper Cretaceous–Eocene, when a reef facies developed in the east. It is possible that this was due to certain regions reaching a depth at which reef formation could occur as a result of Middle Cretaceous volcanic activity.

Except for the Upper Miocene, as has been shown, there are no essential differences between the Sciacca and Iblean domains, and those which do exist are of structural origin. The Sciacca Zone suffered compression from the Miocene to Pliocene, while the Iblean Zone was almost unaffected. At several horizons there are distinct similarities between the Sciacca and Campofiorito Zones. During the Liassic the turbidites of the Campofiorito Zone were fed from the Sciacca Zone. In the Dogger, basalts extended to the most internal parts of the Sciacca Zone, while the nodular limestones developed on the more external parts of the Campofiorito Zone. From the end of the Jurassic to Eocene time, the sequences are the same, a similarity which extends to the Vicari Zone. The breaks in the succession are, however, more numerous and of longer duration in the Sciacca Zone. From the Oligocene and into the Lower Miocene all stages of the transition from calcarenite (Sciacca) to sandy-glauconite beds (Campofiorito) are found (Blondeau *et al.*, 1972).

The Vicari and Campofiorito have different structural locations. The "vicarese" sedimentary klippe are Middle–Upper Miocene and the Campofiorito forms part of their substratum. They show marked differences in Triassic–Upper Jurassic stratigraphy, but the remainder of the sequence is similar, although in the Vicari the Cretaceous sequence is more complete. Note that according to one hypothesis the Vicari Zone is possibly a Middle Cretaceous nappe from the Sciacca Zone.

The resemblances between the Vicari and Sclafani Zones are few and restricted to basic volcanism during the Dogger and to the similarity of material resedimented in the Lower Cretaceous turbidites. While the material forming the Triassic detrital dolomites of Sclafani could have originated from the Vicari Triassic reefs, it could equally well have been derived from the Panormide Zone. According to the hypothesis in which the Vicari Zone was tectonically emplaced during the Cretaceous, the resemblances with the Campofiorito Zone during the Upper Triassic and Jurassic are important.

The relationship of the Sclafani Zone to the Panormide is fairly close. The Panormide reef dolomites of the Triassic supplied debris to the Sclafani Trias, and from the Upper Jurassic to the Upper Cretaceous the Panormide domain was the source of the reef material in the Sclafani turbidites, much as it supplied material along its internal margin (Monte Acci). The occurrence of a mixed facies (radiolarites followed by rudist limestones in place) at Chateau de Termini Imerese (Rangin, 1974, 1975) confirms this point of view. There remains the problem of the Erice unit, which is found structurally in the position of the Sclafani nappe in a region where there is no outcrop of the latter. It contains, from Middle Jurassic to Upper Cretaceous horizons of turbidites with material of panormide origin. It has been assigned to the external margin of the latter zone.

2. *Relation of the Flysch to the External Domain*

The relationship is tenuous, consisting only of the occurrence in the Lower and Upper Cretaceous and in the Eocene of conglomerate horizons. These horizons contain pebbles, some of which were derived from the external domain (Triassic limestone and dolomite, Permian flysch). The conglomerates are found in the most external units of the Monte Soro nappe (Andreieff *et al.*, 1974).

3. *Intraflysch Relationships*

The flysch units belong to two major groups, one consists of an Oligo-Miocene sandy micaceous molasse and comprises the internal units of the Monte Soro nappe, the Reitano nappe (Capizzi and Troina units), and the Peloritan domain; the other consists of a sandy Numidian making up the Argille Scagliose nappe, and perhaps the external units of the Monte Soro nappe and Sclafani Zone. Prior to Middle Eocene two other groups can be recognized, one association consisting of the calcareous, *Aptychus* flysch (Portlandian–Neocomian), sandy flysch (Lower Cretaceous of Monte Soro–Tisirhen–Guerouch), and a thin Upper Cretaceous with conglomeratic horizons found in the Monte Soro nappe and in the Capizzi unit. The other consists of a Lower Cretaceous pelito-quarzitic flysch, a Cenomanian–Turonian sequence with oysters and radiolaria (African facies), and an Upper Cretaceous calcareous flysch found in the Argille Scagliose nappe and the Troina unit. This distinction observed in Sicily in 1963 (Broquet *et al.*, 1963) has been recognized in North Africa and Spain (see Didon *et al.*, 1972). Two distinct series are also known in Calabria. It is thus a general feature of the western Mediterranean, and the hypothesis that the nappes arise from the variation within the same unit (Ogniben, 1960; Durand Delga, 1960; Andrieux, 1972) can no longer be upheld. There is no agreement as to whether the two series were derived from a single basin or from two separate basins, nor as to the position of this basin or basins with respect to the Peloritan domain or its equivalents. Unfortunately, the answers given are seldom based upon objective data. The current hypotheses are that there was a single basin with flysch external to the Peloritano-Kabyle domain (Bouillin *et al.*, 1970; Durand Delga, 1969), or that there was a single sedimentary basin internal to the Kabyle domain (Andrieux, 1972; Delteil *et al.*, 1971), or finally that there were two flysch basins, external and internal, enclosing a discontinuous Peloritano-Kabyle Massif which formed an island arc (Broquet *et al.*, 1963). The question is, however, poorly presented, for each model considers that the Burdigalo-Helveto-Tortonian tectonic phase alone modified the paleogeographic relationships of the different domains.

4. *Relationship of Flysch to the Peloritan Domain*

During part of the Oligocene and the Lower Miocene, sedimentation of the elements later forming the Reitano nappe and those subsequently forming the most internal units of the Monte Soro nappe included the same sandy micaceous molasse, the detrital material of which was from a basement identical to the Peloritan. This molasse, discordant upon thrust structures of the Limestone Chain, is the result of an important tectonic phase and it is possible, if not probable, that the flyschs of the Peloritan domain were tectonically brought together at this time. Prior to that, pebbles originating from the Peloritan domain (in particular Liassic detrital limestones) in the Lower and Upper Cretaceous and Eocene of the Monte Soro nappe or in the external units, are mixed with pebbles of external origin. Phyllite fragments appear in the Eocene of the Capizzi unit. The occurrence of blocks of rudist limestone, and of ophiolites within the Cretaceous limestones flysch ("Argille Scagliose" and Troina unit) poses a problem. While the rudist limestones could be derived from the Panormide domain, or more probably from those of Novara (internal Peloritan), ophiolites are unknown in Sicily.

5. *Interrelationships of the Mesozoic–Eocene Peloritan Units*

The units forming the marginal chain (Longi–Gallodoro and Taormina) show a series of transitional characteristics particularly in the west of the Peloritan domain (Duée, 1969). The facies of the Ali unit itself is not very different from the Longi, and it is probable that all of those may represent diverse aspects of the same domain rather than sedimentation in basins distant from one another. In contrast, the Novara unit shows no analogies with the Peloritan Series. Insofar as there may exist within the basement tectonic contacts of which there is no information, it is possible that the present respective positions of the Novara and other units may be the result of important tectonic transport.

PART II

THE STRUCTURE OF CALABRIA–LUCANIA

IV. INTRODUCTION

The southern Apennines may be subdivided into three natural regions. To the northeast bordering the Adriatic there is the large, calcareous, low-altitude plateau of Apulia. It has a thick, flat-lying limestone sequence and forms the common foreland of the Apennines and Dinarides. Its western limit is marked by the Bradanic depression, which reaches the sea in the Gulf of Taranto. In the center and to the west, the limestone Apennines form a mountainous chain with peaks reaching 2000 m. This chain, formed of limestone massifs, flysch, and local occurrences of greenstones and basement, is built up of a complex of nappes emplaced at various times. To the south lies the crystalline massif of Sila, with a metamorphic basement and a Mesozoic and Tertiary sedimentary cover, on a tectonic greenstone sole. Largely removed by erosion, it is, despite its size, only one of the elements whose origin was internal to that of the nappes. In this segment of the Mediterranean Alpine chain, built up of varied nappes, only the upper Pliocene and Quaternary sediments may be considered as postnappe formation in age.

Beginning with the most external, the easterly domains, and with the lowest zones structurally, the attempt will be made to list the principal complexes rising through the structural pile. The lithologies will be indicated and the attempt made to locate them paleogeographically.

V. THE APENNINE DOMAINS

A. The Neotectonic Foreland, or External Apulia Platform (PF₃)

This is a platform with thick neritic carbonates, lying between the Adriatic and the Neogene Bradano Trough. The principal outcrops are on the Gargano Peninsula, Murge, and Salento.

The oldest horizon recorded in this sequence of 4000 to 6000 m is Upper Triassic, represented by Carnian evaporites and bituminous beds with which are locally associated basic rocks (Punta delle Pietre Nere) followed by dolomites. Jurassic beds are known only in Gargano where three facies types are recognized (Pavan and Pirini, 1966; Martinis, 1962). Along the eastern margin are found dolomites and bioclastic and detrital limestones with or without chert, and marls with lamellibranchs and radiolaria (Lias to Tithonian). The center, the platform *s.s.*, is occupied by bioherms with algae, corals, and ellip-

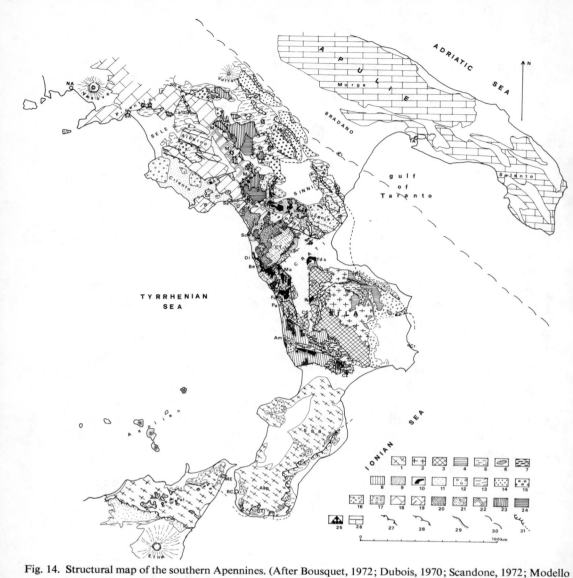

Fig. 14. Structural map of the southern Apennines. (After Bousquet, 1972; Dubois, 1970; Scandone, 1972; Modello strutturale d'Italia, 1:1,000,000). (1) Basement and its Mesozoic and Paleogene sedimentary cover of the Peloritan–Aspromonte–Serra San Bruno; (2–6) *Garnetiferous gneiss and Upper Phyllite unit*—(2) granites, (3) garnetiferous gneiss, (4) phyllites, (5) Rossano–Longobucco Mesozoic and Paleogene, (6) Tiriolo–Martirano–Novara Mesozoic; (7–8) *Augen-gneiss and lower phyllite unit*—(7) augen-gneiss; (8) phyllite; (9–10) *Ophiolitic Series*—(9) Frido, (10) ophiolites; (11–12) *Silentine unit (Cilento–Lucania)*—(11) Cretaceous flysch; (12) Lower Eocene flysch; (13) Argille Scagliose Varicolore unit (A.S.V.); (14–15) *Irpinian Basin Sediments*—(14) Numidian and Serra Palazzo; (15) Gorgoglione; (16) Upper Miocene sediments (Messinian–Tortonian); (17) Aquitanian–Burdigalian clay with blocks; (18–22) *Calabro-Campania internal platform PF₁*—(18) PF$_1$ Pollino–Campo Tenese–Alburno unit (Pollino Alburno facies); (19) PF$_1$ Bulgheria unit; (20) PF$_1$ Verbicaro unit; (21) PF$_1$ Pollino–Campo Tenese–Alburno unit (Campo Tenese facies); (22) PF$_1$ Maddalena–Foraporta unit; (23–24) *Lagonegro–San Fele complex*—(23) L² Upper, proximal Lagonegro unit 2; (24) L¹ Lower, distal Lagonegro unit 1; (25) Intermediate Abruzzi-Campania platform PF$_2$ Monte Alpi; (26) external Apulia platform PF$_3$; (27) thrust of PF$_1$ over Lagonegro; (28) thrust of pre-Apennine units over the Apennine and secondary slices formed in the pre-Apennine units; (29) thrust slices of the Silan basement; (30) thrust front of the nappes in the Pliocene Bradanic foredeep, line Pisticci-Vasto; (31) retrothrust of the A.S.V. during the Miocene. Abbreviations: Am = Amantea; Asp = Aspromonte; B = Bulgheria; Be = Belvedere; C = Coccovello; Cr = Crotone; CS = Cosenza; CT = Campo Tenese; CZ = Catanzaro; Di = Diamante; Fu = Fuscaldo; L = Longobucco; M = Maddalena; Ma = Malvitto; Me = Mezzane; ME = Messina; NA = Naples; Pa = Paola; Pel = Peloritan; Pi = Pisticci; PZ = Potenza; R = Rossano; Ro = Rosa; RC = Reggio Calabria; SA = Salerno; Sa = Sapri; Sc = Scalea; SSB = Serra San Bruno; TA = Taranto; T = Tiriolo; Tds = Terranova di Sibari; V = Verbicaro.

sactinias, while along the western margin oolitic limestones with ellipsactinias, trocholines, and pseudocyclamines of Kimmeridgian to Tithonian age occur.

The Cretaceous (Martinis, in Desio, 1970) in eastern Gargano has a Neocomian sequence of 500 m of fine, marly limestones with chert (radiolarians and tintinids) followed by scaglia (marly limestones with chert, and *Globotruncana*) from Aptian to Senonian. In the west the limestones are bioclastic and detrital with chert and contain *Tintinopsella* radiolaria and orbitolines. They are followed by 200 m of rudist chalk of Cenomanian–Turonian age and of Senonian chalk with orbitoids and siderolites.

In the Murge and Salento there is more than 3000 m of platform carbonate and dolomite of Cretaceous age, distinction being made primarily on the basis of fauna into Barremian–Aptian with *Toucasia*, Cenomanian–Turonian with *Sauvagesia* and *Apricardia*, and Lower Senonian with dicyclines, *Accordiella*, and cuneolines. Slight discordances occur under the Upper Cenomanian and the Senonian.

The platform was probably not continuously submerged during the Cretaceous. The central part emerged almost entirely at the end of the Cretaceous, and only along the margins are neritic and littoral Paleogene deposits found. These consist of limestones with *Nummulites*, discocyclines, alveolines, corals, and algae, extending up to Lutetian and resting discordantly on the Senonian. Basaltic tuffs may be intercalated locally. Oligocene limestones with lepido-cyclines, melobesiae, corals, and molluscs transgress over these beds.

The Neogene sediments are in turn discordant on a limestone relief of older beds. The Neogene in the Gargano Peninsula consists of dolomites and organic calcarenites with a rich Miocene macrofauna. Over this rests a transgressive Upper Pliocene calcarenite. In the northeastern Gargano the Miocene consists of sandy, glauconitic, marly calcarenites locally followed by Upper Miocene gray-brown marls. The sands and limestones of the Middle–Upper Pliocene are everywhere discordant.

The platform is downfaulted to the west toward the Bradanic depression and is not visible at the surface. The stratigraphic succession is known as a result of drilling. The rudist limestones of the Cretaceous are locally covered transgressively by a *Lithothamnion* calcarenite with bryozoans and amphistegines of Lower–Middle Miocene age. These are overlain by Tortonian marls, evaporites of Messinian age, clays and sands of Lower Pliocene age, and terrigenous beds of the Middle–Upper Pliocene.

B. The Neotectonic Foredeep or Bradanic Depression

This foredeep extends from the Molise to the Gulf of Taranto and continues into the Ionian Sea along the curvature of the Calabrian Arc. Its eastern margin and basement consist of Apulia platform deposits. The western margin, formed by the Apennines, was overrun by flysch nappes, slices of Miocene sediments, or even by the thrust front of more internal platform units up to the Pisticci–Vasto line (Carissimo et al., 1963; Ogniben, 1969; Pieri, 1966). The thrust units with their Lower and Middle Pliocene cover were emplaced in the foredeep during the Middle Pliocene, interrupting or perturbing sedimentation there. The Middle and Upper Pliocene and Pleistocene marine sediments mainly occupy the narrow Bradano Channel and the succession of bays eroded into the partially emergent Apennine Ridge (Sant'Arcangelo, Crati, Crotone, and Catanzaro Basins). The basins are filled with conglomerates, sands, and clays from the eroding Apennines. In the Apennines themselves small Pleistocene lacustrine basins became established in grabens.

The emplacement of nappes in the Middle Pliocene reduced the width of the foredeep, which was subsiding. The crustal shortening caused by the Apennine nappes was of the order of a hundred kilometers. Prior to this episode of deformation, the foredeep appears to have been shallower and wider.

During the Upper Miocene–Lower Pliocene a terrigenous fill was derived essentially from the upper Apennine nappes; it now occurs in the basement of the Bradanic depression and as erosional relicts in the Apennines. Along the eastern margin of the Calabrian crystalline, as for example in the Trionto Valley (Magri et al., 1963; Ogniben, 1962) and in Crotonese (Ogniben, 1955), the sequence begins with a sandy conglomerate with Ostrea, Pecten, and Clypeaster, resting unconformably either on the crystalline itself or on its Mesozoic–Paleogene cover. It is followed by Tortonian clays and marls, the Messinian "Gessoso solfifero" made up of tripoli (diatomite), evaporitic limestones, saliferous marls, conglomerates and arkose, gypsum, marls, and sands. Within this Messinian sequence there is evidence of allochthonous material (Argille Scagliose, see below) which appears to have originated east of the site of emplacement. In the Crati Basin and along the Tyrrhenian coast the Messinian lithology is comparable (Perrone et al., 1973), and also contains exotic blocks and slices. The same phenomenon is known in Campania. The sediments, always transgressive over the upper Calabrian nappes, are also discordant in some places over basement now exposed by erosion. Frequent thickness variations indicated by a wedging of the sediments suggests basement instability.

The Upper Miocene is thus characterized by ubiquitous detrital and evaporite sedimentation, recording an important erosive phase correlated with intense folding of the substratum and Apennine nappes. The region appears to have been an archipelago floored by flysch nappes, ophiolites, and crystalline basement, with parts of the platform exposed by erosion.

Removing the tectonic effects of the Upper Miocene reveals a Middle Miocene structure larger than the succeeding one. To the east PF_3 was emergent, and toward the west there occur successively the Molise Trough of unknown width, then platform PF_2. The platform, as will be shown later, serves as the foreland for a Lower and Middle Miocene foredeep, the western margin of the latter being occupied by the more internal Apennine PF$_1$ units and by nappes.

C. The Mesotectonic Foreland or Intermediate Abruzzi–Campania Platform (PF$_2$)

The Abruzzi–Campania platform may be divided into three units which can only be analyzed through tectonic windows or as the result of a few wells. There are, from east to west, the following units:

1. *Frosolone Unit* (*D'Argenio et al., 1973*)

The top part is known in outcrop, and the lower part in borings (AGIP; Pieri, 1966). It consists of a basal Triassic to Liassic of cherty dolomites intercalated with tuffs, with siliceous and black shales containing radiolaria. The rest of the Jurassic is formed of limestones and cherty dolomites, often brecciated, and is followed by cherty limestones and pelagic marls of Lower Cretaceous age and Upper Cretaceous and Paleocene biomicrites with rudists and *Globotruncana*. The Paleogene is represented by detrital scaglia, the Lower Miocene by calcarenites, pelites, and clays. Terrigenous sedimentation began in the Helvetian and continued to the Tortonian in a facies intermediate between platform and pelagic basin type (the latter unknown at outcrop).

2. *Matese–Monte Alpi Unit*

This unit has a discontinuous outcrop along the eastern margin of the Apennine Ridge. The oldest known beds are Triassic and consist of organo-detrital dolomites with algae, lamellibranchs, and gastropods. From place to place there is evidence of hydrocarbon accumulations and of anhydrite lenses. The Lias is composed of microcrystalline dolomites followed by *Lithiotis*, *Paleodasycladus*, and *Orbitopsella* limestones. The Dogger limestones contain algae and foraminifera (*Pfenderina*, *Meyendorfina*), while the Malm, also limestone, is characterized by algae, clypeines, and foraminifera. The Lower and Upper Cretaceous are represented by limestones and dolomitic limestones, with a break during the Albian to Turonian indicated by breccias and bauxite formation. The Paleogene is only present in certain marginal zones so that the Miocene rests upon Jurassic–Lower Cretaceous beds at Monte Alpi (Ortolani and Torre, 1971). The Miocene consists of Burdigalian limestones and calcarenites with algae, oysters, sea-urchins, and foraminifera, followed by marls, sandstones, and conglomerates ranging in age from Helvetian to the base of the Tortonian. The conglomerates indicate synsedimentary and fill naissant grabens. Monte Alpi appears through a window under the flysch nappes, the Lagonegro units, and the most internal platform facies of the Apennines.

3. *Monte Croce di Campagna Unit* (*near Salerno*)

This unit is seen only through a small tectonic window in Campania (Scandone *et al.*, 1967), where the Triassic consists of white, cherty dolomites, with limestones and argillites. The Malm is transgressive, consisting of massive limestones and breccias with *Ellipsactinia*. The Eocene, also transgressive, is also a limestone containing nummulites and alveolines. The Miocene begins with a discordant Burdigalian calcarenite overlain by marls, sandstones, and conglomerates of Helvetian to Tortonian age. This sequence with its many gaps during the Mesozoic marked the western margin of the platform, it was overridden by sedimentary flysch klippe and by the Lagonegro nappes and the most internal part of the platform.

From the Miocene onward the three units had the same type of sedimentary cover. They are most often seen through windows in the Lagonegro and the internal platform PF_1. During the Mesozoic they formed a carbonate platform with transitions into pelagic basins, the Lagonegro Basin (see below) to the west and the hypothetical Molise Basin to the east.

D. The Mesotectonic Foredeep or Irpinian Basin

West of the Abruzzi–Campania platform (PF_2) there existed during the Lower and Middle Miocene a vast basin being filled with sediment and into which nappes flowed from a westerly source. Named the Irpinian Basin by Cocco *et al.* (1972), it includes various formations, argiles à blocs (Grandjacquet, 1967, 1971), paraconglomerates (Ogniben, 1969), Bifurto Formation (Selli, 1958), Pietra san Angelo (Bousquet, 1965), at the base overlain by a rhythmic sequence of detritals, sandstones, and conglomerates, and wildflysch with large sedimentary klippe. The ages of the formations range from the end of the Aquitanian to basal Tortonian. Although this sequence is often spread out in large nappes, the original basin can be reconstructed. It had two principal zones—the eastern, in which the Numidian Sandstone and its cover were deposited, and a western, often invaded by allochthonous slices and wildflysch and in which the detrital series of Gorgoglione were deposited (Selli, 1962).

The Numidian now appears as a wedge seen through the flysch nappes and Gorgoglione Series, in the core of a large Middle Pliocene nappe which projects into the Bradanic Trough. The Wildflysch and Gorgoglione Formations crop out over wide areas involved in nappe movements in the southern Apennines. In the eastern part of the basin, the Lucanian Numidian forms the northeastern corner of the North African Numidian Basin. At the base, both end-Aquitanian (argiles à blocs) and Burdigalian (pelletal sandstone) can be found. Platform material has been recognized in the sandstones (see Ghezzi and Baylis, 1966; Mostardini and Pieri, 1966; Boenzi and Ciaranfi, 1968). The sequence continues with sandy marls, calcarenites, sandstones, pelites, and conglomerates (Ogniben, 1969), ranging in age from Upper Burdigalian and Upper Helvetian to basal Tortonian.

The Wildflysch and Gorgoglione Zone, farthest to the west, can be directly related to the internal nappes overriding the Apennines. The formations themselves generally begin with a clay containing blocks and resting discordantly on the subbasement (internal platform and Lagonegro domain). The blocks, more or less rounded, are disposed in channels within the clay and consist of elements of the substratum (internal platform) pebbles and blocks of argillaceous–calcareous flysch of Paleocene–Eocene age, identical with that in the Canetolo complex of the northern Apennines. Large scales of the internal platform are thrust over this horizon. There follow sandy conglomeratic beds, which are invaded by sedimentary klippe of the platform, of flysch and of peridotites, which often cover zones temporarily exposed to erosion. The Gorgoglione *s.s.* rests transgressively on the flysch nappes. It consists of an alternation of conglomerates with phyllite, granite, and shale clasts and turbidite sandstones whose age ranges from Burdigalian to top-Helvetian–Lower Tortonian.

All these formations of the Irpinian domain have the lower age limit set by *Miogypsina* at the end of the Aquitanian and then by *Globigerinoïdes trilobus* and orbline zones. The facies of the Masseria Luci Formation at the top of the Numidian sequence are similar to those of the Gorgoglione, indicating a homogenization of the sedimentation during the course of the Helvetian.

This large, complex basin was thus a foredeep receiving sediments, which began to submerge the Apennines. The transport of African Numidian sands terminated at the end of the Burdigalian, and the eastern part of the basin was invaded little by little by a detrital facies from the west, and even by nappes. The eastern margin of the basin, PF_2, was also progressively submerged from Burdigalian time onward. During the Tortonian nappes covered the Numidian Series and locally extended onto the PF_2 platform. The Irpinian Basin was, in fact, significantly reduced by the superposition of the western part of the zone over the eastern (the internal platform itself overthrust that part of the basement thus liberated). The greater part of the Irpinian Basin, and to a lesser degree the subbasement, did not begin to move until Middle Pliocene. Given the mechanism, the nature of the subbasement can be analyzed.

E. The Subbasement of the Miocene Foredeep or Irpinian Basin, the Lagonegro–San Fele Complex

The two units which form this complex crop out in a large erosional depression between Lagonegro and San Fele and in two small windows near Salerno at the southern foot of the Picentini Mountains (Campagna and Giffoni

Vallepiana). According to Scandone (1967–1972) there are two principal tectonic units, the lower and upper Lagonegro units, and three facies, the Lagonegro–Sasso di Castalda, Armizzone–Pignola Abriola, and San Fele, respectively.

The lower Lagonegro unit (L¹) begins with 500 m of Upper Triassic, gray, cherty limestones with *Halobia*, ammonites, and radiolaria, which pass progressively into the siliceous shales and radiolarites of the Jurassic. The latter contain some bands of limestone microbreccias (70–80 m thick). Toward the top the radiolarites pass conformally into 400 m of argillites, brown or gray calcilutites, and thin bands of fine-grained quartzite, all of Lower Cretaceous age (galestrino flysch). The beds are rich in iron oxide and flakes of manganese. Locally, the sequence is overlain by siliceous shales (red shales of Pecorone) and argillites with *Globotruncana* microbreccias of Upper Senonian to Paleocene age. All these deposits have a distal depositional character.

In the upper Lagonegro unit (L²) different facies are found, but despite the differences the general succession consists of a Middle Triassic terrigenous series, the Monte Facito Formation, a sandy argillaceous, polychrome flysch sequence, locally clastic, containing blocks of reef limestone. The age ranges from Anisian (*Spiriferina and Anisactinella*) to Ladinian (*Daonella*). Locally, masses of Permian flysch are incorporated. The reefal limestones, surrounded by flank breccias, are clearly interstratified in the pelites. The Upper Triassic, Carnian–Norian, consists of 200 m of cherty limestone and dolomites, passing into 150–250 m of siliceous shales and radiolarites, with numerous microbreccias filling channels and spread out by turbidity currents to which Dogger–Malm age is assigned by, among other fossils, *Protopeneroplis* and *Nautiloculina*. The Galestrino Flysch is a monotonous sequence of dark argillites intercalated with brown calcilutites, marly calcilutites, and occasional microbreccias. On the basis of Berriasian calpionellas it is assigned a Lower Cretaceous age. The Upper Cretaceous–Middle Eocene sequence of red argillites and silicified limestone microbreccias follows, with a Cretaceous fauna of rudist, orbitoid, siderolite debris, and *Globoratalia* and *Nummulites* of Eocene age. The sequence may extend to Stampian–Aquitanian on the basis of *Lepidocylina* and *Miogypsina*. This upper sedimentary sequence is often separated from the lower. The Lagonegro window is covered by Lower Miocene sediments consisting of a thin layer of chaotic argillites or clays and sandstones of the Irpinian Basin.

The upper Lagonegro unit (L²) almost always rests upon the "galestri" of the lower unit (the Pecorone Shales are seldom present). The internal Calabro-Campanian platform is usually thrust over the upper Lagonegro unit (see below). The lower unit supports the Numidian Basin (Ogniben, 1969), while the upper forms part of the wildflysch (shale with blocks) basement of the western Irpinian Basin during the thrusting of the internal platform.

F. The Internal Apennine Margin, or Calabro–Campania Platform (PF₁)

This complex, forming the western margin of the Irpinian Basin, represents a large part of the framework of the southern calcareous Apennines. It consists of large slabs of limestone which can be traced from Rome to central Calabria. The complex, which was defined by D'Argenio (1966–1973) as the Campano–Lucanian platform, can be subdivided into four major units, from east to west, as follows:

1. *Foraporta–Maddalena Unit*

This lowest unit, found along the western margin of the Lagonegro–San Fele inlier, can be further subdivided into the Maddalena and Monte Foraporta subunits. The former consists of 300 m of white, Middle Triassic dolomites overlain by Jurassic perireefal limestones and discordant Maestrichtian–Paleocene calcarenites, which contain rudist fragments, orbitoids, siderolites, and *Globotruncana*. This is followed, discordantly, by nummulitic–alveoline Eocene calcarenites. The top of the succession is formed of polychrome argillites of end-Aquitanian–Burdigalian age. The argillites contain blocks of Mesozoic and nummulitic limestones (Scandone *et al.*, 1968). The subunit is overthrust in the south by a second subunit (Boni *et al.*, 1974) formed of fetid limestones with carbonaceous lenses, chert nodules, and dolomites of Upper Triassic–Lower Liassic age, radiolaria, Middle Liassic limestones and dark or greenish marls, and Upper Liassic yellow, marly limestones, and Dogger limestones with numerous breccia horizons and intraformational conglomerates. The Middle Liassic limestones with nodular chert are affected by slumping. The fauna includes *Paleodasycladus*, vidalines, *Aeolisaccus*, and radiolaria.

2. *The Pollino–Campo Tenese–Alburno Unit*

This second unit, together with the Pallone–Aieta subunit, forms a long band of limestone massifs along the Tyrrhenian Sea margin from Rome to Paola. It is thrust over the Foraporta–Maddalena unit, the Lagonegro complex, and even overrides the Abruzzi–Campania complex. Within the unit two contiguous facies zones are found, one the Pollino–Alburno platform facies found northeast of a line Sibari–Maratea, passes abruptly but without break into the epimetamorphic Campo–Tenese Series. The Pollino–Alburno facies consists of some 3000 m of limestones and dolomites. The Campo-Tenese rocks range in age from Middle Triassic to Upper Aquitanian. The Pallone-Aieta subunit is seen through a tectonic window in the Campo–Tenese unit, where its facies are similar to those of the Pollino–Alburno unit.

In the Pollino–Coccovello–Cervati–Alburno–Picentini massifs, the stratigraphic succession begins with Triassic and Lower Liassic stromatolitic dolomites with occasional gray or buff shales with *Megalodon, Gervillia,* and gyroporellas, followed by Middle Liassic *Paleodasycladus* limestones and Upper Liassic *Lithiotis-Orbitopsella-Paleodasycladus* limestones. The Dogger–Malm, consisting of calcarenites, calcirudites, and oolitic limestones with clypeines and *Chladocoropsis,* is followed by calcarenites, calcilutites, dolomites, and calcioolites with miliolids and cuneolines of Lower Cretaceous age. Middle Cretaceous marls contain orbitolines, and Upper Cretaceous rudist limestones contain cuneolines, dicyclines, ovalveolines, and *Accordiella.* The Paleocene (Trentinara Formation of Selli, 1962) transgresses over the Upper Cretaceous as a green shale containing pseudopebbles of limestone, and is overlain by calcarenites and calcilutites with spirolines. Both the Paleocene and Upper Cretaceous have been subjected locally to karst weathering, and solution hollows are filled with residual brown-red clays. Over the Paleocene lie transgressive calcarenites with lepidocyclines and *Miogypsina* of Upper Stampian–Middle Upper Aquitanian age, in turn covered, often discordantly, by clays and sandstones, of end-Aquitanian–Burdigalian age, the sole of the Irpinian Basin, and internal nappes. This sequence is referred to as the Pollino–Alburno facies, and it passes southwestward abruptly but without break into a sequence referred to as the Campo–Tenese facies.

In the vicinity of Monte Ciagola–Monte Pollino the Campo–Tenese sequence begins with thick, Middle Triassic similar to that found in the upper Lagonegro unit, but absent in the Pollino–Alburno facies. The Middle Triassic consists of an Upper Anisian lens of dolomitic marble (Bousquet and Dubois, 1967), overlain by chloritic and sericitic schists with bands of quartzite and carbonate lenses passing up into dark, marmorized Ladinian limestones. Diabase sills and dikes cut these beds. Beds of dolomite and schistose pelites, assigned to the Carnian, overlie the Middle Triassic. Occasionally, nodular or platy limestones may replace the dolomites. The sequence continues with more Upper Triassic dolomitic marls, then nodular, poorly bedded limestones, tectonically thinned and marmorized, and common intraformational conglomerates in which *Ellipsactinia* and algal balls of Jurassic age have been identified. The Mesozoic sequence is closed by a band of conglomerate or calcarenite, probably of Maestrichtian age. The Eocene, shaly limestones and the transgressive Oligocene–Lower Miocene are preserved in synclinal cores. Clays with beds of fine quartzite and clays with blocks rest discordantly upon a karst surface, but often are tectonized also. The whole sequence is intensely deformed, schistose and metamorphic. It underlies the upper thrust units of the internal platform (Verbicaro) and slices of the internal nappes. The first clearly postmetamorphic sediments are the end-Tortonian–Messinian beds.

A subunit defined by Bousquet and Grandjacquet (1969) as the Pallone–Aieta unit can be seen in the Aieta and Timpone–Pallone windows in the Pollino

Massif. While of the same facies as the Pollino–Alburno unit, it underlies that unit.

3. *The Verbicaro and Bulgheria Sequences*

These form the last two units. The former, the Verbicaro unit (Bousquet and Grandjacquet, 1969), is characterized by beds and nodules of chert breccia (the chert domain of Grandjacquet, 1962) and is the highest external structural unit in the Apennines. It crops out in the coastal Calabrian chain and is well developed in the region of Maratea–Castrovillari–Belvedere. Totally separated from its subbasement below the Upper Triassic, it thrusts over the Campo–Tenese–Pollino unit, and even the Pallone–Aieta subunit.

Stratigraphically, the sequence in the Verbicaro unit begins with dark megalodon dolomites, stromatolites and oncholites, and thick algal masses. There are intercalated yellowish, marly, dolomitic shales which contain rare lenses of gypsum. During the Lower Liassic, dolomites and limestones alternate with green, yellow, or red marls containing megaladons, *Gyroporella*, *Pianella*, and *Triasina*. The Upper Liassic to Malm is made up of cherty limestones and calcarenites, often affected by synsedimentary slumping, with oolitic horizons and intraformational conglomerates. No Lower or Middle Cretaceous is known and the Maestrichtian–Paleocene transgresses over a faulted and eroded basement, depositing coarse calcarenites and oolitic limestones with a debris of rudists and other forms. Locally (Verbicaro), there are intraformational mega-breccias with slabs of chert rounded by the slumping process. Limburgite flows and picritic and teschenitic intrusions are associated with these formations. The breccia is locally preceded by red and green shales, tuff, and hyaloclastites. Overlying the Maestrichtian–Paleocene is an alternation of nummulitic calarenites and calcilutites, and polychromatic marls of Eocene age, traversed by limburgite dikes. The Stampian and Aquitanian, which follows, consists of calcarenites and lenticular, marly limestones which are frequently channelled. At the top of the sequence there is a thin horizon of pelites with thin quartzites and shales with microbreccias of end-Aquitanian–basal Burdigalian age. The whole sequence may be locally folded and schistose (i.e., epimetamorphic) (Buonvicino). It is tectonically overlain by flysch nappes, greenstones, and the crystallines of the internal nappes.

The Bulgheria unit (Scandone *et al.*, 1963) is only known in Monte Bulgheria, and the Isle of Capri is the westernmost indication of the internal platform elements north of the Gulf of Sapri. It contains a dolomitic Middle Triassic followed by a varied Liassic of massive algal limestones intercalated with breccia horizons, cherty crinoidal limestones marls, and marly ammonitic limestones. The Dogger–Malm limestones consist of calcirudites and calcarenites and oolites containing *Ellipsactinia*, tintinids, and crinoids. The transgressive Upper Cretaceous consists of calcarenites with graded rudist fragments, and globotruncanid limestones, and is overlain in stratigraphic continuation by a Maestrichtian–Paleocene scaglia, and alternation red and green marly limestones and calcilutites with *Globotruncana* and *Globorotalia*. Oligo-Miocene (Stampian–Aquitanian) marls and calcarenites and Upper Aquitanian–Burdigalian sandy shales and conglomerates complete the succession.

The Verbicaro and Bulgheria Unit represents the western margin of the Calabria–Campania platform, while the Maddalena–Foraporta unit represents the eastern edge.

The Calabria–Campania complex internal platform (PF$_1$) is broken up into scales, each commonly of different lithology and often of different tectonic facies. All these scales or thrust slices are capped by Upper Aquitanian clays with blocks of reworked elements of the subbasement, and received the first internal nappes represented by an argillo-calcareous flysch. A flysch basin must have existed to the west of this, the most internal platform. The relationship of these sediments with the Verbicaro and Bulgheria units is particularly clear after the Maestrichtian, as will be seen.

G. The Southern Apennines before Miocene Thrusting

Up to the beginning of the Miocene, the southern Apennines *s.s.* consisted of a series of carbonate platforms separated by pelagic basins. From east to west these were the Apulia platform PF$_3$ (Triassic–Recent), a hypothetical Molise Basin (Mesozoic–Miocene), the Abruzzi–Campania platform PF$_2$ (Trias–Burdigalian–Helvetian), the Lagonegro Basin (Trias–Upper Aquitanian), and the Calabro-Campania platform PF$_1$ (Triassic–upper Aquitanian), ending to the west against the flysch basins. In step with the tectonic shortening and the piling up of nappes, basins became established over active structures, e.g., the Irpinian Basin (Upper Aquitanian–basal Tortonian) and Bradanic basin (Upper Tortonian–Recent). Compression of the Apennines began at the end of the Aquitanian and did not end until Middle Pliocene time.

VI. THE PRE-APENNINE DOMAINS

The Apennines are covered by a nappe complex originating in the internal, western domain, and in which occur the Argille Scagliose Varicolore (A.S.V.) and the argillo-calcareous flysch. These are to some extent incorporated in other nappes being at one time the sole of the other nappes and covering their eastern ends. The Silentine unit formed sandy flysch nappes cropping out in Cilento and Lucania. The crystalline Silan unit and the Calabrian ophiolite unit were structured during the middle Upper Cretaceous–Paleocene outside the Apennines and only covered the latter during the Miocene.

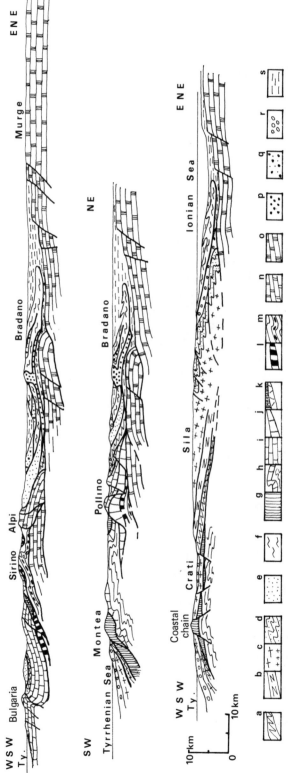

Fig. 16. Structural sections across Calabria and southern Campania (synthesized). (a) Frido Series and ophiolites; (b-d) *Silan basement and Mesozoic–Paleogene cover*—(b) augen-gneiss and phyllite (lower unit); (c) garnetiferous gneiss, granite, and phyllite (upper unit); (d) Mesozoic–Paleogene cover; (e) sandy flysch, Silentine unit; (f) Argille Scagliose Varicolore (A.S.V.); (g-k) *internal Calabria–Campania carbonate platform* PF_1—(g) Verbicaro unit; (h) Campo Tenese (Pollino–Campo Tenese–Alburno) subunit; (i) Pollino–Alburno (Pollino–Campo Tenese–Alburno) subunit; (j) Pollino–Aieta (Pollino–Campo Tenese–Alburno) subunit; (k) Maddalena–Foraporta unit; (l-m) *Lagonegro–San Fele complex*—(l) proximal Lagonegro 2 facies; (m) distal Lagonegro 1 facies; (n) Abruzzi–Campania intermediate platform PF_2; (o) Apulia external platform PF_3; (p) Irpinian unit Numidian–Serra Palazzo facies; (q) Irpinian unit Gorgoglione facies; (r) Upper Miocene sediments; (s) Plio-Pleistocene sediments.

A. The Flysch Basin

Within this basin the argillo-calcareous flysch formed to the east (A.S.V. unit) linked to the Apennine carbonate platforms, while to the west the argillaceous sandy flysch (Silentine unit) was linked to the Silan crystalline basement and an ophiolitic series. The age of the basin ranges from Cretaceous to Eocene.

(1) The argillo-calcareous flysch (Argille Scagliose Varicolore, A.S.V.) or Sicilide complex (Ogniben, 1969), the most external of the central southern Apennine nappes, can be examined in the Molise and Bradano Basins and along the curvature of the Calabrian Arc, as well as in thrusts and in blocks within the Miocene clay with blocks and wildflysch. Despite numerous facies variations the lithological succession consists essentially of Vracono-Cenomanian black or polychrome silty shales and phtanites (overlying possible Eo-Cretaceous schists and quartzites). In Lucania the Nocara Flysch (Ogniben, 1969) consists of shales, sandstones, and conglomerates whose elements were derived from a crystalline massif. The appearance of this facies at the base of the ASV is the result of channel distribution of clastics. The basal beds are followed by a thick (1000 m) polychrome shale series with intercalated greenish *Globotruncana* limestones, which are sometimes cherty and contain flakes of iron and manganese of Middle Upper Cretaceous age as well as Maestrichtian–Lower Eocene marly limestones and microbreccias intercalated with polychrome shales up to 500 m thick (the San Arcangelo members of Ogniben). Locally, coarse calcarenites with rudist debris or a conglomerate with fragments of chert and pebbles or blocks of Mesozoic (cf. Verbicaro) form. This calcareous interval is followed by renewed deposition of polychrome clays (500 m) with some sandy and microbreccia horizons of Eocene age. A transgressive Upper Eocene conglomerate is known in the Molise Basin, followed by younger conglomerates which incorporate platform elements intercalated within Stampian–Aquitanian polychrome shales (Clermonte, 1974), and a cherty calcarenite of Burdigalian age (sediments of the Irpinian Basin). In Lucania the Tusa tuffites (Ogniben, 1969), dark green sandstones with andesitic volcano-detritic debris, cap the succession. They contain a rich, reworked Eocene fauna, but the age of the formation based on the globigerine content is ?Burdigalian (Wezel *et al.*, 1973). The Tusa Formation is older than the Gorgoglione Formation and rests with apparent conformity upon the Argille Scagliose Varicolore, with which it is closely linked.

The unit as a whole seems closely related both in facies and structural position to the Canetolo unit of the northern Apennines. It is the homologue of the "A.S.V." nappe in Sicily. Consideration of the similarity of the Maestrichtian–Eocene facies, and locally, of the Stampian–Aquitanian with the Verbicaro unit suggest that it bordered the western calcareous Apennines. Following initial thrusting over the internal Apennine platform, the unit was broken up in the Irpinian Basin and formed part of the basement of the Gorgoglione Formation, while its front came to be intercalated in the Numidian Basin sediments during the Burdigalian–Helvetian.

(2) The Silentine, or westernmost unit, most commonly consists of a basal argillaceous Cretaceous sequence of Albian–Cenomanian age, followed by two sandy flysch units, one Turonian–Danian the other Paleocene–Lower Eocene, which crop out between Salerno and Sapri in Cilento, and in Lucania or Basilicate between the lower Crati and Agri Valleys. The 200-m basal complex or Crete Nere Formation of Vezzani (1968) and part of the San Venere flysch (Ietto *et al.*, 1965) consist of an alternation of dark or polychrome shales with limestones or fine-grained sandstones with nodules and flakes of manganese. The fauna, which includes *Hedbergella* and *Globigerinelloides*, followed by *Ticinella–Pitonella–Rotalipora* and *Praeglobotruncana*, suggests an age range from upper Albian to basal Cenomanian. The complex is most often separated from its subbasement (Frido Formation) and sometimes includes sedimentary klippe (peridotites, ophicalcites, ophiolitic series, and granite) (Grandjacquet, 1971).

The Cenomanian–Maestrichtian–Danian–Paleocene succession continues with a 600-m sandy, flysch facies, the Saraceno Formation of Vezzani (1968) and Pollica Formation (Ietto *et al.*, 1965), a sequence of sandy, cherty calcarenites, calcilutites, and yellow micaceous marls. Sedimentation continued, either without break (San Mauro) or after transgression (Alidona), with a thick sequence of sandy, conglomeratic flysch with large-scale turbidites interspersed with olistostromes. Some conglomerate horizons up to 30 m thick contain large boulders ranging in composition from granites or microgranites, gneiss, phyllite, basic rocks, marble, and limestone blocks of Mesozoic or Paleocene age. One or two 10–20-m beds of marly limestone occur in the sequence. In Cilento, olistostromes of a sandy scaglia and a variety of blocks invade the sequence (Cocco and Pescatore, 1968). The youngest of the latter elements is Paleocene–basal Eocene. The formation as a whole has a Lower–Middle Eocene age. This unit too, after thrusting over the Apennines, forms part of the basement of the Gorgoglione Formation; previously it was to a certain extent folded up with the Argille Scagliose Varicolore.

B. The Silan Crystalline Complex

The regions of Sila, the coastal chain, the San Bruno, and Aspromonte Mountains are covered by the Silan crystalline complex which corresponds to the Peloritan Massif across the Straits of Messina. An isolated klippe is found of the upper Sinni Valley at Episcopia. Many subunits can be recognized within two major units, the garnetiferous gneiss, granites, and upper phyllites, and augen-gneiss with the lower phyllites.

(1) *The garnetiferous gneiss, granites, and upper phyllites* [the Upper Sila nappe, the phyllites and granites of Dubois' (1970) intermediate nappe, and the granitic and the kinzigitic diorite units of Colonna and Piccaretta (1975)]; this upper unit has a Mesozoic and Eocene sedimentary cover (Longobucco–Rossano, Tiriolo–Martirano, Ionian margin of Serra San Bruno–Aspromonte). The unit has not been subjected to penetrative Alpine deformation except along the sole and represents the highest basement slice in the Calabrian structure. It consists of a slice of garnetiferous-sillimanite-cordieritic gneiss with bands of amphibolite and pyroxenite, which can be traced from Belvedere to Catanzaro and into Serra San Bruno. It is bounded to the east and west by Paleozoic epimetamorphic phyllites, which have been injected by Carboniferous syntectonic granites (Dubois, 1971). At Longobucco, Triassic–Liassic sediments ("anagenites," reddish quartz conglomerates, and marly limestones) and Upper Cretaceous–Lower Eocene sediments rest transgressively on the basement. Between Tiriolo and Amantea, other relics of the sedimentary cover are preserved in a Neogene graben (sandstones with plants, polychrome argillites of ?Triassic–Jurassic age, and Tithonian clypeine limestones). Transgressing over the various tectonic units are Middle Oligocene and then Upper Miocene beds.

Along the eastern margin of the Serra San Bruno–Aspromonte, Triassic–Jurassic and Cretaceous carbonates are discordantly overlain by Upper Aquitanian shales and sandstones, which in turn support allochthonous units (Argille Scagliose Varicolore and Numidian). A klippe in the upper Sinni Valley at Episcopia is made up of garnetiferous gneiss and peridotite with veins of garnetiferous pyroxenite.

(2) The lower and intermediate nappes (in part) of Dubois (1970), the Vaccarizzo unit of De Roever (1972), the phyllites of Fiume Arente and the Paola schists of Bonardi *et al.* (1975), the phyllite metaconglomerate of Colonna *et al.* (1970), the phyllites, mica schists, and augen-gneisses of Colonna and Piccaretta (1975) can all be grouped into the *Lower phyllite, mica schist, and augen-gneiss unit.* The effects of penetrative alpine deformation are evident. The augen-gneiss masses seem to be often tectonically separated from the phyllites and mica schists with which they are normally associated.

In Terme Luigiane, Triassic quartzites and dolomites and Jurassic calcareous shales, red shales, and radiolarites transgress over this unit (Scandone, 1970), but the whole sequence was folded and subjected to epimetamorphism. At Gimigliano, ophiolites are associated (see below) with it. The unit, broken into scales, underlies the preceding unit.

The Silan crystalline complex is thus formed of the two nappe units, the higher made up of an old granulitic basement of metamorphosed Paleozoic sequences overlain by Mesozoic and Paleogene beds. This unit was not affected by Alpine penetrative deformation except along the sole of some reversed faults. It has also been broken up into scales directed toward the northeast in the Paleogene. The oldest deformation with penetrative schistosity occurred during the Carboniferous, the date of granite emplacement (Borsi and Dubois, 1968).

The lower crystalline unit of essentially Paleozoic material, possibly associated with ophiolites, has undergone Alpine penetrative deformation, with schistosity affecting the whole unit.

C. The Ophiolite Complex

Broken up into several thrust sheets, the complex underlies the northern part of the Silan Massif. It crops out in northern Calabria and is exposed in several windows in the coastal chain along the Crati Trough and north of the Catanzaro Trough. Four units of basic rocks can be recognized resting on a sole formed by the Frido Formation.

(1) The highest, the Gimigliano–Reventino unit, is exposed in a tectonic window through the Silan Massif (Dubois, 1970). At Gimigliano, it consists of peridotites overlain by ophicalcites and followed by prasinites. Green and violet sericitic schists intercalated with quartzites containing pebbles of rose quartz (Verrucano facies) and Triassic marbles (Colonna et al., 1972) locally transgressive onto Paleozoic phyllites, seem to thrust the Gimigliano–Reventino unit in an early tectonic episode. In the Coastal chain and in the Crati trough this last formation overlays tectonically the Rose–Fuscaldo unit.

(2) The second, or Rose–Fuscaldo, unit always appears in windows south of the latitude of Cetraro, under the Calabrian crystallines (Quitzow, 1935; Grandjacquet, 1969; Dubois, 1970; De Roever, 1972; Dietrich et al., 1972; Bonardi et al., 1974). It consists of massive or pillowed porphyritic prasinites, locally schistose, overlain by green and violet chlorite and epidote schists, in which marmorized microbrecciated limestones with ankerite can be found, and particularly well represented between Cetraro and Fagnano Castello (the Rota Greca marbles, in part, of De Roever, 1972). The unit has been assigned a Malm–Berriasian age by comparison with, and perhaps because of continuity, with the succeeding unit. At Rose, the schistosity appears concordant with that of the Paleozoic phyllites.

(3) The Mezzane–Laïze–Malvitto unit takes the place of the sequence found in the preceding unit with massive or pillowed porphyritic prasinites (diabase porphyrites of Quitzow) overlain by radiolarites and limestones or microbreccias with calpionellas of Tithonian–Berriasian age. The unit extends from the Sinni Valley in the north (Mezzane) to the Tyrrhenian coast (Laïze near Belvedere, Cetraro) and the lower Crati Valley (Malvitto, Terranova di Sibari). The sedimentary cover is often tectonically separated from the pra-

sinites. At Mezzane (Bousquet, 1962, 1963; Vezzani, 1968; Spadea, 1968) the basal part of the sequence consists of Malm globochaete cherty limestones. The total age range is thus within the Upper Jurassic–Berriasian. Locally, the top of the calpionellid limestones pass to shales, calcareous shales, and quartzites identical to those found in the Frido Formation. The passage beds are rare, for the Frido Formation, as a general rule, is separated from the subbasement, although it is included in numerous tectonic scales.

(4) The lowest, Diamante–Terranova di Sibari, unit is separated from the overlying unit by lenticular, serpentinized peridotites. It consists of glaucophane prasinites, lawsonite, and calc-schists (Quitzow, 1935; Dubois, 1967–1969; De Roever, 1972; Hofmann, 1970; Grandjacquet, 1969), a homologue of the Alpine "schistes lustrés." Of unknown age, it is generally assigned to the Jurassic ophiolitic complex, although an older age cannot be excluded. The unit is broken up and appears in lenses as thin as 100 m or in blocks spared by erosion. It is found in the Lao Valley (Scalea, San Domenica Talao, Mormanno), the upper Sinni Valley, along the Tyrrhenian coast, and in the Crati Valley between Terranova di Sibari and Fontana Tavolara. The glaucophane lawsonite prasinites are either massive or pillowed, and are porphyritic or fine-grained dolerites. The associated schists and calc-schists include intercalated ancient hyaloclastites.

The Frido unit, defined by Vezzani (1960) and made up of schists, quartzites, and microbrecciated limestone (Flysch à quartzites) (Grandjacquet, 1961) acts either as the sole or the tectonic matrix of all the ophiolitic units, much as do the Val Lavagna schists of the upper Ligurian Apennines. The principal outcrops are in the upper Sinni Valley (Torrente Frido), where it rests upon the Verbicaro unit and in the central Calabrian coastal chain as far as the Catanzaro trough. It has a Lower Cretaceous age and has undergone anchimetamorphism. Synsedimentary deformation is common and it has been further affected by synschistosity, folding, and fracture cleavage or, locally, penetrative deformation. In the form of a nappe it rests on the Apennines, emplaced after the deposition of the shale with blocks.

Within this succession of ophiolitic units which are thrust everywhere in northern Calabria, the Mezzane Laïze Malvitto may be assigned to the Upper Jurassic–Berriasian. Of the three remaining, undated units, the Rose–Fuscaldo may be assigned to the Jurassic because of its probable continuity with the first, and the Diamante–Terranova and Gimigliano units to either the Jurassic or an older age. Peridotite is found associated with the Gimigliano unit only. It is possible to attribute to the Mezzane–Laïze–Malvitto unit the crushed slices of peridotite which underlie it. The tectonic contacts which separate the units are seen to be anomalous and post-metamorphic if the sudden changes in metamorphism and deformation grade between units are taken into account.

There is a certain parallelism between the nature and disposition of these units with the Alpine ophiolites, for example, those of the Voltri group in Liguria.

VII. METAMORPHISM AND DEFORMATION

A. Compressional Tectonics

1. *The Apennines s.s.*

The Lower Tortonian conglomerates of Monte Alpi, transgressive over PF_2, have a strongly impressed schistosity in the southwestern part of the massif. The pebbles are flattened and the schistosity is subhorizontal. This deformation, dating from the middle–Upper Tortonian tectonic episode, was probably linked to northerly thrusting and, perhaps, also to E–W displacement.

The metamorphic facies of the Verbicaro and Campo–Tenese units have also been subject to deformation. In the internal domain of PF_1, synschistose folding was associated with the formation of green schists. In the Verbicaro unit only the Buonvicino–Maiera region is affected by intense deformation, while the whole of the Campo–Tenese unit is deformed with all beds from Triassic to Aquitanian being affected. At high horizons, three generations of folds can generally be recognized (synschistose isoclinal folding, refolding of this schistosity by E–W-directed folds with an isoclinal tendency, and folds directed to the northeast with a fracture cleavage along a N110–140° axis). At deeper horizons at San Donato di Ninea and in the coastal chain, the following phases of folding can be recognized: 1) isoclinal folding B_1 with penetrative schistosity and the formation of a quartz–chlorite–sericite assemblage in the pelites and actinote–epidote–chlorite–albite assemblage in the dolerites; the axis of these folds from lineation measurements was probably N20–70°; 2) synschistose isoclinal folds B_2, along a N110–150° axis associated with the formation of sericite in the S_2 plane; 3) concentric B_3 folds tending to be isoclinal along on N70–90° axis; 4) overturned axial plane folds, fracture cleavage, kink bands B_4 (N140–110°) and B_5 (N40–0°). The synschistosity folds are all post-Middle Aquitanian and pre-Upper Tortonian–Messinian.

2. *The Pre-Apennine Domain*

The green schist facies of the Gimigliano–Reventino unit shows evidence of deformation. The prasinites most commonly are in the actinolite epidote ± lawsonite–albite ± glaucophane sodic pyroxene–albite facies. The metabasites usually have a well-defined schistosity with an E–W (N70–110°) lineation

(Dubois, 1970). Conjugate folds deform the early structures giving crenulation and S_2 planes oriented 150 NE 60 and 40 SE 50.

In the Rose–Fuscaldo unit, a lawsonite–blue-green amphibole–pumpellyite facies is found. The prasinites of the unit have locally undergone tectonic thinning with development of penetrative schistosity, and the porphyritic phenocrysts are elongated and show a remarkable E–W lineation (N70–90°). The schistosity associated with the flattening provides planes along which sliding can occur. The structures are deformed by folds and fracture cleavage (overturned). Two conjugate systems can be recognized at Rose, the S_2 planes of schistosity are 40 SE 55 and 150 NE 60. These latter structures are subsequent to the covering of the greenstones by the phyllite and orthogneiss unit.

In the Mezzane–Laïze–Malvitto unit, the green schist facies of massive or pillow basalts are fairly rich in small crystals of lawsonite and locally small blue-green amphiboles, and sodic pyroxenes are present. A prekinematic pumpellyite assemblage can also be found (De Roever, 1972). These rocks are often little deformed save for traces of penetrative schistosity. The sedimentary cover normally attached to the diabase generally shows disharmonic folding, and is separated from its substrate in the southern part of the zone where the unit crops out.

The prasinites in the Diamante–Terranova di Sibari are characterized by two parageneses, the first with glaucophane, lawsonite $\pm$ aragonite $\pm$ jadeite pyroxene shows a synschistose crystal alignment N40–90° (De Roever, 1972; Quitzow, 1935; Dubois, 1967–1969; Hoffmann, 1970). The second, more recent, is a green schist facies. The unit is deformed by concentric folds oriented N40° and N110–140°.

The Torrente Frido series is intimately deformed but does not exhibit a general metamorphic paragenesis. Only small crystals of sericite–chlorite–quartz–albite–ankerite can be observed. Penetrative schistosity can be clearly seen in some pelitic units. This schistosity, which imparts a metallic luster to the pelites, has been refolded by various noncylindrical fold systems. The main phases of deformation probably date from upper Middle Cretaceous and Lower Eocene.

The age of deformation and metamorphism of the ophiolitic series is difficult to discern, for no formations younger than Lower Cretaceous occur. The first sediments resting transgressively over the structures and nappes are the conglomerates, marls, and limestones of the Middle Oligocene and the Upper Tortonian–Messinian series. Additionally, flysch of the Silentine unit locally covers formations of the Frido type, either with Albo-Cenomanian or Lower Eocene horizons. It has been suggested that the principal phase of folding and metamorphism in these units is Upper–Middle Cretaceous (Grandjacquet, 1971; Haccard et al., 1972), an age since confirmed radiometrically.

During the course of the Cenomanian, the basal complex of the Silentine Flysch, in addition, receives sedimentary klippe of peridotite and of the dolerite–radiolarite sequence, which have already been folded. The post-metamorphic shearing of the unit is probably of Upper Eocene age.

The crystalline massifs of Calabria, which were subjected to regional metamorphism at the end of the Paleozoic, suffered Alpine retromorphism.

The Paleozoic patterns had, as a result, epizone metamorphism of phyllite grade, retromorphism of granulitic gneiss to kinzigite, with the syntectonic emplacement of anatectic bodies near the basement-cover boundary resulting in the association of andalusitic cordierite–staurolite $\pm$ sillimanite assemblage in the phyllite, and the injection of microgranites, granophyres, and rhyolites into the sedimentary cover (Dubois, 1971).

Alpine tectogenesis divided the basement into two principal segments—the upper and lower phyllites. The upper unit with garnetiferous gneiss, phyllites, and granite has an unmetamorphosed Mesozoic and Tertiary cover. The unit is broken up into scales and thrust, but retromorphism or mylonitization is restricted to the soles. The lower augen-gneiss, phyllite unit shows development of profound schistosity which has led to the formation of phyllonites (De Roever, 1972). Alpine paragenesis involves quartz–albite–sericite–chlorite–stilpnomelane–tourmaline–hematite–actinolite and incorporates the Triassic–Jurassic cover. The origin of the augen-gneiss is still disputed, but results either from the Carboniferous granites and microgranites (Dubois, 1970; De Roever, 1972) or from Paleozoic and Triassic meta-arkoses, according to Colonna *et al.* (1973) and Bonardi *et al.* (1974).

The Alpine tectonic effects on the lower unit consist of a penetrative schistosity with an elongation seen as a lineation in the felspars (N70–90°) (Dubois, 1970). These initial structures were subsequently folded along an (overturned) axial plane accompanied by a crenulation oriented N145°. This second phase refolded the shear contacts, using the first schistosity. The shearing led to superposition of this unit over the Rose ophiolites, whose deformation is the same. These events may be assigned to the upper Middle Cretaceous. During the Eocene the upper unit overrides the lower unit and the ophiolitic series.

All the metamorphic formations described are presently superposed. They are separated by abnormal contacts which explain sudden jumps in metamorphic grade. In addition, the age of the Alpine paragenesis may be very different in different units. The deformation and metamorphism (green schist) of the Apennine series (Verbicaro–Campo Tenese) dates from the Middle Miocene, while the principal deformation and metamorphism of the pre-Apennine formations is Upper Cretaceous and Upper Eocene. There must, therefore, be two tectonic episodes clearly separated in time, and paleogeographical considerations seem to dicate that they were also separated in space.

For the pre-Apennine units as a whole, three principal stages of compres-

sion occurring during the upper Middle Cretaceous and Eocene may be proposed:

Phase I: Synschistose, synmetamorphic folding along a N40–90° axis (excluding only the greater part of upper Silane unit) with slip along the
planes of schistosity. The lower phyllitic unit is thrust over the Gimigliano–
Reventino unit, and both overlie the Rose–Fuscaldo unit and the southern
edge of the Mezzane–Laïze–Malvitto unit. The direction of thrusting is
not clear, it may be S–N to SE–NW or E–W to NE–SW, but it took place
in the upper Middle Cretaceous.

Phase II: A phase of compression giving rise to large folds and to fracture
cleavage directed N40° and N140° from an E–W stress direction. These
structures redeform the units and the slide planes resulting from Phase I
movements. They presumably date from the Upper Cretaceous–Eocene.

Phase III: Shearing which results in the stacking of the outer units, i.e.,
Diamante Terranova over the Frido, the Mezzane–Laïze–Malvitto, Rose–
Fuscaldo, Gimigliano–Reventino, lower phyllite units over the Diamante–
Terranova and Frido, and the upper phyllite unit over all the preceding.
The principal movements were Eocene, although many local displacements
occurred at other times (Acri Calovetto)(Dubois, 1970). Middle Oligocene
sediments are discordant over all these structures, although the sediments
are preserved principally in the Amantea Tiriolo graben.

In the case of the Apennine units, the tectonic episodes follow one another
during the course of the Neogene. Intense folding and metamorphism are
only known (and perhaps only exist) in the Verbicaro and Campo Tenese
units. Only the latter remained for any lengthy period attached to its basement
(just as its Tuscan homologue did in the Apuane window in the northern
Apennines). The majority of platforms were tectonically separated below the
Triassic from a subbasement not known in outcrop. The structures of the
Campo Tenese unit, therefore, provide partial information on the tectonics of
the Alpine subbasement of the western Apennines. The following phases of
activity are recognized:

Phase I: synmetamorphic–synschistose folds along N20–70° axis.

Phase II: synmetamorphic–synschistose folds along N110–150° axis.

Phase III: concentric folds (tending to isoclinal) along N70–90° axis.

Phases IV and V: folds on an overturned axial plane, along N110–140°
and N0–40° axes.

The folds along a N40° axis, perpendicular to NW–SE Apennine directions,
are probably the result of longitudinal compression with respect to the chain
or the slip of segments of it. The thrusting of PF$_1$ over Lagonegro was accom-

panied by shearing and drag folding also due to a longitudinal, SE–NW directed stress.

The N110–150° folds are parallel to the Apennine trend and to the fronts of the principal thrust sheets; they are the result of SW–NE compression which was the cause of considerable crustal shortening. The N70–90° folds are associated with thrusting and shortening in a N–S direction.

Slide movements involve upper Middle Aquitanian rocks and are pre-Upper Tortonian, as this unit rests upon an erosional unconformity. It is impossible to distinguish between continuous compression and a series of pulses, ending in the Upper Tortonian–Messinian prior to the major shearing movements of the Middle Pliocene.

B. Tensional Tectonics

1. *The Apennine Domain*

The existence of normal faults on the various platforms can be demonstrated. During the Mesozoic and Paleogene, the edges of the platforms were, on several occasions, the site of tensional faulting which controlled sedimentation, the location of basic igneous rocks, and erosion. After Lower Miocene time, normal faults on the platforms in advance of the nappe fronts can also be demonstrated. The nappes are partially fossilized by the clay with blocks and the conglomerates of the Irpinian Basin, and by Tortonian–Messinian and Pliocene–Quaternary sediments. Finally, over the whole Apennine structure, recent normal faults (neotectonic of Bousquet, 1972) permitted the subsidence of the Tyrrhenian domain, the formation of grabens within the nappes, and the tilting of the Apennines towards the Bradanic Trough.

2. *Pre-Apennine Domain*

In the Silan crystalline in particular, tensional phases following the Alpine tectogenesis can be recognized. They are known to be Oligocene in age, for sediments discordantly overlie faults. Upper Miocene rocks are discordant upon those of the Oligocene where it is preserved in grabens, and over the eroded horsts where the substratum is exposed. Finally, the Plio-Quaternary invades still-active grabens where Miocene beds are generally preserved.

VIII. PLUTONISM AND VOLCANISM

Excluding modern volcanicity, the principal evidence of volcanic and plutonic activity found in the various formations of the southern Apennines is briefly indicated below:

A. Burdigalian Cherts

The cherts found with the PF₂ calcarenites (Monte Alpi) and in the Irpinian Basin (Sangro) may be evidence of an important phase of acid volcanism at the beginning of the Miocene which has been recognized elsewhere in the Mediterranean (Didon *et al.*, 1969).

B. Picrites–Teschenites–Limburgites of the Verbicaro Unit (Maestrichtian–Aquitanian)

It is only in this unit that the association of the picrite suite with spilites (= limburgites of Quitzow, 1935) is found. The effusions range from Maestrichtian to Middle Eocene and perhaps as high as Stampian–Aquitanian. They consist of massive or pillow lavas with hyaloclastites interstratified in siliceous pelites with intraformational calcareous breccias (Maestrichtian–Paleocene). Most often, calcareous breccias or oolitic sands contain boulders and fragments of lava, but there are also deposits where the lava has incorporated soft blocks or pebbles of limestone. Synsedimentary deformation of bedding resulting from mixing of lava and carbonate can also be observed. Some soft carbonate pebbles are coated with chloritic-albite and acicular stilpnomelane crystals. The same paragenesis is found at the contact between sills and Eocene calcarenites.

Intrusion of granular picrites and teschenites is found only within the dolomites and cherty limestones of the Verbicaro unit. Its relationship to the lavas is fairly clear. The picrites, dark granular rocks with poikiolitic olivine and pyroxene, biotite and mica pass, in the peripheral regions, to teschenites by enrichment in tabular feldspar and then into limburgites.

C. Paleocene–Eocene Basaltic Tuffs

These occur on the Apulian platform (Murge–Salento).

D. Tusa Tuffites

These tuffites (Ogniben, 1969) of Eocene or Lower Miocene age appear as andesitic sandstones capping the A.S.V. (Argille Scagliose Varicolore) unit. These arenaceous rocks are rich in fresh crystalline andesitic material, plagioclase, biotite, and also chlorite, quartz fragments, orthoclase, granite, garnetiferous gneiss and augite, along with an abundance of fragments of Eocene nummulites. The age of the deposit is Lower Miocene, according to Wezel *et al.* (1973), the most probable time of reworking of the sediment. The deposits pro-

vide evidence of andesitic activity whose paleogeographic location is unfortunately, in doubt.

E. Tuffaceous Ghosts

These ghosts in the Argille Scagliose Varicolore occur in the Maestrichtian–Paleocene microbreccias observed in the Roccagloriosa window (Cilento) and Sasso l'Armi (Lucania). Calcitic pseudomorphs, after plagioclase and chlorite found in the cement of the microbreccias, suggest the presence originally of tuffitic debris.

F. Massive and Pillow Dolerites

These are found in the Paleocene–Eocene flysch of Cilento in the San Mauro Formation (Silentine unit), at Monte Centaurino, and at Tramutola (Migliorini, 1944). These lavas, more or less alkaline, are considered by Cocco and Di Girolamo (1969) to be in their primary position, but are regarded by Dietrich and Scandone (1972) as olistoliths of the Jurassic ophiolite sequence. An analysis of the conditions of deposition favors a primary origin. The homogeneity of the pillows, whether isolated or *en masse*, in the sandstone and hyaloclastites, the concordance between the top of the flow with pillows of decreasing size and a pillow breccia, and the calcareous sandstones with reaction rims at the lava contact, all point to a primary sedimentary and not reworking or a tectonic emplacement.

G. Pthanites

Pthanites in the Vraconian–Cenomanian flysch in the transition zone between the Silentine Series and the Argille Scagliose Varicolore and at the base of the latter is evidence of probable volcanism of Middle Cretaceous age.

H. Massive or Pillow Diabases

These are found in the Jurassic ophiolitic units related to Ligurian or Piedmont units. They are usually metamorphosed, but assigned to a phase of oceanic volcanicity occurring between upper Middle Jurassic and the Tithonian–Berriasian. The lavas are usually found as tectonic scales separated from their subbasement and cover. The presence of lenticular peridotite is perhaps evidence of a partially ultrabasic subbasement. The lavas are principally in the form of metamicrogabbros–metadiabases, often porphyritic, and associated with metahyaloclastites.

I. Diabases

Diabases occur in the Campo Tenese unit and diabase and peridotite occur in the Middle Triassic Gimigliano–Reventino unit. In the schists and limestones of San Donato di Ninea–Aquaformosa of the Campo Tenese unit, Ladinian metabasites and tuffites are found which have been altered to actinolite–epidote–chlorite–albite–chloritoid schists. (The blocks of lawsonite glaucophanite occurring in this area are either erosional relicts or blocks resedimented in the Upper Miocene of the allochthonous Diamante–Terranova di Sibari unit.)

The prasinite of the Campo Tenese unit may be the southern continuation of the Punta-Bianca Argentario series of the northern Apennines.

J. Middle Triassic Tuffites, Diabases, and Pillow Breccias

Blocks of Middle Triassic tuffites, diabases, and pillow breccias were described by Dietrich and Scandone (1972) in the upper Lagonegro unit as occurring in the conglomerates of Monte Facito. They were assigned a Ladinian–Carnian age. The mineral paragenesis consists of plagioclase, calcite, chlorite, epidote, sphene, and apatite. Sanidine, plagioclase, calcite, quartz, chlorite, apatite, and zircon bearing tuffites with *Halobia* are also recorded.

K. Middle Triassic Basic Rocks

Middle Triassic basic rocks occur in the Punta delle Pietre Nere (Gargano) beds.

L. Carboniferous Granites

Carboniferous granites and microgranites found in the Silan crystalline were injected into the basement and Paleozoic cover of the massif (Borsi and Dubois, 1968) during the Hercynian orogenic cycle.

M. Conclusions

If the igneous activity of Hercynian and Plio-Quaternary age is excluded, then igneous activity, plutonic and volcanic, occurred during three principal epochs:

Middle Triassic: diabases and tuffites.
Upper Jurassic: ophiolites, hyaloclastites.
Maestrichtian-Nummulitic: dolerites, tuffites, picrites, teschenites.

To these may be added the minor activity of Cenomanian pthanites and Burdigalian cherts. The problem of the age of the Tusa andesitic tuffites is unsolved, and the geographic localization of the volcanic source is problematic.

IX. GEOGRAPHIC AND TECTONIC EVOLUTION OF THE SOUTHERN APENNINES

The southern Apennines are part of the Alpine structure of the central Mediterranean. They have been subjected to successive tectonic events. The units involved and their geographic position makes them part of the general Alps–Apennines problem; that is, one part of the problem arising from the convergence and collision of the European and African (and Apennine–Austroalpine) blocks. It must be remembered that nowhere in the southern Apennines does basement crop out. The region is one of nappes, and even the Silan crystalline massif, at the top of the structural pile, is only a nappe originating from pre-Apennine basement.

Starting from the Permo-Triassic the attempt can be made to trace the paleogeographic and tectonic evolution of the region through the main orogenic phases. During the Triassic and Jurassic, in particular, a phase of opening of an oceanic region occurred. Small openings, localized in the Upper Triassic, were followed by the main Ligurian–Piedmont–Calabrian trough opening up during the Jurassic. This was probably linked to the opening of the southern part of the North Atlantic (Dewey et al., 1973), and it led to the more or less complete separation of the Apennine–Austroalpine block from the European craton.

During the Cretaceous and Eocene, metamorphism and thrusting of the ophiolitic series and of the cratonic margin resulted from the crushing of the oceanic domain between the two cratonic masses. In Miocene and Pliocene time, folding and metamorphism of the Apennine series with thrusting and a cover of pre-Apennine nappes was the result of the collision of the Apennine–Austroalpine block with the European craton and the rotation of the Corso-Sardinian–Tyrrhenian block.

A. Chronology of Events

1. *Permian to Lower Triassic*

Following in part the paleogeographic model proposed by Dewey et al. (1973), the southern margin of the European platform was formed by an Insubric type zone (of which the residual fragments occur in the Austrian Alps, Insubrian domain, Calabria, the Kabylias, and the Rif). The nucleus of ancient crystalline rocks was partially covered by Paleozoic phyllites which had been injected by Carboniferous and Permian granites and rhyolites (porphyrites). Transgressively overlying these deposits are the Verrucano and an evaporitic series. Towards the open sea to the south and east lay a Permian flysch basin, of which remains now occur in Sicily and Lucania. The African block must also have extended farther to the north and northeast as a thin continental crust.

2. *Middle Triassic*

In the southern Apennines, Middle Triassic sediments (Verrucano), pelites, sandstones, and conglomerates existed in what is now the Campo Tenese unit, and over the phyllites of the Calabrian Massif. In the Lagonegro basin *Halobia* limestones and marls were deposited as well as turbidites. In certain of the units the existence of submarine basic effusions is evidence of tension. At this period the continental margin, in process of disruption, passed to the south and southeast into a pelagic basin broken up by platforms, where evaporites and bituminous beds of the future (PF_2–PF_3) platforms were deposited. To the west, the central part of the Silan crystalline massif was partly emerged.

3. *Upper Triassic–Dogger*

The western and eastern margins of PF_1, and probably the other carbonate platforms as well, were fractured and flexed. Here dolomitic reefs developed with cherty limestones on the slopes (especially Verbicaro, Foraporta). Carbonate deposition was general over the platforms. In the sinking Lagonegro Basin a fine-grained pelagic sedimentation regime became established in the distal part with terrigenous sedimentation in the proximal part. In Sila, over the subsiding upper unit, dolomites and limestones were deposited in the south, anagenites (reddish sandstones and quartz conglomerates) and a detrital calcareous and siliceous sequence in the north. Over the lower phyllite unit a sequence of dolomites and calcareous shales were deposited above the Verrucano facies. These beds may form a transition to the open-water Mezzane limestones, indicating the establishment of a pelagic basin west of the Silan crystalline massif.

4. *Upper Jurassic–Lower Cretaceous*

The ophiolitic basin became distinctive with pillow basalt flows, followed by the deposition of radiolarites and calpionellid limestones. This large basin follows the length of the western Apennine margin from the Alps to Calabria (= Piedmont–Ligurian–Calabrian Basins). Thus, the process leading to the opening of an oceanic basin, which had begun tentatively in the Middle Triassic, extended through the Liassic and Dogger. The relationship of the ophiolite basin to the Silan crystalline massif is not well known. It is possible either to consider that the paleogeographic units of southern Italy are but a continuation of those of the northern Apennines and of the Alpine system *s.s.*, in which case the ophiolites lie west of the Insubric–Silan domain, or that the ophiolitic basin divides the Insubric domain, so that between Lombardy and Tuscany on the one hand and Calabria on the other the relationships are reversed. This would result in the Insubrian domain forming part of the Apennine units in the north while to the south it would be to some extent bound to the European domain. In the first case, thrusting of the Silan basement over the ophiolitic series is toward the west. It is also possible to imagine the coexistence of two ophiolitic basins. Yet another alternative proposed by Dubois (1970) is to consider that the ophiolitic basin separates the Apennines in the north from a Sicilian–African ensemble in the south, of which the Silan crystalline massif represents the frontal zone.

The lower phyllite unit is covered by radiolarites and limestone. The western and eastern margins of PF_1 (Verbicaro–Foraporta) form the continental slopes with reduced sedimentation. In Lagonegro, radiolarites and calpionellid limestones are found. Upon the platforms ($PF_{1,2,3}$) carbonate sedimentation continued. Over the Silan crystallines of the upper unit sporadic reef deposits with clypeines have been recorded.

During the Lower Cretaceous in the ophiolitic trough and Lagonegro trough silty pelites, calcareous microbreccias, and fine-grained sandstones were formed (Galestri Eo-Cretaceous of Lagonegro, Frido Series of the ophiolitic basin).

5. Middle Cretaceous

This was a period of interruption in sedimentation in the Lagonegro Basin and the ophiolitic trough. A large basin of Argille Scagliose Varicolore and of the Silentine unit developed and expanded. Sedimentation began locally with a basal complex (Crete Nere Formation, Vraconian–Cenomanian age) in the flysch basin of eastern Cilento and Lucania (Silentine unit) and by phtanites and siliceous limestones (porcellanites) in the Argille Scagliose Varicolore. The basal complex contains sedimentary klippe of peridotites and of the ophiolitic series.

Over PF_1 the carbonate sedimentation gave way locally to the formation of orbitoline marls, while the marginal slopes remain unchanged. PF_2 however was emergent as is indicated by the formation of bauxites. The bathymetric and paleogeographic variations were probably due to the initiation of movement of certain of the units, reflecting a phase of tectogenesis particularly well marked in the eastern and central Alps, but also known in the western Alps and Calabria. The beginning of tectogenesis in the pre-Apennine domain can be interpreted in one of several ways. The ophiolitic series may have been followed and metamorphosed by uniform subduction to the east and southeast (under the pre-Apennine domain) north and south of the Apennines, with certain slices being obducted along the eastern margin of the flysch basin (sedimentary klippe of the basal complex). Alternatively, the ophiolite series was folded and metamorphosed by east and southeastward subduction under the pre-Apennine Insubric domain (linked to the Apennines) in the north and beyond a transform zone by subduction to the west and northwest under the Silan domain linked to the European block in Calabria. Finally, Dubois (1970) and De Roever (1972) suggest northward thrusting of the Silan crystalline block.

6. *Upper Cretaceous (Including the Maestrichtian Locally)*

Over the platforms carbonate sedimentation continued or returned while a slow, general subsidence was taking place, except on the margins (continental slope deposits). The subsidence continued until the Upper Senonian. Along the eastern margin of southern Sila perireefal deposits formed.

In the flysch basins sedimentation of the sandy flysch of Cilento (Saraceno–Pollica) and Argille Scagliose Varicolore continued. No deposits of this age are known in the ophiolite units or in the lower phyllite unit of Calabria, and it can be inferred that these units were undergoing synmetamorphic–schistose tectogenesis.

7. *Maestrichtian–Paleocene*

Over the platform PF_1, almost totally emergent since the end of Senonian, a progressive transgression occurred during the Maestrichtian covering the Verbicaro (siliceous shales, calcarenites, and breccias associated with basic rocks) and Maddalena units (breccias and calcarenites), while during the Paleocene the transgression extended over the Pollino–Alburno unit (green shales, calcilutites, and calcarenites).

In the Lagonegro Basin, sedimentation recommenced with siliceous argillites, calcarenites, and scaglia in the Maestrichtian and Paleocene. In the Argille Scagliose Varicolore basin calcilutites, calcarenites, varicolored shales, and cherty microbreccias formed, the latter derived from the platform (PF_1). Sandy deposits continued in the Cilento Flysch domain, with the sequences becoming progressively thicker.

This period, one of comparative calm in the southern Apennines, was a period of important tectonic activity in the eastern Mediterranean and the eastern Alps. In the southern Apennines it was marked by a renewal of sedimentation in the basins and on the platforms, with sediments often conglomeratic in unstable zones and coarser turbidite deposits in the basins. The Verbicaro Zone was cut by picrites–teschenites and limburgites. In the ophiolitic units and the lower phyllitic unit no deposits are known from this time, and tectogenesis was probably in progress. It may be envisaged that large scales of crystalline, and of newly metamorphosed ophiolites, were actively thrust over the Frido Formation.

8. *Eocene–Upper Middle Aquitanian*

The transgression begun earlier continued over the platforms, and over the slopes of platform PF_1 Ypresian and Lutetian calcarenites and calcilutites were deposited upon the Verbicaro–Maddalena and Campo–Tenese units. The Priabonian was likewise transgressive. During this time the center of the platform was undergoing karst weathering with the formation of bauxites. The Stampian–Aquitanian was likewise transgressive with the formation of calcarenites.

Eocene calcarenites are also transgressive over various parts of PF_2 and PF_3. In the Lagonegro Basin argillites and calcarenites formed locally. In the Argille Scagliose Varicolore domains the Upper Eocene and Stampian–Aquitanian is marked by deposition of microbreccias, sandstones and argillites, and conglomerates with calcareous elements derived from a platform. The Tusa tuffites, probably formed at this time, contain evidence of an andesitic episode. In Cilento the San Mauro Flysch of middle Lower Eocene rests with local discordance upon the different members of the Silentine unit. It is rich in large crystalline elements of Silan type and has numerous olistostromes and also intercalated alkaline basalt flows. It will be recalled that the limburgites were still being emplaced in the Verbicaro unit.

Over the upper unit of the crystalline Silan, Paleocene–Lower Eocene scaglia and polygenetic conglomerates transgress onto crystalline rocks and the Mesozoic series. They formed scales with basement, probably before Stampian time. The latter is represented by conglomerates and calcarenites which transgress over basement nappes (probably eroded) (Acri Caloveto scale, Dubois, 1970; Ypreso-Lutetian phase in Sicily, Truillet, 1969–1970.)

9. *End-Aquitanian–Middle Tortonian*

The end of the Aquitanian to Middle Tortonian marked the first compressive phase of tectogenesis in the Apennines *s.s.* during which the pre-Apennine domain lost its paleogeographic identity, becoming incorporated, little by little, into the Apennines over which it is thrust. Sedimentation was typically wild-flysch. From the end of the Aquitanian onward, the internal platform PF_1, faulted and temporarily emergent and eroded together with the proximal Lagonegro domain, was covered by clay and blocks derived from PF_1 and the front of the Argille Scagliose Varicolore nappes. The Numidian Sandstone accumulated in the distal part of the Lagonegro domain. A Miocene foredeep, the Irpinian Basin, came into existence. Under stress the platform PF_1 broke up and slices were thrust over the western margin of Lagonegro; over the clay

with blocks, sedimentary klippe slid from the nappe front. During the Burdigalian and Helvetian sandy conglomeratic sedimentation spread over the
backs of the nappes in the south (Gorgoglione) and in front of them, over
PF_1 disrupted by faulting, and over Lagonegro in the north. The Serra Palazzo
Series (Serra Cortina–Serra Palazzo *s.s.*, Masseria Palazzo) was deposited in
the Numidian Basin, and, during the Helvetian, the Gorgoglione facies invaded
the basin (Masseria Luci). Concordant upon the Jurassic beds, a cherty calcarenite formed during the Burdigalian on PF_2 (Monte Alpi). This formation
spread over the Serra Palazzo series and the Argille Scagliose Varicolore in
the Molise Basin. Toward the end of the Helvetian the Irpinian Basin was
partially filled and the western part of the basin, separated from its subbasement
(proximal Lagonegro), broke into scales and was thrust over the eastern part
of the basin (Numidian–Serra Palazzo domain). During the Helvetian–basal
Tortonian, sandstones, conglomerates, clay with blocks, and sedimentary
klippe derived from the nappes were formed over PF_2. These terrigenous
sediments increased to the east at the same time that the PF_2 unit was broken
up by normal faulting. In Calabria during this time, under the pile of pre-
Apennine nappes, the Verbicaro and Campo Tenese subbasement, probably
united to its basement, was subjected to significant shortening and polyphase
metamorphism. All the while erosion was reducing the emerging structures.

10. *Upper Middle Tortonian–Middle Lower Pliocene*

As a result of erosion which exposed numerous units, molassic sediments
of Tortonian–Messinian age and then Lower Pliocene rest discordantly upon
the preceding structures. Such deposits are found over the Argille Scagliose
Varicolore, the Gorgoglione and Frido units, as well as on PF_1 where Triassic
was exposed as a result of faulting and erosion (Campo–Tenese), on the Calabrian crystallines, on PF_2 and PF_3, and are actually buried by nappes in the
Bradanic Trough. During this time the Campo–Tenese unit, in particular,
continued to be folded and up-arched with local readjustments of the tectonic
cover (Verbicaro) taking place. The eastern margin of the Sila was covered
before the Upper Tortonian by retrothrusting of the Argille Scagliose Varicolore, and the same phenomenon occurred on the Valsinni Ridge (the anti-
sicilide movement of Ogniben, 1969) as the result of the folding of the Numidian
and the Argille Scagliose Varicolore nappes. The Lower and Middle Pliocene
are thus transgressive.

Renewed shear of the proximal Lagonegro occurred in the Middle Pliocene, followed by shear of the basal, distal Lagonegro, which permitted overthrusting of the entire Irpinian domain (sediments, nappes, and subbasement)
and of the PF_1 and its nappes, over and beyond PF_2 as well as over the Bradanic
foredeep. During this time, while the front of the Argille Scagliose Varicolore

nappe was forming scales (in thrusting synthetically or antithetically) over the Middle Miocene of the Bradanic Trough, the hinterland around Monte Sirino was folding vigorously, locally blocked by Mt. Alpi. PF$_2$ finished by also breaking into scales which thrust over the foreland. This movement seems to correspond to a sinistral rotation of the southern Apennines, resulting in an overthrusting of the Pliocene foredeep.

The subsiding troughs, especially the Lao, Crati, and Catanzaro (intracontinental rift) troughs were due to rotation linked to an accentuation of the curvature of the Tyrrhenian Arc. Tensional effects are superimposed on folds which run parallel to the arc, such as the Vulture Valsinni, Potenza Sirino, Pellegrino–Palanuda Ridges, and the Calabrian coastal chain. Further, along the inner side of the arc, S–N compression can be detected, in particular along the Sangineto line in the Apennine subbasement (Campo Tenese).

11. *Upper Pliocene–Recent*

The sediments of this time interval rest transgressively on the preceding structures. The main tectonic features are normal faults (Bousquet, 1972; Gueremy, 1972) and flexures. The Apennines seem to be affected by tensional phenomena linked to the completion of the Tyrrhenian Arc, with one margin abrupt or flexured toward the Tyrrhenian Sea and the other a more regular, if faulted, slope toward the Bradanic Trough and the Ionian Sea.

PART III

ATTEMPTED CORRELATION OF SICILIAN AND CALABRIAN UNITS

There are several difficulties in attempting a correlation, not least of which is the period over which correlation is attempted. The displacement of structural units did not occur everywhere at the same time and the succession of events is not always the same, which complicates any scheme of correlation. Nonetheless there is a starting point, the autochthon of the external zone or foreland.

X. CORRELATION OF THE EXTERNAL DOMAINS

Structurally, the Iblean domain equivalent is the Apulian platform, yet there are notable differences in the stratigraphy of these two regions. The Apulian platform during the Upper Cretaceous was a rudistid zone (a Gavrovo-type ridge) passing northeastward to a zone with scaglia intercalated with turbidites of reefal material. The Iblean region at that time belonged to the Scaglia

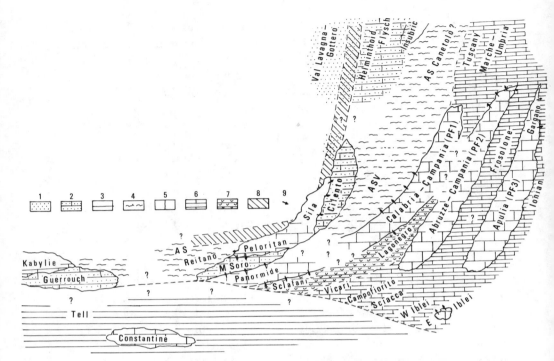

Fig. 17. Tentative paleogeographic reconstruction of the Calabro-Sicilian domain during the Upper Cretaceous, and its relationship to the Apennines and North Africa. (1) Sandy and sandy pelitic series over the ophiolites (Val Lavagna–Monte Gottero–Frido); (2) calcareous flysch and sandy flysch (Helminthoid Flysch, Cilento and Monte Soro Flysch); (3) marls and nodular shales (Tellian facies); (4) Argille Scagliose Varicolore (A.S.V.); (5) reef facies; (6) scaglia; (7) silicified scaglia; (8) ophiolite suture formed at the end of the Cretaceous; (9) source of the crystalline or calcareous detritals.

Zone and the rudistid reefs, the source of turbidite material, occurred only in the east and only at the end of the Cretaceous, as did volcanism.

The Frosolone (or Bradanic) Zone should correspond to the Sciacca and Campofiorito Zones, and the Abruzzi–Campania platform to the Vicari domain. There is a correspondence between the rudistid domain and the Scaglia Basin. To the north, the Abruzzi–Campania platform is continued by the Scaglia domain of Marche and Umbria. Between the two zones the facies interdigitate, so it is not illogical to suppose that a similar passage exists to the south even if, due to allochthonous units and the sea, it cannot be observed. The Tridentine Zone north of Marche and Umbria strikingly resembles the Vicari Zone of Sicily.

The rocks of the Lagonegro and Sclafani Zones are closely similar during the Upper Triassic and Jurassic, and the differences which appear during the Lower Cretaceous attenuate in the Upper Cretaceous, so that there is little difficulty in considering them to be from the same domain. The same agreement exists between the Panormide domain and the Calabro–Campania platform from the Upper Triassic to the Upper Cretaceous.

XI. CORRELATION OF THE INTERNAL DOMAINS

A correlation of the Peloritan domain with the Aspromonte–Serra San Bruno–Sila is acceptable, for there are only two interruptions, the Messina Straits and the Catanzaro Trough. The basement in both is comparable, and the Mesozoic–Eocene of the two domains forms either a reef sequence of the Novara–Stilo type or a sandy Upper Cretaceous–Paleocene Scaglia Series of the Longobucco marginal chain type. Both have the same Oligo-Miocene molasse. Nevertheless, in Sila several units are stacked up over the basement, whereas only one is known in Sicily. In the north the Sila equivalent is the Insubrian domain thrust over the ophiolites at the base of the Upper Cretaceous.

During the Cretaceous the Sila crust provided the detritus of the Cilento Flysch, while the Peloritan basement supplied the Monte Soro Flysch. Although the stratigraphy of the two flysch units is dissimilar, comparable positions with respect the Peloritan–Sila domain along its southern external border (Monte Soro and Capizzi Flysch) and in the northeast (Silentine Flysch) are probable.

It is still necessary to replace the domains thus correlated, external and internal, with respect to the other domains and to locate the "Argille Scagliose" domain. As the external units of the Monte Soro nappe in Sicily appear to be related to the Panormide Zone it seems logical to juxtapose the external province and the Peloritan domain. This results in an ultra-Peloritan position for the "Argille Scagliose," which could explain the presence of reworked blocks of ophiolite in the Upper Cretaceous. In Lucania, in contrast, there is no sign of a possible link between the Cilento Flysch and the Calabro–Campania platform. Here the platform clearly provides the calcareous elements in the Argille Scagliose during the Maestrichtian–Paleocene. It is thus logical to join the two domains, placing the Argille Scagliose Basin between the Calabro–Campania platform and the Sila domain, flanked by the Cilento Flysch.

XII. COMMENT ON POSSIBLE RELATIONSHIP WITH NORTH AFRICA

In North Africa the Rif and Kabyle Massifs are the equivalents of the Peloritan–Sila domain. They are flanked by flysch of the Monte Soro type, the Guerrouch and Tisirhen Flysch, and though not all agree, there is at least a fair presumption that they occupied a position paleogeographically external to the basement. The external domains are represented by the Tellian nappe series and their basement, and by the Baborian units (including the Massylian Series, which appears increasingly like a simple variation of the Tellian Series). The

Tellian and Baborian Series must occupy an external zone corresponding to those of Sicily and Calabria, although there are profound differences. The High Atlas foreland, although very different from the Iblean and Apulian domains, occupies the same relative position. The difference may be explained by the existence of a major discontinuity. It remains to discern the African equivalent of the "Argille Scagliose." The most probable are generally in the north Kabyle, variously named according to the authors (Ziane Flysch, etc.) and often confused with the Massylian Series. There are divergent opinions about the latter too, which is sometimes placed between the Tellian domain and the Guerrouch Flysch, sometimes in either an ultra- or peri-Kabyle location. Further investigation is needed before far-reaching conclusions can be drawn.

XIII. TIME CORRELATIONS

Considering the most recent events (Cretaceous–Tertiary), there is a reasonably good agreement between the Sicilian and Calabro-Lucanian sequences. The same tectonic events may be recognized, though they may not always appear under the same guise. At the base of the Upper Cretaceous the phase of thrusting which affected the Insubrian domain can be seen in Sicily. The sliding of the sedimentary nappes of the marginal chain in the Upper Cretaceous–Paleocene has its equivalent at Longobucco. The discordance between the Middle–Upper Eocene and the Oligo-Miocene of the Peloritan Massif has its equivalent in the Oligo-Miocene discordance in Calabria. The Aquitanian phase in Lucania corresponds in time to the emplacement of the Sicilian Panormid units in the Numidian basin at the same time as those of Calabro-Lucania (Cilento Flysch–Calabro-Campania platform). In both Sicily and Calabria, various units, and in particular the Argille Scagliose, were resedimented several times during the Helvetian and Tortonian. In Calabria the Upper Tortonian was a period of intense folding and metamorphism. In Sicily its equivalent was perhaps the cause of the intense thrusting of the Permian of Lercara Friddi. Messinian resedimentation of the Argille Scagliose is known in both Sicily and Calabria. The shearing of the Abruzzi–Campania platform during the Pliocene corresponds to the thrusting in the Campofiorito and Sciacca Zones. Quaternary movements with sliding of olistostromes (Gela olistostrome, Metaponto nappe) are found in both domains.

REFERENCES

References listed here are additional to those given by Caire in the preceding Chapter.

Accordi, B., La componente traslativa della tettonica dell'Appennino laziale abruzzese, *Geol. Romana*, v. 5, p. 355–406.
Afchain, C., 1961, Observations sur la région de Spezzano Albanese (Calabre) *C.R. Somm., Soc. Géol. France*, p. 287–288.

Afchain, C., 1962, Observations sur la région de Longobucco (Italie méridionale), *Bull. Soc. Géol. France*, (7), v. 4, p. 719–720.

Afchain, C., 1966, La base de la série tertiaire sur le bord de la Calabre ultérieure, *C.R. Somm., Soc. Géol. France*, p. 397–398.

Afchain, C., 1967, Nature du contact entre les "granites" des Serre et les "gneiss fondamentaux" de l'Aspromonte en Calabre méridionale (Italie), *C.R. Somm., Soc. Géol. France*, p. 240.

Afchain, C., 1967, Les argiles écailleuses versicolores et les couches à Huîtres de Motticella (Reggio de Calabre, Italie méridionale), *C.R. Somm., Soc. Géol. France*, p. 329–330.

Afchain, C., 1968, Le témoin calcaire de Sideroni près de Bova Marina (Calabre méridionale), *C.R. Somm., Soc. Géol. France*, p. 329–330.

Afchain, C., 1969, Le substratum des phyllades en Calabre méridionale la fenêtre de Montalbano Jonico (Reggio de Calabre), *C.R. Acad. Sci.*, v. 268, p. 911–912.

Afchain, C., 1970, Présence de Tentaculitidae démontrant l'âge dévonien des niveaux calcaires intercalés dans les "phyllades" du substratum du chaînon calcaire de Stilo-Pazzano (Calabre meridionale, Italie), *C.R. Somm., Soc. Géol. France*, p. 150.

Andrieux, J., 1971, La structure du Rif central, *Notes Mém. Serv. Géol. Maroc*, no. 235, p. 155.

Aruta, L., Buccheri, G., Greco, A., and Sprovieri, R., 1973, Il Siciliano della foce del Belice (Sicilia méridionale), *Riv. Min. Sic.*, no., 136–138, p. 234–239.

Bertolani, M., and Foglia, F., 1975, La formazione kinzigitica della Sila Greca *Boll. Soc. Geol. Ital.*, v. 44, p. 329–345.

Blanc, J. J., 1968, Sedimentary geology of the Mediterranean Sea, in: *Oceanography and Marine Biology. Annual review*, Barnes, H., ed., London: Allen and Unwin, v. 6, p. 377–454.

Blanc-Vernet, L., Chamley H., Froget, C., Le Boulicaut, D., Monaco A., and Robert, C., 1975, Observations sur la sédimentation marine récente dans la région siculo-tunisienne, *Rev. Géol. Médit.*, v. 2, p. 31–48.

Boenzi, F., and Ciaranfi, M., 1968, Sulla presenza di depositi da frane sottomarine nelle argille varicolori dell'Appennino lucano, *Boll. Soc. Geol. Ital.*, v. 87, p. 505–509.

Bonardi, G., Pescatore, T., Scandone, P., and Torre, M., 1971, Problemi paleogeografici connessi con la successione mesozoico–terziaria di Stilo (Calabria meridionale), *Boll. Soc. Nat. Napoli*, v. 80, 14 p.

Bonardi, G., Perrone, V., and Zuppetta, A., 1975, I rapporti tra"metabasiti", "filladi" e "calcisti micacei" nell'area tra Paola e Rose (Calabria), *Boll. Soc. Geol. Ital.*, v. 42, p. 245–276.

Boni, M., Ippolito, F., Scandone, P., and Zamparelli-Torre, V., 1974, L'unità del Monte Foraporta nel Lagonegrese (Appennino meridionale), *Boll. Soc. Geol. Ital.*, v. 43, p. 469–512.

Borsi, S., and Dubois, R., 1968, Données géochronologiques sur l'histoire hercynienne et alpine de la Calabre centrale, *C.R. Acad. Sci.*, v. 266, p. 72–75.

Bouillin, J. P., Durand Delga, M., Gelard, J. P., Leikine, M., Raoult, J. F., Raymond, D., Tefiani, M., and Vila, J. M., 1970, Définition d'un flysch massylien et d'un flysch maurétanien au sein des flyschs allochtones de l'Algérie, *C.R. Acad. Sci.*, v. 270, p. 2249–2252.

Bousquet, J. C., 1962, Age de la série des diabases porphyrites (roches vertes du flysch calabro lucanien Italie méridionale), *Bull. Soc. Géol. France*, (7), v. 4, p. 712–718.

Bousquet, J. C., 1963, *Contribution à l'Etude des Roches Vertes du Nord de la Calabre et du Sud de la Lucanie*, Thèse (3ᵉ cycle), Paris, 145 p.

Bousquet, J. C., 1964, Mise au point sur l'âge du "flysch à quartzites" calabro-lucanien (Italie méridionale), *C.R. Somm., Soc. Géol. France*, p. 55–56.

Bousquet, J. C., 1965, Sur l'allure et la mise en place des formations allochtones de la bordure orientale des massifs calabro lucaniens, *Bull. Soc. Géol. France*, (7), v. 7 p. 937–945.

Bousquet, J. C., 1971, La tectonique tangentielle des séries calcaréo-dolomitiques du Nord Est de l'Apennin calabro lucanien (Italie méridionale), *Geol. Romana*, v. 10, p. 23–52.

Bousquet, J. C., 1972, La tectonique récente de l'Apennin calabro-lucanien dans son cadre géologique et géophysique, *Geol. Romana*, v. 12, p. 1–104.

Bousquet, J. C., and Dubois, R., 1967, Découverte de niveaux anisiens et caractères du métamorphisme alpin dans la région de Lungro (Calabre), *C.R. Acad. Sci.*, v. 264, p. 204–207.

Bousquet, J. C., and Grandjacquet, J. C., 1969, Structure de l'Apennin Calabro-Lucanien (Italie méridionale), *C.R. Acad. Sci.*, v. 264, p. 204–207.

Bousquet, J. C., and Gueremy, P., 1968, Quelques phénomènes de Néotectonique dans l'Apennin Calabro-lucanien et leurs conséquences morphologiques. I: bassin du Mercure et haute vallée du Sinni, *Rev. Geogr. Phys. Geol. Dynam.*, v. 10 (3), p. 225–238.

Bousquet, J. C., and Gueremy, P., 1969, II. L'escarpement méridional du Pollino et de son piedmont, *Rev. Géogr. Phys. Géol. Dynam.*, v. 11 (2), p. 223–236.

Broquet, P., and Mascle, G., 1972, Les grands traits stratigraphiques et structuraux des Monts de Trapani (Sicile occidentale), *Ann. Soc. Geol. Nord.* v. 92, p. 139–146.

Caire, A., Glangeaud, L., and Grandjacquet, C., 1960, Les grands traits structuraux et l'évolution des territoires calabro-lucaniens (Italie méridionale), *Bull. Soc. Géol. France*, (7), v. 2, p. 915–938.

Carissimo, L., d'Agostino, O., Loddo, C., and Pieri, M., 1963, Petroleum exploration by AGIP Mineraria and new geological information in central and southern Italy from the Abruzzi to the Taranto gulf, *IV Congr. Mond. Petrolio* sez., v. 1, no. 27, p. 26.

Chamley, H., 1971, Recherches sur la sédimentation argileuse en Méditerranée, Mém. Sci. Géol. Strasbourg, no. 35, p. 225.

Clermonte, J., 1974, *Etudes Géologiques dans le Bassin Versant du Sangro* (Italie Centro-Méridionale), Thèse, Nancy.

Cocco, E., 1973, Correlazione tra alcune successioni sedimentarie del Cretacico Sup. Paleocene Eocene inf. delle zone interne della "Geosynclinale" sud appenninica, *Boll. Soc. Geol. Ital.*, v. 42, p. 841–860.

Cocco, E., and Digirolamo, P., 1969, Magmatismo hawaïtico nei paraconglomerati terziari del flysch del Cilento, *Mem. Soc. Nat. Napoli*, Suppl. Boll. 78, p. 249–292.

Cocco, E., and Pescatore, T., 1968, Scivolamenti gravitativi (olistostromi) nel flysch del Cilento (Campania), *Boll. Soc. Nat. Napoli*, v. 77, p. 51–91.

Cocco, E., Cravero, E., Ortolani, F., Pescatore, T., Russo, M., Sgrosso, I., and Torre, M., 1972, Les faciès sédimentaires miocènes du Bassin Irpinien (Italie méridionale), *Atti Accad. Pontaniana*, v. 21, p. 1–13.

Colonna, V., and Piccarreta, G., 1975, Schema strutturale della Sila Piccola meridionale *Boll. Soc. Geol. Ital.*, v. 44, p. 3–16.

Colonna, V., and Piccaretta, G., 1975, Metamorfismo di alta pressione bassa temperatura nei micascisti di Zangarona–Ievoli–Monte Dondolo (Sila Piccola, Calabria), *Boll. Soc. Geol. Ital.*, v. 44, p. 17–25.

Colonna, V., and Zanettin-Lorenzoni, E., 1970, Gli scisti cristallini della Sila Piccola 1° I rapporti tra i cosidetti scisti bianchi e le filladi nella zona di Tiriolo, *Mem. Soc. Geol. Ital.*, v. 9, p. 135–156.

Colonna, V., and Zanettin-Lorenzoni, E., 1972, 2° I rapporti tra la formazione delle filladi e la formazione delle pietre verdi nella zona di Gimigliano, *Mem. Soc. Geol. Ital.*, v. 11, p. 261–292.

Colonna, V., Lorenzoni, S., and Zanettin-Lorenzoni, E., 1973, Sull'esistenza di due complessi metamorfici lungo il bordo sud orientale del massiccio "granito" delle Serre (Calabria) *Bull. Soc. Geol. Ital.*, v. 42, p. 801–830.

Cristofolini, R., 1966, Le manifestazioni eruttive basiche del Trias superiore nel sottosuolo di Ragusa (Sicilia sudorientale), *Périod. Min.*, v. 35, p. 1–28.

Cristofolini, R., 1966, Studio petrografico del più profondo livello vulcanico nel sottosuolo (dolomia triassica) di Ragusa, *Atti Accad. Gioenia Sci. Nat. Catania*, v. 18, p. 29–48.

D'Argenio, B., 1966, Zone isopiche e faglie trascorrenti nell'Appennino centro meridionale, *Mem. Soc. Geol. Ital.*, v. 5, p. 279–299.

D'Argenio, B., Pescatore, T., and Scandone, P., 1973, Schema geologico dell'Appennino meridionale (Campania Lucania), Atti del Conv. sul tema "Moderne vedute sulla geologia dell'Appennino," Quad. No. 183, *Accad. Naz. Lincei*.

Delteil, J., Fenet, B., Guardia, P., and Polveche, J., 1971, Géodynamique de l'Algérie nord-occidentale, *C.R. Somm., Soc. Géol. France*, p. 414–417.

De Roever, E. W. F., 1972, Lawsonite albite facies metamorphism near Fuscaldo, Calabria (Southern Italy), Its geological significance and petrological aspects, GVA Amsterdam.

Desio, A., 1970, *Geologia dell'Italia*, Unione Tipografica Editrice Torinese.

Dewey, J. F., Pitman W. C., III, Ryan, W. B. F., and Bonnin, J., 1973, Plate tectonics and the evolution of the Alpine system, *Bull. Geol. Soc. Amer.*, v. 84, p. 3137–3180.

Didon, J., Fernex, F., Lorenz, C., Magné, J., and Peyre, Y., 1969, Sur un niveau remarquable de Silexite dans le néogène inférieur d'Espagne méridionale et d'Italie du nord, *Bull. Soc. Géol. France*, (7), v. 11, p. 849–853.

Dietrich, D., and Scandone, P., 1973, The position of the basic and ultrabasic rocks in the tectonic units of the southern Apennines, *Atti Accad. Pontaniana*, v. 21, p. 1–15.

Dubois, R., 1967, Les glaucophanites a Lawsonite de Terranova di Sibari (Calabre), *C.R. Acad. Sci.*, v. 265, p. 188–191.

Dubois, R., 1969, Le passage latéral des prasinites de Rose Fuscaldo aux épanchements jurassiques de Malvitto et ses conséquences sur l'interprétation de la suture calabro-apenninique, *C.R. Acad. Sci.*, v. 269, p. 1815.

Dubois, R., 1970, Phases de serrage, nappes de socle et métamorphisme alpin à la jonction calabre apennin. La suture Calabro-apenninique, *Rev. Géogr. Phys. Géol. Dynam.*, 12 (3), p. 221–254.

Dubois, R., 1971, Définition d'un socle anté hercynien en Calabre, *C.R. Acad. Sci.*, v. 272, p. 2052–2055.

Dubois, R., and Glangeaud, L., 1965, Grandes structures, microstructures et sens des chevauchements de matériel cristallin à l'extrémité méridionale du massif de la Sila, *C.R. Somm., Soc. Géol. France*, v. 7, p. 229–230.

Durand Delga, M., 1960, Le sillon géosynclinal du flysch tithonique néoconien en Méditerranée occidentale, *Rend. Accad. Naz. Lincei, Cl. Sci. Fis. Mat. Nat.*, (8), v. 19, p. 580–585.

Durand Delga, M., 1961, Le sillon des flyschs du Crétacé supérieur en Méditerranée occidentale, *Rend. Accad. Naz. Lincei, Cl. Sci. Fis. Mat. Nat.*, (8), v. 20, p. 62–66.

Durand Delga, M., 1969, Mise au point sur la structure du Nord Est de la Berbérie, *Publ. Serv. Géol. Algérie*, (N.S.), Bull. 39, p. 89–131.

Emelyanov, E. M., 1972, Principal types of recent bottom sediments in the Mediterranean Sea: their mineralogy and geochemistry, in: *The Mediterranean Sea*, Stanley, D. J., ed., Stroudsburg, Pennsylvania: Dowden, Hutchinson and Ross, p. 355–386.

Ghezzi, G., and Bayliss, D. D., 1964, Uno studio del flysch nella regione calabro-lucana. Stratigrafia, tettonica e nuove idee sul Miocene dell'Appennino meridionale, *Boll. Serv. Geol. Ital.*, v. 84, p. 3–64.

Glangeaud, L., 1962, Paléogéographie dynamique de la Méditerranée et de ses bordures. Le rôle des phases ponto-plio-quaternaires, *Océan. Géol. Géophys. Médit. Occid.*, Villefranche-Sur-Mer, *C.N.R.S.*, p. 125.

Grandjacquet, C., 1961, Le flysch à quartzites des territoires Calabro lucaniens (Italie méridionale), *Bull. Soc. Géol. France*, (7), v. 3, p. 115–120.

Grandjacquet, C., 1963, Schéma structural de l'Apennin Campano-lucanien Italie, *Rev. Geogr. Phys. Géol. Dynam.*, v. 5 (3), p. 185–202.

Grandjacquet, C., 1967, Age et nature du métamorphisme "alpin" en Calabre du Nord, *C.R. Acad. Sci.*, v. 265, p. 1035–1038.

Grandjacquet, C., 1969, Les phases tectoniques et le métamorphisme tertiaire de la Calabre du nord et de la Campanie du sud (Italie), *C.R. Acad. Sci.*, v. 269, p. 1819–1822.

Grandjacquet, C., 1971, Les séries transgressives d'âge Oligo-Miocène inférieur de l'Apennin méridional–conséquences tectoniques et paléogéographiques, *Bull. Soc. Géol. France*, (7), v. 13, p. 315–320.

Grandjacquet, C., 1971, Tectogenèse d'âge anté albien supérieur des séries ophiolifères de Calabre et de Lucanie méridionale—Analyse de leur position structurale, conséquences paléotectoniques, *C.R. Somm., Soc. Géol. France*, p. 436–439.

Grandjacquet, C., and Glangeaud, L., 1962, Structures mégamétriques et évolution de la mer tyrrhénienne et des zones pérityrrheniennes, *Bull. Soc. Géol. France*, (7), v. 4, p. 760–773.

Grandjacquet, C., and Grandjacquet, M., 1962, Géologie de la zone Diamante Verbicaro (Calabre), *Geol. Romana*, v. 1, p. 297–312.

Grandjacquet, C., Glangeaud, L., Dubois, R., and Caire, A., 1961, Hypothèses sur la structure profonde de la Calabre (Italie), *Rev. Géogr. Phys. Géol. Dynam.*, v. 4 (3), p. 131–147.

Gueremy, P., 1972, La Calabre centrale et septentrionale. Guide d'excursion morphologique, *Trav. Inst. Géogr. Reims*, fasc. 10, p. 128.

Haccard, D., Lorenz, C., and Grandjacquet, C., 1972, Essai sur l'évolution tectogénétique de la liaison alpes–apennin (de la Ligurie à la Calabre), *Mem. Soc. Geol. Ital.*, v. 11, p. 309–341.

Heimann, K. O., and Mascle, G., 1974, Les séquences de la série évaporitique messinienne, *C.R. Acad. Sci.*, v. 279, p. 1987–1990.

Hoffmann, C., 1970, Die Glaukophangesteine, ihre stofflichen Äquivalente und Umwandlungsprodukte in Nord Calabrien (Sud Italien), *Contrib. Min. Petrol.*, v. 27, no. 4, p. 283–320.

Ietto, A., Pescatore, T., and Cocco, E., 1965, Il flysch mesozoico terziario del Cilento occidentale, *Boll. Soc. Nat. Napoli*, v. 74, pl 396–402.

Lepvrier, C., 1967, Sur la structure du massif du Chenoua (Algérie), *Bull. Soc. Géol. France*, (7), v. 9, p. 859–864.

Magri, G., Sidoti, G., and Spada, A., 1965, Rilevamento geologico sul versante settentrionale della Sila (Calabria), *Mem. Note Ist. Geol. Appl. Univ. Napoli*, v. 9, p. 5–59.

Martinis, B., 1962, Lineamenti strutturali della parte meridionale della Penisola Salentina, *Geol. Romana*, v. 1, p. 11–23.

Martinis, B., 1962, Osservazioni sulla tettonica del Gargano orientale, *Boll. Soc. Geol. Ital.*, v. 85, p. 45–61.

Mascle, G., 1974, Le système de failles de Comiso, bordure nord-occidentale du haut-plateau ibléen (Sicile), *C.R. Somm., Soc. Géol. France*, p. 134–136.

Migliorini, C., 1944, Affioramento di roccia eruttiva basica nel bacino dell'alto Agri, *Boll. Soc. Geol. Ital.*, v. 62, p. XXXI–XXXII.

Montanari, L., 1964–65, Geologia del Monte Pellegrino (Palerme), *Riv. Min. Sic.*, v. 88, p. 173–197; v. 93, p. 72–106.

Montanari, L., 1966, Geologia dei Monti di Trabia (Sicilia), *Riv. Min. Sic.*, v. 97–99, p. 35–81.

Mostardini, F., Pieri, M., and Pirini, C., 1966, Stratigrafia del foglio 212, Montalbano Jonico, *Boll. Serv. Geol. Ital.*, v. 87, p. 57–88.

Mulder, C. J., 1973, Tectonic framework and distribution of Miocene evaporites in the Mediterranean, in: *Messinian Events in the Mediterranean*, Drooger, C. W., ed., Amsterdam: North-Holland Publ. Co., p. 44–59.

Ogniben, L., 1954, Le "argille brecciate" siciliane, con i rilievi di dettaglio di Grottacalda (Valguarnera, Enna), Passarello (Licata, Agrigento) e Zubbi (S. Cataldo-Caltanissetta), *Mem. Ist. Geol. Univ. Padova*, v. 18, p. 92.

Ogniben, L., 1955, Le argille scagliose del Crotonese, *Mem. Note Ist. Geol. Appl. Napoli*, v. 6, p. 1–72.

Ogniben, L., 1962, Le argille scagliose ed i sedimenti messiniani a sinistra del Trionto (Rosano-Cosenza), *Geol. Romana*, v. 1, p. 255–282.

Ogniben, L., 1969, Schema introduttivo alla geologia del confine calabro-lucano, *Mem. Soc. Geol. Ital.*, v. 8, p. 453–763.

Ogniben, L., 1974, Schema geologico della Calabria in base ai dati odierni, *Geol. Romana*, v. 12, p. 243–585.

Ogniben, L., Parotto, M., and Praturlon, A., 1975, Structural model of Italy, *La Ricerca Scientifica*, Quad. 90, 502 p., 4 maps.

Ortolani, F., 1975, Assetto strutturale dei Monti Picentini, della Valle del Sele e del Gruppo di Monte Marzano–Monte Ogna (Appennino meridionale). Implicazioni idrogeologiche, *Boll. Soc. Geol. Ital.*, v. 44, p. 209–230.

Ortolani, F., and Torre, M., 1971, Il monte Alpi (Lucania) nella paleogeografia dell'Appennino meridionale, *Boll. Soc. Geol. Ital.*, v. 90, p. 213–248.

Paquet, J., 1966, Age de la mise en place des unités supérieures du Bétique de Malaga et de la partie méridionale du Subbétique (transversale de la Sierra de Espuna, province de Murcie, Espagne), *Bull. Soc. Géol. France*, (7), v. 8, p. 946–955.

Pavan, G., and Pirini, C., 1966, Stratigrafia del Foglio 157 "Monte S. Angelo," *Boll. Serv. Geol. Ital.*, v. 86, p. 123–189.

Peronne, D., 1967, *Contribution à l'Etude Sédimentologique de Sondages Sous-Marins*, Thèse (3ᵉ cycle), Sédimentologie, Orsay.

Peronne, V., Torre, M., and Zuppetta, A., 1973, Il Miocene della Catena Costiera Calabra. Primo contributo zone Diamante-Bonifati-Sant'Agata d'Esaro (Cosenza), *Riv. Ital. Paleontol.*, v. 79, p. 157–205.

Pescatore, T., Sgrosso, I., and Torre, M., 1969, Lineamenti di tettonica e sedimentazione nel Miocene dell'Appennino campano lucano, *Mem. Soc. Nat. Napoli*, Suppl. Boll., 78, p. 337–400.

Pescatore, T., and Sgrosso, I., 1973, I rapporti tra la piattaforma campano lucana e la piattaforma abruzzese campana nel Casertano, *Boll. Soc. Geol. Ital.*, v. 42, p. 925–938.

Piccarreta, G., Amodio Morelli, L., and Paglionico, A., 1973, Evoluzione metamorfica delle rocce in facies granulitica nelle Sere nord occidentali (Calabria), *Boll. Soc. Geol. Ital.*, v. 42, p. 861–889.

Pieri, M., 1966, Tentativo di ricostruzione paleogeografico strutturale dell'Italia centro-meridionale, *Geol. Romana*, v. 5, p. 407–424.

Quitzow, W., 1935, Diabas porphyrite und glaucophangesteine in der Trias von Nord Kalabrien, *Nach. Gess. Göttingen Math Phys.*, v. 9, p. 83.

Quitzow, W., 1935, Der Deckenbau des Kalabrischen Massiv und seiner Randgebiete, *Abh. Gess. Win Göttingen Math Phys.*, v. 13, p. 63.

Raoult, J. F., 1967, Chevauchements d'âge éocène dans le dorsale du Djebel Bou Aded (Est de la chaîne numidique, Algérie), *C.R. Acad. Sci.*, v. 266, p. 861–864.

Regione Siciliana, 1962–64, Studi e indagini per ricerche solfifere, *Riv. Min. Sic.*, v. 76–78, 8 p.; v. 79–81, 24 p.; v. 85–87, 28 p.

Renard, V., Pautot, G., Avedik, F., Needham, D., and Melguen, M., 1973, Messina abyssal plain: preliminary results of the *Jean Charcot* polymède II cruise, *Bull. Geol. Soc. Greece*, v. 10, p. 172–173.

Rigo, M., and Barbieri, F., 1959, Stratigrafia pratica applicata in Sicilia, *Boll. Serv. Geol. Ital.*, v. 80, p. 351–442.

Rittmann, A., 1967, Studio geovulcanologico e magmatologico dell'isola di Pantelleria, *Riv. Min. Sic.*, v. 106–108, p. 147–182.

Roda, C., 1965, Il calcare portlandiano a Dasycladacee di M. Mutolo (Reggio Calabria), *Geol. Romana*, v. 4, p. 259–290.

Sancho, J., Letouzey, J., Biju-Duval, B., Courrier, P., Montadert, L., and Winnock, E., 1973, New data on the structure of the eastern Mediterranean basin from seismic reflexion, *Earth. Planet. Sci. Lett.*, v. 18, p. 189–204.

Scandone, P., 1967, Studi di Geologia Lucana la serie calcaro–silico marnosa e i suoi rapporti con l'Appennino calcareo, *Boll. Soc. Nat. Napoli*, v. 76, p. 301–469.

Scandone, P., 1970, Mesozoico trasgressivo nella Catena Costiera della Calabria, *Atti Accad. Pontaniana*, v. 20, p. 1–9.

Scandone, P., 1972, Studi di Geologia Lucana, carta dei Terreni della serie calcareo–silico–marnosa e nota illustrativa, *Boll. Soc. Nat. Napoli*, v. 81, p. 225–293.

Scandone, P., and Bonardi, G., 1967, Synsedimentary tectonics controlling disposition of Mesozoic and tertiary carbonatic sequences of area surrounding vallo di Diano (Southern Apennines), *Mem. Soc. Geol. Ital.*, v. 7, p. 1–10.

Scandone, P., Sgrosso, I., and Bruno, F., 1963, Appunti di Geologia sul Monte Bulgheria (Salerno), *Boll. Soc. Nat. Napoli*, v. 72, p. 19–27.

Scandone, P., Sgrosso, I., and Vallario, A., 1967, Finestra tettonica nella serie calcareo silico marnosa presso Campagna (Monti Picentini, Salerno), *Boll. Soc. Nat. Napoli*, v., 76, p. 247–254.

Selli, R., 1957, Sulla trasgressione del Miocene nell'Italia meridionale, *Giorn. Geol.*, (2), v. 26, p. 1–54.

Selli, R., 1962, Il Paleogene nel quadro della geologia dell'Italia meridionale, *Mem. Soc. Geol. Ital.*, v. 3, p. 737–790.

Spadea, P., 1968, Pillow lava nei terreni alloctoni dell'Appennino lucano, *Atti Accad. Gioenia Sci. Nat. Catania*, v. 20, p. 105–142.

Tanguy, J. C., 1966, *Contribution à la Pétrographie de l'Etna*, Thèse (3e cycle), Paris, 221 p.

Tazieff, H., 1973, *L'Etna et les Volcanologues*, Paris: Arthaud ed., p. 240.

Trevisan, L., 1942, Problemi relativi all'epirogenesi e all'eustatismo nel Pliocene e Pleistocene della Sicilia, *Atti Soc. Tosc. Sci. Nat. Mem.* 51, p. 11–33.

Trevisan, L., and Di Napoli, E., 1938, Tirreniano, Siciliano e Calabriano nella Sicilia sud-occidentale. Note di stratigrafia, paleontologia e morfologia, *Giorn. Sci. Nat. Econ. Palermo*, v. 39 (1937), no. 8, p. 1–39.

Truillet, R., 1970, Etude géologique des Peloritains orientaux (Sicile), *Riv. Min. Sic.*, v. 115–123, p. 1–157.

Vezzani, L., 1966, Nota preliminare sulla stratigrafia della Formazione di Albidona, *Boll. Soc. Geol. Ital.*, v. 85, p. 767–776.

Vezzani, L., 1968, Distribuzione, facies e stratigrafia della formazione del Saraceno (Albiano Daniano) nell'area compresa tra il Mare Jonio ed il Torrente Frido, *Geol. Romana*, v. 7, p. 229–275.

Vezzani, L., 1968, Studio stratigrafico della formazione delle Crete Nere (Aptiano Albiano) al confine calabro-lucano, *Atti Accad. Gioenia Sci. Nat. Catania*, (6), v. 20, p. 189–222.

Vezzani, L., 1968, Rapporti tra ofioliti e formazioni sedimentarie nell'area compresa tra Viggianello, Francavilla sul Sinni, Terranova del Pollino e S. Lorenzo Bellizzi, *Atti Accad. Gioenia Sci. Nat. Catania*, (6), v. 19, Suppl. Sci. Geol., p. 109–144.

Vezzani, L., 1968, La formazione del Frido (Neocomiano Aptiano) tra il Pollino ed il Sinni (Lucania), *Geol. Romana*, v. 8, p. 129–176.

Wendt, J., 1965, Synsedimentäre Bruchtektonik im Jura Westsiziliens, *Neues Jb. Geol. Pal. Mh.*, v. 5, p. 286–311.

Wezel, F. C., and Guerrera, F., 1973, Nuovi dati sulla età e posizione strutturale del flysch di Tusa in Sicilia, *Boll. Soc. Geol. Ital.*, v. 92, p. 193–211.

Chapter 5A

THE GEOLOGY OF THE PELAGIAN BLOCK: THE MARGINS AND BASINS OFF SOUTHERN TUNISIA AND TRIPOLITANIA

P. F. Burollet, J. M. Mugniot, and P. Sweeney

Compagnie Française des Pétroles
Paris, France

I. INTRODUCTION

In a recent synthesis, Burollet (1969, 1973) defined as the Pelagian block a relatively stable area which extended from the Maltese Islands to eastern Tunisia. This area, now largely under water, includes part of Sicily (Ragusa platform) and stretches southward toward the cratonic block of Africa. It is of particular interest in that it bridges the gap between Europe and Africa, between the Atlas Mountains, which may be traced across Africa, and the Apennines of Italy. In combination with the geology of Tripolitania, the relationship of the orogenic belts of North Africa to the Saharan craton can also be studied. These problems have interested Burollet (Petroleum Exploration Society of Libya, 1967), Caire (1973), Castany (1951), Durand Delga (1967), and Rouvier (1973).

Nowhere on the western part of the Pelagian block are *in situ* rocks older than Jurassic exposed at the surface, although tectonically emplaced Triassic rocks are known in Tunisia. As the result of a number of deep boreholes, the existence of a Paleozoic basement is known in southern Tunisia and northern Tripolitania. Beyond the occurrence of Cretaceous, nothing is available on the deeper horizons under the Maltese Islands. The Pelagian block was faulted during the Miocene and parts subsided at the time of subsidence of much of the western Mediterranean. The eastern and northern limits of the block appear to correspond to a major fault structure (Castany 1951) formed in the Pliocene to post-Villafranchian times.

This description of the Pelagian platform and the Ionian Sea is based on the original works of the authors, the C.F.P. (Compagnie Française des Pétroles), and numerous published works, the most important being *The Mediterranean Sea*, edited by D. J. Stanley (1972), papers and maps of the French Petroleum Institute, and the recent report of the Geophysical Observatory of Trieste (Finetti and Morelli, 1973; Morelli *et al.*, 1975).

As far as the geology of Tunisia and Libya is concerned, the reader will refer to the works of G. Busson, P. F. Burollet, and G. Manderscheid, and also to the publications of the Service Géologique de Tunisie and the Petroleum Exploration Society of Libya.

II. PHYSIOGRAPHY

East of Tunisia and north of Tripolitania, the Pelagian platform and the southern part of the Ionian Sea form a very particular area of the Mediterranean (Fig. 1). As opposed to the other Mediterranean basins, the continental shelf here is very wide and the continental slope descends gradually away from the Libyan coast to a depth of -4000 m. There is a strong contrast between the hypsometric curves of this region and those of the Mediterranean as a whole (Fig. 2). We will see later that this is related to the deep-seated crust, an integral part of the Saharan platform.

The Pelagian platform is shallow, but deepens progressively toward the southeast, the bathymetric lines 400 m, 600 m, and even 800 m indicating a wide, uncomplicated furrow, which joins the Gulf of Sidra in the east to the Gabes and Chotts Troughs in the west. Between the Pelagian Islands, Malta, and Sicily, the platform is broken by a graben, the throw of which could be greater than 1000 m (Fig. 1).

To the east, the Pelagian platform is bordered by a vast fault zone, resulting in the easterly deepening of the Ionian Sea. This fault zone is linked to that of the east coast of Sicily and to the Misurata and Bu N'Gem fractures of Tripolitania. It is marked by recent volcanic activity, which is related to that of Etna in Sicily, and Garian and Djebel es Soda in Tripolitania (see Figs. 1, 9). This fault zone is very abrupt and forms a steep slope in the northern part, that is east of the Maltese Plateau. In contrast, south of the 35th parallel, it is less pronounced, resulting in a gradual transition between the Gabes Trough and the Gulf of Sidra slope.

In effect, the southern part of the Ionian Sea is formed by the continental

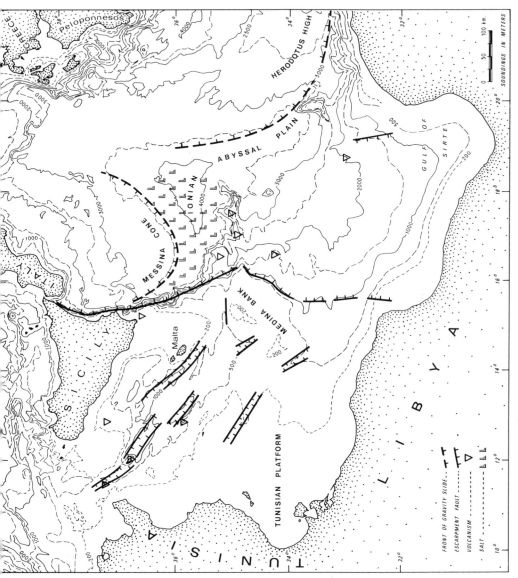

Fig. 1. Physiographic map of the Pelagian and Ionian Seas.

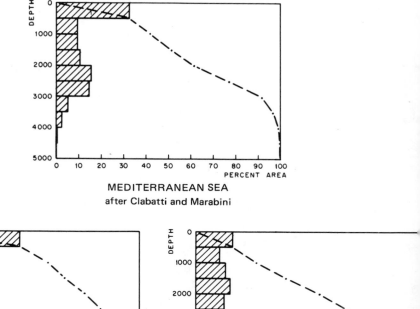

Fig. 2. Hypsometric curves.

slope of the Gulf of Sidra, which deepens progressively from 200 to 2500 m, and is followed by the abyssal plain where water depths are greater than 4000 m. The plain itself is limited to the north by the Cone of Messina with its abundant submarine slides.

At the juncture of the continental slope and the abyssal plain (between latitudes 35°30′ and 34°), the slope is broken by fractures and seamounts, which, in general, are volcanic. This zone is the prolongation of the Malta and Pantelleria Troughs, where a certain number of basic volcanic injections are known along the principal faults.

To the east, the Gulf of Sidra is bounded by the Cyrenaica. The abyssal plain is limited to the northeast by the complex called the Mediterranean Ridge, and continues by way of the Herodotus Trough. This latter feature separates the north of the Cyrenaica from a vast ridge culminating at −1223 m and named Herodotus High.

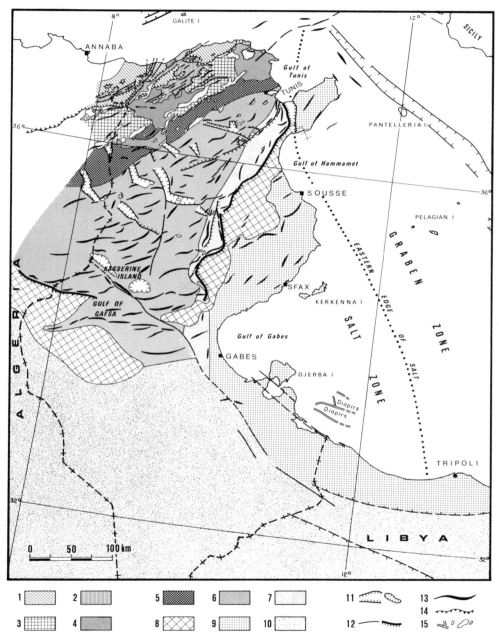

Fig. 3. Tectonic Map of Tunisia. *Northern Alpine Zone:* (1) Numidian nappe; (2) Tellian unit; (3) para-autochthonous and autochthonous zones of the Medjerda; (4) Kechabta–Medjerda Mio-Plio-cene Troughs. *Intermediate ranges of atlasic Tunisia:* (5) diapir area; (6) central and southern Atlas; (7) North–South Axis; (8) peri-atlasic depression; (9) *Eastern shelf*; (10) *Saharan shelf*; (11) graben; (12) faults; (13) anticlines; (14) thrust faults; (15) diapirs.

III. GEOLOGY OF THE SURROUNDING AREAS

The marine area involved in this study is bounded in the west by Tunisia and in the south by Tripolitania. To the north, it is linked to the plateau of Malta and Sicily and the deeper part of the Ionian Sea. These latter areas are considered in other chapters of this volume.

A. Southern Tunisia and Tripolitania (Figs. 3, 4)

Southern Tunisia and Tripolitania form part of the Saharan platform, which received an epicontinental sequence of Cambrian to Mississippian rocks. These rocks were gently folded during the Carboniferous and subsequently eroded, so that the oldest strata occur in the north in the Djeffara. Step faulting during the Pennsylvanian, and particularly in the Permian, caused the area north of the Djeffara to subside and a marine invasion resulted in the local deposition of great thicknesses of sediment (e.g., at Medenine more than 6000 m of Permian are said to occur). This subsidence of the Djeffara was the first step in the development of the Tethys Trough (Fig. 5).

1. *Paleozoic Rocks*

With the exception of the Permian rocks of Djebel Tebaga near Medenine, Paleozoic rocks do not crop out in either Tripolitania or southern Tunisia. They are well known, however, from drilling in the Djeffara Plain and the Hammada el Homra.

Cambrian and Ordovician rocks are generally arenaceous. The following formations, beginning with the oldest, have been described in Libya (P.E.S.L., 1960) and are well known in subsurface.

Hassaouna Formation. This massive, medium- to coarse-grained sandstone forms a continuous sheet across most of the Libyan and southern Tunisian basement. It overlies Precambrian metamorphosed rocks and granites with strong angular unconformity. The sands are cross-bedded and may represent deltaic deposits or tidal deposits on a platform.

Haouaz Formation. This unit consists of fine-grained, well bedded, silty, micaceous, marine sandstones containing many *Tigillites* beds.

Melez Chograne Formation. Consisting principally green or gray shales, this formation contains thin interbeds of very fine-grained micaceous sandstone. Fossils indicate a Caradocian age, and it contains what may be glacial pebbles. In Tunisia it becomes a micro-conglomeratic shale locally known as the Djeffara Formation (Jaeger *et al.*, 1975).

Memouniat Formation. This formation consists of medium-to coarse-grained sandstones of Ashgillian age, and may be a marine periglacial deposit. It becomes finer grained upward, and is overlain by Silurian shales and sandstones.

The Silurian rocks consist of well laminated black shales (Tanezzuft shales) with an increasing component of fine-grained sandstone toward the top (Acacus sandstone). They are unconformably overlain by Lower Devonian sandstones of the Tadrart Formation. There is a progressive progradation of the Silurian rocks from the Fezzan in the south to the Djeffara in the north. Thus, in the southwest of the Murzuk Basin the Tanezzuft shale and Acacus sandstone) are Llandoverian; at Hammada el Homra the shale lithology is Llandoverian and Wenlockian, with the sandstone (Acacus sandstone) being of Lower Ludlovian age. Still farther to the north in the Djeffara Plains, thin euxinic, radioactive, argillaceous beds, associated with argillaceous limestones, represent the Llandoverian and Wenlockian and pass into the Ludlovian. The Acacus sandstone lithology here is of Upper Ludlovian–Pridolian age.

In western Libya and southern Tunisia, three lithological units can be recognized in the Devonian. Their exact age cannot always be precisely determined. The lowermost unit of the Tadrart Formation, a more or less coarse-grained, massive sandstone body with a major regional unconformity at its base, is assigned to the Siegenian by analogy with the facies and stratigraphic position with the Idjerane sandstone in the Algerian Sahara. The Tadrart is overlain by a marine, argillaceous–arenaceous sequence, the Emgayet and Uan Kasa Formations, attributed to the Emsian and Middle Devonian (in part). The Aouinet Ouenine Formation, a highly fossiliferous sequence of marine shale with sandstone intercalations, is then assigned to the Middle (in part) and Upper Devonian.

The latter formation is conformable upon the Uan Kasa Formation in the Ghadames Basin in the north, but it transgresses to the southwest and is diachronous, so that in the latter area the facies assigned to the Aouinet Ouenine Formation may be partly Frasnian in age. Regression occurred before the end of the Devonian and the highest beds are the sandy Oued Tahara member, which contains rare fossils of Strunian age.

The Carboniferous is well defined in western Libya. The M'Rar Formation, a richly fossiliferous, thick, shaly unit with interbedded sandstones is dated by its fauna as Tournaisian and Visean. South of Hammada el Homra, where Carboniferous beds crop out, the top of the M'Rar Formation consists of stromatolitic *Collenia* beds. These are overlain by the shales and sandstones of the Assedjefar Formation containing uppermost Viséan and Namurian faunas. The Moscovian is represented by the limy Dembaba Formation and is followed by continental or lagunal beds of probable Stephanian age of the Tiguentourine Formation.

CHRONO-STRATIGRAPHIC UNITS / LITHO-STRATIGRAPHIC UNITS

SYSTEMS	STAGES	TUNISIAN TROUGH	CENTRAL TUNISIA	EASTERN TUNISIA	SOUTH TUNISIA	LIBYAN JEFFARA	SYRTE	CYRENAICA
HOLOCENE	Holocene and Recent	ALLUVIUM SEBKHAS					CARDIUM BEDS	PANCHINA
PLEIS-TOCENE	U. Quaternary		CALICHE CARAPACE		STROMBUS BEDS / LOESS	GARGARESC LIMESTONE LOESS		
	M. Quaternary	STROMBUS BEDS		RED BEDS	CONGLOMERATE	CONGLOMERATE		
	Villafranchian / Calabrian	RED BEDS	RED BEDS		MARINE SAND / SEGUI			GARET UEDDA
PLIOCENE	Astian / Plaisancian	PORTO FARINA	MARINE PLIOC	SEGUI		MARINE SAND	MARADA	
	Tabianian	RAF RAF	SEGUI		ZARZIS SHALE AND SAND			
MIOCENE	Messinian	CHᵗ ET TEBBALA O. BEL KHEDIM KECHABTA-D-MELAH HAKIMA	SAOUAF	SAOUAF				
	Tortonian		BEGLIA	BEGLIA			DIBA	
	Serravallian		MAHMOUD	MAHMOUD			ARIDA	GIARABUB / CYRENAICA GR.
	Langhian	Numidian Flysch	AIN-GRAB	AIN-GRAB				REGIMA BARDIA FAIDIA SHAHHAT
	Burdigalian	Miocene Transgressive	MESSIOUTA					
	Aquitanian	Glauconitic beds	FORTUNA	FORTUNA			NASAH GROUP	CYRENE LABRAK ALGAL Ls SHAHHAT
OLIGOCENE	Oligocene							

(CAP BON GR. / OUM DOUIL — Eastern Tunisia)

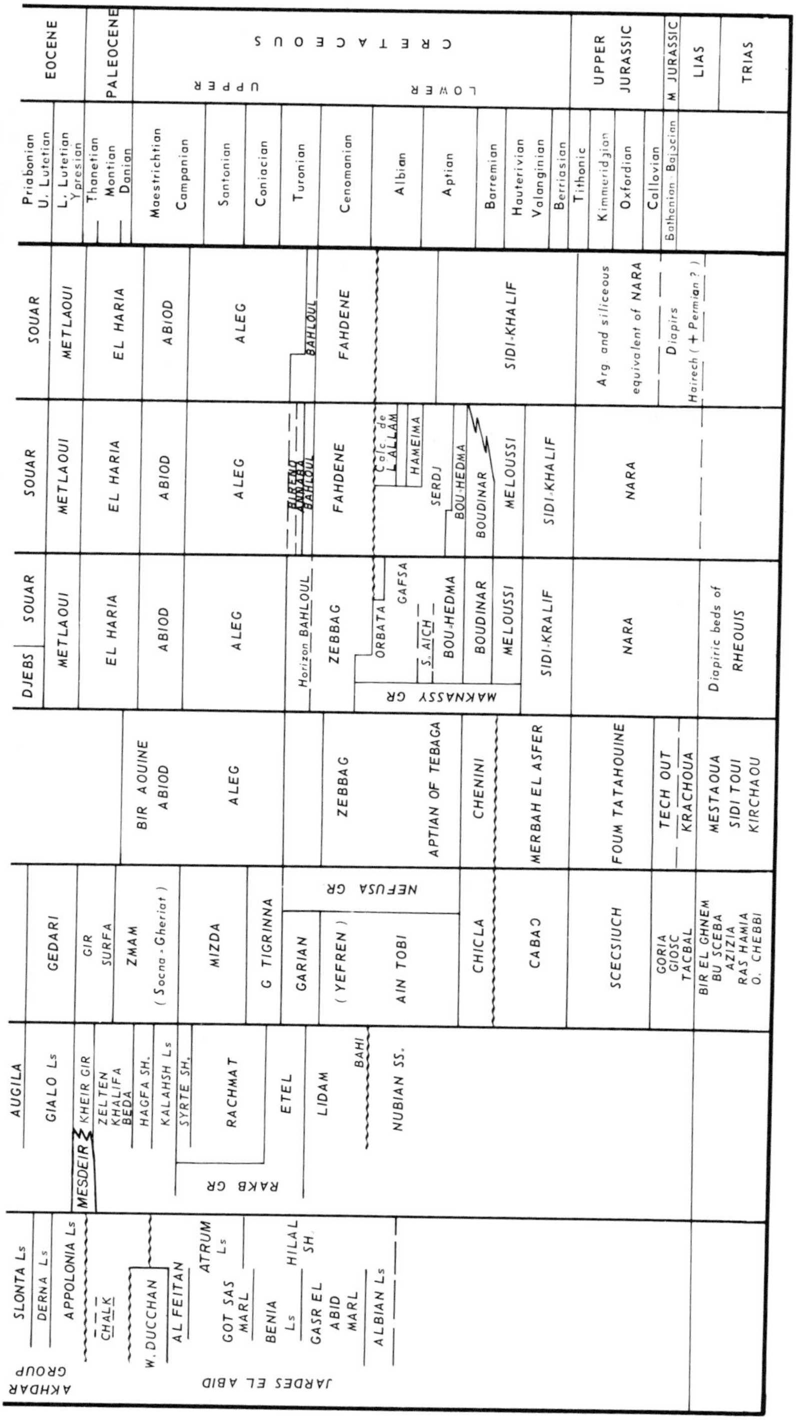

Fig. 4. Correlations of Mesozoic and Cenozoic stratigraphic units in Tunisia and Libya.

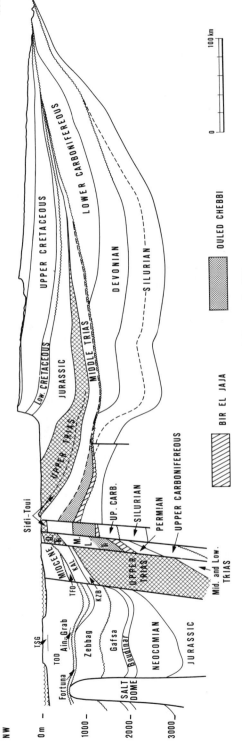

Fig. 5. Tectonic schema across Djeffara and Hammada el Homra.

A fundamental change in the paleogeography of northern Libya occurred during Carboniferous time. Prior to Viséan times the whole Saharan platform, covered by open epicontinental seas, was gently subsiding. At the end of the Viséan the area north of Hammada el Homra was uplifted by prerift epeirogeny and the Carboniferous beds eroded. In the Djeffara, a subsident basin appeared north of this partially faulted flexure, and into this unstable basin Upper Carboniferous and Permian beds were laid down. This trough was the ancestral southern margin of the Mesogean Sea.

Good sections of Upper Carboniferous and Permian beds have been described by Glintzbockel and Rabate (1964) from wells drilled in the Tunisian section of the Djeffara. The Namurian stage begins with an oolitic and bioclastic limestone with Bryozoa, *Millerella*, and *Archaediscus*, overlain by dolomites and argillaceous and sandy limestones with *Millerella*, *Eoschubertella*, *Endothyra*, etc. The Moscovian is represented by oolitic, bioclastic, and argillaceous limestones, with *Fusulina*, *Fusulinella*, *Profusulinella*, *Pseudostafella*, *Eostafella*, *Tetrataxis*, *Endothyra*, etc. At the end of Carboniferous time bioclastic and oolitic, crinoidal limestone with Bryozoa, *Quasifusulina*, *Triticites*, *Bradyina*, *Macroporella*, etc. were laid down.

Lower and Middle Permian are known only in the wells of the Kirchaou and Sidi Toui areas of southern Tunisia. Here bioclastic limestones with *Pseudoschwagerina*, *Schwagerina*, *Rugofusulina*, and *Tricites* were formed. They are overlain by detrital limestones containing *Hemigordius*, *Eoverbeekina*, *Calcitornella*, *Mizzia*, etc., which, toward the top, contain beds of marl and anhydrite. The Upper Permian is more extensive and crops out in Djebel Tebaga. In wells the transgression of the Upper Permian over eroded older Paleozoic beds is dated by the occurrence of *Streblospira* (Fig. 5). In the northern Djeffara the thick sequence of shale with interbedded sandstones and limestones can be subdivided by the occurrence of successive faunal associations:

Polydiexodina, *Yangchiena*, *Dunbarula*, *Parafusulina*, etc. Thickness up to 2900 m.

Codonofusiella, *Schwagerina*, *Dunbarula*, *Verbeekina*, *Geinitzina*, crinoids, algae, *Bellerophon*, etc. Thickness up to 1700 m.

Dunbarula, *Neoschwagerina*, *Yabeina*, *Tetrataxis*, *Osagia*, *Mizzia*, *Epimastopora*, *Gymnocodium*, *Permocalculus*, etc. Thickness up to 1500 m.

In the western Tunisian Djeffara, a reefoid facies with a rich fauna of *Dunbarula* and *Afghanella* zones is known.

The topmost part of the sequence consists of red, unfossiliferous shale which grades imperceptibly into the Lower Triassic.

2. *Mesozoic Rocks* (*Fig. 4*)

During Triassic and Liassic time subsidence of the northern Djeffara Trough continued, but the positive areas both to the west and east of the Djeffara were marked by carbonate sedimentation.

In the southern Djeffara and under the northern Hammada el Homra the succession is well known from subsurface exploration and from outcrops near Sidi Toui, Kirchaou, Azizia, and Garian (Figs. 5 and 6).

The Lower Triassic is represented by the Ouled Chebbi Formation, which consists of interbedded shale and sandstone. This is unconformably overlain by red and brown shales of the Ras Hamia Formation. The formation contains some interbedded sandstone horizons, and at the base yields *Myophoria*, indicating a Middle Triassic age. It passes upward into the Azizia Formation, a sequence predominantly of limestones and dolomites. The lower member of the Azizia Formation marks the end of the Middle Triassic and contains *Enantostreon difforme* Schl., *Ostrea* (*Lopha*) *montis caprilis* Klipst., *Lingula tenuissima* Bronn., and bones of *Nothosaurus*, etc. A rich fauna of the same age has been described from the *Sidi Toui limestone* of Tunisia. The main dolomite member and the associated upper red shale, sand, and dolomite member contains a rich Carnian fauna including *Myophoria inaequicostata* Klip., *M. goldfussi* Alb., *Pleuromyamactroides* (Schl.), *Antijanira* cf. *austriata* (Bittner), *Trachyceras* (*Pr.*) cf. *mandeslochi* (Klip.), etc. In Tunisia the fauna is found in equivalent beds, the Sidi Toui limestone and Rehach dolomite. The Triassic sequence is completed by the Bu Sceba Formation, a sequence of red or violet shales with horizons of fine, cross-bedded sandstone and a few microconglomeratic beds. This formation may be of Carnian age, for near the base it contains a radiolarian bed which has been compared to the Carnian beds of Val Gola in Italy. In the western Djeffara the formation also contains interbedded gypsum and halite.

The assignment of the Bir el Ghnem Gypsum Formation to the Lower Liassic is tentative and based mainly on its stratigraphic position above the Bu Sceba Formation and below the Abregh Formation. It consists of massive gypsum with a few interbedded carbonate bands, but passes laterally to a massive limestone in the eastern Djeffara, locally known as the Bu Gheilan limestone (Fig. 6). The overlying Abregh Formation is also a gypsum and anhydrite unit with interbedded carbonates. At the base there is an important marker bed called the Bu en Niran bed in Libya, the Zmilet Haber limestone in Tunisia, and the "B" reference horizon in Algeria. Its microfacies suggests a Middle Liassic age. The rest of the Abregh Formation extends from Upper Liassic to Bajocian. The equivalent beds in Tunisia comprise the Mestaoua Gypsum Formation.

The Tigi Formation, which rests unconformably upon the beds of the Abregh Formation, has three members. The lowest of these, the Tacbal dolo-

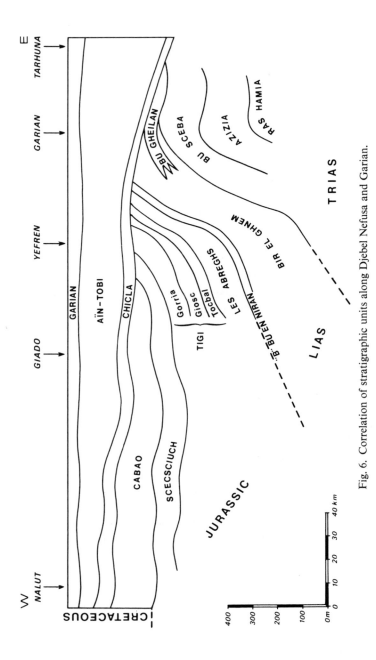

Fig. 6. Correlation of stratigraphic units along Djebel Nefusa and Garian.

mite and limestone has a rich Bathonian fauna (Desio *et al.*, 1960; de Lapparent 1954). In Tunisia the same limestone is called the Krachoua limestone. The middle shale and sandstone member is the Giosc shale. The upper Gorria sandstone member is a white to yellow ferruginous sand with varicolored shale and some lignite. In Tunisia the upper two units are assigned to the Techchout beds. Based on the macrofauna and microfauna the uppermost Jurassic unit ranges in age from the end of the Bathonian to the Kimmeridgian. This unit, the Scecsiuch Formation, consists of an alternation of limestone, dolomitic limestone, sandstone, and shale and crops out in the Djebel Nefusa cliffs in northern Tripolitania.

The Cretaceous begins with Wealden or Neocomian beds of the Cabao Formation in Tripolitania or Merbah el Asfer Formation in Tunisia. The beds consist of alternations of gray sandstone, green silt, and green shale above red basal beds. Along the Djebel Nefusa scarp all the formations are progressively cut out by the discordant Chicla Sandstone Formation of probable Barremian age, and near Garian the Chicla sandstone rests upon the Lower Liassic Bu Gheilan limestone (Fig. 6).

In the northwestern part of the Tunisian Djeffara, Triassic, Jurassic, and Lower Cretaceous rocks are considerably thinner and finally wedge out (Busson, 1967). This is due to stratigraphic thinning, hiatuses, and unconformities. From Upper Triassic times onward this province is the site of almost exclusively carbonate sedimentation.

Lower and Middle Triassic red sandstones rest unconformably on the Upper Permian beds of Tebaga. Above the unconformity, there is a progressive onlap of younger formations from south to north. Near Kef el Aneba and Djebel Zebassa, a yellow dolomite overlies Triassic sandstones and its basal beds have yielded Triassic fossils; it is a dolomitic equivalent of the Rehach dolomite, the lower evaporite, and the Mestaoua gypsum. North of Zemlet en Negueb, the Jurassic Krachoua limestone has onlapped Lower Triassic sands, and in the Tebaga area near the Medenine–Toujane road, Aptian limestone directly overlies the Permian series with angular unconformity.

In the northeastern Djeffara, thick Upper Permian and Triassic sequences have been penetrated by boreholes. The Triassic here consists mainly of evaporites, including a thick halite found in piercement domes in the Gulf of Gabes (Fig. 3).

The Upper Cretaceous, consisting mainly of limestone, dolomite, green clay, and gypsum represents an epineritic platform facies, which is well developed around the Djeffara in the Tunisian Dahar, the Libyan Djebel Nefusa, and in the eastern part of Hammada el Homra. Paleocene and Eocene beds, however, are absent in and around the Djeffara. They do occur southeast of Hammada el Homra and thicken in the Syrte Basin. An open marine facies along the coast passes to a restricted shaly environment in the center of the

Syrte Basin. On highs within the basin reefoid facies occur. This gently subsiding platform and embayment is a prolific oil province (Fig. 3). North of
Djeffara and Djerba Island in the Gulf of Gabes and below the Pelagian Sea,
seismic and drilling exploration shows the existence of Paleocene and Lower
Eocene series, with the same facies as in eastern Tunisia.

B. Eastern Tunisia and the Pelagian Sea

Eastern Tunisia and the Pelagian Sea form a vast, stable platform which
progressively, but unevenly, subsided during the Mesozoic and Tertiary. The
western limit of the area is a major lineament called the north–south axis,
extending more or less from Gabes to Tunis, bordering the Atlas of central
Tunisia. This latter area was an unstable platform which, at the end of the
Tertiary, became an area of folding of the intermediate peri-Alpine chain type
(Figs. 3, 4).

The Jurassic series in Tunisia consists mainly of carbonates. The subsident
trough of the northern Djeffara (Permian and Triassic) migrated a little to the
north, to Chott Fedjedj, while the whole of Tunisia was covered by an open
sea with an irregular floor. At several places over the north–south axis condensed sequences occur. The sea deepened northwestward. The "Ammonitico
rosso" facies of red nodular limestones is found northwest of a line from
Aouareb (southwest of Kairouan) to the vicinity of Aures in Algeria. It sometimes occurs in rocks of Toarcian age, sometimes in the Bajocian, but most
commonly in Callovian–Oxfordian rocks. Farther to the northwest the facies
become more pelagic with radiolarites common, and, in the Ghardimaou area,
Upper Jurassic jasper beds are found (Glaçon and Rouvier, 1970).

During the Lower Cretaceous the unstable platform of central Tunisia
occupied the boundary zone between Saharan clastic, alluvial deposits from
the southwest and the open marine facies and the open marine conditions in the
northwest. The north–south axis played a dominant role at the edge of the
stable eastern platform. Shoal facies and condensed sequences are common
(Burollet, 1956) and recently Biely et al. (1973b) have found in the Enfidaville
area an Aptian series condensed to several centimeters. Wells drilled through
the eastern platform found some tuffs, diabases, spilites, etc., of Neocomian
to Upper Cretaceous age (de Raaf and Althuis, 1952; Burollet, 1956; Bajanik,
1971). During the Aptian the Gafsa Trough received great thicknesses of
sediment, which are the northern counterpart of those laid down in the Chott
Fedjedj during the Neocomian and Jurassic. These are good examples of
typical "precursory" basins for intermediate chains.

In northern Tunisia, the Lower Cretaceous began with deposition of thick
shales and some interbedded sandstones. In the upper Lower Cretaceous a
few limy beds contain ammonites and belemnites (Sidi Khalif Formation,

Berriasian to Aptian). In central Tunisia (Kasserine Island) the marine facies of Sidi Khalif is restricted to the Berriasian and Valanginian. The Meloussi and Boudinar Formations consist of mostly sandy, neritic, lagoonal, or deltaic beds. North of Kasserine Island and in the Central area (Bargou, Maktar, Tadjerouine), reef limestones of the Serdj Formation of Aptian age occur. The Serdj limestone is overlain by the infraneritic Hameima Formation or by the lower shaly part of the Fahdene Formation (Fig. 7). To the south (Maknassy and Gafsa), the Aptian and Albian stages consist of dolomites, sands, clays, and evaporites of the Gafsa Formation; *Orbitolina* and *Choffatella* are found.

Along the north–south axis, erosion followed uplift at the end of the Lower Cretaceous, and Upper Albian or Cenomanian sediments frequently overlie older rocks (Burollet, 1956; Richert, 1971; Biely *et al.*, 1973a). East of the north–south axis, several highs are partially bare of Cretaceous sediments (Nasr Allah, Hamman Zriba, etc.) (Bouju and Burollet, 1975; M'Rabet and Dufaure, 1974).

It is probable that the frequent occurrence of extrusive Triassic in northwest Tunisia is principally due to the absence of resistant carbonate rocks in the overlying Jurassic and Cretaceous sediments of the diapiric area. Such carbonates exist in the central and eastern Atlas. Incompetent cover has allowed the halokinetic processes to begin as soon as the overburden (Upper Cretaceous and Eocene) was sufficient. The proximity of tectonic areas to the northwest accentuated stratigraphical factors.

The Upper Cretaceous of northern Tunisia is represented by four formations—the Fahdene, Bahloul, Aleg, and Abiod—and the basal part of a fifth, the El Haria Formation. The dark gray shales of the Fahdene Formation are of Upper Albian and Cenomanian age, dated by an abundant fauna with ammonites, echinoids, and a planktonic microfauna. Some radiolarian micrites are included in the Upper Albian. The thinly laminated, euxinic, argillaceous limestones of the Bahloul Formation represent the Lower Turonian. The Upper Turonian, Coniacian, Santonian, and Lower Campanian was a period of shale deposition assigned to the Aleg Formation. Everywhere in Tunisia the Aleg shale is overlain by chalky limestones of the Abiod Formation, ranging in age from Campanian to Maestrichtian. The overlying the El Haria Shale Formation is Maestrichtian at the base, while the upper part of the formation extends into the Tertiary.

General facies trends can be observed in Tunisia. The beds of the Fahdene Bahloul and Aleg Formations change southward from open marine to neritic limy shelf facies. Near Thala, Kasserine, and Sbeitla the Turonian is bioclastic limestone (Bireno limestone), overlying shales (Annaba), and the foliated Bahloul horizon. The Aleg shale is absent from the center of Kasserine Island where the last marine deposition is that of the Turonian upper Zebbag dolomite. Along the north–south axis the Upper Cretaceous may be reduced or partially

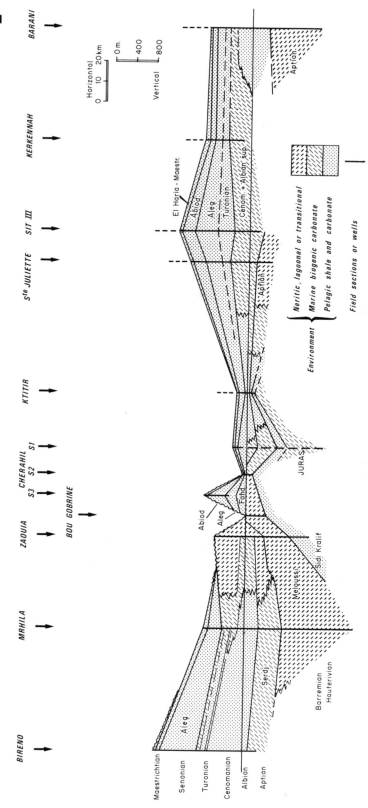

Fig. 7. Cretaceous stratigraphic correlation profile in central and eastern Tunisia.

absent as at Djebel Siouf, north of Sidi Nasr Allah. In other places north of the north–south axis, reefoid facies may occur, surrounded by shales of the Fahdene or Aleg Formations, as at Potinville where Cenomanian and Turonian reefs occur (Fig. 7).

Locally around the central (Kasserine) island and along the north–south axis, the Abiod Formation is transgressive over Jurassic or Lower Cretaceous rocks. During the Upper Cretaceous the paleogeography established earlier persists and is accentuated. A considerable area of Kasserine Island remained emergent after the Turonian. In the Gafsa area, an important NW–SE orientated flexure marks facies and thickness limits. This is probably one of the rare basement faults of the southern Atlas. In northwestern Tunisia the unstable character of the Tellian Trough is shown by Senonian microbreccia facies found in the allochthonous units (Glaçon and Rouvier, 1964; Durand Delga, 1967).

The Paleocene and Eocene are represented by the shales of the El Haria Formation, the Metlaoui Limestone Formation, and the shales of the Souar Formation. The Metlaoui limestone is a massive carbonate unit with local chert bands. It ranges in age from Thanetian to Lower Lutetian, although Lutetian is not present everywhere. The lower part, Thanetian–Lower Ypresian, may include glauconitic and phosphatic beds, the latter sufficiently thick around Kasserine Island to be commercially exploited. In restricted intracratonic basins (Gafsa–Metlaoui Gulf, and southeast of the north–south axis) it may contain a high percentage of evaporites, mostly gypsum. The main Ypresian limestone exhibits three facies. These are a coquinoid facies extending southward from Kasserine Island, a Nummulitic facies north and east, and a pelagic facies with globigerinids in eastern and northern Tunisia. In eastern Tunisia and in the Gulf of Gabes, the boundary between the nummulitic and pelagic facies trends NW–SE. It is independent of local thicknesses and subsidence. There are no marine Tertiary deposits on the central Kasserine Island. Along the north–south axis (e.g., near to Rheouis or Meheri Zebbeus) there are rapid thickness changes and some depocenters are related to salt migration.

In the north, between Kef and Bizerta, a high ridge formed, over which the Upper Lutetian transgressed directly onto older rocks (Glaçon and Rouvier, 1972). To the northwest are allochthonous Tellian facies, which represent an unstable trough, the original position of which is not known. The lower unit (Ed Diss) consists of a dark shale series containing yellow spheres and enclosing the *Globigerina* limestone with numerous cherts. The upper unit (Adissa) is a microbrecciated marly limestone (see work of Rouvier).

The Middle to Upper Eocene Souar shale is present over most of northern and eastern Tunisia. It becomes very thick in the northwestern trough and may include black and yellow weathered dolomitic boulders in the Tellian zones. Coquina and coquinoid limestone are frequently interbedded with this shale, particularly close to the edge of Kasserine Island.

Tectonic activity increased in Oligocene time and resulted in the deposition of the thick Fortuna Sandstone Formation in eastern central Tunisia. The lower part is shaly with interbedded sandstones and is often fossiliferous, containing echinoids and *Nummulites* (*N. fichteli*, *N. vascus*, and *N. incrassatus*). At Cap Bon, the lower part is calcareous with *Nephrolepidina tournoueri*, *N. t.* var. *simplex*, *Eulepidina formosoides*, *E. levis*, *Heterostegina sp.*, *Operculina sp.*, etc. The upper part of the Fortuna (Upper Oligocene and Aquitanian) is a massive sandstone, often cross-bedded and becoming very coarse grained at the top, with gravel and round quartz pebbles. It may contain silicified wood and vertebrate remains. There is a hiatus at the top of the massive sandstones which is often indicated by red continental beds (Messiouta). The source of the thick, coarse-grained sandstones is not known. It may be due to reworking of Lower Cretaceous deposits from large eroded anticlines or uplifts of pre-Saharan Tunisia. Current directions indicate transportation from SW to NE. The aerial distribution of the Fortuna sandstone differs from previous Paleogene sedimentation patterns, reflecting a new tectonic framework. In central Tunisia, along the north–south axis, a continuous mantle of Fortuna sandstone is present, varying in thickness from 400 to 1200 m. Eastward it grades into a finer and more marine facies. It is not present in southern and southwestern Tunisia. In northern Tunisia it is absent from a narrow NE–SW-trending band, and north of this high thin littoral glauconitic beds immediately underlie the Lower Miocene (Souk el Khemis, Beja).

In contrast, further to the northwest, Numidian flysch deposits up to 4000 m thick crop out along the Tunisian coast and in the forest-covered Khoumirie and Mogod hills. The wedges of Oligocene flysch are thrust over different tectonic units. The existence of a fairly fine-grained, marly and micaceous sandy Oligocene in the Galite Archipelago leaves the question of the origin (or origins) of Tunisian and Algerian Numidian Flysch unanswered (Durand Delga 1956a, b). The source of the Numidian Flysch is thought to be a now submerged area northwest of continental Tunisia. However, according to some authors the flysch may be derived from the Fortuna sandstone of Cap Bon by longitudinal transportation of contourite type.

The Miocene history of Tunisia is complex. The succession of events in the central and eastern Atlas as well as in the eastern plains has been described by Castany (1951), Burollet (1956), Biely *et al.* (1973a), Biely (1973), and others in the following terms:

Emersion and sometimes erosion, red beds locally (Messiouta).

Transgression marked by a marine detrital limestone (Ain Grab limestone) usually Upper Burdigalian but sometimes basal Langhian.

Mahmoud (marine) shale, Langhian to Lower Serravallian.

Beglia sands, Upper Miocene.

Paralic Saouaf Series, Upper Miocene.

Segui Series (continental), Mio-Pliocene.

The recommencement of the deposition of thick sand bodies, with the Beglia sands, corresponds to a new uplift of the source areas. The Saouaf-type subsident basins, containing thick lagoonal–paralic series, were formed at the same time as the terminal Miocene tectonic phase, particularly noticeable along the north–south axis and the scarp at the limit of the eastern platform. Salt migration was induced by the movement, and is seen in many eastern Tunisian seismic profiles. To the northwest, the Lower Miocene, and sometimes a thin Middle Miocene, follows the Oligocene or directly overlies older beds. The displacement, to their present locations, of the Numidian and Tellian Units, as well as their break-up into wedges and the overthrusting of areas of the Medjerda, took place toward the beginning of the Upper Miocene.

Thick, post-orogenic molasse and lagoonal beds developed in the Medjerda and Kechabta troughs during the Upper Miocene. The lateral variations suggest an unstable depositional environment. The Miocene terminates with Messinian beds, often of lagoonal or brackish water facies. A widespread tectonic phase affected the Tunisian Atlas at the end of the Miocene, bringing with it emersion and post-Messinian erosion. In the coastal areas, the Lower Pliocene rests transgressively on different Miocene series, often with angular discordance (Raf Raf, Cap Bon, Monastir, etc.).

The acid and basic volcanism of Nefza was subsequent to the displacement of the nappes, and several radiometric dates show a Miocene age for the acid rocks, and an Upper Messinian to Pliocene age for the basic rocks (Bajanik, 1971; Durand Delga, 1956*a, b*; Glaçon and H. Rouvier, 1972). The troughs of the Pelagian Sea and the main features east of Malta correspond to collapse, which began in the Upper Miocene and continued through the Pliocene and Quaternary (see Fig. 9).

The Lower Pliocene was marked by strong local subsidence in northern and eastern Tunisia (Ghar el Melah, Hammamet, Kuriates Islands, etc.). In the Bizerta region there was a spectacular reversal of the main topographical features. The mountain massif which surrounded the post-nappe Miocene trough (Lac de Bizerte, Jebel Ichkeul, etc.) sank rapidly during the Pliocene, falling more than 2000 m between the beginning of the Pliocene and the present. This was accompanied by a major tectonic phase at the end of Villafranchian in the whole of the Atlas and eastern Tunisia. Although the tectonic lines already marked in the Atlas were definitively emphasized, those of eastern and northern Tunisia are asymmetrical to the north. Could this be due to a general slippage of the region, on the salt, toward the vacuum caused by the sinking of the western Mediterranean? It would appear, anyway, to show a "plis de couverture" style. More difficult to explain are the orthogonal (perpendicular to folding) rifts of the Tunisian Atlas. They were certainly formed late, and they have the orientation of the African rifts and those of the Pelagian platform. We are inclined to connect them with the terminal Miocene or Pliocene phase, but when one considers the displacement of structural axes, they are also intimately linked with the main post-Villafranchian phase (Richert, 1971; Burollet, 1973).

The history of the Pleistocene is characterized by rejuvenation. A negative zone lies immediately east of the north–south axis, a recent basin filled by alluvium or by salt pans. It has the same significance as the Chotts Trough, south of the Saharan Atlas hinge line (Flexure Saharienne) in southwestern Tunisia and southern Algeria. In eastern Tunisia the Sfax–Kerkennah uplift, the eastern counterpart of Kasserine Island, is characterized by the divergence of the Pleistocene drainage pattern and by the bathymetric curves on the Pelagian platform. Coque and Jauzein (1967) have underlined the effect of the recent movements: "On land, the latest movements of the Tunisian orogeny were so recent that there was not always sufficient time for erosion to remove the morphostructural effects of these movements. Collapse and subsidence disrupted the existing hydrographic patterns and created closed basins. These features were complicated by other endoreic conditions peculiar to the palaeo-climatology of certain Quaternary stages. When the soil structure is favorable, the aridity of the summers facilitates the removal of wind-transportable material which may accumulate in the near vicinity of the depression (or Sebkha), forming an eolian ridge on the downwind side."

Along the coasts, the most recent orogenic movements have greatly disturbed the Quaternary marine shorelines, particularly the oldest ones. The *Strombus bubonius* shoreline, also identified as the Neotyrrhenian or Tyrrhenian III shore, extends along a considerable length of the Tunisian coast. The normal elevation of maximum transgression is between 5 and 8 m elevation in those areas where the coastal area has not been deformed by recent tectonic activity. In places, however, it occurs at different elevations, between 0 and 32 m above sea level, and occasionally even below sea level (Monastir, Mahdia, Sfax, etc.). Locally two different Tyrrhenian terraces were described. Calcareous sand dunes are connected with these shorelines and are often encrusted by caliche. More recently, the Flandrian stage is represented by some abrasion surfaces located near the 2-m level, and post-Flandrian dunes are known along the east coast.

IV. STRUCTURAL INTERPRETATION OF THE PELAGIAN PLATFORM AND THE IONIAN SEA

The platform area is well known from geological studies on shore, from wells drilled in the area, and from onshore and offshore seismic work (Fig. 8). The deeper parts have been explored only by geophysical methods, the principal

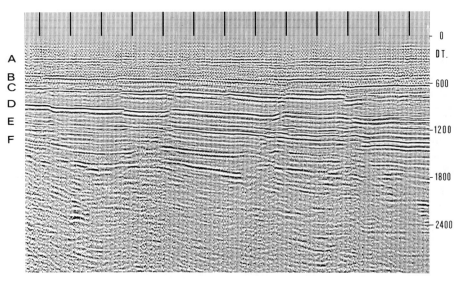

Fig. 8. Example of an E–W seismic profile south of Kerkennah Archipelago. Horizon E: Oligocene. Horizon F: Middle Eocene. Lower limit of good reflections would represent top of Mesozoic series. (For deep-water profiles, see Finetti and Morelli, 1973.)

work being the seismic profiling of the *Meteor* 1969–1971 campaign (refraction and reflection, Closs and Hinz, 1972; Hinz, 1973), the OGST 1971 campaign (reflection and Flexotir, Finetti and Morelli, 1972, 1973), and the French campaigns of 1972, 1973 (reflection and Flexichoc IFP, CNEXO, CFP, ELF, SNPA).

A. The Pelagian Platform

The Pelagian platform is divided into the Tunisian platform, the Sicilian platform and Malta Horst, and the Medina Bank by the large NW–SE Pantelleria and Malta grabens (Burollet, 1962; Finetti and Morelli, 1972) (Fig. 9).

The interpretation of seismic lines leads to the conclusion that the Mesozoic series studied in the wells drilled continues in the Tunisian and Sicilian offshore area and throughout the platform. The top of the Mesozoic is situated at a depth of approximately 2000–2500 m. Above the Mesozoic, the Tertiary series has a relatively regular thickness in the platform zones, displaying open-marine facies (Fig. 9). In late-Miocene or post-Miocene time impressive graben subsidence occurred with the development of a thick Plio-Quaternary sequence which in places reaches as much as 1000 m. A number of deep wells have been drilled on the platforms, three northeast of Malta, one near Lampedusa, and several in the Tunisian zone, but the results are not generally available.

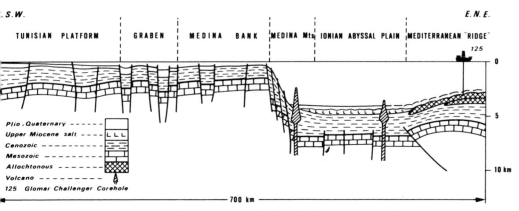

Fig. 9. Schematic section across the Pelagian platform and the Ionian Sea.

B. The Ionian Sea

The Ionian Sea is made up of several structural units. The central, abyssal Ionian plain is bounded to the north by the Messina Cone with its gravity slides, to the east by the western termination of the Mediterranean Ridge, to the south by the Gulf of Syrte (Biju-Duval *et al.*, 1974), and to the west by the faulted margin of the Pelagian block. Eastward it connects with the Herodotus Trough in which subhorizontal sediments up to 5000 m thick (Sancho *et al.*, 1973) are known. The abyssal plain is characteristically flat and is floored by a thin Plio-Quaternary series over a late Miocene evaporitic (halite) sequence, which may reach a thickness of 0.5 sec (two-way travel time) although no salt domes are known. According to Finetti and Morelli (1973) the thickness is less consistent than that found in the eastern Mediterranean and decreases toward the southwest part of the basin and the south Adriatic Sea.

Below the salt beds, seismic reflections lead one to believe that the deeper series are the equivalent of the Tertiary and Mesozoic series on the plateau; the top of the Mesozoic series is marked by strong reflections which could correspond to carbonate series identical to those on the Pelagian platform. In the southern part of the Ionian Sea certain Tertiary and Cretaceous seismic horizons can be followed quite easily and correlated with the known section in the Syrte Basin. The reflections attributed to the Mesozoic continue under the Messina Cone to the north and across the Gulf of Syrte to the south, where they are affected by numerous faults through which volcanic material has been injected.

Although it cannot be seen in the seismic records, the authors agree with Mulder (1973) in believing that the Mesozoic section continues under the so-called Mediterranean Ridge, including the Herodotus High north of Cyrenaica.

The development of the evaporitic series and the relative thickening of the Plio-Quaternary series in the Ionian abyssal plain prove that the subsidence occurred recently, in Upper Miocene to Recent time. A great deal of volcanism accompanied the subsidence, and its most remarkable mainfestation is east of Medina Bank where volcanics form the E–W-trending Medina Mountains. Thus, if the Ionian Sea is the result of late foundering (Upper Miocene and probably post-Miocene), the Tertiary and Mesozoic series should occur in facies similar to those of the Pelagian platform and Libyan shelf. The Bouguer gravity map of the Ionian Sea shows the highest positive anomaly found in the whole Mediterranean Sea, +310 mgal (Finetti and Morelli, 1973). The basement in the Gulf of Syrte is known to consist of Precambrian metamorphic and igneous rocks overlain by an early Paleozoic sequence. The Mesozoic of the Pelagian platform indicates a stable shelf tectonic framework.

The interpretation of Finetti and Morelli (1973), the one accepted here, is that the deep Ionian Sea is underlain by continental crust which foundered after late Miocene times. That crust is probably partly oceanized and associated with a mantle density anomaly. Consequently, the Mesozoic and Tertiary series, up to the end of the Miocene, should present facies similar to those on the Pelagian platform and Libyan shelf. This tilted and foundered platform is partially overthrust by the Tyrrhenian and Hellenic Arcs and by recent gravity slides of the Messina Cone and of the so-called Mediterranean Ridge. It represented part of the pre-Upper Miocene foreland of the Alpine system, north of the Saharan craton, and south of the Apulian and central Adriatic carbonate shelves ("Promontoire Africain" of E. Argand).

V. RECENT SEDIMENTATION

The most prominent characteristic of the area studied is the lack of permanent rivers between the Gulf of Tunis (Medjerda River) and the Nile. Mention may be made of the Wadi es Sebt, north of Sousse in eastern Tunisia, which appeared in 1969 during the dramatic floods and is still flowing. The permanent flow is small and could disappear following several dry years. In eastern Tunisia, the absence of permanent rivers is emphasized by the semiendoreic character of the hydrographic pattern. Wadis coming from the central Tunisian Atlas mountains flow irregularly between long, dry periods and run into large salt-pans or sebkhas, acting as decantation basins. In very rare cases, the overflow may reach the sea but it is poor in sediments. Wadi Gabes and some streams of Djeffara are exceptions, for they, like the Tripolitanian Wadis, do not cross any sebkha.

As a result of the existing drainage pattern, the Pelagian platform, particularly around the Kerkennah Archipelago, receives practically no terrigenous sediments. It is covered by *Posidonia* lawns or mats in shallow parts and *Caulerpa* lawns in deeper areas (below 40 m). The Holocene mud and the biogenetic fragments are reworked by currents and by tides (up to 3 m near Gabes). In the deeper parts of the Gulf of Gabes there is muddy sand with sponges and reworked algal concretions, or relatively clean bioclastic sands (detritique cotier) (Blanc, 1972; Blanc-Vernet, 1974). North of Djerba Island oolites have been described (Coque and Jauzein, 1967) in Tyrrhenian deposits. More recent oolitization has also been suggested. According to Fabricius (1973) the ooids of the eastern Mediterranean are relicts of interglacial or postglacial warm periods more recent than the Tyrrhenian (30,000–20,000 yr and 7000–5000 yr). A recent stage of oolitization may have occurred during the Atlantic optimum (Blanc-Vernet, 1974).

Off the coast of Tripolitania clastic sediments from the continent are periodic due to seasonal stream flow and to sandstorms. The rate of sedimentation may be as high as 10 to 50 cm/1000 yr (Blanc-Vernet, 1974). In the deeper parts of the Pelagian platform at depths of -200 to -600, fine, sandy deposits are progressively replaced by silts (coarse aleurites) and fine aleuritic muds.

On the slope and the upper part of the rise, recent deposits are mainly aleuritic pelitic muds, and in the deep Ionian Sea sediments consist of foraminiferal and nannoplankton oozes. In some places of the Ionian Sea, the lutite fraction may be rich in magnesian–calcite, especially near the so-called Mediterranean Ridge (Muller, 1973).

Mineralogical studies made by Emelyanov (1972) show that on the Pelagian platform and along the Tripolitanian shelf, the supply of terrigenous clastics comes from the southwest or south; in the deep Ionian sea, on the contrary, the main supply comes from the Messina Cone. During the late Quaternary, variations in climate are neatly reflected in the faunal assemblages of the Ionian abyssal plain (Ryan, 1972). During cold events, sedimentation rates were high (up to 200 cm/1000 yr); during warming episodes, stagnant conditions gave place to sapropelic muds.

REFERENCES

Allan, T. D., and Morelli, C., 1971, A geophysical study of the Mediterranean Sea, *Boll. Geofis. Teor. Appl.*, v. 3 (50), p. 99–134.

Argand, E., 1924, La tectonique de l'Asie, *C.R. XIII Int. Geol. Congr.*, Liège (1922), p. 169–371.

Auboin, J., 1973, Paléotectonique, tectonique, tarditectonique et néotectonique en Méditerranée moyenne: à la recherche d'un guide pour la comparaison des données de la géophysique et de la géologie, *C.I.E.S.M.*, *Assemblée*, Athènes (1972), v. 22, no. 2a, p. 105–117.

Bajanik, S., 1971, Volcanisme en Tunisie, *Anno Mines Géol.*, Tunis, no. 25, p. 63.

Biely, A., 1973, Le problème de la conception du Miocène nord tunisien, in: *Livre Jubilaire de M. Solignac*, *Ann. Mines Géol.*, Tunis, v. 28, p. 257–261.

Biely, A., Rakus, M., Robinson, P., and Salaj, J., 1972, Essai de subdivision lithostratigraphique du Trias de la Tunisie Orientale, *Notes Serv. Géol. Tunisie*, no. 7, p. 27–28.

Biely, A., Burollet, P. F., and Ladjmi, T., 1973a, Etude géodynamique de la Tunisie et des secteurs voisins de la Méditerranée, *XXIII Congr. C.I.E.S.M.*, Athens (1972), *Notes Serv. Géol. Tunisie.*

Biely, A., Memmi, L., and Salaj, J., 1973b, Le Crétacé inférieur de la région d'Enfidaville— Découverte d'Aptien condensé, in: *Livre Jubilaire de M. Solignac, Ann. Mines Géol.*, Tunis, v. 26, p. 169–178.

Biju-Duval, B., Letouzey, J., Montadert, L., Courrier, P., Mugniot, J. F., and Sancho, J., 1974, Geology of the Mediterranean Sea basins, in: *The Geology of the Continental Margins*, Burk, C. A., and Drake, C. L., eds., New York: Springer-Verlag, p. 695–721.

Blanc, J. J., 1972, Observations sur la sédimentation bioclastique en quelques points de la marge continentale de la Méditerranée, in: *The Mediterranean Sea: A Natural Sedimentation Laboratory*, Stanley, D. J., ed., Stroudsburg, Pa.: Dowden, Hutchinson and Ross, p. 225–240.

Blanc-Vernet, L., 1974, Microfaune de quelques dragages et carottages effectués devant les côtes de Tunisie (Golfe de Gabès) et de Libye (Tripolitaine), *Géol. Médit.*, v. I (1), p. 9–26.

Bouju, J. P., and Burollet, P. F., 1975, Dynamique de la Sédimentation en Tunisie, *Congr. Int. Sédiment. Nice*, Extr. Publ. Congrès, v. 6, p. 16–21.

Burollet, P. F., 1956, Contribution à l'étude stratigraphique de la Tunisie centrale, *Ann. Mines Géol.*, Tunis, no. 18, p. 350.

Burollet, P. F., 1962, Signification géologique du Détroit de Sicile, *C.R. 87e Congr. Soc. Savantes*, Poitiers, p. 848–853.

Burollet, P. F., 1969, Petroleum Geology of the Western Mediterranean Basin, *Joint Conf. Inst. Petrol. and Amer. Assoc. Petrol. Geol.*, Brighton.

Burollet, P. F., 1973, Importance des facteurs salifères dans la tectonique tunisienne, in: *Livre Jubilaire de M. Solignac, Ann. Mines Géol.*, Tunis, no. 26, p. 111–120.

Burollet, P. F., and Byramjee, R. S., 1974a, Evolution géodynamique néogène de la Méditerranée occidentale, *C.R. Acad. Sci. Paris*, v. 278, p. 1321–1324.

Burollet, P. F., and Byramjee, R. S., 1974b, Réflexions sur la tectonique globale, Exemples africains et méditerranéens, *Notes Mém., Compagnie Française des Pétroles*, Paris, no. 11, p. 71–120.

Burollet, P. F., and Manderscheid, G., 1967, Le Dévonien en Libye et en Tunisie, *Int. Symp. on the Devonian System*, Calgary, Alberta, p. 285–302.

Burollet, P. F., and Rouvier, H., 1971, La Tunisie, *UNESCO, Tectonique de l'Afrique (Sciences de la Terre, 6)*, p. 91–99.

Busson, G., 1967, *Le Mesozoïque Saharien*, 1ère partie: *L'extrème Sud Tunisiens*, Paris: C.N.R.S., 194 p.

Byramjee, R. S., Biju-Duval, B., and Mugniot, J. F., 1975, Petroleum potential of deep water areas of the Mediterranean and Caribbean Seas, *9th World Petrol. Congr.*, Tokyo, Panel 5, p. 299–312.

Caire, A., 1970, Tectonique de la Méditerranée Centrale, *Ann. Soc. Géol. Nord*, v. 90, p. 307–346.

Caire, A., 1973, The Calabro-Sicilian Arc, in: *Gravity and Tectonics*, De Jong, K. A., and Scholten, R., eds., New York: J. Wiley and Sons, Inc., p. 157–173.

Castany, G., 1951, Etude géologique de l'Atlas Tunisien oriental, *Ann. Mines Géol.*, Tunis, no. 8, p. 632.

Castany, G., Gobert, E. G., and Harson, L., 1956, Le Quaternaire marin de Monastir, *Ann. Mines Géol.*, no. 19.

Ciabatti, M., and Marabini, F., 1973, Hypsometric researches in the Mediterranean Sea, *C.I.E.S.M. Assemblée*, Athènes (1972), v. 22, fasc. 2a, p. 178–180.

Closs, H., and Hinz, K., 1972, Seismische und bathymetrische Ergebnisse von Mediterranean Rücken und Hellenischen Graben, *Aegean Symp.*, Hannover.

Comte, D., and Dufaure, P., 1973, Quelques précisions sur la stratigraphie et la paléogéographie tertiaires en Tunisie Centrale et Centro-orientale du Cap Bon à Mezzouna, in: *Livre Jubil. M. Solignac, Ann. Mines Géol.*, Tunis, no. 26, p. 241–256.

Conant, L. C., and Goudarzi, G. H., 1967, Stratigraphic and tectonic framework of Libya, *Bull. Amer. Assoc. Petrol. Geol.*, v. 51 (5), p. 719–730.

Coque, R., and Jauzein, A., 1965, Essai d'une carte néotectonique de la Tunisie au 1/1,000,000, *Rev. Géogr. Phys. Géol. Dynam.*, Paris.

Coque, R., Jauzein, A., 1967, The geomorphology and Quaternary geology of Tunisia, in: *Petrol. Expl. Soc. Libya: Guidebook to the Geology and History of Tunisia*, p. 227–257.

Desio, A., Rossi Ronchetti, C., and Vigano, P. L., 1960, Sulla Stratigrafia del Trias in Tripolitania e nel Sud-Tunisino, *Riv. Ital. Paleontol. Stratigr.*, Milan, v. 66 (3), p. 273–322.

Durand Delga, M., 1956a, Sur les formations ante-miocènes de l'archipel de la Galite (Tunisie), *C.R. Accad. Sci. Paris*, v. 243, p. 389–392,

Durand Delga, M., 1956b, L'évolution de l'archipel de la Galite au Néogène et au Quaternaire, *C.R. Acad. Sci. Paris*, v. 243, p. 507–509.

Durand Delga, M., 1967, Structure and geology of the northeast Atlas Mountains, in: *Petrol. Expl. Soc. Libya: Guidebook to the Geology and History of Tunisia*, p. 59–83.

Emelyanov, E. M., 1972, Principal types of recent bottom sediments in the Mediterranean Sea: Their mineralogy and geochemistry, in: *The Mediterranean Sea: A Natural Sedimentation Laboratory*, Stanley, D. J., ed., Stroudsburg, Pa.: Dowden, Hutchinson and Ross, p. 355–386.

Fabricius, F., 1973, Pleistocene and early Holocene Ooids in Mediterranean Coastal Waters, *C.I.E.S.M. Assemblée*, Athènes (1972), v. 22, fasc. 2a, p. 162–164.

Fabricius, F., and Schmidt-Thome, P., 1972, Contribution to recent sedimentation on the shelves of the Southern Adriatic, Ionian, and Syrtis Seas, in: *The Mediterranean Sea: A Natural Sedimentation Laboratory*, Stanley, D. J., ed., Stroudsburg, Pa.: Dowden, Hutchinson and Ross, p. 333–343.

Faculty of Science, University of Libya, 1971, *Symposium on the Geology of Libya*, Gray, C., ed., p. 522.

Fairbridge, R. W., 1972, Quaternary sedimentation in the Mediterranean region controlled by tectonics, paleoclimates and sea level, in: *The Mediterranean Sea: A Natural Sedimentation Laboratory*, Stanley, D. J., ed., Stroudsburg, Pa.: Dowden, Hutchinson and Ross, p. 99–113.

Finetti, I., and Morelli, C., 1972, Wide scale seismic exploration of the Mediterranean Sea, *Boll. Geofis. Teor. Appl.*, v. 14 (56), p. 291–342.

Finetti, I., and Morelli, C., 1973, Geophysical exploration of the Mediterranean Sea, *Boll. Geofis. Teor. Appl.*, v. XV (60), p. 261–341, 14 maps, Trieste, Udine.

Glaçon, G., and Rouvier, H., 1964, Sur l'existence de Sénonien à microbrèches calcaires sur les feuilles au 50.000è de la Calle et de Fernana (Tunisie), *C.R. Somm., Soc. Géol. France*, p. 133–135.

Glaçon, G., and Rouvier, H., 1970, Découverte de Jurassique dans les Monts de la Moyenne Medjerda (Tunisie Sept.), *C.R. Acad. Sci. Paris*, v. 270, p. 3007–3009.

Glaçon, G., and Rouvier, H., 1972, Age des mouvements tectoniques majeurs en Tunisie Septentrionale, *C.R. Acad. Sci. Paris*, v. 274, p. 1257–1260.

Glangeaud, L., 1956, Corrélation chronologique des phénomènes géodynamiques dans les Alpes, l'Apennin et l'Atlas nord-africain, *Bull. Soc. Géol. France*, (6), v. VI, p. 867–891.

Glintzboeckel, C., and Rabate, J., 1964, Microfaunes et Microfaciès du Sud Tunisien, *Int. Sediment. Petrogr. Series*, Leyden: E. J., Brill ed., v. 7, 45 p.

Hassan, M., and Massa, D., 1975, Some geochemical and sedimentological aspects of the Lower Silurian of North Africa, *Congr. Int. Sédiment.*, Nice, Extr. Publ. Congrès, v. 1, p. 97–102.

Hinz, K., 1973, The results of refraction and reflection seismic surveys of the *F.S. Meteor* in the Ionian Sea, *C.I.E.S.M. Assemblée*, Athènes (1972) v. 22, fasc. 2a, p. 97–98.

Illies, J. H., 1970, Die grossen Graben: harmonische Struktur in einer disharmonisch strukturierte Erdkrüste, *Geol. Rundschau*, Stuttgart, v. 59 (2), p. 528–552.

Jaeger, J., Bonnefous, J., and Massa, D., 1975, Le Silurien en Tunisie; ses relations avec le Silurien de Libye nord-occidentale, *B.S.G.F.*, (7), v. XVII, (1), p. 68–76.

Klemme, H. D., 1958, Regional geology of circum-mediterranean region, *Bull. Amer. Assoc. Petrol. Geol.*, v. 42, p. 477–512.

Laffitte, R., 1939, Les plissements post-nummulitiques dans l'Atlas Saharien, *Bull. Soc. Géol. France*, (5), v. IX, p. 135–159.

Laffitte, R., and Dumon, E., 1948, Plissements pliocènes et mouvements quaternaires en Tunisie, *C.R. Acad. Sci. Paris*, v. 227, p. 138–140.

Lapparent, A. F., (de), 1952–1954, Stratigraphie du Trias de la Jeffara (Extrême Sud tunisien et Tripolitaine), *C.R. 19ᵉ Congr. Géol. Int.*, Alger (1952) [published 1954, *Assoc. Serv. Géol. Africa*, fasc. 21, p. 129–134].

Lort, J. M., 1971, The tectonics of the Eastern Mediterranean: A geophysical review, *Rev. Geophys. Space Phys.*, v. 9 (2), Washington.

Massa, D., and Beltrandi, M., 1975, Sédimentologie du Silurien de Libye Occidentale, *Congr. Int. Sédiment.*, Nice, Extr. Publ. Congrès, v. 1, p. 113–117.

Massa, D., Termier G., and Termier, H., 1974, Le Carbonifère de Libye Occidentale, Stratigraphie, Paléontologie, *C.F.P. Notes Mém.*, Paris, no. 11, p. 139–206.

Morelli, C., Bellemo, S., Finetti, I., and Visintini, G. de, 1967, Preliminary depth contour maps for the Conrad and Moho discontinuities in Europe, *Boll. Geofis. Teor. Appl.*, v. 9 (34), p. 1–48.

Morelli, C., Gantar, G., and Pisani, M., 1975, Bathymetry, gravity and magnetism in the Strait of Sicily and in the Jonian Sea, *Boll. Geofis. Teor. Appl.*, v. 17, 39–58.

Mirabet, A., and Dufaure, P., 1975, Nouvelles données sur la série crétacée du Djebel Hallouf: Application à l'Axe Nord-Sud (Tunisie centrale), *C.R. Séances, Acad. Sci. Paris*, v. 280, p. 9–12.

Mulder, C. J., 1973, Tectonic framework and distribution of Miocene evaporites in the Mediterranean, in: *Messinian Events in the Mediterranean*, Drooger, C. W., ed., Amsterdam: North-Holland Publ. Co., p. 44–59.

Muller, J. W., 1973, Precipitation and diagenesis of carbonates in the Ionian deep sea, *C.I.E.S.M. Assemblée*, Athènes (1972), v. 22, fasc. 2a, p. 162.

Petroleum Exploration Society of Libya, 1960, *Names and Nomenclature Committee—Lexique stratigraphique*, International IV Afrique, Paris: C.N.R.S., 62 p.

Petroleum Exploration Society of Libya, 1967, *Guidebook to the Geology and History of Tunisia*, Ninth Annual Field Conference.

Petroleum Exploration Society of Libya, 1970, *Geology and History of Sicily*, Twelfth Annual Field Conference, 291 p.

Poncet, J., 1970, La "catastrophe" climatique de l'automne 1969 en Tunisie, *Ann. Géogr.*, Paris, (79), v. 435, p. 581–595.

Raaf, J. F. M., de, and Althuis, S. P., 1952, Présence d'ophites spilitiques dans le Crétacé des environs d'Enfidaville, in: *Atlas Tunisien Oriental et Sahel*, 19ᵉ Congr. Géol. Int., Alger (1952), Monogr. Région., 2ᵉ série, Tunisie, no. 6, p. 127–137.

Richert, J. P., 1971, Mise en évidence de quatre phases tectoniques successives en Tunisie, *Notes Serv. Géol. Tunisie*, no. 34, p. 115–125.

Rouvier, H., 1973, Nappes de charriage en Tunisie Septentrionale: preuves et conséquences paléogéographiques, in: *Livre Jubilaire M. Solignac, Ann. Mines Géol.*, Tunis, no. 26, p. 33–47.

Ryan, W. B. F., 1972, Stratigraphy of Late Quaternary sediments in the eastern Mediterranean, in: *The Mediterranean Sea: A Natural Sedimentation Laboratory*, Stroudsburg, Pa.: Dowden, Hutchinson and Ross, p. 149–169.

Ryan, W. B. F., Hsü, K. J., *et al.*, eds., 1973, *Initial reports of the Deep Sea Drilling Project*, v. XIII, Washington, D.C.: U.S. Gov. Printing Office, 514 p.

Sancho, J., Letouzey, J., *et al.*, 1973, New data on the structure of the Eastern Mediterranean Basin from seismic reflection, *Earth Planet. Sci. Lett.*, v. 18, p. 189–204.

Stanley, D. J., ed., 1972, *The Mediterranean Sea: A Natural Sedimentation Laboratory*, Stroudsburg, Pa.: Dowden, Hutchinson and Ross, p. 765.

Van Bemmelen, R. W., 1972, Geodynamic models—An evaluation and a synthesis, in: *Developments in Geotectonics*, v. 2, Amsterdam–London–N.Y.: Elsevier, 267 p.

Woodside, J., and Bowin, C., 1970, Gravity anomalies and inferred crustal structure in the Eastern Mediterranean Sea, *Bull. Geol. Soc. Amer.*, v. 81, p. 1107–1122.

Zarudski, E. F. K., 1972, The Strait of Sicily. A geophysical study, *Rev. Geogr. Phys. Géol. Dynam.*, Paris, 2ᵉ sér., v. 14 (1), p. 11–28.

Zarudski, E. F. K., and Rossi, S., 1973, Medina and Cyrene seamounts in the Southwesterly Ionian Sea, *C.I.E.S.M. Assemblée*, Athènes (1972), v. 22, fasc. 2a, p. 100–102.

Chapter 5B

THE GEOLOGY OF THE PELAGIAN BLOCK: THE EASTERN TUNISIAN PLATFORM

J. Salaj

Service Geologique
Tunis, Tunisia
and Geologicky ústav Dionýza Štura
Bratislava, Czechoslovakia

I. INTRODUCTION

The eastern platform of Tunisia, the Tunisian section of the Pelagian block, is one of four major tectonic units which can be seen in a northwest–southeast traverse across the country (Fig. 1). In the northwest lies the Alpine zone with the Numidian nappes, Tell units, and the autochthon and para-autochthon of Hedil and Medjerda. This zone is bordered immediately to the south by a zone of diapirs, but is separated from it by a major fault. The two together formed a sediment-filled foredeep in Miocene time. The importance of this fault was noted by Jauzein (1967) who called it the Teboursouk suture (the Teboursouk thrust of Rouvier, 1973). Jauzein considered it to be the eastern limit of the zone affected by Upper Pliocene–post-Villafranchian tangential thrust faulting. Detailed study of a section of the fault (Bajanik and Salaj, 1971) showed it to be a zone varying in width from a few dozen meters to a kilometer, and trending northeast–southwest. The zone of diapirs marks the northern edge of the intermediate chains of the Tunisian Atlas. These chains terminate to the south against the Saharan platform, and to the east against the Tunisian platform, the

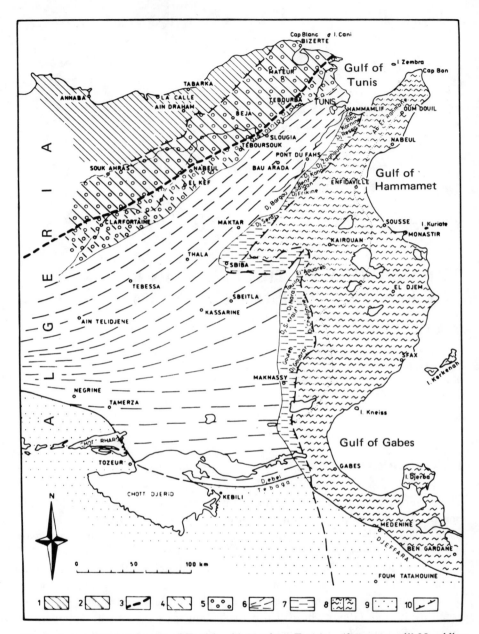

Fig. 1. A tectonic sketch map of Tunisia. (*I*) *Northern Tunisia—Alpine zone:* (1) Numidian nappes. Tellian units, and para-autochthon of Hedil; (2) Medjerda para-autochthon and autochthon; (3) thrust zone of Teboursouk; (4) Miocene foredeep (also occurring in intermediate chains). (*II*) *Tunisian Atlas—intermediate chains:* (5) diapiric zone; (6) central and southern Atlas; (7) north–south axis. (*III*) *Platforms:* (8) eastern platform (western part of Pelagian block), structural depressions, and elevations of eastern platform; (9) Saharan platform; (10) thrusts.

edge of which is formed by a tectonic feature known as the north–south axis (Fig. 1). At Sbiba there is a westward jutting salient in the north–south axis; north and west of this point the autochthonous Atlas tends to have structures striking northeast–southwest, while to the south the influence of the Saharan platform seems to have imparted an east–west structural trend.

To the south, the Saharan platform, at the edge of the African craton, forms the third unit. The fourth unit is the eastern platform, whose western limit is marked by the north–south axis. It is an anticlinal zone overthrust by the Atlas in the region of Zaghouan. The axis is a feature of long standing, for as will be seen in the following pages it formed an important facies division throughout the later Mesozoic between the mobile zone of the Atlas Trough and the Chotts and the more stable Tunisian platform. Only during the Cenozoic did its importance as a structural feature diminish. During the Permian the Tunisian platform is indistinguishable from the region to the west.

The north–south axis is in two distinct segments, linked by a curve from Djebel Touila to Maknassy (Fig. 1). This region is important for it suggests that much of the area included by Burollet (1973) in the central and southern Atlas in reality belongs to the eastern platform. The area west of Djebel Touila–Maknassy, as far as Tebessa–Tamerza, has been described by Burollet (1956, 1967) as the Kasserine Island, for its positive role particularly during the Senonian. No marine beds younger than Turonian occur in the central region of the "island." The eastern platform is characterized in its tectonic stability from the Upper Paleozoic into the Cenozoic. Condensed sedimentary sequences with numerous lacunes are common. At the end of the Miocene and during the Lower Pliocene part of the platform was emergent, forming a barrier separating the eastern and western Mediterranean. This barrier was, however, submerged during the course of the Upper Pliocene.

In the following section an attempt will be made to outline the stratigraphy of Tunisia with particular reference to the Pelagian block. The Cretaceous, in particular, is well enough known to permit the construction of facies distribution maps with a certain degree of confidence. From these maps certain broad paleogeographic conclusions can be drawn, permitting an overall picture of the evolution of the area as a whole.

II. STRATIGRAPHY

In the following section the stratigraphy of eastern Tunisia will be reviewed, system by system, and an attempt will be made to develop an interpretive paleogeographic model of the evolution of the block as a whole. However, only the Cretaceous is exposed adequately to allow reasonably detailed interpretations.

A. Triassic

1. *Regional Stratigraphy*

On the eastern Tunisian platform (Fig. 2, zone 2) as well as along the north–south axis no Triassic rocks are known *in situ*, but allochthonous rocks of this age do occur. The consequence is that it is difficult to get a clear understanding of the stratigraphy.

The most varied section, seen in Djebel Rheouis, has been studied by Castany (1952), Burollet and Dumestre (1952), and Burollet (1963, 1967, 1973). In this diapir Burollet (1973) recognized seven lithogical units:

(7) Upper evaporites

(5–6) Yellow, relatively fossiliferous, dolomitic limestones and silts, 60 m

(4) Middle evaporites, 850 m

(3) Black dolomites (Upper Ladinian–Carnian), 100 m

(2) Lower silts of variable thickness

(1) Lower evaporites, 200 m

The Triassic facies are basically of Germanic type. Horizon 3 has yielded ostracods and foraminifera comparable to those found in the Upper Ladinian–Lower Carnian of Bechateur (cf. Salaj and Bajanik, 1972*b*; Biely and Rakus, 1972*b*), and it may be that the lower evaporites belong to the lower rather than the Middle Triassic, as proposed by Burollet (1973).

The dolomitic limestones of the Rhaetic are not present in the above section, but have been found in sections at El Attaris above a variegated gypsum, which may be compared to the upper evaporites (cf. Castany, 1952), and in the Djebel Fkirine (Salaj and Stranik, 1970, 1971*a*), where both Lower and Upper Rhaetic were recognized in a thickness of 83 m of limestone. In the region Hairech–Djebel Ichkeul, *Megalodontes* of Norian–Rhaetian age was found (Fig. 2). In the Cape Bon well CB-1, Bonnefous (1967) and Burollet (1973) indicated the presence of Upper Triassic dolomitic limestones and argillaceous dolomites. The data suggest that true open marine conditions occur in the Ladinian of Cape Bon, but do not appear until Rhaetic times in the Djebel Fkirine section. In this the Triassic succession differs from the typically Germanic facies.

In the Tello-Tunisian Trough (Fig. 2, zone 1) four units can be recognized. The lower consists of a detrital sandstone comparable in age and lithology with the Buntsandstein as at Djebel Baouala and Oued Melleg, but including thin marine dolomites with lamellibranchs, and capped in places by a conglomerate and volcanics (Djebel Bechateur). It is followed by blackish limestones of Upper Ladinian to Lower Carnian in age, and therefore equivalent to the

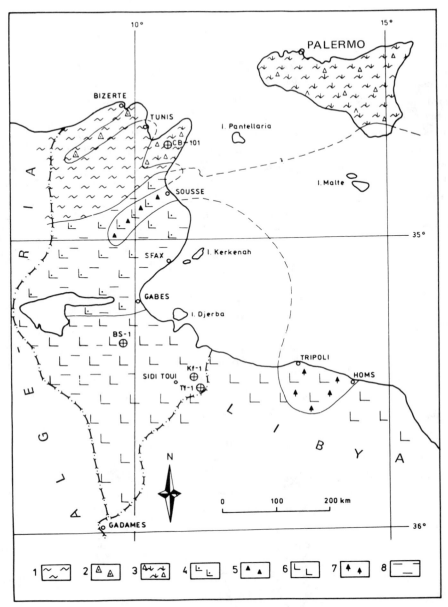

Fig. 2. A sketch map of possible Triassic facies distribution. (*I*) *"Pre"-Tello-Tunisian Trough:* (1) predominantly detrital beds (= Germanic facies); (2) mixed facies, detrital beds, and dolomitic limestone (= modified Germanic facies + Alpine); (3) occurrence of Norian–Rhaetian dolomitic limestone (Alpine facies). (*II*) *Central and southern Atlas and eastern platform:* (4) Middle–Upper Triassic saliferous series; (5) occurrence of ?Norian–Rhaetian dolomitic limestones. (*III*) *southern Tunisia–northern Tripolitania:* (6) Lower and Middle Triassic (Upper Triassic usually absent); (7) Upper Triassic dolomites followed by shales and sandstones; (8) region emergent during much of the Lower and Upper Triassic; ⊕ location of boreholes.

Muschelkalk. Overlying this is a gypsiferous series with a basal conglomerate and some volcanics (Keuper equivalent), followed by dolomitic limestones probably representing the Rhaetic (Bajanik, 1971; Salaj and Bajanik, 1972b). The trend of the Tunisian Trough continues into Sicily, but volcanism there does not occur until Bajocian time.

The Triassic of southern Tunisia (Fig. 2, zone 3) has been extensively studied by Pervinquière (1912), Domerque et al. (1952), and Busson (1967), and may be examined in the Kef Touareg in the Djeffara. The microfauna, of Alpine type, is known from wells Tt-1 (Bir Ben Tartar) and L-687 (Sidi Toui), and gives evidence of the existence of a Muschelkalk facies of Julian (Middle Carnian), Cordovolian (Lower Carnian), and Campilian (Glintzboeckel, 1956; Salaj, 1969).

2. *Outline of Paleogeography*

Given the absence of *in situ* outcrops in the northern half of the country a paleogeographic reconstruction is frankly speculative. Apart from southern Tunisia there is still very little evidence for the existence of Lower Triassic beds in the central and southern Atlas, or in what may be termed the Tello-Tunisian "pretrough" in the north. It is possible that the lowest horizon of Cape Bon may be Lower Triassic in age, and that certain of the patchily developed redbeds under the saliferous series of the central and southern Atlas may also be Lower Triassic in age.

The most continuous and thickest section of Triassic is developed in the south, and has been exhaustively described by Busson (1967).

In the north the attempt has been made to distinguish a relatively more subsident "pre"-Tello-Tunisian Trough from the zone encompassing the central and southern Atlas (Fig. 2). The deposits in this trough are of a modified Germanic type; that is, they consist of continental detrital redbeds, with intercalated yellow and black fossiliferous marine limestones and evaporites with synsedimentary volcanism. The volcanics occur both below and above the Muschelkalk. They differ from the typical Germanic facies by the association of blackish dolomitic limestones. The oldest of these is Ladinian in age, the youngest are assigned to the Rhaetic. There are many similarities between these beds and the succession known in Sicily.

In the central and southern Atlas a Middle to Upper Triassic saliferous series rests upon continental–lagoonal beds. However, uplift and erosion in pre-Liassic time leaves the succession thin where it has not been removed altogether. Only in a narrow zone striking southwestward from Sousse is a series of dolomitic limestones of probable Norian to Carnian age preserved.

The most complete Triassic sections in southern Tunisia and northern Tripolitania are known from boreholes in the Djeffara. The Upper Triassic here differs from that to the north by being represented by the massive Rehach (or Azizia) dolomites and their associated clays and anhydrite.

B. Jurassic

1. *Regional Stratigraphy*

There have been many studies of the Jurassic of the eastern platform of Tunisia (Baltzer, 1895; Pervinquiere, 1903; Spath, 1913; Solignac, 1927; Castany, 1951, 1952, 1955; Dubar, 1953; Burollet, 1956; Bonnefous, 1967, 1972; Bismuth *et al.*, 1967; Floridia and Massin, 1969; Biely 1969; Biely and Rakus, 1969, 1972*a*; Rakus and Biely, 1971; Rakus, 1973; Gaudant *et al.*, 1972; Salaj, 1972), as well as mapping on a 1:50,000 scale of the sheets La Goulette (Bujalka *et al.*, 1972), Grombalia (Bujalka *et al.*, 1971), Bou Ficha (Johan and Krivy, 1969), Zaghouan (Bajanik *et al.*, 1968), Djebel Fkirine (Stranik and Mencik, 1970), and Enfidaville (Bajanik *et al.*, 1973). To this may be added the lithological interpretation of numerous wells (CB-1, CB-101, Kt-2, SO-1, ABK, CF-2, CF-1, and G-10) on the Tunisian platform by Bonnefous (1967, 1972). A discussion of the Jurassic in the southern part of the country may be found in Busson (1967). A summary of Jurassic data is given by Bonnefous (1967, 1972) and by Bismuth *et al.* (1967).

Lias. In view of the problem concerning Triassic outcrops, the Rhaetic–Hettangian boundary raises some problems. It may be supposed that the Hettangian continues the pattern of limestone deposition from the Rhaetic of Djebel Fkirine and south of Cape Bon (Rakus and Biely, 1971; Rakus, 1973). The Hettangian–Lower Sinemurian, usually represented by some 300–350 m of massive limestones of the Oust Formation *s.s.* (Rakus, 1973), may exceptionally reach as much as 500 m in thickness. In Djebel Nara in central Tunisia (well SO-1) and in the region of Chott Fedjedj (wells CF-1, CF-2), and at Zemlet El Beida (well ZB-1) the limestones are strongly dolomitized (cf. Bonnefous, 1967).

The limestone is a biosparite with shallow-water dasyclydacean algae, some of which have been recognized in Morocco (Levy, 1966). Until the end of the Lower Sinemurian the North–South Axis was a zone of subsidence, and only with the beginning of the Upper Sinemurian are the first positive movements felt in the region of Djebel Oust, Djebel Fkirine, Djebel Zaghouan, Hammam Djedidi, etc. (cf. Rakus and Biely, 1971), which lead to the formation of a stable shoal area (Fig. 3).

The Upper Sinemurian and Lower Pliensbachian are thus represented by a condensed sequence of 50 cm to 2 m, always accompanied by the development of a ferruginous crust, phosphate, and glauconite. The limestone is an intra-

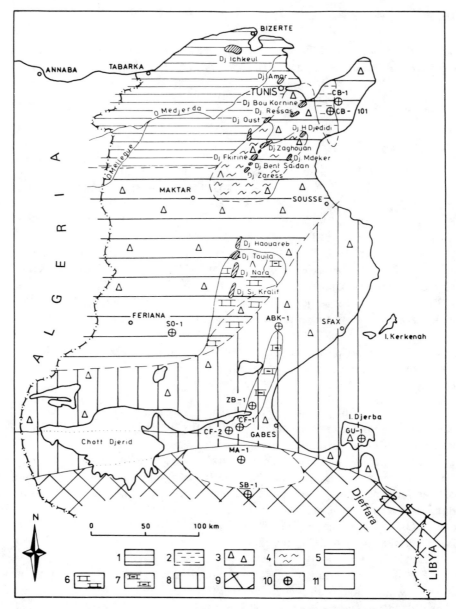

Fig. 3. Sketch map of Jurassic facies distribution. (1) Subsiding pelagic facies of Tunisian Trough; (2) red argillaceous radiolarian facies of Oxfordian (Tunisian Trough facies); (3) littoral Lower and Middle Lias (algal mats); (4) condensed Upper Sinemurian–Carixian sequence; (5) pelagic Callovian, dolomitized Oxfordian–Lower Portlandian, pelagic Upper Portlandian; (6) neritic Callovian, dolomitized Oxfordian–Lower Portlandian, pelagic Upper Portlandian; (7) dolomitized Kimmeridgian–Lower Portlandian; (8) (i) subsiding pelagic Oxfordian with terrestrial detritus, (ii) subsiding littoral Kimmeridgian–Lower Portlandian with *Pseudocyclammina*, (iii) littoral Upper Portlandian with *Iberina*; (9) littoral–brackish-water facies of Djeffara Jurassic; (10) wells; (11) Matmata High erosion; ..... basin margin.

biosparite. The sequence seems to be incomplete, for according to Rakus and Biely (1971) the Lower Domerian was a period of nondeposition and current scouring. Breaks in sedimentation are clearly indicated by the ferruginous crusts of which there may be two to five.

Northwest of this zone of condensed sedimentation (Djebel Zaress, Djebel Oust, Djebel Djedidi; Fig. 3) is a subsiding zone in which the sequence is represented by 15–20 m of organodetrital limestones with black chert. The Pliensbachian part of the succession of dark gray limestones, in places pseudo-nodular, is followed by the Domerian, which is here present in the form of gray, sublithographic limestones with marly partings.

Along the north–south axis and on the eastern Tunisian platform the Upper Domerian–Toarcian is represented by a gray, marly limestone with a cream colored patina, alternating with olive-green marls (Castany, 1951; Biely and Rakus 1972a). Based on the existence of hard ground (ferruginous crust) below the Toarcian sequence, Castany (1951) supposed the existence of a gap in the Lower Toarcian. Subsequently Rakus (1964) showed that the succession was complete at Djebel Bou Kornine near Hammam-Lif, while Bonnefous and Massin (1964), Bonnefous and Rakus (1965), and Bonnefous (1967) showed the presence of both Lower and Upper Toarcian at Dj. Kornine, and Floridia and Massin (1969) showed that all the Toarcian was present at Hammam Djedidi. Similarly a supposed break at Zaghouan (Castany, 1951) must be discounted because of the recognition there of an Upper Domerian and Lower Toarcian fauna (Biely and Rakus, 1972a).

Farther south along the north–south axis the marly limestone alternation, in places with red calcareous nodules and marls, contains intercalations of limestone rich in fish remains. The continuity of sedimentation and of facies argues against any stratigraphical break during the Toarcian. The microfacies show a biomicrite with short, algal filaments, rare radiolaria, and various species of *Lenticulina*.

In central Tunisia (Souima well SO-1) the Upper Liassic is absent and the same is true in the region of Chott Fedjedj (cf. Bonnefous, 1967). Both locations, however, lie west of the edge of the Tunisian platform.

Middle Jurassic. The Aalenian is represented by 20–34 m of well-bedded, gray limestones with chert at Hammam Lif. The succession contains some rare marl intercalations. In some places the Upper Aalenian is condensed to about 30 cm (Djebel Fkirine), but it is rich in ammonites which permit recognition of the complete succession. The limestone is a biomicrite with short, algal filaments and lamellibranchs and is relatively rich in radiolaria.

Three facies have been recognized in rocks of Bajocian age:

(a) a gray-black, finely banded and folded limestone full of filaments, 5–10 m thick, found in the western part of Zaghouan. The same facies is

found in the immediately overlying Bathonian, and above the red Oxfordian limestones (Biely, 1969).

(b) a red nodular limestone found by Biely (1969), 1–5 m thick, in the
cliffs east of Djebel Zaghouan. These are overlain by finely banded, blackish Bathonian–Callovian limestones.

(c) yellowish, nodular, argillaceous limestones found at Hammam Lif.
This facies extends from the Aalenian to the Bathonian, and when interbedded with marls into the Callovian (Rakus, 1973).

Within the Bathonian there are two conglomeratic horizons made up of
organo-detrital limestone pebbles (of Middle Liassic age), evidence of exposure
and erosion during the Bathonian of regions which had begun to emerge at
the beginning of the Domerian. This is presumably related to volcanic activity
in Sicily.

Over the eastern platform the same facies can be found, but the succession
is between 150–230 m thick, while in the subsiding zone in the south from Sfax
extending westward under the Chotts, the same sequence is much thicker and
totals some 1380 m in well ABK (Bonnefous, 1967). Local variations occur on
the platform; for example, at Djebel Saidane the reddish radiolarian limestones
and red marls occur prior to the Upper Callovian horizon known as "ammonitico rosso" (Mencik et al., 1970).

Upper Jurassic. The Oxfordian and the upper part of the Callovian are
represented by red nodular limestones. The presence of Upper Callovian in the
Djebel Zaghouan section, first indicated by Spath (1913), has been fully confirmed by the rich ammonite fauna found by Biely and Rakus (1969) despite
assignment to the Upper Oxfordian by Pervinquière (1903), Solignac (1927),
Castany (1951, 1955), and Bonnefous (1967, 1972). A little to the south in
Djebel Bent Saidane, Mencik et al. (1970) found the Upper Callovian *Reineckia*
horizon at the base of the "ammonitico rosso," indicating that the onset of the
"ammonitico rosso" phase regionally began in Upper Callovian time and
continued through the Oxfordian to the base of the Kimmeridgian. This facies,
from 15–56 m thick, is widely distributed along the north–south axis and occurs
on the platform at Hammam Djedidi (Floridia and Massin, 1969), at Djebel
Mdeker (Bonnefous, 1967) (Fig. 3), and it has been recognized in wells on Cape
Bon.

Off the platform in southern-central Tunisia, in well SO-1, reddish limestones 70 m thick and attributable to the Oxfordian have been found (Bonnefous, 1967), and Oxfordian in an oolitic limestone facies has been reported
from Djebel el Attaris (about 5 km south of Djebel Nara and 15 km north of
Djebel Sidi Kralif).

Along the northern and central parts of the north–south axis the Kimmeridgian and Lower Portlandian are represented by a pelagic facies consisting

of greenish or yellowish argillaceous limestones and marls 10–25 m in thickness. The fauna includes crinoids, radiolaria, and other microfossils. In the Upper Portlandian calpionellids make their appearance. A reef facies of Upper Portlandian reaching a thickness of 300–390 m is found at Djebel Ressas, Djebel Mdeker, Djebel Zaghouan, Hamman Djedidi, and Hammam Zriba. This facies consists of organodetrital calcarenites and calcirudites with corals, rudists, algae, and foraminifera, especially trocholinids. At most of the above-named localities a transition between the two facies can be found. Unfortunately along the southern part of the north–south axis only a dolomitic facies is present.

South of the Tunisian platform, in the Chotts, the Upper Portlandian is represented by a littoral facies consisting of sandy oolitic limestones, dolomitized to a greater or lesser degree, and accompanied by evaporites. This was a zone of strong subsidence, for more than 1000 m of littoral sediments is recorded in wells at Djerba (Guellala 1) and north and west of Gabes (ABK-1, SO-1, and CF-2). South of the Chotts lies the Djeffara, where the Jurassic has been described by Busson (1967). Here between the Oued Tatahouine and Djebel Tebaga the Jurassic forms a condensed sequence of dolomites and evaporites laid down in a shallow sea close to the continental margin. Very often lagoonal beds and horizons of continental beds are intercalated, and the uppermost Jurassic, Kimmeridgian–Portlandian, presents a facies similar to the Wealden facies of the Purbeckian–Lower Cretaceous.

2. Outline of Paleogeography

The attempt has been made in Fig. 3 to summarize the distribution of the principal Jurassic facies, and to relate the development of the Jurassic to eastern Tunisia and surrounding regions.

The distributions shown in Fig. 3 form the basis from which an attempt can be made to deduce something of the paleogeographic conditions. Along the edge of the Sahara platform in the south, a littoral to brackish-water environment developed but was interrupted locally by the Matmata High which was undergoing erosion. To the north in the region of the Chotts lay a zone of subsidence. Here, a pelagic facies predominated, although there were some terruginous horizons representing an influx of terrestrial material.

Central Tunisia to the north was also a zone of pelagic sedimentation, although subsiding at a slower rate than the Chotts. The eastern platform is clearly defined only in the north and south by the north–south axis on which a reef developed. Still farther to the north lay the subsiding zone of the Tunisian trough with its pelagic deposits. The facies pattern is most clearly emphasized in the Middle and the Upper Jurassic, for during the Lias the central region including the eastern platform and the region of the Chotts was a shallow-water area in which algal mats developed.

C. Cretaceous

1. *Lower Cretaceous Stratigraphy*

The Lower Cretaceous in Tunisia is well developed and must be described stage by stage before a general synthesis can be attempted. Sedimentation was continuous from the Jurassic into the Cretaceous along the north–south axis at the western limit of the Tunisian platform. Toward the north the sediments are of a neritic aspect. The northern part of the platform was invaded five times by flysch wedges, once each in the Upper Valanginian, the Upper Hauterivian, and the Gargasian, and twice during the Barremian.

Berriasian. The Berriasian along the north–south axis is represented by three different facies which are most easily located by reference to the facies of the underlying Portlandian (Fig. 3). For example, above the mixed reef and calpionellid facies of the Portlandian in the vicinity of Djebel Zaghouan (Figs. 4, 5) and Djebel Djedidi and Djebel Bent Saidane (Fig. 6), the 15–20 m of Berriasian consists of greenish or dark gray marls with intercalations of sublithographic limestones (with calpionellids), and by yellowish organodetrital limestones in which crinoids predominate. However, above the sublithographic limestones of the Portlandian the Berriasian consists of a 16.5-m alternation of gray marls and sublithographic calpionellid limestones with abundant trocholinids. These limestones, compared with the limestones of Zaghouan, are less detrital with fewer quartz grains. The calpionellid limestones of the Portlandian of Oued Guelta and Djebel Zarrez are followed by an alternation of pelagic marls and sublithographic limestone, the paleontology of which has been studied by Colom *et al.* (1953), Salaj (1972), and Memmi and Salaj (1973). This facies continues as far south as Djebel Chaabet et Attaris (Bonnefous, 1965, 1972; Memmi, 1967; Memmi and Salaj, 1973). Although the Jurassic–Cretaceous passage is continuous, dolomitic limestones followed by gray marls, pelagic limestone, and ochreous marly limestones develop (the Sidi Kralif Formation of Burollet, 1956). The rich fauna found in the marls has been studied by Arnould and Saget (1951) and Memmi (1967).

The Chotts to the south have been penetrated by numerous oil borings (Bismuth *et al.*, 1967; Memmi and Salaj, 1973), which indicate that the Jurassic–Cretaceous boundary lies within beds of a littoral, neritic facies with sandy oolitic limestones, followed by lagoonal-type gypsiferous shales, and gypsum with sandstone intercalations containing lagoonal ostracods, of a Wealden facies. The Barremian, Aptian, and Albian are widely represented in this marginal region of the Sahara platform by a continental sandstone facies. This facies at the end of the Barremian–Aptian extended onto the Tunisian platform as far north as the latitude of Kairouan.

East of the north–south axis on the platform, the Berriasian is present in

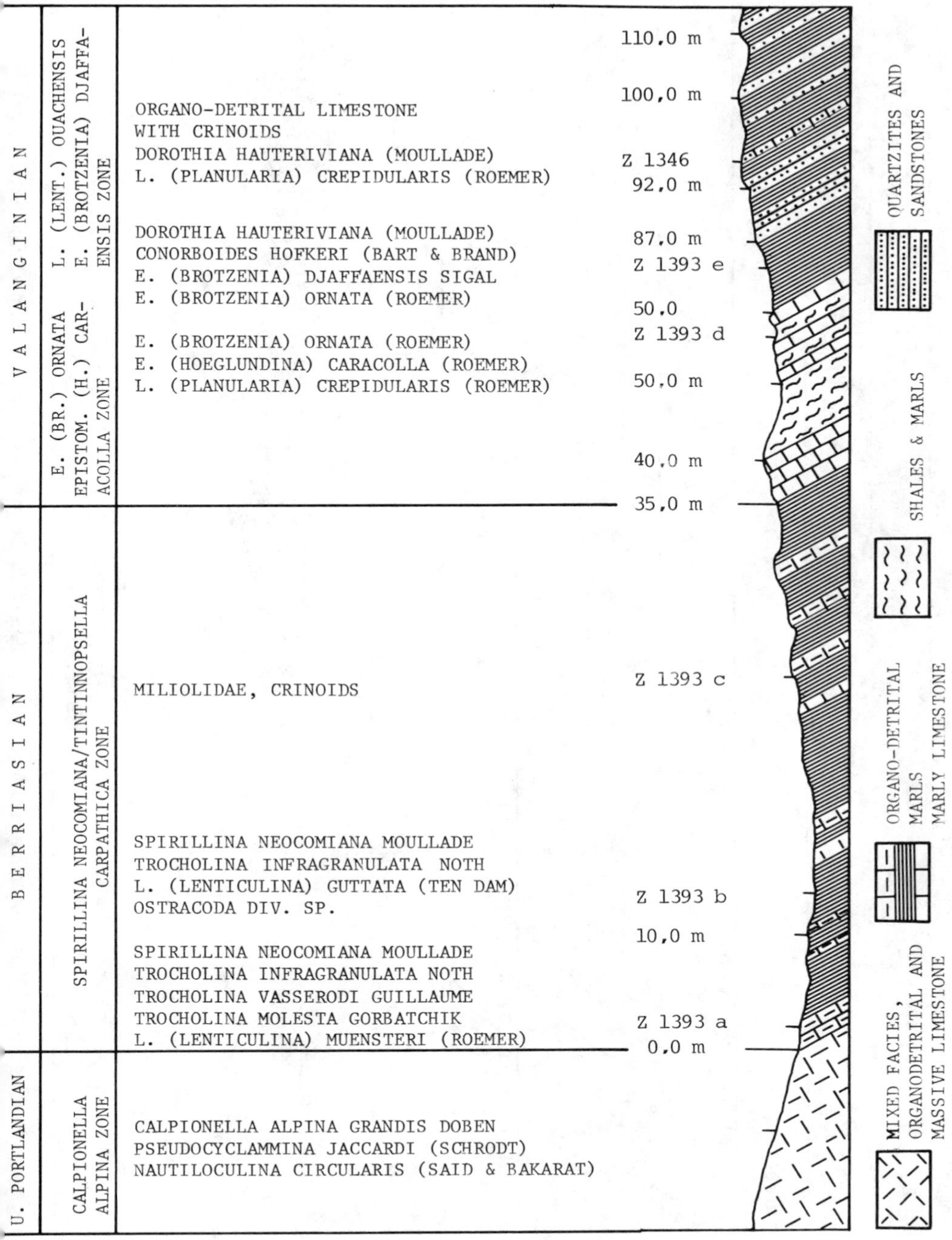

Fig. 4. Biostratigraphical and lithological section of the Temple des Eaux, near Zaghouan.

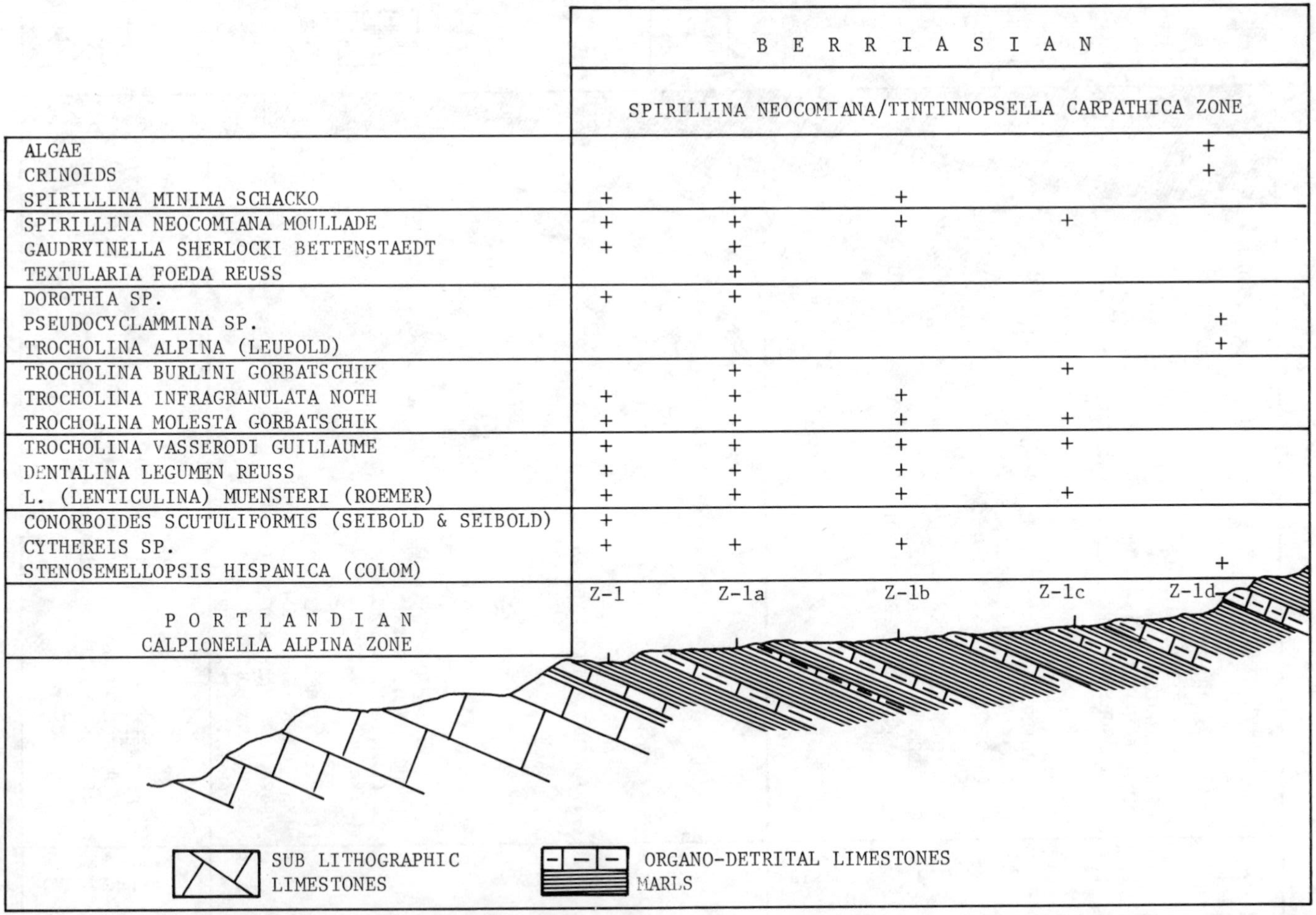

Fig. 5. Biostratigraphical and lithological section of the Berriasian at Col du Vent, near Zaghouan.

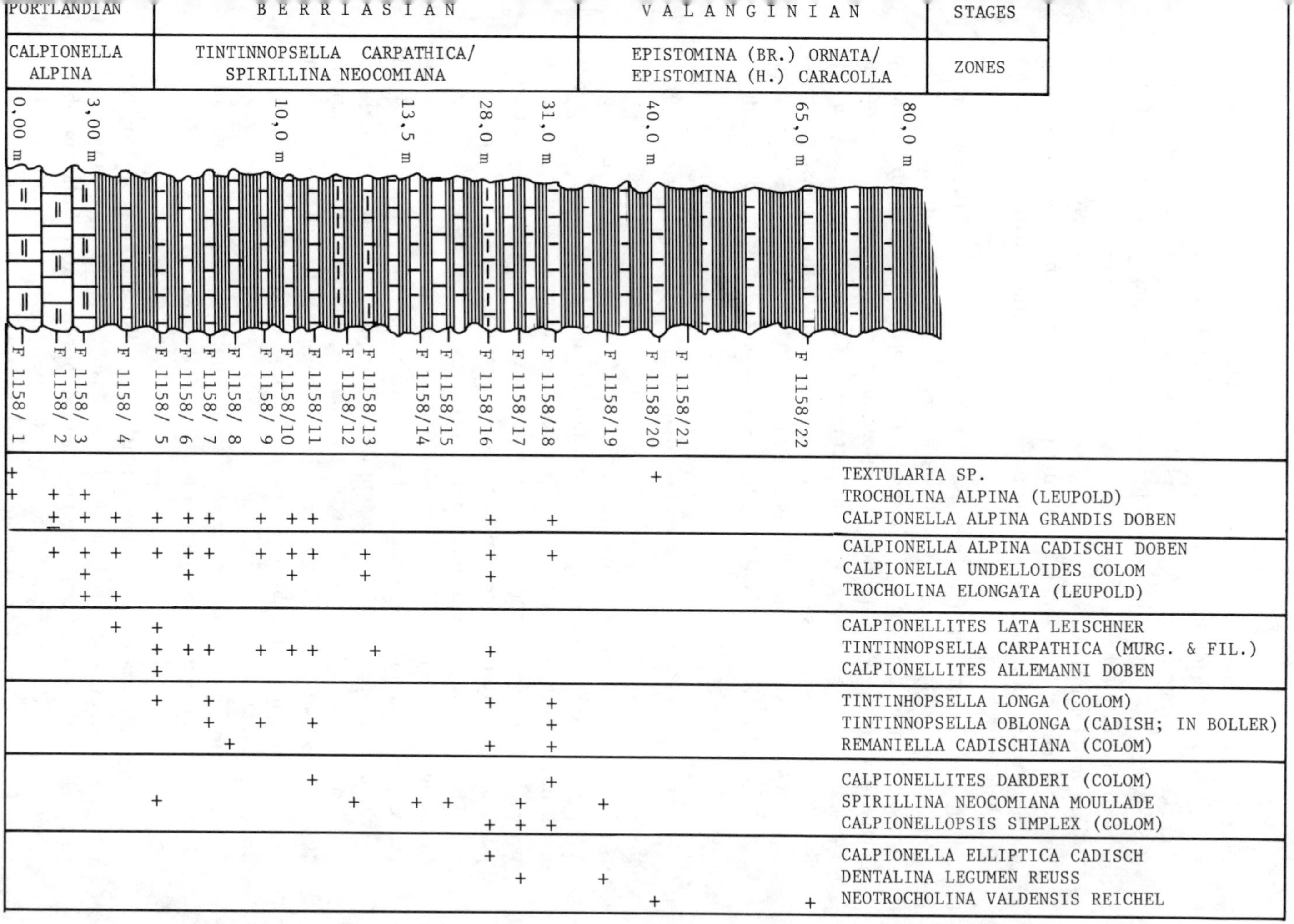

Fig. 6. Biostratigraphical and lithological section of the Berriasian in the eastern part of Djebel Bent Saidan, Fkirine sheet. The distribution of calpionellids is illustrated. Sampling sites and numbers are recorded in the archives of the Geological Survey of Tunisia.

either a pelagic or an organodetrital facies. The pelagic facies has been described in Djebel Mdeker (Bonnefous, 1956; 1972; Bajanik *et al.*, 1973), and although the limestones are dolomitized, they still contain an identifiable fauna (Bonnefous, 1965). Organodetrital limestones alternating with marls, a facies comparable with that occuring near Zaghouan, occurs in Djebel el Messella-Sidi Salem (Johan and Krivy, 1968; 1969). The facies variations found in the Portlandian and Berriasian are indications of slight tectonic movements linked not only to the north–south axis, but also to the northern part of the eastern Tunisian platform.

In the authors' opinion, because of the closer affinity of the ammonites (Wiedmann, 1971) and other faunal elements (Sigal, 1965) with those of the Jurassic than to those of the Valanginian, the so-called Berriasian should be included with the Portlandian. The upper limit of the two substages of the Tithonian may be fixed by the disappearance of *Calpionella* ex. gr. *alpina* (cf. Salaj, 1975). This is consistent with the second solution of the problem proposed at the Berriasian Colloquium (Neuchâtel, Lyon, 1973) [see also Busnardo *et al.* (1965) and Memmi and Salaj (1973)]; it is also consistent with a change in the facies pattern.

Valanginian. Beginning with the Valanginian *s.s.* sedimentation in both northern and eastern Tunisia took place in relatively deep water according to the interpretation of the cephalopod facies (cf. Castany, 1951). During the Valanginian and Hauterivian the Tunisian Trough (cf. Bolze *et al.*, 1952), immediately north of the platform, reached its maximum extent. During this period, the anticlinal zone of the north–south axis, which had played an important role up to this time, was engulfed in the trough.

The stratigraphic sequence of the Valanginian consists of two units, a lower, marly sequence 60 to 150 m thick with limestone and marly limestone intercalations (ammonite fauna studied by Memmi, 1965), and an overlying flyschoid sequence from 200 to 300 m thick. The limit with the Hauterivian is roughly marked by intercalations of fossiliferous, marly limestones (Memmi, 1965; Bajanik *et al.*, 1973). The least detrital section in the whole of northern and eastern Tunisia is that found on the northwest flank of the Dj. Mecella anticline (Bujalka *et al.*, 1971; Bolze and Sigal, 1964; Floridia and Massin, 1969), where the thickness of the Upper Valanginian is reduced. Apparently a slight elevation within the basin of sedimentation was responsible for this reduction.

The greater part of the eastern platform, and of the Pelagian block generally, was gradually covered during the Berriasian and Lower Valanginian by relatively shallow water. The zone emerged during upper Valangian–basal Hauterivian and was eroded, the detrital material being transported northwestward into the "sillon tunisien" (Tunisian Trough).

Hauterivian. The basal part of the Hauterivian continues the flyschoid facies found in the Upper Valanginian. It contains a few marly limestones and marl horizons in a thickness of some 40 m, and is best exposed in Oued Guelta along the north–south axis (Fig. 7) (Bajanik *et al.*, 1968). This section is followed by a blackish, sublithographic limestone sequence with intercalations of marl. The Upper Hauterivian consists predominantly of marls with some marly limestone bands. The top of the sequence is often marked by a limestone horizon.

The preceding facies characterizes the northern part of the north–south axis, while toward the south (locality of Oued Guelta, Zaghouan) a few sandstone intercalations appear in the Upper Hauterivian, and yet farther to the south (Djebel Fkirine, Djebel Bahalil) the whole of the Upper Hauterivian becomes flyschoid (Mencik *et al.*, 1970), with rare intercalations of marly limestone. On the platform itself, e.g., Hammam Djedidi (Johan and Krivy, 1968) and elsewhere, the Hauterivian is thinner, the flysch facies essentially is absent, and is replaced by limestones, marly limestones, and marls (Biely *et al.*, 1973).

Barremian. Along the north–south axis in the region of Zaghouan, the Lower Barremian is represented by a total of 10 to 80 m of massive, blackish limestones with gray marl intercalations (cf. Fig. 8), followed by 30–40 m of marls (Bajanik *et al.*, 1968). The sequence is thicker (150–250 m) and more marly, though containing argillaceous limestone and quartzite intercalations between Djebel Mecella and Djebel Ressas. The sandstones, however, disappear before Djebel Ressas and Djebel Bou Kornine (in the northeast) are reached. In the vicinity of Djebel Fkirine a subreef organodetrital limestone facies, with massive and nodular stromatoparoid limestones, shell banks, and many crinoids (Stranik *et al.*, 1970; Remack-Petitot, 1971) replaces the argillaceous facies. In a northwesterly direction the section thins by condensation from 40–60 m to 4 m.

On the eastern platform the Lower Barremian consists of alternating pelagic limestones and marls. It is not very thick where seen in Djebel Mdeker (Biely *et al.*, 1973), ranging between 6–30 m and is a condensed sequence formed on a shoal area.

The lithological sequence of the Upper Barremian found in the north–south axis zone around Djebel Zaghouan–Djebel Fkirine–Djebel Zaress is given below:

(d) blackish platy bedded limestone, 4–18 m

(c) upper flyschoid series, 10–150 m

(b) alternation of marly limestones and marls, 50–75 m

(a) lower flyschoid series, which makes its appearance before the end of the Lower Barremian, 40–150 m

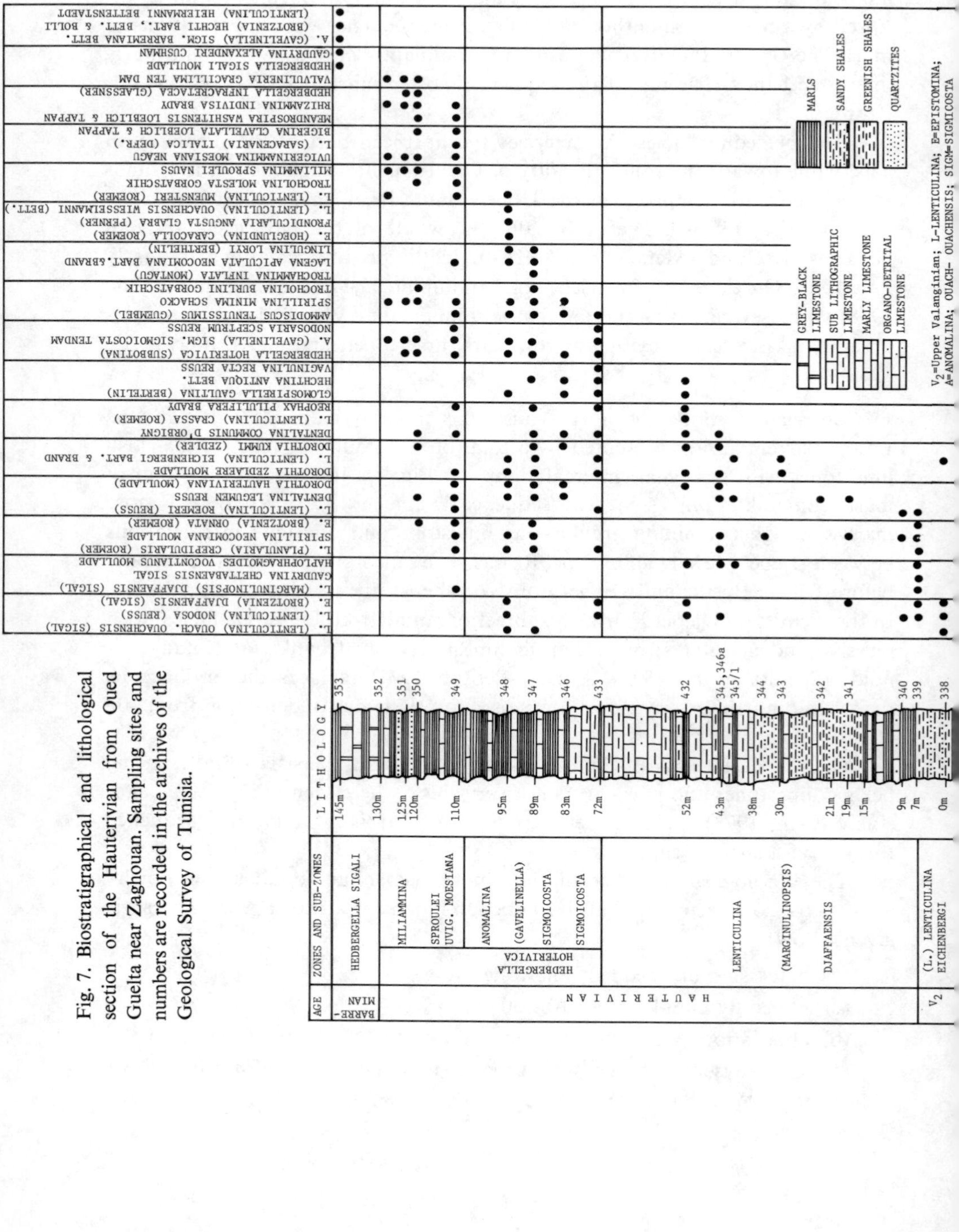

Fig. 7. Biostratigraphical and lithological section of the Hauterivian from Oued Guelta near Zaghouan. Sampling sites and numbers are recorded in the archives of the Geological Survey of Tunisia.

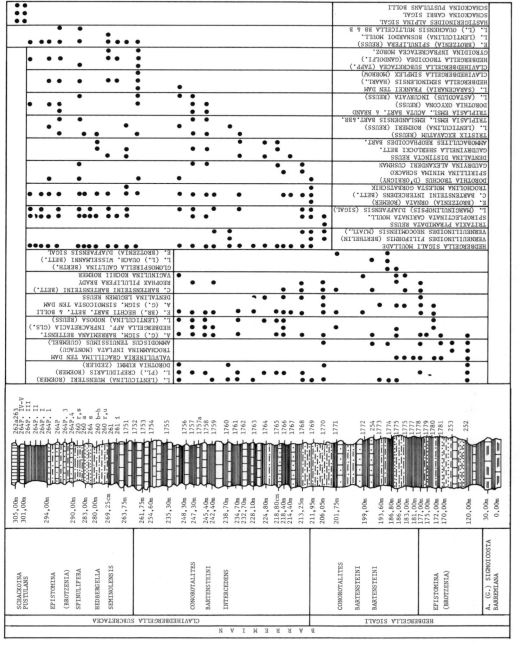

Fig. 8. Biostratigraphical and lithological section of the Barremian from Djebel Ech Chama, near Zaghouan. Sampling sites and numbers are recorded in the archives of the Geological Survey of Tunisia.

Thus, in this region the thick massive limestone horizon which separates the Lower and Upper Barremian at Djebel Oust is missing (Jauzein, 1957, 1967; Busnardo and Memmi, 1972; Memmi and Maamouri, 1974).

On the eastern platform, there is a flysch horizon (Biely et al., 1973) and a maximum of 100 m of limestone–marl alternations, topped by a characteristic clear, gray-banded limestone 10 m thick, and it appears that horizons c and d of the Zaghouan section are absent. This gap may be due to nondeposition or, more probably, to deposition and subsequent erosion during a short emergence prior to the deposition of the Aptian. This appears probable for the Albian rests on different levels of the Barremian (Biely et al., 1973).

Aptian. The best Aptian exposures along the north–south axis are in the vicinity of Zaghouan, where both the Lower (Bedoulian) and Upper (Gargasian) are represented by marls and intercalations of marly limestones and whitish limestones (Castany, 1951; Solignac, 1927; Bajanik et al., 1968; Salaj, 1972). There are also good exposures at Djebel Fkirine, Djebel Ben Saidan, Djebel Zaress, Djebel Ressas, and Bou Kornine (Mencik et al., 1970; Bujalka et al., 1971; Bujalka et al., 1972). The ammonite fauna has been studied by Stranik et al. (1970), the gastropods by Mongin (1971), and the echinoids by Remack-Petitot (1971). Subdivision of the zones is possible because of the rich ammonite fauna, and comparisons can be made with the foraminiferal zones (Stranik et al., 1970; Salaj, 1972; Maamouri and Salaj, 1974).

The top of the Gargasian is developed as a flysch facies, with the arenaceous content diminishing to the north. East of Djebel Ressas no quartzite bands are found. The Aptian thickness along the north–south axis varies between 170–300 m (Bujalka et al., 1971). Thinning is related to the eastern platform, which south of Khanquet and Hadjadj formed a relatively stable high (Biely et al., 1973). Over the eastern platform the Aptian is represented by 20–30 cm of glauconitic, phosphatic, highly fossiliferous limestone. Examination of the fauna (Biely et al., 1973) shows that both the Bedoulian and Gargasian are present except at Djebel Azreg, where the only two ammonites collected are both Bedoulian. The top of the Aptian has a ferruginous crust, suggesting nondeposition and exposure.

A reduced thickness is also found to the east and northeast in the wells at Enfidaville and Cape Bon. Burollet (1956) links this to orogenic movements along the north–south axis as far as Djebel Serdj. The movement also results in the absence of the Albian (Castany, 1951; Burollet, 1956). However, this is not general along the axis (Stranik et al., 1970) for near Zaghouan a glauconite Upper Albian horizon is found, thus confirming the shallow-water deposition mentioned above. At Djebel Rhezala the Lower Albian is also missing. In other places there is no break in sedimentation between Aptian and Albian (Djebel Mehjoul, Kef en Nsoura near Enfidaville). At Djebel Azreg any break must have been short, for the condensed Bedoulian has a ferruginous crust, but is overlain by Gargasian (Biely *et al.*, 1973).

The condensed sequence can also be seen on the south flank of Djebel Oust, where only 4 m is found (Memmi and Maamouri, 1974). Thus, it seems clear that tectonic movements along the north–south axis affected the whole sedimentation on eastern platform. Farther to the south at Djebel Garci (Biely *et al.*, 1973) and Djebel Fadeloun (Castany, 1951), the Aptian consists of 5–20 m of organo-detrital limestones above Barremian flysch. Somewhat dolomitized calcareous beds in the Upper Aptian are widespread in the south, and in the Chotts and Medenine areas rest transgressively upon Wealden sandstones.

Albian. During the Albian, the eastern platform, Saharan platform, and Tunisian Trough begin to develop as distinctive units. Over the eastern platform, a stable area, the Lower Albian is absent and a thin Upper Albian is generally transgressive. In the developing Tunisian Trough in central Tunisia there accumulated a thickness of marls (Fahdene Formation). To the south lay a zone of neritic facies. The Saharan platform remained a region of shallow-water sedimentation which included evaporites.

Along the north–south axis (Djebel Bou Kornine–Djebel Zaress) as well as on the eastern platform, the Albian is formed by an alternation of limestones and marls. Marls sometimes predominate at the base followed by platy Vraconian limestones. Where the top of the Albian is formed by a flysch facies the Aptian–Albian boundary is a bed of glauconitic sandstone at the base of a marl limestone sequence. Where the glauconitic bed is absent or poorly developed, the limit is well marked by a marl–clay alternation above marl–sandstone horizons of the Aptian (Bujalka *et al.*, 1971). Around Zaghouan the Lower Albian consists of greenish marls, a facies widespread in eastern Tunisia (Memmi and Maamouri, 1974; Salaj and Bajanik, 1972a). Where flysch is absent in the Upper Aptian, the lower limit of the Albian is determined by a horizon with barite concretions (see Fig. 9) at Djebel Donamess (Djebel Fkirine sheet). It must also be noted that in some regions the Albian is thin and transgressive. The Middle and Upper Albian total 100 m on Djebel Abid, but the base is not visible. The thickness is less in the region of Djebel Fadeloun near Kairouan where about 50 m of limestones and marl are found (Fig. 7). This characteristic persists into the Upper Cretaceous where a condensed sequence appears to have formed in a shoal zone. At the latter locality (Djebel Fadeloun near Kairouan), the Albian (Fig. 10), represented by a neritic limestone and marl facies, overlies reefal sediments (coral and orbitoline limestones).

The lagoonal–neritic zone, extending from Sfax to Gabes, is characterized during the Albian by limestones, dolomites, sandy marls, and some evaporites. On the Saharan platform itself, the sequence of limestones, dolomites, and evaporites together with sandstones, is thin and incomplete.

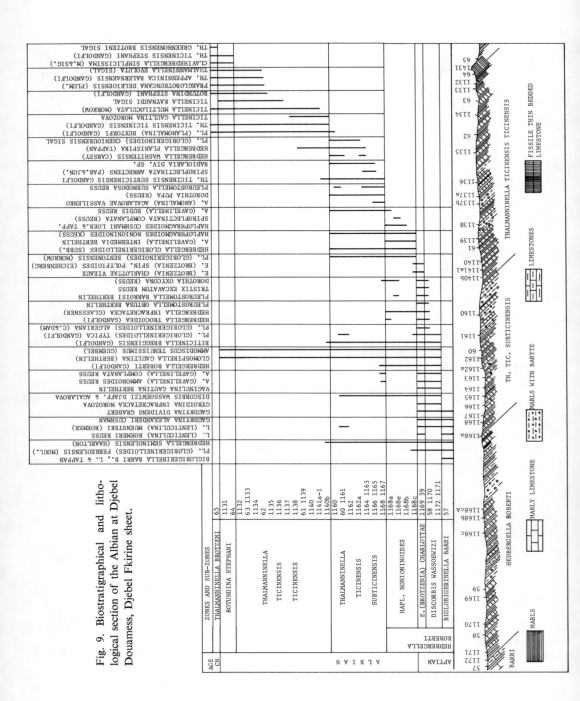

Fig. 9. Biostratigraphical and lithological section of the Albian at Djebel
Douamess, Djebel Fkirine sheet.

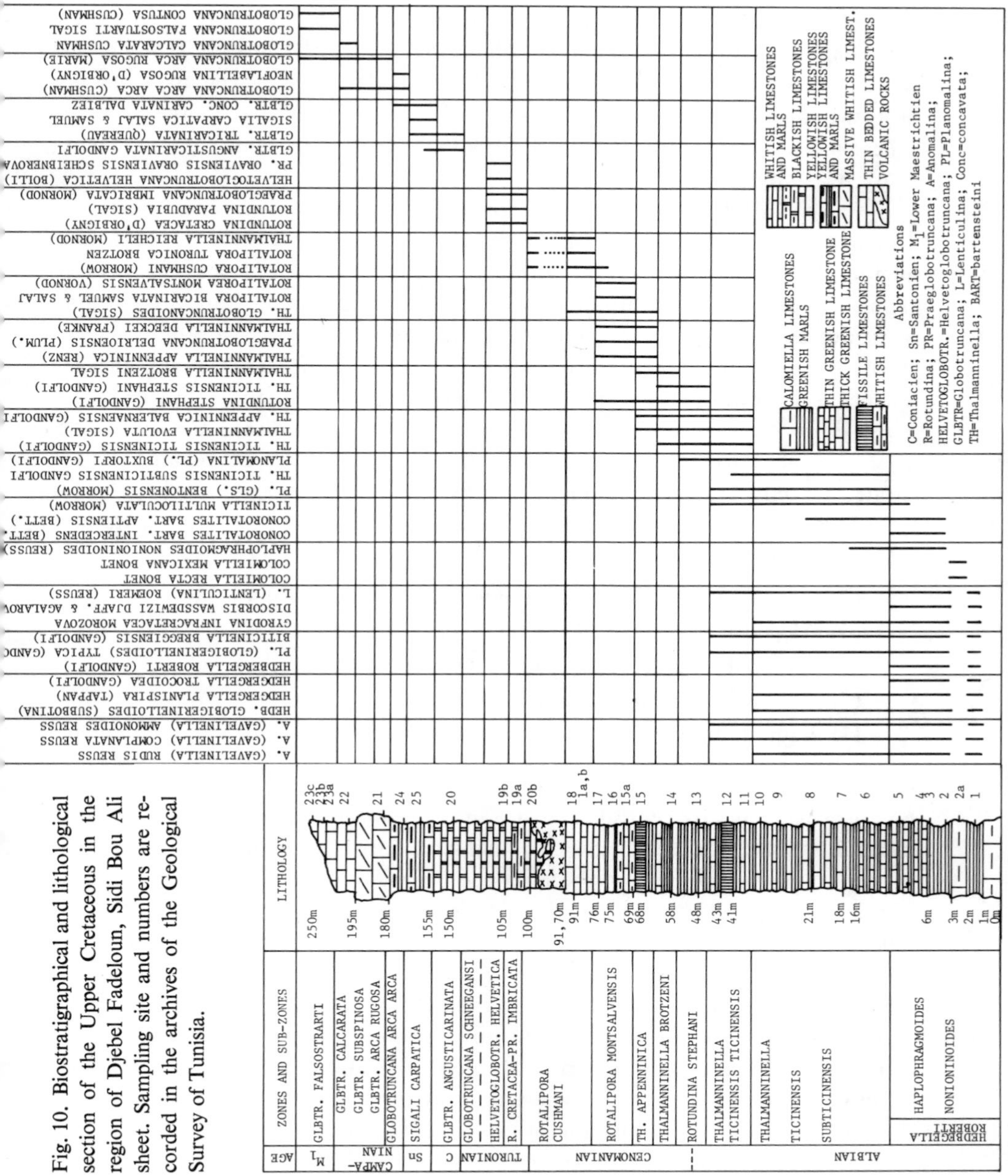

Fig. 10. Biostratigraphical and lithological section of the Upper Cretaceous in the region of Djebel Fadeloun, Sidi Bou Ali sheet. Sampling site and numbers are recorded in the archives of the Geological Survey of Tunisia.

2. *Upper Cretaceous Stratigraphy*

Pervinquière (1903) considered three principal facies divisions—northern, central, and southern—for the Upper Cretaceous, with particular reference to the Senonian. His northern facies, which included the Tunisian Trough was subsequently modified by Solignac (1927) who described a Tellian subfacies near the Algerian border and a Tunisian facies in the eastern part of the country. These divisions are still valid though modified in detail by later work (Jauzein, 1967; Castany, 1951).

Facies variations are confined principally to the Tunisian Trough in the Cenomanian–Conacian interval. From the Santonian onward over the whole region, the facies is monotonous although thickness variations occur. Northern Tunisia, that is the eastern platform, north–south axis, and Tunisian Trough, was covered by pelagic marls and limestones, with some minor variations; in the Tellian facies hard, black marls absent in eastern Tunisia occur and the sequence is somewhat richer in detrital material. The central and southern facies consist primarily of bioclastic beds. There is a complete disappearance of the flyschoid facies found in the Lower Cretaceous, and in the Cenomanian limestones the only detrital component is a few fine quartz grains.

The lithostratigraphic sequence set up by Burollet (1956) from outcrops in central Tunisia is given below:

(5) Abiod Formation, Campanian–Lower Maestrichtian

(4) Aleg Formation *s.s.*, Coniacian–basal Campanian

(3) Bireno Formation, Middle–Upper Turonian

(2) Anaba Formation, Lower Turonian

(1) Bahloul Formation, Upper Cenomanian–basal Turonian — Fahdene Formation, Albian–Middle Cenomanian

Cenomanian. In central Tunisia and the western part of the Tunisian Trough ("sillon tunisien") the Cenomanian consists of a marly sequence assigned by Burollet (1956) to the Fahdene Formation. The sequence in eastern Tunisia and the north–south axis at Djebel Douamess, and in the region of Djebel Fkirine and Djebel Ressas (Bujalka *et al.*, 1971) consists of Lower Cenomanian marls similar to these of the Fahdene Formation, followed by an alternation of platy limestones and marls with marly limestones. A band of ochreous limestone with brachiopods in this sequence has been recognized by Jauzein (1967) as a good reference horizon. The Middle Cenomanian consists of gray limestones and marls. The Upper Cenomanian characteristically has a 5–10 m hard band of whitish limestones within the marl sequence. The Bahloul Formation begins at the top of the Cenomanian and continues into the Turonian as a series of platy, bedded, brownish limestones. The type section, proposed as a hypostratotype, is near Pont du Fahs.

In the area of Djebel Fkirine and Hammam Djedidi the base of the Cenomanian is formed of marls followed by a limestone/marl alternation. The
limestones of the Bahloul Formation are restricted to the basal Turonian;
the same is true in the vicinity of Zaghouan where the sequence of lower marls
and marls and limestone alternations is some 15–20 m thick. In the same general
region deeper water bituminous limestones, which persist to the Lower Coniacian, occur (Edjehaf facies).

In the region of Enfidaville, the Cenomanian consists of 4–8 m of marls
overlain by brownish, marly limestones and marls. In the region of Djebel
Fadeloun (Fig. 10) the Cenomanian consists of alternations of marly limestones
and marls with only two levels of platy limestones 2–3 m thick. However, from
Baudinar and Sfax in the north to the Chotts in the south there is a bioclastic
limestone facies; farther south on the Saharan platform the sequence is replaced by dolomitic limestones and gypsiferous marls.

Turonian. The facies of the Turonian are widely developed in the Zaghouan
region. The Lower Turonian contains the top part of the Bahloul Formation,
which consists of blackish marls and limestones. The Middle and Upper Turonian Bireno Formation consists of an alternation of marls and nodular
limestones, with marl increasing (Bujalka et al., 1971). Elsewhere in the general
region of Zaghouan and Fkirine a deeper water (Edjehof facies, Menčik et al.,
1970), succession occurs comprising bituminous, blackish, well-bedded sublithographic limestones.

The Turonian of Djebel Abid and Djebel Bayada on the eastern platform,
chosen by Salaj (1970, 1972) as a parahypostratotype (Salaj 1973, 1974a), ap-
proaches lithogically the subdivision of Burollet (1956), with the basal Bahloul
Formation followed by the Bireno Formation, for the marl sequence of the
Lower Turonian Anaba Formation is not found here. The top of the Turonian
consists of greenish marls with some limestone intercalations similar to the
base of the Aleg Formation. South of Hammam Djedidi (Johan and Krivy,
1969) the same lithological divisions can be made, although here some repre-
sentation of the Anaba marls does occur. These outcrops present a sequence
comparable with those found in northern Tunisia (cf. Salaj and Bajanik, 1972;
Massin and Salaj, 1970).

A second type of Turonian is the condensed sequence found at Kef en
Nehzal and Kef en Nsoura in the region of Enfidaville (Fig. 11), where some
10–30 m of massive nodular limestones, comparable to those of the Bireno
Formation (Biely et al., 1973) occur. They overlie limestones of the Bahloul
Formation which here only represent the topmost Cenomanian. A level of
volcanic rocks of Lower Turonian age (cf. Bajanik, 1971) is found near Sidi el
Abid (Sidi Bov Ali, Sheet 1:50,000).

There was no break in sedimentation between Cenomanian and Turonian;

Fig. 11. Biostratigraphical and lithological section of the condensed Upper Cretaceous section at Kef en Nsoura, Enfidaville sheet. Sampling sites and numbers are recorded in the archives of the Geological Survey of Tunisia.

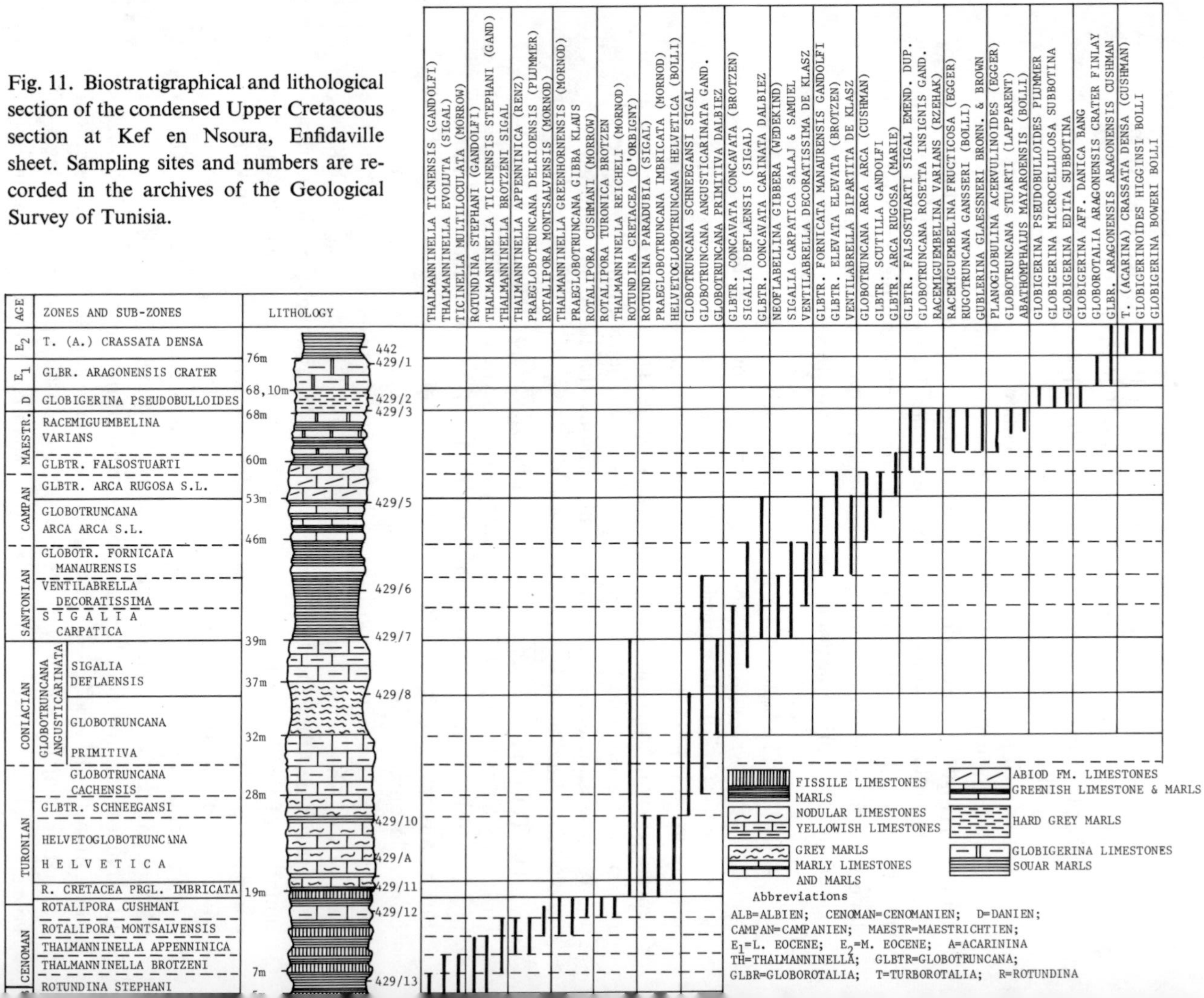

the limit is usually placed within the Bahloul Formation (Burollet, 1956) throughout eastern-central and northern Tunisia. The Bahloul Formation represents sedimentation in a fairly deep environment. Where water depths were shallower (e.g., Pont du Fahs, Salaj, 1974a) the Bahloul Formation is either not developed or is reflected by rare, blackish, platy limestone intercalation. Here the succession contains an Upper Cenomanian and Lower–Middle Turonian rudistid horizon with *Hippurites* within marls of a pelagic sequence. A reef facies also develops east of Djebel Bou Kornine (Djebel Mokta), and here too there appears to be no break between Cenomanian and Turonian (Bujalka *et al.*, 1971; Maurin and Tlatli, in Burollet,1975).

In the south, in the Chotts from Sfax, Boudinar, and Kasserine, the Turonian is principally represented by organodetrital limestone and dolomites with *Hippurites*. These give way still farther south on the Saharan platform to dolomites and evaporites which overlie the marls and dolomites of the Cenomanian.

Coniacian. Over most of the length of the north–south axis the Coniacian is represented by an alternation of marly limestones 150–200 m in thickness. In some places, e.g., Djebel Fkirine, the lower part of the Coniacian is formed of a hard, yellowish limestone above which the marl sequence passes into the Santonian without break. The best development is in the section named a hypostratotype, in the region of Pont du Fahs.

On the eastern platform in the vicinity of Enfidaville, a condensed sequence can be seen at Kef en Nehzal where there is only 10-m yellowish basal Coniacian limestone, while at Kef en Nsoura (Fig. 11) the limestone is separated into two bands by 5 m of marl. However, at Djebel Abid–Djebel Bayada (Salaj, 1974a) the sedimentation is not condensed, and since it is characteristic of that found over northern Tunisia it was chosen as a parahypostratotype.

In the south, north of the Chotts the Coniacian to Lower Santonian is represented by dolomites and organodetrital limestones or oolites containing rudists (Bismuth *et al.*, 1967). Still farther south in the Chotts and on the Saharan platform, this facies passes into gypsiferous marls with bands of dolomite (Busson, 1967).

Santonian. The Santonian beds, forming the upper part of the Aleg Formation (Burollet, 1956) are of uniform facies along the north–south axis and in northern Tunisia generally. The lower part consists of an alternation of marly limestones and marls continuing from the Upper Coniacian, followed by marls with sparse limestone intercalations. Thickness variations studied by Bujalka *et al.* (1971) are controlled by a high in the vicinity of Djebel el Mecella–Djebel Ressas, uplift occurring during the Santonian and Lower Companian, producing a wide emergent zone with erosion and reworking of Turonian and Lower Senonian and with transgression of Campanian on to Cenomanian, which at Hammam Zriba (Johan and Krivy, 1969) rests directly upon Portlandian limestones.

In the Santonian of Djebel Ech Chama, chosen as a hypostratotype by Salaj (1973), shallow-water deposition is indicated by a thin glauconitic marls horizon, presumably related to the movements mentioned above. Here, the thickness is of the order of 100–150 m. Toward the top the number of marly limestone intercalations increases and the sequence passes into the alternation characteristic of the Lower Campanian Lower Abiod Formation. Farther south the effects of tectonic movement disappear and at Djebel Fkirine there is no reduction in thickness and no glauconitic horizon.

East of the north–south axis on the eastern platform, the effect of movement is seen in the condensed sequence of greenish, marly limestones and marls (about 30 m, Bajanik et al., 1973) in the Kef en Nehzal section near Enfidaville and the 7 m of greenish marls in the Kef en Nsoura section. However, at Djebel Abid–Djebel Bayada the sediments have their normal thickness of about 250 m with a facies comparable to that found in the "sillon tunisien," or at Djebel Fkirine.

The gap in sedimentation (Santonian–Lower Campanian) is clearly related to movement affecting the eastern platform and presumably linked to movements which affect northern Libya/Cyrenaica (Röhlich, 1974; Žert 1974) along the northern margin of the Sahara platform, and may be precursors of events along the southern edge of the Alpine orogenic system in the Neogene (Klitzch, 1970, 1971; Röhlich, 1974). Thus, it seems that the eastern platform must also belong to the southern border of the Tethys geosyncline as well as the eastern margin of the western Mediterranean represented by the "sillon tunisien."

In the Chotts the Santonian consists of marls, but there may sometimes be a gap with transgressive Campanian (Bismuth et al., 1967). Over a wide area north of the Chotts ("Kasserine Island") Lower Senonian beds are absent. The Santonian, neritic in Central Tunisia and on the Pelagian block, is here represented by organodetrital and shelly limestones (see Fig. 7).

Campanian. The Abiod Formation (northern facies), consisting of two limestone horizons separating three sequences of limestone/marl alternations, covers the time span from Campanian to basal Maestrichtian (Burollet, 1956; Crampon, 1971). (The upper of these limestones and the beds above it represent basal Maestrichtian.) This formation, widely developed in the "sillon tunisien," is absent on the platform. Along the north–south axis the northern facies is best seen in the region of Djebel Fkirine (Mencik et al., 1971) where 110–120 m are exposed, about twice the thickness seen in the region of Zaghouan (Tunisian facies, cf. Fig. 7).

At Djebel Rorfa (Grombalia sheet) Upper Campanian limestones (Tunisian facies) are transgressive over the Albian–Cenomanian (Bujalka et al., 1971). The basal limestones are sandy and this transgressive limestone facies may also be observed in the region of Djebel Bou Kornine and Djebel Zaina (Bujalka et al., 1972).

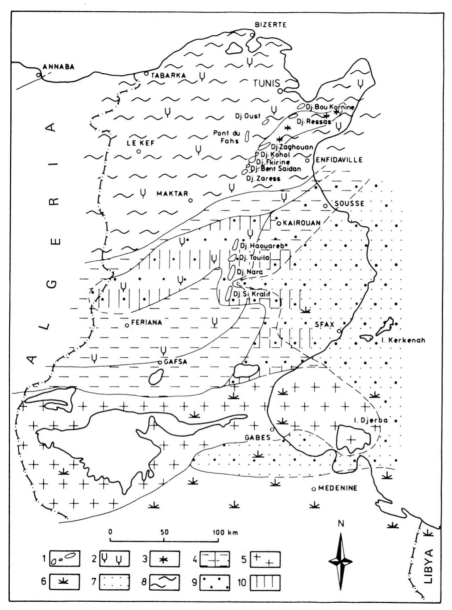

Fig. 12. Sketch map of the facies distribution during the Berriasian and Valanginian. In the figure separate symbols are used for each stage so that facies changes can be discerned by the superposition of symbols. *Berriasian:* (1) Jurassic massifs of north–south axis; (2) pelagic sublithographic limestones and marls; (3) organodetrital limestones formed in shoal areas; (4) micaceous silty shales formed on platform; (5) intracratonic lagoonal facies (including Valanginian); (6) "Wealden" facies of the continental Sahara. *Valanginian:* (7) continental zone of Sahara and eastern platform; (8) facies of Tunisian Trough, mostly marls; (9) micaceous, silty shales formed on platform; (10) epineritic facies—dolomites of top Valanginian–Lower Hauterivian.

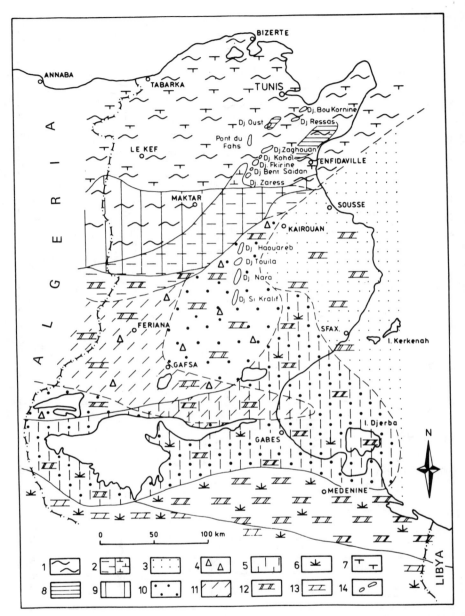

Fig. 13. Sketch map of the facies distribution during the Hauterivian to Aptian. In the figure separate symbols are used for the Hauterivian–Lower Barremian and Upper Barremian–Aptian, so that facies changes may be discerned by the superposition of symbols. *Hauterivian–Lower Barremian:* (1) pelagic—marl and limestone facies of the Tunisian Trough; (2) neritic organodetrital facies of margin of Pelagian block, (a) Hauterivian, (b) Lower Barremian; (3) continental–subcontinental facies of Pelagian block; (4) epineritic zone carbonates of the Pelagian block; (5) epineritic and lagoonal facies some intercalations of "Wealden" facies; (6) epicontinental "Wealden" facies. *Upper Barremian–Aptian:* (7) pelagic—marl and limestone facies and flyschoid facies of the Tunisian Trough; (8) shoal area of condensed sedi-

The Campanian on the eastern platform is best known in the region of Enfidaville, where the Abiod Formation consists of limestones underlain and overlain by a condensed alternation of limestones and marls still reflecting some effect of earlier tectonic movements (see the Kef en Nsoura, Kef en Nehzal section). The limestone and the upper limestone/marl sequence belong to the Lower Maestrichtian. In the vicinity of Djebel Abiod–Djebel Bayada, the upper alternation was eroded during the Maestrichtian (Salaj, 1972) for it is overlain by a thin (1 cm) Danian marl and then by uppermost Paleocene–Lower Eocene.

Maestrichtian. The Lower Maestrichtian of the northern facies (sillon Tunisien) includes the top part of the Abiod Formation (Burollet, 1956). Along the greater part of the north–south axis the Abiod Formation is represented by a single limestone and an upper and lower marl/limestone alternation, i.e., Lower Maestrichtian, followed by the El Haria Formation. Thickness variations of the El Haria in the Grombalia area show the increase from 22 m (Djebel Zouaine) to 100 m (Djebel Diouane) (cf. Bujalka *et al.*, 1971).

Over the eastern platform the Maestrichtian is represented by 5–10 m of gray marls above the limestone (top of the Abiod Formation), followed by 15–25 m dark gray and greenish marls, which include both Upper Maestrichtian and Danian (Salaj *et al.*, 1973). Around Enfidaville the sequence is still more condensed with 5 m of Abiod Formation followed by 15 m of greenish limestones and marls; however, there is a break in the succession which includes the top of the Maestrichtian (Salaj *et al.*, 1973). Condensed sections are to be found at Kef en Nehzal and Kef en Nsoura (Fig. 11).

North of the Chotts, as far as Boudinar and Sfax, both the Campanian and Maestrichtian are represented by organodetrital limestones with large foraminifera and some marl intercalations (central and southern facies). These units are also known on the Saharan platform (Busson, 1967).

3. *Outline of Cretaceous Paleogeography*

The better exposure and more complete knowledge of the Cretaceous makes it possible to illustrate the facies distribution at four different stages (Figs. 12–15).

In the lowest stage, from the Berriasian to the Barremian, a Wealden facies covers the northern margin of the Saharan platform (Fig. 12) passing into a lagoonal facies occupying the area of the Chotts. To the north, in the Tunisian Trough, the Berriasian is represented by sublithographic limestones, which, during the Valanginian, give way to marls and marly limestones with a sandy flysch horizon at the top of the Valanginian and basal Hauterivian.

---

mentation (emergent during Aptian); (9) littoral zone with subreefal limestones; (10) continental–subcontinental facies; (11) epineritic lagoonal facies; (12) epineritic marine limestone and dolomite facies of the Pelagian block; (13) dolomite facies succeeding the "Wealden" of the Saharan platform; (14) Jurassic massifs of the north–south axis.

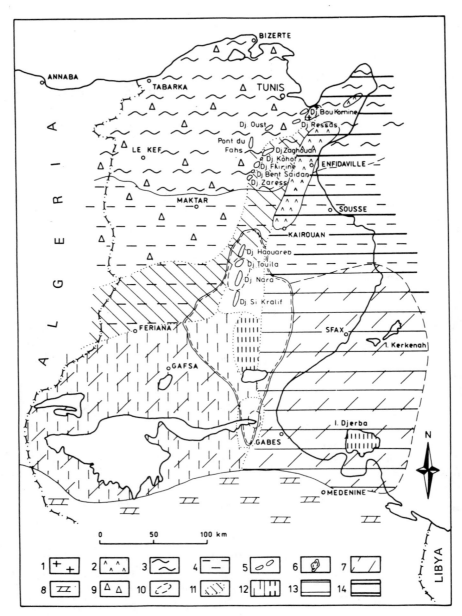

Fig. 14. Sketch map of the facies distribution during the Albian to Turonian. In the figure separate symbols are used for the Albian and Cenomanian–Turonian, so that changes of facies may be discerned by the superposition of symbols. *Albian:* (2) zone of condensed sedimentation prior to Upper Albian transgression; (3) pelagic—marls and marly limestones of the Tunisian Trough; (4) pelagic marls; (5) Jurassic massifs of the north-south axis; (7) laguno-neritic zone with dolomites, evaporites, and marls; (8) epineritic carbonates of the Sahara platform. *Cenomanian–Turonian:* (1) reef facies (rudists—Vraconian, Cenomanian, and Lower–Middle Turonian); (9) pelagic, marly facies of the Tunisian Trough; (10) region where Cenomanian is not represented; (11) transition zone between Tunisian Trough and Saharan platform and Pelagian block; (12) epineritic—lagoonal carbonates and evaporites

For the first time the eastern platform is clearly distinguished from the central Tunisian zone by its facies. Much of the platform was exposed to erosion while in central Tunisia generally, silty and micaceous shales were deposited. Along the southern half of the north–south axis the emersion ended with the development of an epineritic facies. In the north organodetrital limestones formed around highs and are underlain by Portlandian reef limestones.

During the greater part of the Hauterivian–Aptian (Fig. 13) interval the conditions remained basically the same; however, the Upper Aptian was marked by a transgression which flooded the eastern platform and formed a sequence of dolomites, coral, and orbitolinoid limestones. The same transgression led to the replacement of continental conditions on the Saharan platform by dolomite formation. The Chotts region, as usual, formed a transition zone between the two.

The major transgression of the Cenomanian–Turonian resulted in a link with the Gulf of Guinea through Nigeria and deposited a blanket of limestone and dolomite on the northern margin of the Saharan craton. Associated with this carbonate material are marls with echinoids and gypsum.

The Chotts remained a laguno-neritic zone, with, during the Albian, an emergent zone centered on the southern part of the north–south axis (Fig. 14). This represents a major change in the paleogeographic situation, with the Chotts subsiding less than the platform to the south. The deposits are, therefore, basically similar to those of earlier periods with marls, some dolomites and limestones, and evaporitic horizons. During the Albian to Turonian pelagic marls and some limestone formed in the Tunisian Trough, while the eastern platform remained a zone of relatively thin or condensed deposition.

In the Senonian of the Chotts marls, sometimes gypsiferous, replaced the calcareous facies of the Saharan platform, although calcareous conditions with the development of shelly limestones are also represented. The Tunisian Trough in the north formed a relatively narrow zone (Fig. 15), with maximum subsidence in the northwest where hard, blackish marls predominate.

The north–south axis forms a well-defined zone, characterized by either nondeposition or a condensed platform deposition, with the eastern platform too a relatively positive area, characterized by condensed sequences with numerous gaps in the succession except for a small area east of Maktar (Fig. 15).

D. Paleocene

1. Stratigraphy

Along the north–south axis and on the Tunisian platform, the Upper Cretaceous–Paleocene boundary lies within the El Haria Formation (Upper

(heavy symbols, evaporites predominate); (13) neritic marls and shelly limestones of the southern part of the Pelagian block; (14) pelagic marls and limestones of Pelagian block, condensed sequence. ----- zone emergent and eroded prior to Cenomanian transgression.

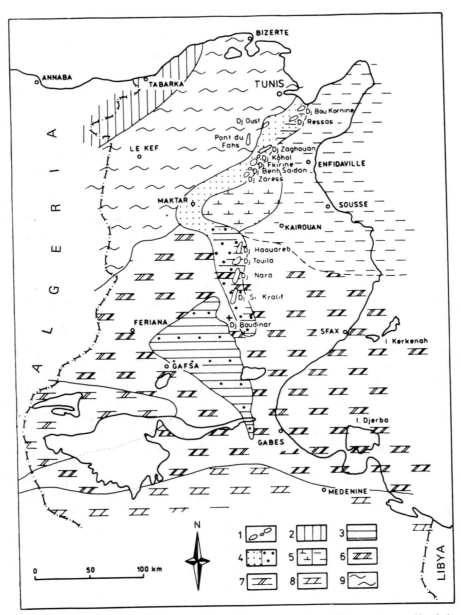

Fig. 15. Sketch map of the facies distribution of the Senonian. (1) Jurassic massifs of the north–south axis; (2) pelagic, marly facies of most rapidly subsiding segment of the Tunisian Trough; (3) zone where lower part of Senonian absent prior to Upper Campanian; (4) non-subsiding zone of north–south axis, condensed sequence; (5) condensed sequence, (a) with numerous gaps, (b) greater subsidence in the west; (6) neritic marls, organodetrital and shelly limestones; (7) neritic subsiding trough of Gafsa; (8) marls, bioclastic shelly and dolomitic limestones of the Saharan platform; (9) pelagic limestone and marls facies of the Tunisian Trough.

Maestrichtian and Paleocene *s.s.*). Cuvillier *et al.* (1955) recognized two provinces in Tunisia, separated by a line from Thala to Bizerte. Northwest of the line sedimentation, continuous from the Cretaceous to the Paleocene, is complete, while to the southeast, even if the Danian is not always absent as originally thought (Jauzein 1957, 1967), there are numerous gaps in the Upper Danian and Montian (Johan and Krivy, 1969; Bujalka *et al.*, 1971; Salaj *et al.*, 1973). As tectonic movement made itself felt after Danian time in most cases, it is sometimes difficult to establish whether a gap is due to nondeposition or secondary erosion.

Lithologically the Danian and Montian are represented by marly sandstones and greenish marls. The Upper Montian marly sandstones are fossiliferous and glauconitic along the north–south axis, the condensed sequence (45 cm thick) showing the effects of tectonic movements in eastern Tunisia. In the same area, e.g., Djebel Fkirine, the Upper Paleocene transgresses over Maestrichtian.

Over a wide area of the Tunisian platform Paleocene is absent, both in the region of Enfidaville and over most of the Grombalia sheet, and here Eocene rests on Upper Maestrichtian. However, there are areas where sedimentation was continuous even though the sequence is condensed.

E. Eocene

1. *Stratigraphy*

Along the north–south axis and over the Tunisian platform the Eocene is represented by two formations, the Metlaoui Formation (Ilerdian and Cuisian-Ypresian, and Lower Lutetian) and the Souar Formation (top of Lower Lutetian to Priabonian). The lower unit consists of globigerinal or nummulitic limestones which pass, by intercalation, into the marls of the Souar Formation. The facies and thicknesses of the two formations are variable and the boundary between them is not precisely defined (Figs. 16, 17).

In the region of Bou Kornine, Djebel Ressas, and Zaghouan, along the north–south axis the Metlouai Formation consists of 30–80 m of chalky, whitish, globigerine limestones, which become brownish in some regions and represent both the Ilerdian and the Cuisian. The transition to the 300–700-m thick Souar marls, represents the Middle and Upper Eocene. These marls are in places sandy and glauconitic, and toward the top, in the region of Zaghouan, contain sandstone bands and a lens of gypsum (Mencik *et al.*, 1968; Salaj, 1970).

In places on the Tunisian platform near Enfidaville, white chalky globigerine limestones 10–20 m thick (Metlouai Formation) rest directly on beds of the Abiod Formation (Djebel Bayada); they are 100 m thick in others and overlie the El Haria Formation (Djebel Mdeker). The Souar Formation, which completes the Eocene, consists of marls within which one or two limestone

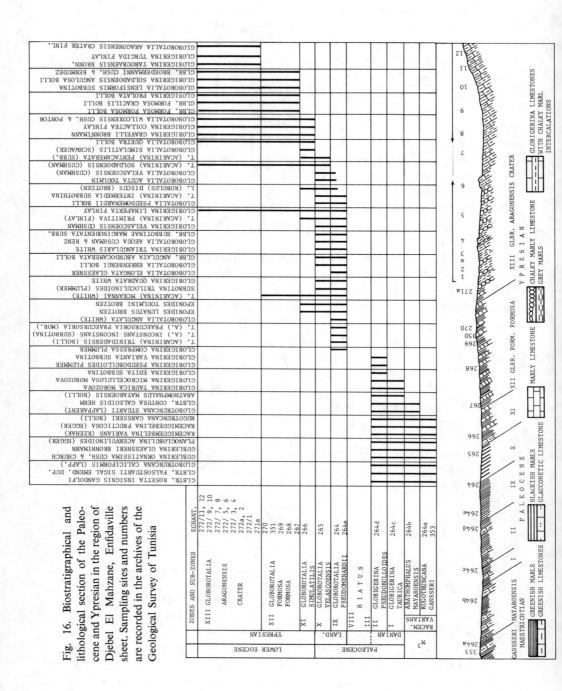

Fig. 16. Biostratigraphical and lithological section of the Paleocene and Ypresian in the region of Djebel El Mahzane, Enfidaville sheet. Sampling sites and numbers are recorded in the archives of the Geological Survey of Tunisia

Fig. 17. Biostratigraphical and lithological section of the Paleocene and Eocene between Djebel El Touijine and Kef El Haouach, Djebel Fkirine sheet. Sampling sites and numbers are recorded in the archives of the Geological Survey of Tunisia.

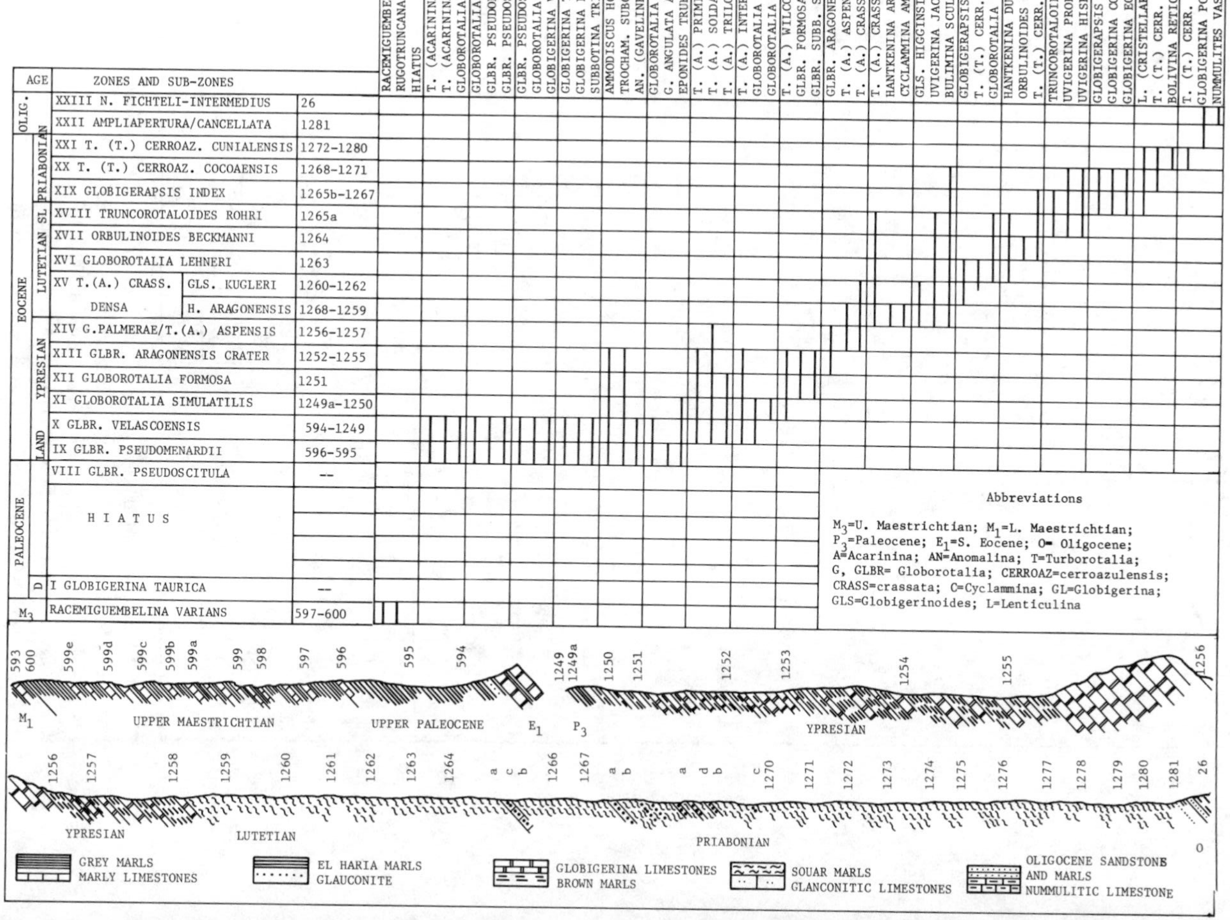

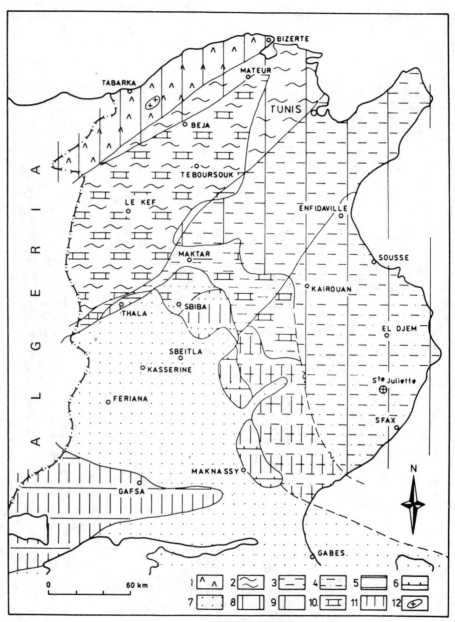

Fig. 18. Sketch map of the facies distribution during the Paleocene and Lower and Middle Eocene. In the figure different symbols are used for the Paleocene and the Lower and Middle Eocene, so that facies changes may be discerned from the superposition of symbols. *Paleocene:* (1) El Haria Formation marls with dolomitic limestone and two limestone horizons in zone of subsidence (Formation also includes Maestrichtian); (2) El Haria Formation marls and marly limestone (facies originally described by Burollet); (3) El Haria Formation condensed gray marls and greenish limestone facies (some breaks in succession); (4) El Haria Formation commonly condensed (here Cretaceous in part and Danian phosphates and Montian phosphates are missing); (5) Metlaoui facies of Upper Paleocene; (6) Metlaoui Formation lagoonal facies with phosphates; (7) emergent zone. *Lower and Middle Eocene:* (8)

horizons occur. On the southern part of the eastern platform, between Kairouan–Sousse and Sfax, the Upper Eocene El Haria Formation is transgressive with the Danian and Montian absent (cf. Fig. 18).

F. Oligocene–Lower Miocene

1. Stratigraphy

Beds of Oligocene age, widely developed over the Tunisian platform and north–south axis, were assigned to a "Cherichera" facies by Castany (1951) (see also Burollet, 1956). This term encompasses a variety of lithologies, the initial brownish gypsiferous clays follow the Eocene beds and pass into a flyschoid facies containing some organodetrital nummulitic limestones followed by the Fortuna Sandstone and the Messiouta Formation (Burollet, 1967), consisting of variegated, coarse sandstones of continental origin. The Fortuna Sandstone, formerly assigned to the Upper Oligocene, actually extends from the Aquitanian to the Lower Langhian, i.e., it is primarily of Miocene age. The Messiouta Formation, regarded by Schoeller (1933) as Aquitanian, is here regarded as Upper Burdigalian, that is, as the continental facies equivalent of the top part of the Fortuna Sandstone. This relationship is seen in the comparison between the outcrops at Cape Bon where the continental facies is absent and the Fortuna Sandstone is followed by marine Langhian beds, and Enfidaville where the Messioua Formation rests on some 3 m of Aquitanian sandstone, the rest of the Fortuna Sandstone probably having been removed by erosion.

The lower part of the Fortuna Formation, according to Cassan et al. (1973), can be divided into a number of cycles. The initial regressive cycle, A1, is characterized by marine deltaic sediments. The second cycle, A2, consists of coarser-grained sediments and at the highest level, the coarse sandstones of torrential-fluvial origin form cycle B. Cycles A1–A2 are assigned to the Aquitanian and Lower Burdigalian, cycle B is mainly Upper Burdigalian, but extends into the lowest Langhian. Thus the Messiouta Formation is the continental facies equivalent to the topmost and coarsest part of the Fortuna Sandstone, although it is conceivable that it may extend down to the A2 horizon.

In northernmost Tunisia the Oligocene is represented by the Numidian facies (Fig. 19). The Numidian is primarily a flysch sequence; it consists of lower marls (Lower Oligocene) followed by sandstones of flysch type. At the base of the flysch glauconitic levels have been found (Batik, personal communication; location 2.5 km southwest of Djebel Choucha), indicating that deposition did not occur in deep water. The absence of nummulites has made determination of the age of the Numidian, and its correlation with deposits of the Tunisian Trough from which it was separated by an emergent ridge, difficult.

---

black, marly limestone facies with bituminous beds; (9) white, marly limestones; (10) nummulitic limestone facies; (11) lagoonal facies with shelly limestones and gypsiferous horizon; (12) littoral organodetrital limestone facies.

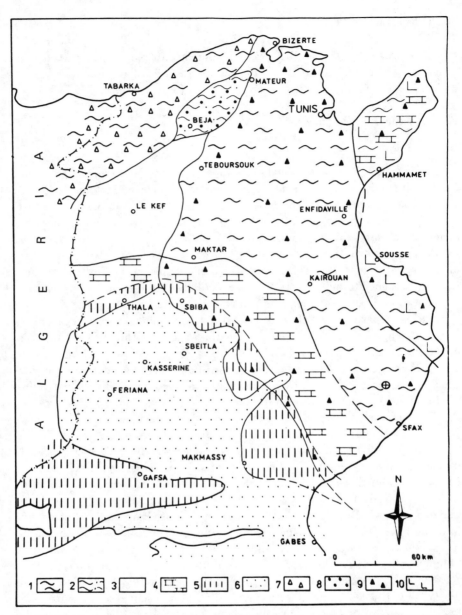

Fig. 19. Sketch map of the facies distribution during the Upper Eocene and Oligocene. (1) Souar Formation marls; (2) Souar Formation formed on shoals with glauconitic horizons; (3) emergent zone Eocene, rarely recorded; (4) shelly, marly limestone or Cherahil facies; (5) Faid Formation clay or clayey limestone with gypsum horizons (Djebs Formation when gypsum dominant); (6) continental area; (7) Numidian sandstone facies (Oligocene); (8) Bejoua facies–Miogypsina limestone; (9) Cherichira brownish marl and Nummulitic limestone facies; (10) Cherichira facies in Lower Oligocene and Upper Oligocene lepidocycline beds.

The Oligo-Miocene outcrops on Cape Bon, at Korbous, Ain Octor, and Djebel Sidi Abd er Rahmane consist predominantly of shale, sandy marl, and sandstone horizons, with limestone bands which prove invaluable for dating purposes (Hooyberghs and Lajmi, 1974; Compte and Dufaure, 1973; Bismuth *et al.*, 1972). One of the thicker limestones at Korbous belongs to basal Aquitanian, and lies just below cycle A1 of the Fortuna Sandstone, which is here 650 m thick. The section of brownish marls, sandy marls, sandstones, and limestones below the Korbous Limestone is Lower Oligocene. This lower Stampian section is well developed near Enfidaville by about 150 m of the same lithology; however, the Fortuna Sandstone is represented by about 3 m of pelletal sandstone of Aquitanian age. The Djebel El Meugtella section, where at the top a sandstone horizon contains a conglomeratic level with large sandstone pebbles resting directly on fine-grained sandstones, suggest that the Fortuna Sandstone has been removed by erosion.

The Messiouta Formation can be found directly overlying coarse sandstones of Fortuna Sandstone cycle B. It contains shelly, marine horizons, indicating the close association of marine and continental beds widespread southwest of Kairouan (Biely *et al.*, 1972). However, still farther to the south in the region of Sousse, richly fossiliferous marls with Lepidocyclines found in well SS2 (Dolle, 1952) replace the arenaceous beds with hard, gray-white reef limestones and soft, gray-blue sandy marls. This suggests that the marine zone was not distant from the emergent zone.

G. Miocene

1. *Stratigraphy*

The Lower Miocene has been discussed in the preceding section. The Middle Miocene is absent along the north–south axis (Castany, 1951), but is widespread over the Tunisian platform, where outcrops have been described by Salaj and Stranik (1971*b*), Biely *et al.* (1972), Cassan *et al.* (1973), Compte and Dufaure (1973), and Ballais (1973). Over most of the area rocks of Middle Miocene age have been assigned to units considered chronostratigraphic, the Ain Grab, Oued Hammam, and Mahmound (Hooyberghs, 1973; Hooyberghs and Lajmi, 1974). The lithologies are related to the tectonic movements affecting Cape Bon. Difficulties which arose because two limestones of similar lithology but different age, were assigned to the Ain Grab Formation were resolved when it was recognized that the "Ain Grab" prevalent over most of the region is cut out at Cape Bon. The upper "Ain Grab" limestone at Cape Bon (not found elsewhere) overlies the Oued Hammam marls in some places, and in others rests directly on the Fortuna Sandstone (Salaj and Stranik, 1971*b*). The marls assigned to the Mahmoud Formation cover both limestones; thus, in time-stratigraphic terms both the Upper "Ain Grab" limestone and the Oued Hammam marls correspond to the lower part of the Mahmoud Formation elsewhere. The succession indicates a reversal in the tectonic history of this region, with transgression from the region of Cape Bon to the north rather than from the south. The general character of the "Ain Grab" limestones permits their correlation with the Al Jaghlred Formation of Middle Miocene* age (Helvetian–Langhian) in Libya and Egypt (Annoscia, 1969).

The Mahmoud Formation in the Saouaf syncline, defined by Biely et al. (1972), consists of a series of marls 5–110 m in thickness with a glauconitic horizon at the base. It corresponds to the basal part of Burollet's (1956) Oum Douil Formation, and in terms of age extends from the top of the Langhian through most of the Serravallian (Wiman, 1974).

The top of the Miocene consists of a sandy sequence (Salaj and Stranik, 1971b) which can be regarded as equivalent to the second member of the Oum Douil Formation [and the probable equivalent of the Beglia Formation defined by Burollet (1956) in southwestern Tunisia]. The Beglia Formation was regarded as continental Tortonian by Solignac (1931), but later attributed to the Pontian (Solignac, 1931). Upper Miocene sands and sandstones 50 to 300 m thick follow the Mahmoud Formation without a break in sedimentation, although to the south they may rest upon much older rocks. The beds have a rich vertebrate fauna, and can be dated as topmost Serravallian and Tortonian (Biely et al., 1972). The Beglia Formation represents a new sedimentary cycle, cycle C3 of Cassan et al. (1973), and reflects orogenic movements of Upper Serravallian age (Salaj and Stranik, 1971b; Bajanik and Salaj, 1971, 1972, 1973; and Burollet, 1971). These littoral beds at Cape Bon pass laterally into beds attributed to the Saouaf Formation of Biely et al. (1972) and southward into the Segui Formation of Burollet (1956). Lithologically the Saouaf beds at Cape Bon are formed of sands, sandstones, marls, and clays, and are sometimes gypsiferous and lignitic. The lignites, rare and thin in the lower part, become more numerous (with 5 to 11 horizons) and thicker (may reach 3 m) near the top. Several cycles can be recognized, each beginning with a thin sand or breccia with gastropods and oysters, usually resting on a lignite, and representing shallow, marine incursions.

Toward the south the Saouaf Formation is replaced by the continental Segui Formation of variegated clays, of variable but sometimes great thickness. They are well developed between Gafsa and Gabes.

* The zone Borelis melo and the subzone Metrarabdotos vinassi are recognized in Tunisia for the first time (Salaj, unpublished data) through foraminifera. Bryozoans (Cyclostomata and Cheilostomata) are represented by Idmidronea atlantica (Forbes) Johnston, Neillia tenella (Lamarck), Holoporella polythele (Reuss), Grayporella cervicornis (Pallas). In addition to Borelis melo (Fichtel and Moll), Elphidium gr. fichtelianum (d'Orbigny).

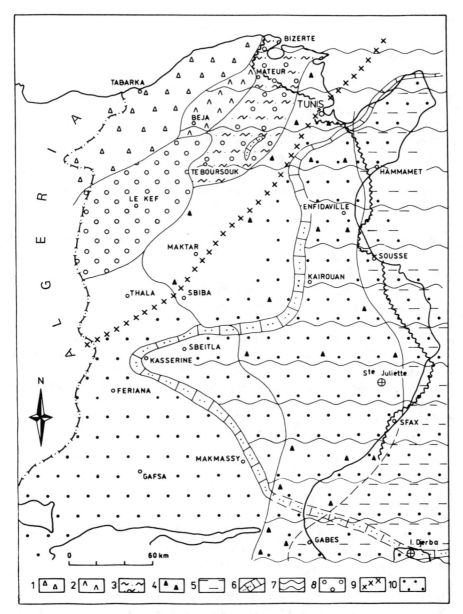

Fig. 20. Sketch map of the Neogene facies distribution. (1) Aquitanian–Burdigalian in Numidian Zone of continuous sedimentation emergent in Serravallian; (2) Lower Aquitanian limestones, Upper Aquitanian foraminiferal marls; (3) Aquitanian–Burdigalian sandstone, passing to a marly facies; (4) Aquitanian–Burdigalian Fortuna Sandstone facies, Burdigalian Messiouta facies; (5) Aquitanian marine facies, Burdigalian–Helvetian–Langhian phosphate, nonmarine; (6) western limit of Ain Grab Formation; (7) Burdigalian–Tortonian marl facies during Serravallian, conglomerate intercalations; Langhian marls only Serravallian on the eastern platform; (8) Upper Miocene Oued Mellah, Kechabta, Oued Bel Khedine Formations; (9) Axis of emergent zone during Upper Miocene; (10) Upper Miocene Beglia and Saouaf Formations; 〜〜〜 Limit of Pliocene marine transgression; ⊕ location of boreholes.

H. Continental Mio-Pliocene

1. *Stratigraphy*

There are marine beds of Mio-Pliocene age along the north–south axis, and it is presumed that the sedimentation of the Mahmoud Formation during the Langhian–Serravallian was general in northern Tunisia (Bajanik and Salaj, 1972; Kujawski, 1969; Biely and Salaj, 1971; Caire *et al.*, 1971) and along the north–south axis. Tectonic uplift during the Upper Serravallian–Lower Tortonian in northern Tunisia (Bajanik and Salaj, 1971; Burollet 1971; Glaçon and Rouvier, 1972) resulted in the emergence and erosion of the Miocene beds along the north–south axis. Sedimentation began later, presumably in the uppermost Miocene with the formation of lacustro-continental deposits well known near Zaghouan (Bajanik *et al.*, 1968; Planderova *et al.*, 1969) and near Pont du Fahs where they transgress on to Lower Oligocene (Bajanik *et al.*, 1968). Elsewhere they rest on the beds of the Beglia Formation (Soauaf Formation absent). These lacustro-continental beds cannot be older than Upper Messinian and are presumed to represent the Lower Pliocene. Lithologically they consist of up to 350 m of whitish and reddish cross-bedded sands and sandstones, often with ripple marks, and inclusions of gray, or gray-green sandy clays, and rare conglomeratic horizons. This, the lower series, is followed by an upper sequence of conglomerates, clays, and shales, which is variable in thickness and extent (Bajanik *et al.*, 1968). Lying upon the lower series at Pont du Fahs, it transgresses older beds at Zaghouan. It is followed by a sequence of clays and gypsiferous marls.

I. Marine Pliocene

1. *Stratigraphy*

In the region of Nabeul, the topmost part of the Oum Douil Formation (the Saouaf Formation) is represented by sandstones and marls, the latter sometimes gypsiferous of Upper Tortonian–Lower Messinian age (Hooyberghs and Lajmi, 1974) [considered as Serravallian–Upper Tortonian (Wiman, 1975)]. This is followed by a succession of sandstones, clays, and conglomeratic horizons of Pontian age, which contain marls with a marine microfauna. The microfauna suggests that the sequence continues at least as high as Plaisancian.

J. Cenozoic Paleogeography

Three sketch maps of facies distribution through time are shown (Figs. 18–20). The patterns for the Paleocene, Eocene, and Oligocene are similar in showing a wide emergent zone in the southwest, broken only by an east–west trough which broadens westward. This trough, with Gafsa near its northern margin, received lagoonal sedimentation, with the occasional development of gypsum in the Upper Eocene. The eastern platform remained a relatively positive area characterized by a thin succession of marls and limestones, which become almost entirely marls by Middle Eocene time.

The Tunisian Trough received continuous sedimentation in its north-western part, but during the course of the Upper Eocene a zone of emergence or nondeposition developed around Le Kef, restricting its former southern part. During the Neogene this zone became part of an axis of uplift which extended to the coast near Tunis.

III. REVIEW OF PALEOGEOGRAPHY AND PRINCIPAL TECTONIC EVENTS

As indicated earlier, data on the pre-Cretaceous section are relatively incomplete. Consequently, in the following account all rocks prior to the Cretaceous will be considered together.

A. Pre-Cretaceous

The Lower Paleozoic basement of the eastern platform is known only through borings near Kirchaou and Kasbah Leguine along the margins of the Djeffara. Upon these rocks rests a transgressive Permo-Carboniferous section beginning with Namurian, followed by a regressive cycle beginning in the Permian. The paleogeographic interpretation of Glintzboeckel and Rabaté (1964) suggests a markedly subsiding marine trough (Tebaga), separated by a barrier reef (Bir Soltane) from a lagoonal zone to the south-southeast along the border of the Saharan platform (Kasbah Leguine–Kirchaou). Biely et al. (1973) have demonstrated that during the Upper Carboniferous the Saharan border was formed by a more or less faulted flexure, south of which was an uplifted region undergoing erosion. The Permian section in the trough reaches a thickness of several thousand meters (Busson, 1967). Conditions on the eastern platform are largely unknown, but a zone of emergence may be assumed, for if Upper Carboniferous was deposited in places it was certainly removed either before the Permian or together with the Permian before the general Triassic transgression.

During the Triassic another trough began to develop in northern Tunisia where beds of a modified Germanic facies were deposited (Fig. 2). This trough was separated from Upper Triassic limestones and dolomites of an Alpine facies (Cape Bon well CB-1) (Bonnefous, 1972; Biely et al., 1973), along a line extending roughly from Hammamet to Tripoli. The Permian trough of the

Djeffara, which was inactive during the Triassic and Lower Liassic, became enlarged and somewhat displaced to the north and was the site of deposition of beds which have a neritic facies. In contrast, the trough to the north received pelagic sediments. Its eastward extent was restricted as effects associated with formation of the north–south axis began to be felt. In the north a shoal area with a condensed succession formed during the Carixian and Aalenian, breccias formed during the Bajocian and Bathonian, and a reef–organodetrital facies formed during the Portlandian and Berriasian (Figs. 3, 12).

B. Cretaceous

During the course of the Valanginian–Barremian the Tunisian Trough, which lay in roughly the same position as the proposed Triassic trough, expanded to include the northern part of the eastern platform. The pattern of sedimentation was strongly influenced by the introduction of terruginous material from the southern part of the eastern platform (south of a line from Kairouan to Sousse). The Chotts region was undergoing depression and receiving deposits of a lagoonal "Wealden" type during the Neocomian. During the Aptian this gave way to a succession of neritic limestones. It seems that the northern part of the eastern platform probably became emergent and was briefly subjected to erosion during the Aptian, and a condensed sequence similar to the sequence seen in Djebel Oust (Biely et al., 1973) formed.

From the Aptian to the Turonian, when the north–south axis was most active, the eastern platform was clearly defined as a stable area often marked by condensed sedimentation. To the west lay a less stable area now occupied by the Tunisian Atlas (see Biely et al., 1973) and also the Tunisian trough lay farther to the north. It is interesting to note that the north–south axis was marked by sporadic volcanic activity from Neocomian to Upper Cretaceous time (Burollet, 1956; Bajanik, 1971). This basically tectonic division influenced the sedimentation pattern as late as Senonian time (Pervinquière, 1903; Solignac, 1927).

Solignac (1927, p. 194–195) recognized two facies in the Senonian of northern Tunisia which he believed corresponded to different tectonic environments. In the present article the definitions of Solignac, and of Pervinquière will be expanded. Three facies are recognized, one of which may be split into two subfacies:

(1) Tunisian facies (faciès tunisois) is characteristic of the eastern part of the Tunisian Trough, the northern part of the eastern platform, and the intervening part of the north–south axis. This was a tectonically active area with shallow-water sedimentation, breaks in the succession, and condensed sequences at one or more horizons. In some places, nevertheless,

the succession is complete. An important characteristic of this facies is the occurrence of a single limestone in the Abiod Formation which in places is transgressive. Senonian of this facies is found in Sicily (cf. Castany, 1956), and it may be traced in the Ionian Sea to near the coast of Cyrenaica by following a reflecting horizon at the top of the Senonian (Finetti and Morelli, CIESM symposium, in Biely et al., 1974). In Cyrenaica the same facies with limestones predominating has been described by Röhlich (1974), Klen (1974), and Žert (1974), who report an incomplete section.

(2) North of the Tunisian Trough two Senonian facies can be recognized:

(2a) Tellian facies, as defined by Solignac (1927), corresponds to a zone of subsidence near the Algerian frontier and links with the Tellian units (cf. tectonic map of Burollet, 1973). The beds consist of hard, blackish Santonian and Lower Campanian marls, which are often brecciated. Here the Abiod Formation has two limestone horizons.

(2b) Northern facies here reinstituted after the original definition of Pervinquière (1903) was replaced by the Centro-Tellian facies (Castany, 1951; Jauzein, 1967). This facies characterizes the Tunisian Trough and is linked with the autochthon of Medjerda and the region north of the central and southern Atlas (see Burollet, 1973 tectonic map). As in (2a) there are two limestone horizons in the Abiod Formation.

(3) A central and southern facies was originally defined by Pervinquière (1903) as occupying the unstable zone of the southern part of the eastern platform south of a line from Sfax to Medenine and the intracratonic trough of the Chotts. It even reaches onto the Saharan platform (see Fig. 15). The beds laid down consist of marls, sometimes gypsiferous, and of shelly and organodetrital limestones with orbitoids (the latter unrepresented in the other facies), the Abiod Formation equivalents with only one limestone horizon. The succession has gaps and in places is transgressive.

C. Paleogene

The southern part of the Pelagian block was emergent during the Lower to Middle Paleocene, while the northern part was covered by shallow seas in which a condensed section formed. During the Middle Paleocene, in particular, the northern area must have formed an archipelago. During the course of the Upper Paleocene the islands were submerged as part of a general north-to-south transgression. West of the north–south axis subsidence was general and continuous, although a transgression can be noted in the Upper Paleocene in the area of Pont du Fahs (Salaj et al., 1973) not far distant from the north–south axis.

The Eocene of the eastern platform suggests that the conditions of the late Paleocene continued; although condensed sequences do occur sedimentation is dominantly pelagic with globigerine limestones and marls common. In the zone Le Kef–Teboursouk–Mateur, however, despite continued pelagic sedimentation the passage of globigerine limestones to nummulitic limestones suggests a diminished water depth. This is consistent with the occurrence of breaks in the succession and glauconitic horizons in the Upper Eocene. Over much of the region Upper Eocene is lacking. This distribution may be as much due to nondeposition as to subsequent erosion.

During the Oligocene, the zone of uplift, first active in the Eocene and which may be traced from Le Kef–Mateur toward Sicily, continued to play an important role. It formed the southern limit of the Tunisian Trough which, although shallow, continued to subside and to receive marly beds similar to the Cherichera facies (see Fig. 19). Much of this uplifted zone was temporarily resubmerged by the Upper Oligocene to Aquitanian transgression. In the vicinity of Bizerte a reworked Upper Eocene to Middle Eocene microfauna is found, indicating the emergence and erosion of an uplift (anticlinal) zone during the Lower and Middle Oligocene. The basal beds of the Upper Oligocene contain pebbles of light colored Lower Eocene limestone (Crampon, 1971; Bajanik and Salaj, 1972), an indication that the Upper Oligocene is autochthonous and not allochthonous (Biely *et al.*, 1974).

The source of the detrital material deposited in the Tunisian Trough and Tello-Atlas region must ultimately have been the Saharan platform. Some may have been derived from the Saharan part of Algeria, the rest from the southeast, perhaps from the basement of the Pelagian block (cf. Burollet, 1967; Bishop, 1975). Recently Gaudette *et al.* (1975) have demonstrated that detrital zircon from the Oligocene beds has an absolute age of 1740 ± 100 m.y. and suggest that this is most readily explained by a source on the African craton. The zone Cape Bon–Hammamet–Sousse, where limestones and marls with Lepidocyclines formed, represents the border of the Pelagian block. It was a shallower area and, in consequence, accumulated a thinner section.

D. Neogene

The arenaceous sediments which began to form in the Upper Oligocene continued locally into the Lower Langhian (Fig. 20). They mark a general regression, excluding only the region of Beja-Mateur–Teboursouk–Bizerte, where marls continued to be deposited. This region formed a new foredeep in front of the former Tunisian Trough which was being uplifted.

Toward the end of the Upper Burdigalian–Upper Langhian there is evidence of renewed transgression over the eastern platform. In the region of Cape Bon this appears to have occurred in two phases, an earlier general phase and a later, Lower Langhian phase represented only in Cape Bon (Hooy-berghs and Lajmi, 1974). Tectonic uplift during the Upper Langhian led to the emergence of certain regions of Cape Bon, and this is reflected in the transgressive relation over eroded horizons of the Ain Grab Formation. After a period of calm with widespread formation of marls, major tectonic activity in northern Tunisia during the Serravallian–Tortonian led to further deposition of detritals from the erosion of the emergent areas. Elsewhere, in central and eastern Tunisia, these movements were less marked, but post-tectonic formations are widely distributed.

During the Upper Miocene and Pliocene a wide area between Thala and Tunis began to emerge, resulting in a differentiation of the sedimentation pattern between northern Tunisia (and Sicily) and the eastern platform or Pelagian block. The latter was characterized by limestone formation (Biely *et al.*, 1974) during the Miocene; similar limestones are found in Cyrenaica and the Ionian Sea. The Pliocene, nearly everywhere, is represented by continental beds.

Post-Pliocene tectonic movements have an important effect on the morphology (Castany, 1951; Burollet; 1956; Jauzein, 1967). It seems clear, however, that the principal faults were formed much earlier, as for example the Senonian limit between the eastern platform and central and southern Tunisia. Thus, the evolution of facies can be seen as a consequence of the geodynamic evolution of the region, and the establishment of the different zones of sedimentation give a clear idea of the evolution of Tunisia in time and space.

IV. SOME STRATIGRAPHIC PROBLEMS

Some stratigraphic questions still unresolved concerning the geodynamic and Mesozoic paleogeography of Tunisia are given below:

The basal Triassic on the eastern platform and Tunisian Trough (or pre-Trough) is unknown. It is, therefore, uncertain whether the Paleozoic is present or if the Triassic rests on crystalline basement rocks. At Djebel Ichkeul, where the Upper Triassic is metamorphosed with beds regarded as Jurassic, it is possible that there may also be metamorphosed Paleozoic beds. It is curious that in the zone of Triassic diapirs, the Triassic has not been found clearly associated with Jurassic beds. Nearly everywhere the Triassic is in contact with Upper Albian–Lower Cenomanian beds. Can some contacts be transgressive rather than tectonic (Pont du Fahs area)? However, no quartz grains or Jurassic pebbles have ever been found in the Upper Albian–Lower Cenomanian beds.

If the tectonic penetration occurred during the Upper Jurassic–Lower Cretaceous it is tempting to associate it with other tectonic events, in particular, the activity of the north–south axis with its crown of Upper Jurassic reefs. It also would account for detrital grains in the Portlandian and Berriasian beds of the north–south axis.

The presence or absence of the Lower Cretaceous then becomes an important question. Was the tectonism related to the diapir emplacement also responsible for the absence of Lower Cretaceous; i.e., were there emergent regions, an archipelago in the Tunisian Trough? This is still not known in certain regions of the diapir zone.

A minor point is raised by the discovery of radioloria in Lower Eocene globigerine limestones from near El Kef. Given that the beds are not of deep-water origin, it seems possible that ecological conditions favorable to the development of radiolaria may be linked to volcanic activity.

Wiman (1975) recently raised doubts about the age of the Saouaf Formation. Considered as Messinian, he suggested it was Upper Serravallian–Tortonian. Independently, Hooyberghs and Lajmi (1974) found an Upper Tortonian planktonic microfauna in the basal part of the formation, yet Salaj and Stranik (1971b) record the presence of a foraminiferan known in the Sarmatian and correlated with the Messinian.

A possible correlation of the Tunisian section with Malta can be proposed:

	Malta	*Tunisia*
Tortonian	Greensand Formation	Beglia Formation
Langhian–Serravallian	Blue clay Formation	Mahmoud Formation
Upper Burdigalian–Langhian	Upper globigerine limestone	Ain Grab Formation
	Lower globigerine limestone	

REFERENCES

Aguilar, J. P., and Thaler, L., 1974, Problèmes de corrélations entre échelles stratigraphiques continentales et marines d'après l'exemple du Miocène inférieur, *Résumé Comm., VI Colloq. Africain Micropaléontol.*, Tunis (1974).

Annoscia, E., 1969, The Bryofauna of Mesomiocenic Al-Jaghbub Formation in Eastern Cyrenaica/Libya, *Proc. Third African Micropaleontol. Colloq.*, Cairo, p. 37–94.

Arnould, D., and Saget, S., 1951, Les Ammonites pyriteuses du Tithonique supérieur et du Berriasien de Tunisie centrale, *Ann. Mines Géol.*, Tunis, no. 10.

Bajanik, S., 1971, Volcanisme en Tunisie, *Ann. Mines Geol.* Tunis, no. 25, p. 5–63.

Bajanik, S., and Biely, A., 1969, Carte géologique de la Tunisie a 1/50,000, feuille no. 43: Enfidaville manuscript, *Serv. Géol. Tunisie*, Tunis.

Bajanik, S., and Salaj J., 1971, Données nouvelles sur la zone de cicatrice de la région d'Oued Zarga (Tunisie septentrionale), *Notes Serv. Géol. Tunisie*, Tunis, no. 32, p. 3–23.

Bajanik, S., and Salaj, J., 1972, Néogène sous faciès Bou Sefra entre Teboursouk et Medjez el Bab (Tunisie septentrionale), *Notes Serv. Géol. Tunisie*, Tunis, no. 39, p. 121–135.

Bajanik, S., and Salaj, J., 1973, Quelques remarques sur la tectonique du Néogène entre Teboursouk et Medjez el Bab, *Livre Jubilaire M. Solignac, Ann. Mines Géol.*, Tunis, v. 26, p. 121–124.

Bajanik, S., Biely, A., Mencik, E., Salaj, J., and Stranik, Z., 1968, Carte Géologique de la Tunisie au 1/50,000, feuille no. 35: Dj. Zaghouan et notice explicative, *Serv. Géol. Tunisie*, Tunis.

Bajanik, S., Biely, A., and Salaj J., 1973, Carte géologique de la Tunisie au 1/50,000, feuille no. 43: Enfidaville et notice explicative, *Ed. Serv. Géol. Tunisie*, Tunis.

Ballais, J. L., 1973, Le Burdigalien de Mezzouna (Sud Tunisien), *Livre Jubilaire M. Solignac, Ann. Mines Géol.*, Tunis, v. 26, p. 263–266.

Baltzer, A., 1895, Versteinerungen aus dem tunisischen Atlas, *Nj. Jahrb. J. Min. Geol.*, Stuttgart, v. 1, p. 105–107.

Biely, A., 1969, Bajocien sous faciès "Ammonitico rosso" au Djebel Zaghouan, *Notes Serv. Géol. Tunisie*, Tunis, no. 30, p. 11–16.

Biely, A., and Rakus, M., 1969, Sur l'âge de la base de l'"Ammonitico rosso" dans la Dorsale tunisienne, *Notes Ser. Géol. Tunisie*, Tunis, no. 31, p. 37–39.

Biely, A., and Rakus, M., 1972a, Remarques stratigraphiques sur le Toarcien au Dj. Zaghouan *Notes Serv. Géol. Tunisie*, Tunis, no. 40, p. 95–101.

Biely, A., and Rakus, M., 1972b, Analyse critique des données sur l'age du Salifère en Tunisie septentrionale, *Notes Serv. Géol. Tunisie*, Tunis, p. 35–48.

Biely, A., and Salaj, J., 1971, L'Oligo-Miocène du Béjaoua oriental (Tunisie septentrionale), *Notes Serv. Géol. Tunisie*, Tunis, no. 34, p. 71–107, pls. VI–XII.

Biely, A., Rakus, M., Robinson, P., and Salaj, J., 1972, Essai de correlation des formation miocènes au Sud de la Dorsale tunisienne, *Notes Serv. Géol. Tunisie*, Tunis, no. 38, p. 73–92.

Biely, A., Memmi, L., and Salaj, J., 1973, Le Crétacé inférieur de la région d'Enfidaville. Découverte d'Aptien condensé, *Livre Jubilaire M. Solignac, Ann. Mines Géol.*, Tunis, v. 26, p. 169–178.

Bishop, W. F., 1976, Geology of Tunisia and adjacent parts of Algeria and Libya, *Amer. Assoc. Petrol. Geol.*, v. 59, p. 413–450.

Bismuth, H., Bonnefous, J., and Dufaure, P., 1967, Mesozoic microfacies of Tunisia. Guide Book to the Geology and History of Tunisia, *Petrol. Expl. Soc. Libya*, p. 159–214, pls. 1–27.

Bismuth, H., Lazaar, A., Lorenz, C., and Rakus, M., 1972, Reconnaissance géologique de l'île de Zembra (Golfe de Tunis, Tunisie), *C.R. Acad. Sci. Paris*, v. 275, sér. D, no. 25, p. 2807–2810.

Bizon, G., 1967, Contribution à la connaissance des Foraminifères planctoniques d'épire et des îles Ioniennes (Grèce occidentale) depuis le Paléogène supérieur jusqu'àu Pliocène, *Publ. Inst. Franç. Pétrole*, Paris, 142 p.

Bolze, J., Burollet, P. F., and Castany, G., 1952, Le Sillon Tunisien, *XIXe Congr. Géol. Int. Alger, Monogr. Region.*, Tunis, sér. 2, no. 5.

Bolze, J., and Sigal, J., 1964, Précisions sur la stratigraphie et les Foraminifères du Crétacé entre Zaghouan et le Ressas (Tunisie orientale), *C.R. Somm., Soc. Géol. France*, Paris, v. 1, p. 31–32.

Bonnefous, J., 1965, Livret guide de l'excursion géologique dans le massifs Jurassiques de la Tunisie du nord-est, *Rapport Int. S.E.R.E.P.T.*, Tunis.

Bonnefous J., 1967, Jurassic stratigraphy of Tunisia: A tentative synthesis (Northern and central Tunisia, Sahel and Chotts areas), in: *Guide Book to the Geology and History of Tunisia*, Tripoli: *Petrol. Expl. Soc. Libya*, p. 109–190.

Bonnefous, J., 1972, *Contribution à l'Etude Stratigraphique et Micropaléontologique du Jurassique de Tunisie* (*Tunisie Septentrionale et Centrale, Sahel, Zone des Chotts*) Thèse, Paris, p. 397.

Bonnefous, J., and Massin, J. M., 1964, Étude géologique des massifs jurassiques de la Tunisie du nord-est et du centre, *Rapport Int. S.E.R.E.P.T.*, Tunis.

Bonnefous, J., and Rakus, M., 1965, Précisions nouvelles sur le Jurassique du Djebel Bou Kornine d'Hammam Lif (Tunisie), *Bull. Soc. Géol. France*, Paris, (7), v. 7, p. 855–859.

Bujalka, P., Johan, Z., Krivy, M., Rakus, M., and Vacek, J., 1971, Notice explicative de Grombalia, feuille no. 29, Carte géologique de la Tunisie à 1/50,000, *Serv. Géol. Tunisie*, Tunis, 93 p.

Bujalka, P., Rakus, M., and Vacek, J., 1972, Notice explicative de la Goulette, feuille no. 21, Carte géologique de la Tunisie, *Serv. Géol. Tunisie*, Tunis, 71 p.

Burollet, P. F., 1956, Contribution à l'étude stratigraphique de la Tunisie centrale, *Ann. Mines Geol.*, Tunis, no. 18, p. 1–350.

Burollet, P. F., 1963, Trias de Tunisie et Libye. Relations avec le Trias européen et saharien, *Colloque sur le Trias, Mém. B.R.G.M.*, Paris, no. 15, p. 482–494.

Burollet, P. F., 1967, General geology of Tunisia, in: *Guide Book to the Geology and History of Tunisia*, Tripoli: Petrol. Expl. Soc. Libya, p. 51–58.

Burollet, P. F., 1971, Remarques géodynamiques sur le nord-east de la Tunisie, *C.R. Somm., Soc. Géol. France*, Paris, fasc. 8, p. 411–414.

Burollet, P. F., 1973, Importance des facteurs salifères dans la tectonique tunisienne, *Livre Jubilaire M. Solignac, Ann. Mines Géol.*, Tunis, no. 26, p. 111–120.

Burollet, P. F., 1975, Géologie et sédimentologie de la Tunisie, *IX Congr. Int. Sédimentol.* Nice, Excursion 15, 112 p.

Burollet, P. F., and Dumestre, A., 1952, Le Djebel Rheouis. Remarques sur un percement diapirique. *XIXe Congr. Géol. Int. Alger, Monogr. Région.*, Tunis, sér. 2, no. 6, p. 121–125.

Busnardo, R., and Memmi, L., 1972, La série infracrétacée du Djebel Oust (Tunisie), *Notes Serv. Géol. Tunisie.*, Tunis, no. 38, p. 49–61.

Busnardo, R., Le Hégarat, G., and Magné J., 1965, Le stratotype du Berriasien, *Colloque sur le Crétacé inférieur*, Mém. B.R.G.M., Paris, no. 34, p. 5–33.

Busson, G., 1967, Mesozoic of southern Tunisia, in: *Guide Book to the Geology and History of Tunisia*, Tripoli: Petrol. Expl. Soc. Libya, p. 131–151.

Caire, A., 1973, Les liaison alpines précoces entre Afrique du Nord et Sicile et la place de la Tunisie dans l'arc tyrrhenien, *Livre Jubilaire M. Solignac, Ann. Mines Géol.* Tunis, no. 26, p. 87–110.

Caire, A., Maamouri, M., and Stranik, Z., 1971, Contribution à l'étude structurale de la région des Hedil (Tunisie septentrionale) et comparaison avec Tell Algerien, *Notes Sérv. Géol. Tunisie*, Tunis, no. 36, p. 30.

Cassan, J. P., Mathieu, M., Tchimichkian, G., Cazes, M., and Dondon, J., 1973, Contribution à l'étude de l'Oligo-Miocène tunisien. Origine des apports et milieu de dépôt des séries detritiques, *Livre Jubilaire M. Solignac, Ann. Mines Géol.*, Tunis, v. 26, p. 233–239.

Castany, G., 1951, Etude géologique de l'Atlas tunisien oriental (Thèse), *Ann. Mines Géol.*, Tunis, no. 8, p. 1–632, pl. A–u, AA–AG, pl. I–XXVIII, 1 carte géol.

Castany, G., 1952, Atlas tunisien oriental et Sahel, *XIXe Congr. Géol. Int. Alger, Monogr. Région.*, Tunis, sér. 2, no. 6, p. 5–83.

Castany, G., 1955, Les extrusions jurassiques en Tunisie, *Ann. Mines Géol.*, Tunis, n. 14, p. 72.

Castany, G., 1956, Essai de synthèse géologique du territoire Tunisie–Sicile, *Ann. Mines Géol.*, Tunis, v. 16, p. 1–101.

Colom, G., Castany, G., and Durand-Delga, M., 1953, Microfaunes pélagiques dans le NE de la Berberie, *Bull. Soc. Géol. France*, Paris, sér. 6, v. III, p. 517–534.

Compte, D., and Dufaure, P., 1973, Quelques précisions sur la stratigraphie et la paléogéographie tertiaires en Tunisie central et centro-orientale, du Cap Bon a Mezzouna, *Livre Jubilaire M. Solignac, Ann. Mines Géol.*, Tunis, v. 26, p. 241–256.

Crampon, N., 1971, *Etude Géologique de la Bordure des Mogods, du Pays de Bizerte et du Nord des Hedil* (Tunisie Septentrionale), Thèse, Nancy, 3 v.

Cuvillier, J., Dalbiez, F., Glintzboeckel, G., Lys M., Magné J., Perebaskine, V., and Le Rey, M., 1955, Etudes micropaléontologiques de la limite Crétacé–Tertiaire dans les mers mesogéennes, in: *Proc. IV World Petrol. Congr.*, Rome (1955), Sect. 1/D, p. 517–544.

Dolle, M., 1952, Les Lepidocyclines de l'Oligocène de Sousse, *XIXe Congr. Géol. Int. Alger.*, *Monogr. Région.*, Tunis, sér. 2, no. 6, p. 139–140.

Domergue, C., Dumon, E., De Lapparent, A. F., and Lossel, P., 1952, Sud et Extrême Sud tunisiens, *XIXe Congr. Géol. Int. Alger, Monogr. Région.*, Tunis, sér. 2, no. 7.

Dubar, G., 1953, Gisements liassiques de la Dorsale tunisienne (Dj. Saidane et Dj. Zaghouan), *Bull. Soc. Géol. France*, Paris (16), v. 3, p. 354–356.

Durand-Delga, M., 1967, Structure and Geology of the Northeast Atlas Mountains, in: *Guide Book to the Geology and History of Tunisia*, Amsterdam: Petrol. Expl. Soc. Libya, p. 59–83.

Floridia, S., and Massin, J. M., 1969, Contribution à l'étude des extrusions jurassiques de la Tunisie septentrionale. Etude géologique et minière du sector minier d'Hamman Djedidi, *Bull. Soc. Géol. France*, Paris, (7), v. 11, p. 82–97.

Fournie, D., and Pacaud, M., 1973, Esquisses sédimentologiques et paléogéographiques sur le Crétacé inférieur de Tunisie du Berriasien au Barremien, *Livre Jubilaire M. Solignac, Ann. Mines Géol.*, Tunis, no. 26.

Gaudant, J., Rakus, M., and Stranik, Z., 1972, Leptolepis (Poisson teleosteen) dans le Toarcien de la Dorsale tunisienne, *Notes Serv. Géol. Tunisie*, Tunis, no. 38, p. 5–19, 3 pls.

Gaudette, H. E., Hurley, P. M., Fairbairn, H. W., and Lajmi, T., 1975, Source area of the Numidian flysch of Tunisia as suggested by detrital zircon ages, *Geol. Soc. Amer.* (Abstract), p. 1.

Glaçon, G., and Rouvier, H., 1972, Age des mouvements tectoniques majeurs en Tunisie septentrionale, *C.R. Acad. Sci., Paris*, v. 274, p. 1257–1260.

Glintzboeckel, C., 1956, Présence de Trocholines dans le Trias de Tunisie, *C.R. Soc. France*, Paris, p. 238–240.

Glintzboeckel, C., and Rabaté, J., 1964, Microfaune et microfaciès du Permo-Carbonifère du sud tunisien, *Int. Sediment. Petrograph. Ser.*, Leiden: E. J. Brill, 36 p.

Hooyberghs, H. J. F., 1973, Les Foraminifères planctoniques de la formation de l'Oued Hammam, une nouvelle unité lithologique in Tunisie d'âge langhien inférieur, *Livre Jubilaire M. Solignac, Ann. Mines Géol.*, Tunis, v. 26, p. 319–332, 2 pls.

Hooyberghs, H. J. F., and Lajmi, T., 1974, Oligo-Miocène et Pliocène du Cap Bon, Livret-guide des excursions du VIe Colloque Africain de Micropaléontologie, *Ed. Serv. Géol. Tunisie*, Tunis, p. 81–88.

Jauzein, A., 1957, Notice explicative de la feuille géologique au 1/50,000 de Bir M'Cherga, *Publ. Serv. Geol. Tunis*, no. 28.

Jauzein, A., 1959, Notice explicative de la feuille géologique au 1/50,000 de Maktar no. 53, *Publ. Serv. Géol.*, Tunis.

Jauzein, A., 1960, Notice explicative de la feuille géologique au 1/50,000 de Mansour, *Publ. Serv. Géol.*, Tunis.

Jauzein, A., 1967, Contribution à l'étude géologique des confins de la Dorsale Tunisienne, *Ann. Min. Géol.*, Tunis, no. 22, p. 1–475.

Johan, Z., and Krivy, M., 1968, Géologie de la structure anticlinale Dj. Messella–Sidi-Salem, *Notes Serv. Géol.*, Tunis, p. 26.

Johan, Z., and Krivy, M., 1969, Carte géologique de la Tunisie à 1/50,000, feuille no. 36, Bou Ficha et notice explicative, *Ed. Serv. Géol. Tunisie*, Tunis, 84 p.

Klen, L., 1974, Geological map of Libya 1:250,000, Shat: Benghazi NI 34-14. Explanatory booklet, Praha-Tripoli: Industrial Research Centre, p. 5–56.

Klitzsch, E., 1970, Die Strukturgeschichte der Zentralsahara—Neuerkenntnisse zum Bau und zur Palaogeographie eines Tafellandes, *Geol. Rundschau*, Stuttgart, v. 59, no. 2, p. 459–572.

Klitzsch, E., 1971, The structural development of parts of North Africa since Cambrian time, *Symp. Geol. Libya*, Tripoli, p. 253–262.

Kujawski, H., 1964, Sur un niveau gisement danien en Tunisie orientale, *C.R. Somm., Soc. Géol. France*, p. Paris, p. 413.

Kujawski, H., 1969, Contribution à l'étude géologique de la région des Hedil et du Béjaoua Oriental, *Ann. Mines Géol.*, Tunis, v. 24, p. 1–281.

Levy, J., 1966, Neomizzia (dasycladacée) nouveau genre de Lias du Maro, *Rev. Micropaléontol.*, Paris, v. 9 (1), p. 37–39.

Maamouri, A. L., and Salaj, J., 1974, Subdivision microbiostratigraphiques du Crétacé inférieur du Djebel Oust (Tunisie septentrionale), *VIᵉ Colloq. Africain Micropaléontol.*, Tunis (1974).

Massin, J., and Salaj, J., 1970, Contribution à l'étude stratigraphique de la région Nebeur (Tunisie septentrionale), *Bull. Soc. Géol. France*, Paris, (7) v. XII, p. 818–825.

Memmi, L., 1965, Sur quelques Ammonites du Valanginien de l'Oued Guelta, Tunisie, *Bull. Soc. Géol. France*, Paris, (7), v. 7, p. 833–838.

Memmi, L., 1967, Succession de faunes dans le Tithonique supérieur et le Berriasien du Djebel Nara (Tunisie Centrale), *Bull. Soc. Géol. France*, (7), v. IX, p. 267–272.

Memmi, L., 1969, Eléments pour une biostratigraphie de l'Hauterivien du "Sillon Tunisien," *Notes Serv. Geol.*, Tunis, v. 31, p. 41–50.

Memmi, L., and Maamouri, A. L., 1974, Crétacé inférieur du Djebel Oust, Livret-guide des excursions du VIᵉ Colloque Africain de Micropaléontologie, *Ed. Serv. Géol. Tunisie*, Tunis, p. 31–39.

Memmi, L., and Salaj, J., 1973, Le Berriasien de Tunisie. Succession de faunes d'Ammonites, de Foraminifères et de Tintinoidiens, *Colloque sur la limite Jurassique–Crétacé*, Lyon-Neuchâtel (September, 1973), *Mémoire du B.R.G.M.*, no. 86, p. 58–67.

Mencik, E., Salaj, J., and Stranik, Z., 1970, Notice explicative de la feuille géologique à 1/50,000 du Dj. Fkirine, *Manuscript Serv. Géol.*, Tunis.

Meuter F. De, and Symons, F., 1973, Etude quantitative des Foraminifères de l'Oued Hammam, Tunisie nord-orientale: une déscription à l'aide de variables structurelles dites Varimau, *Livre Jubilaire M. Solignac, Ann. Mines Géol.*, Tunis, v. 26, p. 337–343.

Mongin, D., 1971, Quelques Gastéropodes nouveaux dans l'Aptien supérieur de Tunisie (Henchir Demane el Bekr; région du Dj. Fkirine) Pl. I, *Notes Serv. Géol. Tunisie*, Tunis, no. 34, p. 19–42.

Moullade, M., 1966, Etude stratigraphique et micropaléontologique du Crétacé inférieur de la "fosse Vocontienne," *Doc. Lab. Geol. Fac. Sci. Lyon.*, Lyon, v. 15, p. 1–369.

Pervinquière, L., 1903, Etude géologique de la Tunisie Centrale. Dir. Trav. Publ. Carte géol. Tunisie, Paris: *F.R. de Rudeval ed.*

Pervinquière L., 1912, Sur la géologie de l'Extrême-Sud tunisien et de la Tripolitaine, *Bull. Soc. Géol. France*, Paris, (4), v. XII, p. 143–193.

Planderova, E., 1971, Contribution à l'étude palynologique des sédiments tertiaires de la Tunisie, *Geologicke Prace, Spravy*, Bratislava, v. 56, p. 199–216.

Planderova, E., Biely, A., Mencik, E., and Stranik, Z., 1969, Pollens des séries continentales de la cuvette de Zaghouan, *Notes Serv. Géol. Tunisie*, Tunis, no. 30, p. 38–43, 4 pls.

Rakus, M., 1964, Le Jurassique au Djebel Ressas, *Rapport Inedit., Serv. Géol. Tunisie*, Tunis.

Rakus, M., 1973, Le Jurassique au Djebel Ressas (Tunisie septentrionale), *Livre Jubilaire M. Solignac, Ann. Mines Géol.*, Tunis, no. 26, p. 137–147.

Rakus, M., and Biely, A., 1971, Stratigraphie du Lias dans la Dorsale tunisienne, *Notes Serv. Géol. Tunisie*, Tunis, no. 32, p. 45–63.

Remack-Petitot, M. L., 1971, Précisions stratigraphiques sur des Echinides du Crétacé inférieur de Tunisie, *Notes Serv. Géol.*, Tunis, no. 4, p. 5–17.

Robinson, P., 1975, Neogene continental reck units of Tunisia, *Proc. VI Congr.* Bratislava, p. 415–419.

Röhlich, P., 1974, Geological map of Libya 1:250,000, Sheet: Al Bayada, NI 34-15 Explanatory booklet, Praha-Tripoli: Industrial Research Centre, 70 p.

Rouvier, H., 1973, Nappes de charriage en Tunisie septentrionale: preuves, conséquences paléogéographiques, *Livre Jubilaire M. Solignac. Ann. Mines Géol.*, Tunis, no. 26, p. 33–47.

Salaj, J., 1969, Quelques remarques sur les problèmes microbiostratigraphiques du Trias, *Notes Serv. Géol. Tunisie*, Tunis, no. 31, p. 5–23.

Salaj, J., 1970, Quelques remarques sur les problèmes de microbiostratigraphie du Crétacé supérieur et du Paleogène, *IV Colloq. Africain Micropaléontol.*, Nice, p. 357–374.

Salaj, J., 1972, Contribution à la microbiostratigraphie du Mésozoique et du Tertiaire de la Tunisie septentrionale, *V Colloq. Africain Micropaléontol.*, Addis Ababa.

Salaj, J., 1973, Proposition pour des Néostratotypes du Crétacé supérieur (en vue de la zonational des régions de la Thetys), *Livre Jubilaire M. Solignac, Ann. Mines Géol.*, Tunis, no. 26, p. 219–222.

Salaj, J., 1974a, Microbiostratigraphie du Crétacé supérieur de la région de Pont du Fahs, *Livret-guide des Excursions du VIe Colloq. Africain Micropaleontol.*, Tunis, p. 41–49.

Salaj, J., 1974b, Proposition pour les hypostratotypes du Danien et du Paléocène (localité type de la formation El Haria: le Kef, Tunisie septentrionale). *Notes Serv. Géol. Tunisie*, no. 41, p. 91–100.

Salaj, J., 1974c, Microbiostratigraphie du Senonien supérieur du Danien et du Paléocène de la région du Kef, *Livret-guide des Excursions du VIe Colloq. Africain Micropaleontol.*, Tunis, p. 51–57.

Salaj, J., 1974d, Contribution à la microbiostratigraphie des hypostratotypes tunisiens du Crétacé supérieur, du Danien et du Paléocène *VIe Colloq. Africain Micropaléontol.*, Tunis.

Salaj, J., 1975, Quelques remarques sur le Miocène sous faciès Bou Sefra de Tunisie septentrionale, *N.Jb. Geol. Palaont. Mh.*, v. 7, p. 412–423.

Salaj, J., and Bajanik, S., 1972a, Contribution à la stratigraphie du Crétacé et du Paléogène de la région de l'Oued Zarga, *Notes Serv. Géol.*, Tunis, no. 38, p. 63–71.

Salaj, J., and Bajanik, S., 1972b, Essai de subdivision lithostratigraphique du Trias de la Tunisie septentrionale, *Notes Serv. Géol. Tunisie*, Tunis, p. 27–33.

Salaj, J., and Maamouri, A. L., 1970, Remarques microbiostratigraphiques sur le Senonien supérieur de l'Anticlinal de l'Oued Bazina (Region de Beja, Tunisie septentrionale) *Notes Serv. Géol.* Tunis, no. 32, p. 65–78.

Salaj, J., and Samuel, O., 1966, Foraminifera der Westkarpaten-Kreide (Slowakei) *Geol. Ustav D Stura*, Bratislava, p. 1–292, pls. 1–48.

Salaj, J., and Stranik, Z., 1970 Découverte du Rhétien dans l'Atlas tunisien oriental, *C.R. Acad. Sci. Paris*, v. 271, p. 2087–2089.

Salaj, J., and Stranik, Z., 1971a, Rhétien dans l'Atlas tunisien oriental, *Notes Serv. Géol. Tunisie*, Tunis, no. 32, p. 37–44.

Salaj, J., and Stranik, Z., 1971b, Contribution à la stratigraphie du Miocene du synclinal de Saouaf (Region du Djebel Fkirine, Tunisie orientale), *Notes Serv. Géol. Tunisie*, Tunis, no. 32, p. 79–82.

Salaj, J., Bajanik, S., Mencik, E., and Stranik, Z., 1973, Quelques problèmes relatifs au Paleocène du Sillon tunisien et de l'Atlas oriental, *Livre Jubilaire M. Solignac, Ann. Mines Géol.*, Tunis, no. 26, p. 223–231.

Salaj, J., Batik, P., Maameri, N., and Maamouri, A. L., 1974a, Le Senonien supérieur et le Paleogene de la région des Hedil, *Livret-guide des Excursions du VIe Colloq. Africain Micropaleontol., Ed. Serv. Géol. Tunisie*, Tunis, p. 59–66.

Salaj, J., Pozaryska, K., and Szczechura, J., 1974b, Foraminifera, zonation and subdivision of the hypostratotypes of Tunisia. *Acta Paleontol. Polonica*, Warsaw, Vol. 27, no. 2, p. 127–190.

Schoeller, H., 1933, Présence de l'Aquitanien en Tunisie, *C.R. Soc. Géol. France*, Paris, fasc. 2, p. 158–159.

Sigal, J., 1965, Etat des connaissances sur les Foraminifères du Crétacé inférieur, Colloque sur le Crétacé inférieur, Mém. B.R.G.M., Paris, no. 34, p. 489–502.

Solignac, M., 1927, Etude géologique de la Tunisie septentrionale, *Publ. Serv. Géol. Tunisie.*

Solignac, M., 1931, Le Pontien dans le Sud tunisen, *Ann. Univ. Lyon*, Lyon, Nouv. Sér., fasc. 48.

Spath, L. F., 1913, On Jurassic Ammonites from Jebel Zaghouan, *Proc. Q. J. Geol. Soc. London*, v. LXIX, (838), p. 540–581.

Stranik, Z., and Mencik, E., 1970, Carte géologique de la Tunisie à 1/50,000, feuille du Djebel Fkirine, *Ed. Serv. Géol. Tunisie*, Tunis.

Stranik, Z., Mencik, E., Memmi, L., and Salaj, J., 1970, Biostratigraphie du Crétacé inférieur de l'Atlas tunisien oriental, Conference on African Geology (December, 1970), *Proc. Congr. Ibadan.*

Verhoeve, D., 1973, Les Foraminifères benthiques du Pliocène (Astien) du sondage de Korba (+5 à −200 m), Cap. Bon, Tunisie, *Livre Jubilaire M. Solignac, Ann. Mines Géol.*, Tunis, v. 26, p. 345–377.

Wiedmann, J., 1971, Zur Frage der Jura (Kreide-Grenze), *Ann. Inst. Geol. Publ. Hungarici*, v. LIV, fasc. 2, p. 149–154.

Wiman, S. K., 1974, Micropaleontology of Miocene sediments exposed near El Haorearia, Tunisia, *VIe Colloq. Africain Micropaleontol.*, Tunis.

Wiman, S. K., 1975, Middle Miocene foraminifera from the Island of Zembra, Gulf of Tunis, Tunisia, *Proc. VI Congr. Bratislava*, p. 437–441.

Žert, B., 1974, Geological map of Libya 1:250,000, sheet: Darnah NI 34-16, Explanatory booklet, Praha-Tripoli: Industrial Research Centre, p. 5–49.

Chapter 5C

THE GEOLOGY OF THE PELAGIAN BLOCK:
THE MALTESE ISLANDS

Hugh Martyn Pedley

Department of Biology and Geology
The Polytechnic of North London
Holloway, London, U.K.

Michael Robert House and Brian Waugh

Department of Geology
The University
Hull, U.K.

I. INTRODUCTION

The Maltese Archipelago is 45 km in length and comprises Malta, the most southerly island, which is 27 km long, and Gozo, the more northerly island, which is 14.5 km long. Comino, in the Comino Straits separating the two larger islands, is only 2.5 km across, and the adjacent Cominoto is even smaller (Fig. 1). The islands trend northwest to southeast, and they lie 93 km due south of the Ragusa Peninsula of Sicily on the southern end of the shallow submarine shelf which extends under the separating Malta Channel. The shelf continues on to the coast of Tripoli and Tunisia, but reaches depths in excess of 600 fm 15 km southwest of Malta. The shelf between Malta and Sicily is mostly less than 50 fm.

The Maltese Islands are composed of Tertiary limestones and marls with

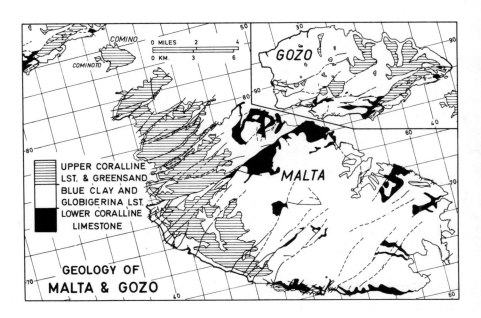

TABLE I

Mid-Tertiary Rock Succession of the Maltese Islands[a]

UPPER CORALLINE LIMESTONE FORMATION *up to 162 m*		TORTONIAN	M I O C E N E
GREENSAND FORMATION *0-12 m*		SERRAVALLIAN	
BLUE CLAY FORMATION *0-65 m*			
GLOBIGERINA LIMESTONE FORMATION *23-207m*	UPPER *UPPER PHOSPHORITE*	LANGHIAN	
	MIDDLE *LOWER PHOSPHORITE*	BURDIGALIAN	
	LOWER	AQUITANIAN	
LOWER CORALLINE LIMESTONE FORMATION *over 140 m*		CHATTIAN	OLIGO-CENE

[a] Lithostratigraphy mainly after Murray (1890); chronostratigraphy after Felix (1973).

very subsidiary Quaternary deposits. The succession is a simple "layer-cake" of Lower and Upper Coralline Limestones with intervening soft Globigerina Limestone and Blue Clay (Table I). This succession is correlated with the top of the Oligocene and the Miocene (Felix, 1973).

The distinctive formations in the succession have contrasting lithologies, and this is reflected in the characteristic topography and vegetation which they produce. The Lower Coralline Limestone is responsible for forming spectacular cliffs, some reaching 140 m in height, which bound the islands, especially to the west. Inland this unit forms barren gray limestone-pavement plateaus. The succeeding Globigerina Limestone, which is the most extensive formation on the islands, forms a broad, rolling landscape. The soil is meager, but it is intensively cultivated, and hill slopes on it are densely terraced. The Blue Clay forms the most fertile bedrock on the islands, especially where springs seep from the overlying Upper Coralline Limestone and lush vineyards occur. The Greensand forms part of the Upper Coralline Limestone complex, and the Limestone forms massive cliffs and limestone pavements with karstic topography similar to the Lower Coralline Limestone. It makes the highest tabular hills and mesas, and these reach their highest point for the islands of 253 m on the cliffs due south of Rabat, Malta (4567).*

The Victoria Lines Fault divides a southerly area of Malta, which is relatively unfaulted and gently folded, from a northern area, which is cut up into faulted horsts and grabens with a northeasterly trend. Southern Gozo is also much faulted.

Geological mention of the islands goes back to Classical times, with the observation of fossils here made by Xenophanes of Colophon in the Sixth Century B.C. Serious study dates from the work of Commander Spratt (Spratt and Forbes, 1843; Spratt, 1852), Leith Adams (1864, 1870, 1879), and especially John Murray (1890). More recent reviews have been given by Reed (1949, p. 7–15), House et al. (1961), and a volume on the geology was published by Hyde (1955). Recent contributions are by Vossmerbäumer (1972) and Felix (1973).

This report includes information from an unpublished report on the geology and gravity of the islands by Durham University, written in 1955 by K. C. Dunham, M. R. House, A. A. Wilson, M. H. P. Bott, and others for the British Petroleum Co. Ltd., and also an unpublished thesis by J. C. Wigglesworth (1964). We are indebted to them for permission to use some of their work. Other information results from six-inch mapping by M. R. House, field work by all authors, and an unpublished thesis (Pedley, 1975).

---

* Four-figure grid references locate the southwest corner of the kilometer grid square containing the feature named. The grid used is that appearing on the two-inch map of Malta and Gozo, G.S.G.S. 3859, Edition 4, 1954. This grid system is used on Figs. 1, 2 and 3.

II. STRATIGRAPHY

A. Introduction

The current terminology for the mid-Tertiary rocks of the Maltese Islands essentially dates from Murray (1890), but Cooke (1893a, b, 1896a, b) gave further details on the Globigerina Limestone and Blue Clay, Felix (1973) and Giannelli and Salvatorini (1975) have revised the biostratigraphy, and Pedley *et al.* (1976) give interpretative work on the paleoenvironment.

Despite the early acceptance of a lithostratigraphy, the biostratigraphy and chronostratigraphy, the correlation of the Maltese sequence with the Tertiary "stages" especially, have remained a source of continued debate. Fuchs (1874), Gregory (1891), and Cooke (1896a) all recognized the mid-Tertiary age of the rocks and attempted correlations with other areas, but it is the proposals of Bather (*in* Trechmann, 1938) which are the most detailed attempts at precision. He placed the Lower Coralline Limestone in the Tongrian Stage, the Globigerina Limestone in the Aquitanian and Burdigalian, and the overlying formations in the Helvetian and Tortonian, thus recognizing that both Oligocene and Miocene sediments were represented. However, on the basis of the pectinid faunas, Roman and Roger (1939) maintained that a Lower and Middle Miocene age could be ascribed to the entire succession. House *et al.* (1961) referred strata up to the top of the Lower Globigerina Limestone to the Aquitanian, the rest of the Globigerina Limestone and the Blue Clay to the Burdigalian, and the Greensand and the Upper Coralline Limestone to the Helvetian and Tortonian. A more restricted range was envisaged by Eames *et al.* (1962) on the evidence of foraminiferal studies. They considered all the strata to be of Lower Miocene age, the Lower Coralline Limestone being Aquitanian and overlying formations Burdigalian. The most recent correlation, also based on Foraminifera, is that of Felix (1973), the details of which are given in Table I.

Studies of the fauna of these Oligo-Miocene sediments have been made by Gregory (1891), Stefanini (1908), Roman and Roger (1939), Eames and Cox (1956), Wigglesworth (1964), and Felix (1973). But much remains to be done.

The distribution of the formations on Malta is given on two accompanying maps (Figs. 2 and 3), which are the first to map a subdivision of the Globigerina Limestone.

B. The Oligo-Miocene Succession

The sequence of rock units of limestones and associated marls represents a succession of sediments deposited within a variety of shallow water marine environments. In many respects these resemble the mid-Tertiary limestones occurring in the Ragusa region of Sicily and North Africa. Recent paleo-

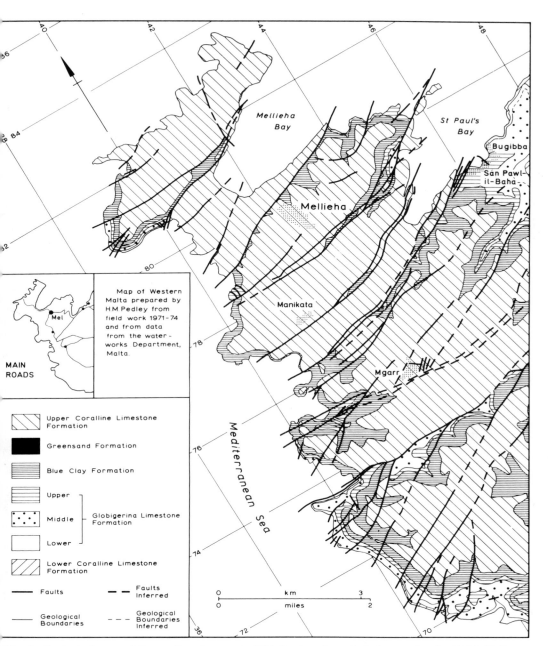

Fig. 2. Geological map of the northwestern part of Malta, based on six-inch mapping by H. M. Pedley.

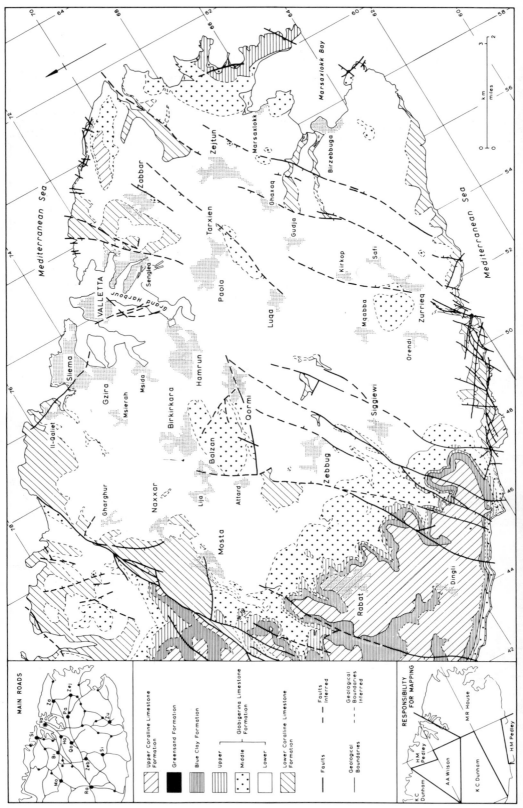

Fig. 3. Geological map of central and southern Malta, based on six-inch mapping by M. R. House, K. C. Dunham, A. A. Wilson, and H. M. Pedley.

magnetic and volcanological evidence from Sicily (Barberi *et al.*, 1974) demonstrates that the Africa–Europe plate boundary passes through northern
Sicily. Consequently, it would appear that Malta was part of a mid-Tertiary
Tethyan carbonate platform, extending from southern Sicily to North Africa,
with Malta situated toward the leading edge of the African plate.

That Malta has been a region of continued carbonate sedimentation for a
considerable period prior to the Miocene is illustrated by a deep borehole
sunk by the British Petroleum Co. Ltd., at Naxxar (4974), Malta. Commencing
at the top of the Lower Coralline Limestone, the hole terminated at a depth of
3000 m in dolomites, which carried spores of Lower Cretaceous affinity. Higher
Cretaceous and Eocene rocks were also dolomitized limestones. The uppermost
650 m, of shelly limestones and subordinate shales, was referred to the Oligocene, and this is in agreement with the correlation of Felix (Table I).

1. *Lower Coralline Limestone Formation*

This formation is exposed to 140 m in the vertical cliffs of southwest Gozo,
and is seen to over 100 m in the sections between Fomm ir-Rih (4073) and
Benghisa Point (5862) of western and southern Malta. Inland exposures are
mostly associated either with valley–gorge sections (*wieds*), as in southern
Malta, or with faulted inliers (Figs. 1, 2, 3). The upper part is exploited in
quarries around Attard, Malta (4873), and elsewhere, often under the name
zonqor.

The lowest exposed horizons consist of pale yellow biomicrites composed
largely of the tests of benthonic foraminiferans, and these are particularly well
seen around the base of the cliffs at Ghar Lapsi, Malta (4864). They are succeeded by massive, coarse-grained limestones, containing abundant algal rho-
doliths. These strata can be recognized throughout Malta and Gozo, the rhodo-
liths being composed of the coralline algae *Lithothamnion* and *Archaeolitho-
thamnion*. Benthonic foraminifera, including *Heterostegina* and *Nummulites*,
gastropods, pectinids, and echinoids are also present. In some western areas
of Malta at this horizon, there is a local biohermal development where the
algal rhodolites are associated with abundant hexacorals. Coarse bioclastic,
cross-bedded limestones, containing the thick-tested *Scutella subrotunda*, overlie
the algal limestones. The top of these beds, and of the Lower Coralline Limestone, is marked by a prominent echinoid horizon, the *Scutella* Bed of Roman
and Roger (1939), and this can be recognized both on Malta and on Gozo.
In eastern Malta, the cross-bedded facies is replaced by fine-grained, yellow
limestones containing abundant bryozoans, *Echinolampas*, *Lepidocyclina*, and
Terebratula. These pass transitionally upward into the Globigerina Limestone
and frequently have no *Scutella* Bed at their top.

A depositional setting, encompassing a sheltered, open marine situation, through reefal to offshore bar environment is envisaged for the formation of these deposits.

2. *Globigerina Limestone Formation*

   This formation covers large areas of central and southern Malta and Gozo (Fig. 1), and it shows marked variations in thickness. The thinnest sequence is probably near Fort Chambray, Gozo (3686), where it is about 23 m, but about 207 m is exposed in southern Malta around Marsaxlokk (5965). In general, the formation is thinnest around the Comino Straits, and probably thickest in the Valletta Basin (where only the Lower Globigerina Limestone is now preserved). This is the formation which provides the beautiful golden building stone, or *franka*, with which most Maltese buildings are made.

   Lithologically, the formation consists of yellow to pale gray, fine-grained limestones, almost wholly composed of the tests of globigerinid planktonic foraminiferans. Several phosphorite horizons are developed within the formation and consist of beds, frequently less than 0.5 m in thickness, of amoeboidal-shaped pebbles and nodules of dark brown to black collophanite. Two of these can be traced throughout the islands and enable the formation to be subdivided and mapped in three divisions (House *et al.*, 1961).

   Lower Globigerina Limestone. This subdivision is composed of massive bedded, pale yellow, globigerinid biomicrites, and often exhibits a characteristic honeycomb weathering witnessing to syndepositional trace fossils. Maximum thicknesses in excess of 100 m are inferred in the Valletta Basin of eastern Malta. On Gozo a maximum of nearly 40 m occurs in the extreme northwest of the island, but this division seems locally to have thinned out completely in northwestern Malta (4078). Fossils are locally abundant and include the molluscs *Chlamys* and *Flabellipecten*, the echinoids *Schizaster* and *Eupatagus*, pteropods such as *Cavolina*, and extensive thalassinoidean burrow systems.

   The highest stratum included here is the lower phosphorite conglomerate bed. This is approximately one meter thick and is characterized by a planar top and irregular base. In many western areas of Malta the phosphatic pebbles have been cemented by concretionary development of phosphate. Derived phosphatized mollusc casts and corals are common in the bed, together with the teeth of the sharks *Carcharodon megalodon*, *Odontaspis*, *Isurus*, and *Hemeprestes*, and waterworn bones and tusks of aquatic Mammalia.

   Middle Globigerina Limestone. Overlying the lower phosphorite horizon are the white to pale gray, marly limestones of the Middle Globigerina Limestone. Their whiteness normally allows them to be readily mapped as distinct from the units above or below, except where those are bleached; some uncertainties surround some of the isolated patches of Middle Globigerina Limestone shown on the maps here (Fig. 3). This unit shows similar thickness variations to the preceding unit. The Middle Globigerina Limestone is absent in eastern Gozo, a fact first recognized by Wigglesworth (1964). In western and northern Gozo it may reach 15 m in thickness. On Malta it is at its thinnest in the northernmost part of the island around Marfa, where it may be less than 15 m. It progressively thickens to the southeast and reaches 110 m near Delimara Point (5965). Since the Middle Globigerina Limestone is not found in the Valletta Basin area, it is impossible now to say whether the center of deposition was similar to that of the lower unit or not. Around San Leonardo (6070), the Middle Globigerina Limestone is progressively cut out by the basal Upper Coralline Limestone unconformity.

The dominant megafossils of this unit are the echinoids *Brissopsis* and *Schizaster*, the bivalves *Chlamys* and *Flabellipecten*, thalassinoidean burrows, and remains of the turtle *Tryonyx* and the crocodile *Tomistoma*.

A widespread phosphorite horizon terminates the Middle Globigerina Limestone and consists of 0.5 m, or thereabouts, of reworked phosphorite pebbles, mollusc casts, corals, echinoids, shark teeth, and casts of the nautiloid *Aturia aturi*. Thalassinoidean burrows extend up to 0.75 m below this bed and they are usually infilled with reworked phosphorite pebbles. There are other phosphatic pebble horizons within the main body of the Middle Globigerina Limestone, and also a khaki chert horizon, and all can be mapped locally (House *et al.*, 1961, p. 28).

Upper Globigerina Limestone. The upper division of the Globigerina Limestone is a tripartite sequence (Morris, 1952), comprising upper and lower divisions of yellow biomicrites and a middle division of gray marls. It does not show the same pattern of thickness variations as the Lower and Middle Globigerina Limestones. The maximum thickness of about 20 m occurs in northeastern Gozo, northern Malta, and the Delimara area (6064), while the strata thin rapidly toward southwest Gozo and are absent, as a result of post-Miocene erosion, in east-central Malta. The fauna is sparse, although *Schizaster eurynotus*, the gastropod *Epitonium*, and burrow systems do occur, and the pteropod *Vaginella* is common in the upper yellow limestone.

It is believed by Felix (1973) that the limestones were deposited in water depths ranging from 40–150 m. The accumulation of sediments with such a high planktonic foraminiferal content on a shallow shelf area is probably due to the position of Malta during the Miocene on a mid-Tethyan submarine rise.

3. *Blue Clay Formation*

The Blue Clay Formation comprises a sequence of alternating pale gray and dark gray banded marls, with lighter bands containing the highest proportion of carbonate. This lithology persists throughout the island, although in regions flanking the Comino Straits the upper part of the succession contains clays which are uniformly dark gray in color, lacking banding, and yielding abundant limonite and goethite concretions.

The maximum thickness noted is approximately 75 m at Xaghra, which is situated in north Gozo. Marked thinning occurs toward the south and east, but mostly there the formation has been removed by erosion, and at San Leonardo (6070) the Blue Clay is absent as a result of pre-Upper Coralline Limestone erosion. In Gozo the formation increases in thickness from 10 m along the southern coast to over 60 m in the north. A depositional high in the region of the Comino Straits is again apparent.

Fossils are most frequent in the upper horizons, particularly in northern Malta and southeastern Gozo. They include the corals *Flabellum*, *Stephanophyllia*, and *Balanophyllia*; molluscs such as *Aturia aturi*, *Sepia*, *Flabellipecten*, and *Chlamys*; echinoids; and the pteropod *Vaginella*. Foraminifera are abundant throughout, with species of *Globigerina* and *Orbulina* being the most common. The remains of marine vertebrates, including *Phoca*, dugongs, and many fish, are also present.

A muddy, open marine environment is clearly envisaged, the fauna indicating water depths of up to 150 m for the lower part of the formation, shallowing to less than 100 m in the upper part.

4. *Greensand Formation*

The Greensand Formation is composed of thickly bedded, coarse, glauconitic, bioclastic limestones in which, in unweathered sections, the green and black glauconite grains are readily discernible. Usually, however, due to the release of limonite upon weathering and oxidation of the glauconite, the rock possesses a characteristic orange-brown color. Early workers were clearly impressed with similarities to the English Cretaceous greensands. In Malta and eastern Gozo the contact with the underlying Blue Clay is frequently sharp, but the upper contact appears transitional due to the incorporation of glauconite into the lower beds of the Upper Coralline Limestone as a result of bioturbation.

The maximum development is at Il Gelmus (3189), Gozo, where 11 m can be measured, with thicknesses of 7 m in northwestern Gozo. Throughout Malta the formation, if restricted to the main glauconitic beds, is usually less than one meter thick and shows extensive reworking and assimilation into the overlying strata. The developments of *Heterostegina* rock are here considered as belonging to the basal Upper Coralline Limestone, but this giant foraminiferan is common in the Greensand. The molluscs *Chlamys*, *Cardium*, *Glycimeris*, and *Ostrea* occur, and this is a famous level for the echinoids *Clypeaster altus, C. marginatus, Echinolampas pignatarii,* and others. Shark teeth, the remains of dugongs, manatees, dolphins, and whales have been recorded, and bioturbation is common throughout. Deposition in a shallow water environment is envisaged.

5. Upper Coralline Limestone

A Middle Miocene, Tortonian age, is given to this, the youngest of the Tertiary formations of Malta, by Felix (1973). The formation is extensively developed, particularly in western Malta, Comino, and east-central Gozo, where it displays a wide range of lateral and vertical facies variations. A maximum thickness of approximately 100 m of strata is present in a lithological sequence, which can be divided into three divisions.

The lower part of the succession depicts the complex association of related facies types. In Gozo, coarse bioclastic limestones containing bivalves and echinoids are dominant, together with some foraminiferal micrites. This facies, representing a fairly high energy environment, passes eastward into a north–south trending coralline algal bioherm, in which the constituent algal rhodolites are mostly composed of *Lithophyllum.* Encrusting bryozoans and the brachiopods *Terebratula, Aphelesia, Megathiris, Argyrotheca,* and *Megerlia* are common. To the east of the bioherm, which is best developed in western Malta and eastern Gozo, micritic limestones with an echinoid–bivalve fauna represent deposition within an open shelf–lagoon environment.

The overlying sediments are mainly coarse-grained, bioclastic and oolitic limestones, in which discrete *Lithophyllum* rhodolites are common. Patch reefs, composed of an association of coralline algae, molluscs, and corals occur in western Malta and throughout Gozo. The molluscs include the bivalves *Cardium, Glycymeris, Lithophaga, Chama,* and *Gastrochaena,* and the gastropods *Turritella* and *Strombus,* while *Tarbellastraea* is the most common coral. At this horizon in Comino and northwest Malta large scale cross-bedded limestones, with an easterly dip to the foresets, are thought to indicate a minor phase of tidal delta sedimentation, or accumulation on offshore bars or shoals.

The highest group has a restricted distribution and is limited to the northwestern parts of Malta and isolated outcrops on Comino and northeastern Gozo. They consist of cross-bedded, oolitic, pelletoidal, and bioclastic limestones in which scour channels and ripple marks are locally important. Micritic limestones, with a stromatolitic algal layer containing gypsum crystals, succeed these in Malta. Deposition in very shallow water is clearly indicated by these various facies, ranging from shallow subtidal through possible intertidal and supratidal environments.

C. Quaternary Deposits

Quaternary deposits are of local extent only and consist of isolated cave and fissure infills and sediment veneers.

The earliest are the Pleistocene bone deposits of various cave systems of Malta, particularly that of Ghar Dalam Cave (5765) (Cooke and Smith-Woodward, 1893). Dwarf hippopotami, pygmy elephants, and swans are recorded from the oldest deposits, while later horizons have horse and deer remains. The giant dormouse is the typical rodent in fissure infills. The presence of so many land quadrupedal animals is taken as evidence that there was land communication between Sicily and Malta at this period.

Later deposits, which invariably possess a distinct red color, include alluvial fan deposits, calcreted breccias and conglomerates, and caliche soil profiles. At Wied Maghlek (4865) in southern Malta, alluvial fanglomerates and associated soil horizons occur along the flanks of the Maghlek Fault. Similar fans also occur in the Pwales Valley (4577), west of San Pawl il-Bahar, Malta. A caliche soil profile, capped by a red carbonate horizon at Marfa Point (3982), Malta, yields terrestrial gastropods (Cooke, 1896c). Trechmann (1938) has discussed the Quaternary history of the islands.

D. Comparison of the Maltese Miocene Succession with Adjacent Areas

The isolated nature of the Maltese Archipelago, together with the development of faunal elements often peculiar to the islands, has made correlation with other areas difficult. The reader is referred to Felix (1973) for further discussion on this subject.

Shallow water marine sequences of a similar nature and age to the Maltese succession occur in several adjacent areas, the nearest of which is Sicily to the north. Here, the northern and central outcrops of Miocene strata are mainly of marls and sandstones, but in the Ragusa region of southeast Sicily a thick sequence of shallow-water marine carbonates occurs. These are rich in coralline algae, bivalves, gastropods, cephalopods, echinoids, and Foraminifera, including species which are common to those occurring on Malta.

The island of Lampedusa, lying about 220 km to the northwest of Malta, provides a further shallow-water carbonate succession equally rich in molluscs and coralline algae. These limestones probably correlate with the upper parts of the Maltese succession.

To the south of the Maltese Archipelago, the North African coastal region shows the development of Miocene strata, much of which are, however, of a clastic nature. In the Sirte Basin of Libya, a mixed clastic-carbonate shoreline sequence of complex sedimentary facies can be recognized (Selley, 1969), passing northward into shallow-water carbonates. In the latter, coralline algae are again well represented and the invertebrate faunas also have much in common with the Maltese Miocene succession.

III. STRUCTURE

The main structural features of the Maltese Islands have been clear for a century or more, and the recently completed mapping on the six-inches-to-the mile scale (Figs. 2 and 3) has added only detail. Recent reviews have been given by Hyde (1955), Vossmerbäumer (1972), and Felix (1973, p. 12–16).

Normal faulting predominates and the strata are only gently flexured. A prominent series of horst and graben structures, of east-northeast trend, characterizes Malta north of the major Victoria Lines Fault, which crosses the island from Fomm ir-Rih (4073) to near Madalena Tower (5276). Northern Malta is also the largest structurally depressed region. Another structural low is the Valletta Basin, the center of which is just southwest of Valletta (5571). In southeastern Malta, around Marsaxlokk Bay (6064), there is a structural shelving to the southeast, and this seems to be the lowest point on the islands (estimates depend, however, on thicknesses inferred for the Globigerina Limestone). On Malta, the highest elevations of the Lower Coralline Limestone are found near Naxxar (4975) and near the southwest coast (4468, 4865). Gozo shows a general tilt to the northeast, and the southern parts are much affected by faulting. The highest structural elevation of the Lower Coralline Limestone on the Maltese Islands appears to be on the southwest coast of Gozo (2987, 3286).

There is evidence that structural movements are still in progress. Earthquakes were recorded in 1659, 1693, 1740, 1811, 1856 (Hyde, 1955, p. 111), and March, 1972. At St. Paul's Bay and St. George's Bay, submerged prehistoric cart tracks are recorded. The stratigraphical evidence indicates contemporary mid-Tertiary faulting and movement, but the main period of faulting was probably post-Miocene and, in view of the fresh nature of the major fault scarps, quite recent. The evidence of Pleistocene conglomerate caught up in the Maghlek Fault has not been confirmed in our work. Discrete subsidence structures, the "circular faults" of former authors, have been demonstrated to have been activated several times since their initiation (Pedley et al., 1976).

A. Victoria Lines Fault

This important normal fault forms a bold scarp across the islands (4073 to 5276), which is perhaps the most striking single topographic feature of the islands. In one area it brings the Upper Coralline Limestone down to the north against Lower Coralline Limestone on the southern, upthrown side. The throw of the fault varies from about 200 m near the Bingemma Syncline (4473) (Morris, 1952) in the west, to about 100 m in the east near Madalena Tower. Hobbs (1914) gave an analysis of the fault plane and the relations of the strata on either side to it. He showed that the fault plane changed its strike quite frequently while maintaining a general east-northeast trend.

B. Northern Malta

The horst and graben blocks of northern Malta are topographically indicated by prominent ridges and valleys. The main units, from north to south are: Marfa Ridge, Mellieha Valley, Mellieha Ridge, Mizieb Valley, Bajda Ridge, St Paul's Valley, Wardija Ridge, and Bingemma Valley (the latter developed only against the western part of the Victoria Lines Fault).

These structures are not simple. The fault fractures are often compound, and sharp stratal flexures in both valleys and ridges occur. Approximate figures for the throw of these faults were given by House *et al.* (1961), and they have been discussed by Vossmerbäumer (1972). The fault of greatest throw would appear to be that forming the southern margin of the Marfa Ridge where the Upper Coralline Limestone is brought down to the south against the Lower Coralline Limestone, a displacement of over 100 m.

C. Central and Southern Malta

South from the Victoria Lines Fault a different tectonic situation is apparent. Horst and graben structures do not occur. While there are many normal faults they almost entirely have a northeast, rather than east-northeast trend, and structurally they frequently form a step fault system, especially the group which scissors in to the western end of the Victoria Lines Fault. While it has often been thought that the Maltese fault pattern must be part of a conjugate system, it is only here that there is significant evidence of the northwesterly trending element. This is provided by the Maghlek Fault (4865), which runs parallel with the coast, and erosion has only preserved a sliver of Upper Coralline Limestone faulted against Lower Coralline Limestone, with Globigerina Limestone, Blue Clay, and Greensand caught up between fault planes at Halk It-Tafel. The Upper Coralline Limestone is just overturned at one place near here.

South of the Victoria Lines Fault, however, flexuring provides the main tectonic effect. The Durham University Survey (House *et al.*, 1961) confirmed the general observation of E. B. Bailey that a structural high passes south from the Victoria Lines Fault toward Ghar Lapsi (4764) but demonstrated that this gave several culminations which would carry the top of the Lower Coralline Limestone higher than 120 m above sea level (at 4273, 4975, 4468, 4865). In a broad area around and inland from Valletta, and centering upon the head of Grand Harbour (5571), the top of the Lower Coralline Limestone is below sea level, reaching over 40 m below. The only other area where this occurs significantly is between Marsaskala (6068) and Benghisa (5862), where it would appear that the Lower Coralline Limestone shelves to the southeast at a rate of 50 m/km.

D. Gozo

The regional dip on Gozo is to the northeast so that the Lower Coralline Limestone, which forms vertical cliffs over 120 m in height in southwestern Gozo, between Dwejra (2789) and near Sannat (3286), shelves down to be over 20 m below sea level between Marsalforn (3392) and east of San Blas Bay (3990). But there is much flexuring, and both southeast of Zubbug (3290) and around Mgarr (3787), the top of the Lower Coralline Limestone is below sea level.

South from a line joining Dwejra (2790) to the easternmost point of the island (4087), Gozo is affected by a complex fault belt, the irregular trend of which is approximately east–west, contrasting with the patterns on Malta (Vossmerbäumer, 1972, p. 31). The "circular fault" structures (Hyde, 1955; Pedley et al., 1976) often form fulcra for the fault complexes. The faults are all normal, but Rizzo (1932) and others have noted that the smooth striations and slickensides shown on the well-exposed fault plane surfaces are inclined, not vertical, and examples occur in which their angle can be seen to decrease, suggesting rotational movement.

IV. GEOPHYSICAL SETTING

Early geophysical surveys around the Maltese Islands were done on *H. M. Submarine Talent* in 1950 (Cooper *et al.*, 1952), when 46 pendulum stations were occupied in the area of Malta and Pantelleria. This work was extended and interpreted by Harrison (1954, 1955) and by M. H. P. Bott in an unpublished report for the British Petroleum Co. Ltd.

The regional situation suggests a congruence of the high Bouguer anomalies (above 20 mgal) of the shelf between Malta and southeast Sicily, values reaching +110 mgal on the Ragusa Peninsula of Sicily (Wunderlich, 1965). There is evidence for mass surplus beneath Malta, where the basement depth seems unlikely to exceed 1.8 km.

The Durham University gravity survey by M. H. P. Bott established an important positive anomaly of 69 mgal centered near Zabbar. This formed the southeast end of a high extending to others at Naxxar and Il Ghallis, with another, separated from these, near Ghar Lapsi. All follow the structural situation, but in the case of the Zabbar anomaly (5670), the tectonic culmination is slight and more fundamental control seems probable; associated to the southeast is also the anomalous transgressive Upper Coralline Limestone outlier of San Leonardo. The northwest trend of the main anomaly belt coincides with the trend of the northeast and southwest coasts of Malta, which have usually been thought to be fault-bounded. It would appear that any faulting may reflect basement structure. In southern Malta and northern Gozo there is a sharp fallaway of the Bouguer anomalies, and this has been interpreted as marking the limits of the Malta–Ragusa shelf basement structure.

REFERENCES

Adams, A. L., 1864, Outline of the geology of the Maltese Islands, *Ann. Mag. Nat. Hist.*, v. 14, p. 1–11.

Adams, A. L., 1870, *Notes of a Naturalist in the Nile Valley and Malta*, Edinburgh: Edmonston and Douglas.

Adams, A. L., 1879, On the remains of Mastodon and other vertebrates of the Miocene Beds of the Maltese Islands, *Quart. J. Geol. Soc. London*, v. 35, p. 517–531.

Barberi, F., Civetta, L., Gasparini, P., Innocenti, F., Scandone, R., and Villari, L., 1974, Evolution of a section of the Africa–Europe Plate boundary: paleomagnetic and vulcanological evidence from Sicily, *Earth Planet. Sci. Lett.*, v. 22, p. 123–132.

Cooke, J. H., 1893a, On the occurrence of concretionary masses of flint and chert in the Maltese limestones, *Geol. Mag.*, v. 20, p. 157–160.

Cooke, J. H., 1893b, The marls and clays of the Maltese Islands, *Quart. J. Geol. Soc. London*, v. 49, p. 117–128.

Cooke, J. H., 1896a, Contributions to the stratigraphy and palaeontology of the Globigerina Limestones of the Maltese Islands, *Quart. J. Geol. Soc. London*, v. 52, p. 461–462.

Cooke, J. H., 1896b, Notes on the Globigerina Limestone of the Maltese Islands, *Geol. Mag.*, v. 33, p. 502–511.

Cooke, J. H., 1896c, Notes on the "Pleistocene Beds" of the Maltese Islands, *Geol. Mag.*, v. 32, p. 201–210.

Cooke, J. H., and Smith-Woodward, A., 1893, The Har Dalam Cavern, Malta, *Proc. Roy. Soc. (London)*, v. 54, p. 273–283.

Cooper, R. I. B., Harrison, J. C., and Willmore, P. L., 1952, Gravity measurements in the eastern Mediterranean, *Phil. Trans. Roy. Soc. London*, Ser. A, v. 244, p. 533–559.

Eames, F. E., and Cox, L. R., 1956, Some Tertiary Pectinacea from East Africa, Persia and the Mediterranean region, *Proc. Malac. Soc.* v. 32, p. 1–68.

Eames, F. E., Banner, F. T., Blow, W. H., and Clarke, W. J., 1962, *Fundamentals of Mid-Tertiary Correlation*, Cambridge: Cambridge University Press.

Felix, R., 1973, *Oligo-Miocene Stratigraphy of Malta and Gozo*, Wageningen: H. Veenman and Zonen, B. V.

Fuchs, T., 1874, Das Alter des Tertiarschichten von Malta, *Sitz. K. K. Akad. Wiss. Wien.*, v. 70, p. 92–105.

Giannelli, L., and Salvatorini, G., 1975, I foraminiferi planctonici dei sedimenti Terziari dell'arcipelago Maltese. II Biostratigrafia di: "Blue Clay", "Greensand" e "Upper Coralline Limestone", *Atti. Soc. Tosc. Sc. Nat.*, Mem. Ser. A. v. 82, p. 1–24.

Gregory, J. W., 1891, The Maltese fossil Echinoidea and their evidence on the correlation of the Maltese rocks, *Trans. Roy. Soc. Edinburgh*, v. 36, p. 585–639.

Harrison, J. C., 1954, Gravity measurements on Malta and at Tunis, *Geophys. Suppl. Roy. Astronom. Soc.*, v. 6, p. 604–609.

Harrison, J. C., 1955, An interpretation of gravity anomalies in the eastern Mediterranean, *Phil. Trans. Roy. Soc. London*, Ser. A, v. 248, p. 283–325.

Hobbs, W. H., 1914, The Maltese Islands: a tectonic–topographic study, *Scot. Geograph. Mag.*, v. 30, p. 1–13.

House, M. R., Dunham, K. C., and Wigglesworth, J. C., 1961, Geology of the Maltese Islands, in: *Malta: Background for Development*, Bowen-Jones, H., Dewdney, J. C., and Fisher, W. B., eds., p. 24–33, University of Durham, Newcastle.

Hyde, H. P. T., 1955, *Geology of the Maltese Islands*, Malta: Lux Press.

Morris, T. O., 1952, *The Water Supply Resources of Malta*, Malta: Government Printing Office.

Murray, J., 1890, The Maltese Islands with special reference to their geological structure, *Geograph. Mag.*, v. 6, p. 449–488.

Pedley, H. M., House, M. R., and Waugh, B., 1976, The geology of Malta and Gozo, *Proc. Geol. Ass.* v. 87 (3), p. 325–341.

Reed, F. R. C., 1949, *The Geology of the British Empire*, London: Arnold.

Rizzo, C., 1932, *Report on the Geology of the Maltese Islands*, Malta: Government Printing Office.

Roman, F., and Roger, J., 1939, Observations sur la faune de Pectinides de Malte, *Bull. Soc. Géol. France*, Ser. 5, v. 9, p. 59–79.

Selley, R. C., 1969, Near-shore marine and continental sediments of the Sirte Basin, Libya, *Quart. J. Geol. Soc. London*, v. 124, p. 419–460.

Spratt, T. A. B., 1852, *On the Geology of Malta*, Malta: Valletta.

Spratt, T. A. B., and Forbes, E., 1843, Geology of the Maltese Islands, *Proc. Geol. Soc. London*, v. 4, p. 225–230.

Stefanini, G., 1908, Echini Miocenici di Malta, *Boll. Soc. Geol. Ital.*, v. 27, p. 435–483.

Trechmann, C. T., 1938, Quaternary conditions in Malta, *Geol. Mag.*, v. 75, p. 1–26.

Vossmerbäumer, H., 1972, Malta, ein Beitrag zur Geologie und Geomorphologie des Zentral-mediterranen Raumes, *Würzburger Geog. Arb.*, v. 38, p. 1–212.

Wigglesworth, J. C., 1964, The Tertiary stratigraphy and echinoid palaeontology of Gozo, Malta, Unpublished Ph.D. thesis, University of Durham.

Wunderlich, H. G., 1965, Der gegenwärtige orogenetische Zustand der Appennin-Halbinsel, *Tectonophysics*, v. 1, p. 495–516.

INDEX